T0320749

BOUNDARY CONFORMAL FIELD THEORY AND THE WORLDSHEET APPROACH TO D-BRANES

Boundary conformal field theory is concerned with a class of two-dimensional quantum field theories that display a rich mathematical structure and have many applications, ranging from string theory to condensed matter physics. In particular, the framework allows discussion of strings and branes directly at the quantum level.

Written by internationally renowned experts, this comprehensive introduction to boundary conformal field theory reaches from theoretical foundations to recent developments, with an emphasis on the algebraic treatment of string backgrounds. Topics covered include basic concepts in conformal field theory with and without boundaries, the mathematical description of strings and D-branes, and the geometry of strongly curved spacetime. The book offers insights into string geometry that go beyond classical notions.

Describing the theory from basic concepts, and providing numerous worked examples from conformal field theory and string theory, this reference is of interest to graduate students and researchers in physics and mathematics.

ANDREAS RECKNAGEL is a member of staff at the Department of Mathematics, King's College London. His research centres on quantum field theory in two dimensions, in particular conformal field theories and their applications to strings and branes. He is also interested in topological field theory and the relation of non-commutative geometry to quantum field theory.

VOLKER SCHOMERUS is a Professor and scientist at the Theory Group of DESY, Hamburg. He has worked intensively in quantum field theory, symmetries in physics, non-commutative geometry and string theory, for which he has received several distinctions. He serves on the editorial board of several prestigious journals.

CAMBRIDGE MONOGRAPHS ON MATHEMATICAL PHYSICS

General Editors: P. V. Landshoff, D. R. Nelson, S. Weinberg

M. Le Bellac *Thermal Field Theory*[†]

Y. Makeenko *Methods of Contemporary Gauge Theory*

N. Manton and P. Sutcliffe *Topological Solitons*

N. H. March *Liquid Metals: Concepts and Theory*

I. Montvay and G. Münster *Quantum Fields on a Lattice*[†]

L. O'Raifeartaigh *Group Structure of Gauge Theories*[†]

T. Ortín *Gravity and Strings*

A. M. Ozorio de Almeida *Hamiltonian Systems: Chaos and Quantization*[†]

L. Parker and D. Toms *Quantum Field Theory in Curved Spacetime: Quantized Fields and Gravity*

R. Penrose and W. Rindler *Spinors and Space-Time Volume 1: Two-Spinor Calculus and Relativistic Fields*[†]

R. Penrose and W. Rindler *Spinors and Space-Time Volume 2: Spinor and Twistor Methods in Space-Time Geometry*[†]

S. Pokorski *Gauge Field Theories, 2nd edition*[†]

J. Polchinski *String Theory Volume 1: An Introduction to the Bosonic String*

J. Polchinski *String Theory Volume 2: Superstring Theory and Beyond*

J. C. Polkinghorne *Models of High Energy Processes*[†]

V. N. Popov *Functional Integrals and Collective Excitations*[†]

L. V. Prokhorov and S. V. Shabanov *Hamiltonian Mechanics of Gauge Systems*

A. Recknagel and V. Schomerus *Boundary Conformal Field Theory and the Worldsheet Approach to D-Branes*

R. J. Rivers *Path Integral Methods in Quantum Field Theory*[†]

R. G. Roberts *The Structure of the Proton: Deep Inelastic Scattering*[†]

C. Rovelli *Quantum Gravity*[†]

W. C. Saslaw *Gravitational Physics of Stellar and Galactic Systems*[†]

R. N. Sen *Causality, Measurement Theory and the Differentiable Structure of Space-Time*

M. Shifman and A. Yung *Supersymmetric Solitons*

H. Stephani, D. Kramer, M. MacCallum, C. Hoenselaers and E. Herlt *Exact Solutions of Einstein's Field Equations, 2nd edition*[†]

J. Stewart *Advanced General Relativity*[†]

J. C. Taylor *Gauge Theories of Weak Interactions*[†]

T. Thiemann *Modern Canonical Quantum General Relativity*

D. J. Toms *The Schwinger Action Principle and Effective Action*[†]

A. Vilenkin and E. P. S. Shellard *Cosmic Strings and Other Topological Defects*

R. S. Ward and R. O. Wells, Jr *Twistor Geometry and Field Theory*

E. J. Weinberg *Classical Solutions in Quantum Field Theory: Solitons and Instantons in High Energy Physics*

J. R. Wilson and G. J. Mathews *Relativistic Numerical Hydrodynamics*

[†] Issued as a paperback

Boundary Conformal Field Theory and the Worldsheet Approach to D-Branes

ANDREAS RECKNAGEL

King's College London

VOLKER SCHOMERUS

DESY, Hamburg

CAMBRIDGE
UNIVERSITY PRESS

CAMBRIDGE
UNIVERSITY PRESS

University Printing House, Cambridge CB2 8BS, United Kingdom

One Liberty Plaza, 20th Floor, New York, NY 10006, USA

477 Williamstown Road, Port Melbourne, VIC 3207, Australia

314-321, 3rd Floor, Plot 3, Splendor Forum, Jasola District Centre, New Delhi - 110025, India

103 Penang Road, #05-06/07, Visioncrest Commercial, Singapore 238467

Cambridge University Press is part of the University of Cambridge.

It furthers the University's mission by disseminating knowledge in the pursuit of education, learning and research at the highest international levels of excellence.

www.cambridge.org
Information on this title: www.cambridge.org/9780521832236

© A. Recknagel and V. Schomerus 2013

This publication is in copyright. Subject to statutory exception and to the provisions of relevant collective licensing agreements, no reproduction of any part may take place without the written permission of Cambridge University Press.

First published 2013

A catalogue record for this publication is available from the British Library

Library of Congress Cataloging in Publication data
Recknagel, Andreas, author.
Boundary conformal field theory and the worldsheet approach to D-branes / Andreas Recknagel, King's College London, Volker Schomerus, DESY.
pages   cm – (Cambridge monographs on mathematical physics)
Includes bibliographical references and index.
ISBN 978-0-521-83223-6 (hardback)
1. Conformal invariants.   2. Quantum field theory.   3. D-branes.   I. Schomerus, Volker, author.   II. Title.
QC174.52.C66R43   2014
530.14′3 – dc23   2013038661

ISBN 978-0-521-83223-6 Hardback

Cambridge University Press has no responsibility for the persistence or accuracy of URLs for external or third-party internet websites referred to in this publication, and does not guarantee that any content on such websites is, or will remain, accurate or appropriate.

To our parents

to Elena

und Hans Recknagel zum Angedenken

Contents

Introduction

The aim of this book is to give an introduction to two-dimensional boundary conformal field theory (CFT) and an overview of its various applications to D-branes in string theory.

The study of two-dimensional scale invariant quantum field theories (QFTs) has a long history. Applications to important problems in various branches of physics are so numerous that CFT has established its position as one of the leading techniques in modern theoretical physics. Its first significant success was the exact computation of critical exponents for second-order phase transitions in two-dimensional statistical systems, such as the critical Ising model and many extensions thereof. Surface critical phenomena are still among the key applications, but many less obvious ones have joined.

The applications most relevant for us are those in string theory. To this date, string theory clearly offers the most promising candidate for a fundamental theory of quantum gravity and an intriguing approach to the unification of all four known interactions. Strings may be considered as one-dimensional quantum systems or, equivalently, two-dimensional statistical systems. Among all such systems, CFTs are singled out since they solve the string theoretic equations of motion.

Let us also mention that two-dimensional CFTs offer a very fruitful laboratory for QFT. Many models of CFT can be solved non-perturbatively, mostly aided by their infinite-dimensional symmetries. This offers unique insights into the very nature of QFT, the importance of non-perturbative effects and dualities between theories with seemingly different field content. In addition, most CFTs also possess perturbations which may break scale invariance but preserve integrability, so that analytic expressions for many quantities become available.

The analysis of boundary conditions is a natural problem in physics. All realistic two-dimensional statistical systems possess boundaries and hence their full theoretical understanding clearly requires a good control of boundary conditions. For two-dimensional CFT, the study of boundaries was pioneered by Cardy in a series of papers, in particular [98, 100]. Once again, the presence of powerful infinite-dimensional symmetries has led to many exact results on boundary critical exponents and correlation functions.

Boundary CFTs are in fact more directly applicable to "real" physical situations than their relatives on closed surfaces. Many processes in three space dimensions are dominated by scattering in the s-wave channel, where the relevant quantities depend on time and a radial coordinate. Therefore, QFTs on the half-plane appear naturally. Quantum impurity scattering (the Kondo effect) is the most famous example.

In string theory, boundaries enter through the description of open strings. But it was not until the mid nineties and the discovery of branes that the whole relevance of the plethora of boundary conditions for string theory was fully realised. At low spacetime energy, p-branes appear as supergravity solutions, which describe stable objects whose mass and charge is distributed along $(p + 1)$-dimensional hypersurfaces in spacetime. Beyond the low-energy regime, supergravity needs to be replaced by string theory, and so a natural question to ask was how to describe branes in string theory. For a large class of branes, those that became known as D-branes, the answer was given by Polchinski in [375]: D-branes are objects on which open strings can end. The "D" in D-branes stands for Dirichlet boundary conditions, which force the open-string endpoints to stay within the brane worldvolume. String theory contains many different branes, which are characterised by their dimension and additional data to be described extensively below. All these data must be encoded in the choice of boundary conditions.

The importance of D-branes for our understanding of string theory, and perhaps even for the development of modern theoretical physics, can hardly be overestimated. After about 25 years of perturbative calculations, branes have made some non-perturbative features of string theory accessible. This "string revolution" has led to the conviction that string theories which previously were considered as independent are in fact merely different realisations of a single underlying theory, M-theory, see [437, 453] but also the earlier pre-D-brane proposals in [173]. The discussion of non-perturbative aspects is intimately related to the observation that string theory is more than a theory of one-dimensional objects travelling in a (quantised) spacetime, and that its consistent formulation requires the inclusion of higher-dimensional extended objects as well, namely membranes or D-branes.

As a "spin-off", D-branes have triggered much progress in the understanding of dualities in (supersymmetric) gauge theories in various dimensions, see e.g. [238, 329, 418] for some early results, and they continue to do so. Arguably the most profound development originating from brane physics, one that has influenced thousands of papers on many aspects of gauge theories and string theory in various dimensions, is Maldacena's intriguing anti-de Sitter (AdS)/CFT correspondence between (conformal) gauge theories and string theory in (asymptotically) AdS backgrounds, see [260, 343, 456] and the many excellent reviews that were written later.

Although branes have also led to new insights into 11-dimensional M-theory, the status of this theory is still rather unsettled. Therefore, this text focuses on D-branes proper as ingredients of modern string theory, the existence of which is well established, and which can be investigated in a rigorous way.

In the original formulation of D-branes by Polchinski and co-workers [119], "non-standard" boundary conditions on open strings arising from T-dualities are an essential ingredient. It is this point where the worldsheet description of string theory in terms of two-dimensional CFT comes into play. For a long time, "post-revolutionary" string literature was almost exclusively concerned with flat targets or simple variations thereof, where branes can easily be treated with methods from classical spacetime geometry. But subsequent developments revealed that naive intuitions drawn from this classical picture can be unreliable, and also that there are other kinds of branes (in particular non-BPS branes) which are not easily described in spacetime terms. Here, the worldsheet approach to D-branes, i.e. boundary CFT, proves much more effective, since it does not depend on target-space symmetries like spacetime supersymmetry. The worldsheet methods lead to a more general picture of D-branes, which appear simply as conformal boundary conditions. Geometric ideas like the interpretation of D-branes as hyperplanes in the target are no longer an essential ingredient and may sometimes in fact be misleading, even in rather simple situations.

Given the huge amount of work that has been invested into the development of the basic techniques and their applications, we have clearly been forced into many (painful) omissions. These include M-theory and the AdS/CFT correspondence, which will not be discussed, partly because, so far, worldsheet methods are only of indirect use for their study. This is not entirely true for the AdS/CFT correspondence, but the examples that can be analysed with the worldsheet techniques, such as string theory in AdS_3 without Ramond–Ramond flux, are quite exceptional. Even in this limited class of backgrounds, analytical techniques are needed to supplement the algebraic approach to be described below, because non-compactness of the curved AdS space renders spectra continuous. This provides new challenges, which, so far, have only been overcome in specific examples, model by model. In other words, in the context of the so-called non-rational CFT that describes non-compact string backgrounds, there are few model-independent constructions and results, in sharp contrast to the situation for rational CFTs that we shall treat in much detail. The interested reader can find an introduction to non-rational CFT e.g. in [409]. There is little doubt that much more technology will be developed in the future, simply because of the large number of important problems that non-rational CFT can be applied to.

There are many other applications of boundary CFT in physics and mathematics that we will only mention in passing. Among the most famous ones is the study of the Kondo effect with methods of boundary CFT that was carried out by Affleck and Ludwig [3, 4, 336]; see also [167] for related work and [458] for a discussion of the quantum wires with boundary CFT methods. Another somewhat different application concerns quantum systems with dissipation, see [91]. The connection to boundary CFT can be seen as follows: consider an open string and concentrate on one of its endpoints. Its motion will feel the presence of the rest of the string because energy can flow from the endpoint into the oscillatory modes of the string. Viewed from the endpoint, this appears as dissipation as described by the well-known Caldeira–Legget model of dissipative quantum mechanics [85]. None of these applications will be described below, but the underlying theoretical tools will be fully developed.

Let us now briefly describe the scope of this book and the organisation of its chapters, and make some suggestions on how to read them, depending on background and interests.

Chapter 1 treats free field theories on surfaces with boundaries, aiming to introduce branes without invoking abstract CFT concepts. We focus on uncompactified free bosons for the most part, but also treat fermionic theories as well as bosons compactified on a circle. To enrich the situation, non-vanishing B-fields are included from the start, which will in particular show that non-commutative geometry appears naturally in the context of boundary CFT and branes.

Chapter 2 aims at linking these field theoretic results to notions from string theory. After a lightning review of closed string theory in the first section, we turn to open strings and D-branes and try to establish a "dictionary" between worldsheet data and quantities encountered in the effective physics of strings and branes. These translation rules extend to models beyond the free field situation. We will also review some relatively technical aspects, because they are crucial in the construction of string compactifications or lead to interesting physical interpretations.

We then turn to a more general and also more abstract approach, which allows "branes" to be treated in more general situations than string theory in flat targets. To set the stage, Chapter 3 recapitulates CFT on the plane in the approach of Belavin, Polyakov and Zamolodchikov. We discuss symmetry algebras and their representations, then fields and correlation functions, including conformal Ward identities and non-linear constraints arising from sewing relations and modular invariance.

Chapter 4 contains a detailed exposition of CFT on (genus zero) surfaces with boundary, building upon the approach to CFT on the plane reviewed in the previous chapter. In the first two sections, we discuss the defining data of conformal boundary conditions and the ingredients that are new compared to CFTs on the full plane, in particular boundary fields. Section 4.3 introduces the boundary state formalism as an alternative, and computationally very efficient, description of boundary CFTs. We discuss the precise relation to the upper half-plane language used before, and also show how the examples from Chapter 1 fit into the general language. The fourth section is again devoted to non-linear constraints, some of which can be viewed as immediate generalisations of the sewing relations encountered on the plane, some of them, called Cardy constraints and exploiting modular transformations, arise naturally from the boundary state formalism. Their study provides a direct method to compute the spectrum of a boundary CFT. The collection of additional material at the end of Chapter 4 addresses crosscap states, orbifolds and a few other topics.

Chapter 5 deals with a subject that is, at least qualitatively, easier to discuss within a geometric rather than the worldsheet approach, namely D-brane moduli. We first review how to implement perturbations by arbitrary boundary fields, including relevant ones, and mention some general results like the g-theorem. We then turn to deformations by marginal boundary fields and give a criterion sufficient to show that certain operators indeed provide moduli of boundary conditions. Discussions of concrete examples and applications to string theory, including topology changes and condensation processes of brane configurations, conclude the chapter.

Chapter 6 can be seen as a "case study": we discuss the $SU(2)$ Wess–Zumino–Witten (WZW) model and CFTs derived from it via orbifold and coset constructions, and show how to apply some of the general machinery developed in Chapters 3 to 5. In particular, we will study condensation processes of stacks of branes, and it will turn out that the necessary techniques are the same as those needed for a CFT treatment of the Kondo effect in condensed matter physics. Non-commutative worldvolumes will make another prominent appearance.

Chapter 7 is more exclusively motivated by string theory, and shows how the worldsheet formulation allows the generalisation of the notion of D-branes to string vacua formulated without recourse to a target space description, by giving a construction of BPS boundary states for Gepner models. Using the general framework of Chapter 4 and the translation rules established in Chapter 2, this task is conceptually straightforward; but the implementation of string theoretic consistency conditions (in particular the GSO projection) is somewhat technical. Gepner models provide an exact CFT description of special points in the moduli space of string compactifications whose large-volume limits correspond to Calabi–Yau manifolds in weighted projective space. We explain this relation after a review of Gepner's algebraic construction of the bulk theories, and we also discuss the geometric content of Gepner model boundary states, highlighting where the CFT results refine preconceptions based on the picture from classical geometry. The chapter concludes with an introduction to the basics of an algebraic method that allows the study of topological branes in Landau–Ginzburg models and on Calabi–Yau manifolds in a very efficient way, namely matrix factorisations.

Chapters 5 to 7 use the general formalism developed in Chapters 3 and 4, and at times language from string theory as reviewed in Chapter 2. Nevertheless, readers familiar with with boundary CFT and brane physics may have no difficulty diving directly into any of the last three chapters if they so wish.

Chapters 3 and 4 can be read as a stand-alone introduction to (boundary) CFT in the approach by BPZ and Cardy. Here and there, motivations from string theory may be felt to lurk under the surface, but they are not crucial to understanding the methods explained in these two chapters.

The opening chapters try to provide a gentle introduction to the subjects of boundary conditions and branes for readers who have some background knowledge in QFT, but little previous exposure to abstract CFT in two dimensions or to string theory.

In our research work and on the long road towards completing this book, we have greatly profited from stimulating discussions with, and valuable suggestions from, many friends and colleagues, too many to list individually. We are grateful to all of them.

London and Hamburg, November 2012

1

Free field theory with boundaries

Our aim in the first chapter is to explain the microscopic theory of strings and branes in flat backgrounds. The corresponding two-dimensional worldsheet theories can be solved with elementary methods. This will not only give us a chance to meet most of the relevant quantities from two-dimensional field theory but also to pass from worldsheet to target space concepts without much technical struggle.

Our exposition begins with an analysis of free bosonic fields on the half-plane. After a few comments on the classical action and possible boundary conditions we shall solve the model. The solution will enable us to calculate bulk operator products and correlation functions. Moreover, we shall take a first look at a few central concepts, such as gluing conditions, boundary states, bulk and boundary fields and operator products. Once free bosonic fields are dealt with, we shall perform a similar analysis for fermions. Throughout the entire chapter, some familiarity with free field theory on the complex plane will be assumed.

1.1 Free bosonic field theory

This section is devoted to the simplest two-dimensional Euclidean quantum field theory, namely a theory of free bosons. Since the model is easily solved, it allows an unobstructed approach to the main features of conformal field theory (CFT) in the presence of boundaries. The choice of boundary conditions has many interesting consequences. For example, it influences the correlation functions of bulk fields, i.e. of fields that are inserted at points in the interior of the worldsheet. Furthermore, a new set of excitations emerges that can only exist at the boundary. The precise spectrum of such boundary excitations and the associated correlations also depend on the boundary condition. In order to render our framework sufficiently rich, we shall work with $D \geq 1$ bosonic fields. Multi-component bosonic fields are also relevant for string theory where the number of bosonic fields is given by the dimension of spacetime.

1.1.1 Solution of free bosonic field theory

We want to study free bosonic field theories with a D-dimensional Euclidean target space $\mathbb{R}^D$. In order to spell out a concrete model we shall fix a constant symmetric matrix $g_{\mu\nu}$ and an antisymmetric matrix $B_{\mu\nu}$ with $\mu, \nu = 1, \ldots, D$.

These will also be referred to as *metric* and *B-field*. The D-component bosonic field $X : \Sigma \to \mathbb{R}^D$ is assumed to live on a strip $[0, \pi] \times \mathbb{R}$ or, equivalently, on the upper half-plane

$$\Sigma = \{z \in \mathbb{C} \,|\, \Im z \geq 0\}.$$

These two realisations of the worldsheet are related by the exponential map. The action of our theory is then given by the following quadratic functional,

$$S(X) = \frac{1}{2\pi\alpha'} \int_\Sigma d^2z \, (g_{\mu\nu} + B_{\mu\nu}) \, \partial X^\mu \bar\partial X^\nu, \qquad (1.1)$$

where α' is a constant referred to as string tension in the string theory context. It is important to notice that for constant B the worldsheet action can be rewritten in the form

$$S(X) = \frac{1}{2\pi\alpha'} \int_\Sigma d^2z \, g_{\mu\nu} \partial X^\mu \bar\partial X^\nu + \frac{1}{\pi\alpha'} \int_\mathbb{R} dx \, B_{\mu\nu} X^\mu \partial_x X^\nu, \qquad (1.2)$$

where the second term involving the B-field is a pure boundary term and we have used the decomposition $z = x + iy$, i.e. the coordinate x parametrises the boundary of Σ. Hence, the constant B-field does not affect the dynamics in the interior of Σ ("in the bulk"), but it provides a linear background $A_\mu(X) = B_{\mu\nu} X^\nu$ that couples only to the boundary values of the field X. If we ignore the bulk term for a moment, the boundary term alone may be interpreted as the action of a particle that moves through $\mathbb{R}^D$ in the presence of a constant magnetic field $B_{\mu\nu}$. Since the boundary coordinate x plays the role of time now, $\partial_x X^\mu$ is the velocity. If we continue to focus on the boundary, but add the bulk term back into our setup, the latter may be considered as a bath of oscillators that drags energy from the boundary. In such an interpretation, the boundary theory (1.2) becomes the starting point for the study of dissipative quantum mechanics in the presence of a magnetic background field.

The complete description of our two-dimensional field theory requires us to specify boundary conditions for X. We shall not attempt to discuss the most general such boundary conditions. With later applications in mind, we demand

$$(g_{\mu\nu}\partial_y X^\nu(z,\bar z))_{z=\bar z} = (iB_{\mu\nu}\partial_x X^\nu(z,\bar z))_{z=\bar z} \quad \text{for} \quad \mu = 1,\dots,d, \qquad (1.3)$$

$$(\partial_x X^\mu(z,\bar z))_{z=\bar z} = 0 \quad \text{for} \quad \mu = d+1,\dots,D. \qquad (1.4)$$

For $B = 0$, the first line reduces to Neumann boundary conditions for the first d components of X. A non-vanishing B-field gives rise to a deformation of Neumann boundary conditions. This has a number of interesting effects, which we shall address below. Through the second relation, we impose Dirichlet boundary conditions for the remaining $D - d$ fields, i.e. those components of X are constant along the boundary. The value of X^a at $z = \bar z$ shall be denoted by

$$X^a(z,\bar z)_{z=\bar z} = x_0^a \quad \text{for} \quad a = d+1,\dots,D.$$

The constants x_0^a define a d-dimensional hyperplane V in $\mathbb{R}^D$ through the equations $X^a = x_0^a$. From time to time we might refer to the hyperplane as a "brane". The terminology shall be explained in detail in Chapter 2.

The free bosonic field theory (1.2) with boundary conditions (1.3) and (1.4) is easy to solve. One possibility is to spell out an explicit formula for the 2-point functions from which more general correlators can then be computed via Wick's theorem. Here we shall follow an alternative operator approach.

Solution for Dirichlet boundary conditions

For the transverse directions $a = d+1, \ldots, D$, the fields X^a are constructed through the following formula for the general solution of the two-dimensional Laplace equation $\partial\bar{\partial}X^a(z,\bar{z}) = 0$ with Dirichlet boundary conditions,

$$X^a(z,\bar{z}) = x_0^a + i\sqrt{\frac{\alpha'}{2}} \sum_{n \neq 0} \frac{\alpha_n^a}{n} \left(z^{-n} - \bar{z}^{-n} \right). \tag{1.5}$$

In the quantum theory, the objects α_n^a become operators obeying the relations

$$[\alpha_n^a, \alpha_m^b] = n\, g^{ab}\, \delta_{n,-m}, \quad (\alpha_n^a)^\dagger = \alpha_{-n}^a. \tag{1.6}$$

The commutation relations for α_n^a ensure that the field X^a and its time derivative possess the usual canonical commutator. Reality of the bosonic field X^a is encoded in the behaviour of α_n^a under conjugation. In comparison to the solution on the full complex plane, i.e. a bulk theory with periodic boundary conditions, the construction of X^a involves a single set of modes α_n^a. While α_n^a and $\bar{\alpha}_n^a$ are independent on the complex plane, they are related by a reflection on the boundary when we pass to the half-plane.

Similar to the bulk theory, the operators α_n^a for $n \neq 0$ act as creation and annihilation operators on the Fock space $\mathcal{H}_0^a = V_0$, which is generated by α_n^a with $n < 0$ from a unique ground state $|0\rangle$ subject to the conditions

$$\alpha_n^a |0\rangle = 0 \quad \text{for} \ \ n > 0. \tag{1.7}$$

This construction of the state space $\mathcal{H}_0^a$ along with the formula (1.5) provides the complete solution for any component of the bosonic field that satisfies Dirichlet boundary conditions.

Before we turn to the boundary condition (1.3), let us briefly remark that the operators α_n^a can be obtained as Fourier modes

$$\alpha_n^a = \frac{1}{2\pi i} \int_C z^n\, J^a(z)\, dz - \frac{1}{2\pi i} \int_C \bar{z}^n\, \bar{J}^a(\bar{z})\, d\bar{z}. \tag{1.8}$$

Here, C is a semi-circle in Σ centred around the point $z = 0$, and $J^a, \bar{J}^a$ denote the usual chiral currents

$$J^a(z) = i\partial X^a(z,\bar{z}) = \sqrt{\frac{\alpha'}{2}} \sum_n \alpha_n^a\, z^{-n-1},$$

$$\bar{J}^a(\bar{z}) = i\bar{\partial}X^a(z,\bar{z}) = -\sqrt{\frac{\alpha'}{2}} \sum_n \alpha_n^a\, \bar{z}^{-n-1}.$$

Because the operators α_n^a can be constructed from the U(1) current J^a of the theory, the algebra with relations (1.6) is often referred to as U(1) current algebra. From the explicit formulas for the currents we read off that they obey

$$J^a(z) = -\bar{J}^a(\bar{z}) \tag{1.9}$$

all along the real line $z = \bar{z}$. This relation is equivalent to the Dirichlet boundary condition and it tells us that the two sets of conserved currents in our theory are identified along the real line. Hence, there is only a single set of currents living on the boundary, while there are two sets throughout the bulk of the worldsheet.

Solution for Neumann boundary conditions

Let us now repeat the above free field theory analysis for the directions subject to the boundary condition (1.3). The fields $X^i, i = 1, \ldots, d$, are once more constructed using the general solution of the wave equation

$$X^i(z, \bar{z}) = \hat{x}^i - i\sqrt{\frac{\alpha'}{2}} \alpha_0^i \ln z\bar{z} - i\sqrt{\frac{\alpha'}{2}} B^i{}_j \alpha_0^j \ln \frac{z}{\bar{z}}$$
$$+ i\sqrt{\frac{\alpha'}{2}} \sum_{n \neq 0} \frac{\alpha_n^i}{n} \left(z^{-n} + \bar{z}^{-n}\right) + i\sqrt{\frac{\alpha'}{2}} \sum_{n \neq 0} \frac{B^i{}_j \alpha_n^j}{n} \left(z^{-n} - \bar{z}^{-n}\right),$$
$$\tag{1.10}$$

where summation over $j = 1, \ldots, d$ is understood. We have also raised one index of the matrix B with the help of the metric g. In passing to the quantum theory, $\hat{x}^i, \alpha_n^i$ become operators satisfying

$$[\alpha_n^i, \alpha_m^j] = n\, G^{ij}\, \delta_{n,-m}, \qquad\qquad [\hat{x}^i, \alpha_n^j] = i\sqrt{2\alpha'}\, G^{ij}\, \delta_{0,n}, \tag{1.11}$$
$$[\hat{x}^i, \hat{x}^j] = i\, \Theta^{ij}. \tag{1.12}$$

Furthermore, they obey the reality properties $(\hat{x}^i)^\dagger = \hat{x}^i$ and $(\alpha_n^i)^\dagger = \alpha_{-n}^i$. The commutation relations involve new structure constants G^{ij} and Θ^{ij} which are obtained from the background fields through

$$G^{ij} = \left(\frac{1}{g+B}\right)_S^{ij}, \qquad \Theta^{ij} = \left(\frac{\alpha'}{g+B}\right)_A^{ij}. \tag{1.13}$$

Here, S or A mean that the expression in brackets gets symmetrised or antisymmetrised, respectively. By construction, G is a symmetric $d \times d$ matrix that shall serve as a new metric in many of the formulas to be derived.

Note that the matrix Θ vanishes if and only if the B-field vanishes. A non-zero Θ implies that the "centre of mass coordinates" $\hat{x}^i$ no longer commute. We shall see below that this has very interesting consequences. Readers who are unfamiliar with free field theory on the boundary are invited to set $B = 0$ upon first reading. This simplifies many of the subsequent formulas. In particular, the metric G^{ij} then agrees with the metric g^{ij} in the action. Note that the modes α_n^i

generate D copies of the same U(1) current algebra which we met in the case of Dirichlet boundary conditions. When we work with $B \neq 0$, the algebra generated by the modes α_n^i does not change, but we work with a set of generators whose commutators contain G rather than g simply because these turn out to be more convenient.

Once more, the operators α_n^i act as creation and annihilation operators but now there is a d-parameter family of ground states $|k\rangle$, which are parametrised by a momentum $k = (k_i)_{i=1,\dots,d}$,

$$\alpha_0^i |k\rangle = \sqrt{2\alpha'}\, G^{ij} k_j |k\rangle . \tag{1.14}$$

If we denote the associated Fock spaces by $\mathcal{V}_k$, the state space $\mathcal{H}^B$ for the first d directions can be written as a direct integral $\mathcal{H}^B = \int_k d^d k\, \mathcal{V}_k$. On this state space we can also represent the position operators $\hat{x}^i$ as simple shifts of the momentum,

$$\exp(ik_i'\, \hat{x}^i) |k\rangle = e^{\frac{i}{2} k \times k'} |k + k'\rangle ,$$

where the vector product $\times$ is defined through $k \times k' = k_i \Theta^{ij} k_j'$. The fields we shall consider below involve only exponentials of $\hat{x}^i$ and not $\hat{x}^i$ itself.

From X^i we obtain the chiral currents J^i and $\bar{J}^i$ in the same way as above

$$J^i(z) = i\, \partial X^i(z,\bar{z}) = \sqrt{\frac{\alpha'}{2}} \sum_n (1+B)^i_{\ j}\, \alpha_n^j\, z^{-n-1} , \tag{1.15}$$

$$\bar{J}^i(\bar{z}) = i\, \bar{\partial} X^i(z,\bar{z}) = \sqrt{\frac{\alpha'}{2}} \sum_n (1-B)^i_{\ j}\, \alpha_n^j\, \bar{z}^{-n-1} . \tag{1.16}$$

These currents obey a linear boundary condition all along the real line $z = \bar{z}$, which is equivalent to the condition (1.3), namely

$$J^i(z) = \left(\frac{1+B}{1-B}\right)^i_{\ j} \bar{J}^j(\bar{z}) =: \left(\Omega_B \bar{J}\right)^i (\bar{z}) . \tag{1.17}$$

In this case, there appears a non-trivial orthogonal $d \times d$ matrix Ω_B that rotates the anti-holomorphic fields before they are identified with their holomorphic counterparts. It replaces the simple sign that we found in equation (1.9) for the directions satisfying Dirichlet boundary conditions. When $B = 0$, i.e. in the case of ordinary Neumann boundary conditions, the left and right moving currents are glued trivially, $J^i(z) = \bar{J}^i(\bar{z})$ along the boundary. The map Ω_B also shows up in the formula

$$\alpha_n^i = \frac{1}{2\pi i} \int_C z^n\, J^i(z)\, dz + \frac{1}{2\pi i} \int_C \bar{z}^n\, \left(\Omega_B \bar{J}\right)^i (\bar{z})\, d\bar{z} , \tag{1.18}$$

which is used to obtain the oscillators α_n^i from the local fields J^i and $\bar{J}^i$. These remarks complete our solution of the worldsheet theory. We can now start to study some properties of the solution.

1.1.2 Bulk fields and their properties

Our discussion of the solution begins with a few comments on fields that can be inserted at points in the interior of the upper half-plane. Such fields are referred to as bulk fields. After a few brief remarks on the so-called Virasoro fields, we shall explain how to obtain the vertex operators and compute their operator products and the one-point functions in the presence of the boundary.

The stress tensor T

Along with the chiral currents J^μ and $\bar{J}^\mu$, there is another very important pair of chiral fields, namely the components T and $\bar{T}$ of the stress tensor. The holomorphic field T is obtained from the chiral currents J^μ (the index μ is taken from $\mu = 1, \ldots, D$) through the prescription

$$T(z) := \frac{1}{\alpha'} \sum g_{\mu\nu} :J^\mu J^\nu: (z) := \frac{1}{\alpha'} \lim_{w \to z} \left(g_{\mu\nu} J^\mu(w) J^\nu(z) - \frac{\alpha'}{2} \frac{D}{(w-z)^2} \right).$$

$$(1.19)$$

Here, we introduced conformal normal ordering $:\cdot:$ defined by the limiting procedure on the right-hand side. For the anti-holomorphic partner $\bar{T}$ we employ the same construction with currents $\bar{J}^\mu$ instead of J^μ. The boundary conditions (1.9, 1.17) for the chiral currents imply that the two components of the stress tensor coincide along the boundary $z = \bar{z}$,

$$T(z) = \bar{T}(\bar{z}). \qquad (1.20)$$

The relation (1.20) can be seen to prevent worldsheet momentum from leaking out across the boundary of Σ. Technically, it allows us to construct the following modes,

$$L_n := \frac{1}{2\pi i} \int_C z^{n+1} T(z)\, dz - \frac{1}{2\pi i} \int_C \bar{z}^{n+1} \bar{T}(\bar{z})\, d\bar{z}. \qquad (1.21)$$

The elements L_n can also be expressed in terms of oscillators α_n^μ through the so-called Sugawara construction,

$$L_n = \frac{1}{2} \sum_{n=-\infty}^{\infty} G_{ij}^{-1} \, {}^\circ_\circ \alpha_{m-n}^i \alpha_n^j {}^\circ_\circ + \frac{1}{2} \sum_{n=-\infty}^{\infty} g_{ab} \, {}^\circ_\circ \alpha_{m-n}^a \alpha_n^b {}^\circ_\circ, \qquad (1.22)$$

where G^{-1} denotes the inverse of G and it appears in the expression for L_n because of the factors $(1 \pm B)$ in equations (1.15, 1.16). The symbol ${}^\circ_\circ \cdot {}^\circ_\circ$ stands for operator normal ordering, i.e. it instructs us to move all the annihilation operators $\alpha_n^\mu, n \geq 0$, to the right of the creation operators $\alpha_n^\mu, n < 0$. The modes (1.22) obey the defining relations of the Virasoro algebra

$$[L_n, L_m] = (n-m)L_{n+m} + \frac{c}{12} n(n^2-1)\delta_{n+m,0}, \qquad (1.23)$$

with central charge $c = D$. The appearance of this algebra is the most characteristic feature of scale invariant two-dimensional field theories (see also Chapter 3). We shall often refer to T and $\overline{T}$ as Virasoro fields.

For later reference let us stress that the formula (1.22) simplifies when we set $B = 0$. Since this choice implies $G = g^{-1}$, the expression for L_n assumes the more standard form

$$L_n = \frac{1}{2} \sum_{n=-\infty}^{\infty} g_{\mu\nu} {}^{\circ}_{\circ} \alpha^{\mu}_{m-n} \alpha^{\nu}_{n} {}^{\circ}_{\circ} . \tag{1.24}$$

The same Sugawara construction is used in theories on the full complex plane to obtain the modes L_n and $\overline{L}_n$ of the left- and right-moving Virasoro algebras from the left- and right-moving modes α^{μ}_n and $\overline{\alpha}^{\mu}_n$ of the U(1) currents, respectively. In a bulk CFT on the full plane, the left and right movers are independent, but this is no longer the case when we consider a theory on the half-plane and impose, say, Dirichlet or Neumann-type boundary conditions along the boundary: in the theory on the half-plane, only a single copy (1.21) of the Virasoro algebra acts on the state space of our theory. It will always be clear from the context, i.e. from whether the L_n acts on a state from the boundary theory or from a theory on the full plane, which Virasoro generators are meant.

Bulk vertex operators

Let us now turn to another family of bulk fields that we shall use to probe the effect of boundary conditions on the bulk. These "tachyon vertex operators" are defined by the expression

$$\phi_{k,k}(z, \bar{z}) := \,:e^{ikX(z,\bar{z})}: \,= \sum_{n=1}^{\infty} \frac{i^n}{n!} :(kX)^n:(z, \bar{z}) . \tag{1.25}$$

Here the "momentum" $k = (k_\mu)$ is a vector with D components, as is X. The nth-order normal ordered product is defined recursively by subtracting lower-order terms where two of the fields have been "contracted". Very explicit formulas can be found e.g. [377], but the procedure is familiar from Wick's theorem. Let us, however, emphasise that the contractions occurring in equation (1.25) are computed as two-point functions (propagators) of the free boson bulk theory on the full plane, i.e. using

$$\langle X^\mu(z_1, \bar{z}_1) X^\nu(z_2, \bar{z}_2) \rangle_{\text{bulk}} = -\alpha' g^{\mu\nu} \ln|z_1 - z_2| .$$

Bulk correlators were also used to define the normal ordering in the formula (1.19) for the bulk stress energy tensor.

The bulk free boson propagator differs from the propagator on the upper half-plane,

$$\langle X^\mu(z_1, \bar{z}_1) X^\nu(z_2, \bar{z}_2) \rangle_{\text{bdy}} = -\alpha' g^{\mu\nu} \ln|z_1 - z_2| + \alpha' g^{\mu\nu} \ln|z_1 - \bar{z}_2|$$

$$-\alpha' G^{\mu\nu} \ln|z_1 - \bar{z}_2|^2 - \Theta^{\mu\nu} \ln \frac{z_1 - \bar{z}_2}{\bar{z}_1 - z_2} , \tag{1.26}$$

by terms which are regular in the upper half-plane $\Im z > 0$ and become singular only along the boundary. In writing down equation (1.26), we have promoted G and Θ to $D \times D$ matrices by filling up with zeroes.

For some computations, in particular for evaluating correlation functions of bulk fields in the boundary theory, it is advantageous to rewrite the bulk vertex operator $\phi_{k,k}(z, \bar{z})$ in terms of the oscillator modes that occur in the boundary theory, i.e. of the α_n^μ and the $\hat{x}^i$ from the expansions (1.5) and (1.10) above. In particular, we want to use operator normal ordering as in the formula (1.22), i.e. we move creation operators (including, by convention, $\hat{x}^i$) to the left of the annihilation operators. Using the commutation relations leads to the expression

$$\phi_{k,k}(z, \bar{z}) := \frac{1}{|z - \bar{z}|^{\alpha'(k^2/2 - k \diamond k)}} \, {}^\circ_\circ \, e^{ikX(z,\bar{z})} \, {}^\circ_\circ \,. \tag{1.27}$$

We use $kk' = g^{\mu\nu} k_\nu k'_\mu$ as before, and we have introduced the new product

$$k \diamond k' = G^{ij} k_i k'_j \tag{1.28}$$

for the momentum components along directions with Neumann-type boundary conditions. In the case of pure Dirichlet boundary conditions, i.e. $d = 0$, the prefactor diverges with an exponent $\alpha' kk = \alpha' k^2$. For $d = D$ and $B = 0$, on the other hand, the product $k \diamond k$ is given by $k \diamond k = k^2$ and hence the exponent becomes $-\alpha' k^2$. Note the relation of the exponent of $|z - \bar{z}|$ to the second and third term in the propagator (1.26).

Correlation functions of the fields (1.25) will have singularities when two of the bulk fields come close to each other,

$$\phi_{k,k}(z, \bar{z}) \, \phi_{k',k'}(w, \bar{w}) = |z - w|^{\alpha' kk'} \phi_{(k+k'),(k+k')}(w, \bar{w}) + \cdots, \tag{1.29}$$

where the dots stand for terms that are less singular in the limit $z - w \to 0$. This can be worked out from equation (1.27), and one finds that the leading singularities agree with those in the operator product expansions (OPEs) of the vertex operators (1.25) in the free field theory in the full complex plane. Note, in particular, that the $|z - w|$-exponent in the expansion (1.29) neither depends on the B-field nor on the parameter d, i.e. there is no dependence on the boundary condition. On the other hand, the leading singularities in bulk operator products encode complete information about the metric g that determines the bulk dynamics.

The fact that the singularities in OPE are independent of the boundary conditions does not, however, imply that correlators of bulk fields on a surface with a boundary are the same as on the complex plane. For example, correlators in the boundary theory diverge whenever one of the bulk fields approaches the boundary.

One-point functions and boundary states

The simplest correlators one can consider are one-point functions of bulk fields. They are trivial in a CFT on the whole plane, but non-trivial if the worldsheet has a boundary. The expression (1.27) makes it easy to calculate these quantities for the tachyon vertex operators,

$$\langle \phi_{k,k}(z,\bar{z}) \rangle_{x_0} = \delta^{(d)}(k_i) \frac{e^{ik_a x_0^a}}{|z - \bar{z}|^{\alpha' k^2/2}} . \tag{1.30}$$

The delta function for momenta along the Neumann directions appears because the $\exp\{i\,k\hat{x}\}$ contained in $\phi_{k,k}$ shifts the momentum by k, so that only $k_i = 0$ survives taking the vacuum expectation value.

From the set of all these one-point functions we can recover the parameters x_0^a, i.e. the boundary values of the bosonic field's position are completely specified by the one-point functions of the bulk vertex operators. We shall see in Chapter 6 that to some extent a similar statement remains true for branes in curved backgrounds. The exponent in the denominator on the right-hand side of equation (1.30) involves the (left- and right-moving) *conformal dimensions*

$$h(\phi_{k,k}) = \frac{\alpha'}{4} k^2 = \bar{h}(\phi_{k,k}) \tag{1.31}$$

of the bulk field $\phi_{k,k}$. It describes the behaviour of the vertex operator under rescalings $z \to \lambda z$ of the coordinates z of the worldsheet. Note that the exponent in the denominator of equation (1.30) only involves the sum of h and $\bar{h}$.

Another interesting quantity is the following one-point function of descendants fields

$$\langle :J^{\mu}(z)\bar{J}^{\nu}(\bar{z})\phi_{k,k}(z,\bar{z}): \rangle_{x_0} = \delta^{(d)}(k_i) \frac{\Omega_{(B,d)}^{\mu\nu}\, e^{ik_a x_0^a}}{|z - \bar{z}|^{2+\alpha' k^2/2}}, \tag{1.32}$$

where we have extended the $d \times d$ matrix Ω_B defined in equation (1.17) to a $D \times D$ matrix $\Omega_{(B,d)}$ such that the gluing conditions (1.9) and (1.17) may be written in a single equation

$$J^{\mu}(z) = \Omega_{(B,d),\nu}^{\mu}\, \bar{J}^{\nu}(\bar{z}), \quad \text{where} \quad \Omega_{(B,d)} = \begin{pmatrix} \Omega_B & 0 \\ 0 & -1_{D-d} \end{pmatrix}. \tag{1.33}$$

For $k = 0$, the one-point function (1.32) can be obtained through differentiation of the propagator (1.26).

Since there is an infinite number of bulk fields, there are infinitely many one-point amplitudes that one may compute. The information about all bulk one-point functions may be stored in a single object, the so-called boundary state $\| B, q; x_0 \rangle\rangle$. We shall think of this object as a product

$$\| B, q; x_0 \rangle\rangle \equiv \| N(B) \rangle\rangle \otimes \| D(x_0) \rangle\rangle \tag{1.34}$$

of a boundary state $\| N(B) \rangle\rangle$ for the first d components of our bosonic field and another factor $\| D(x_0) \rangle\rangle$ that deals with the remaining ones. We have used the notation $N(B)$ and $D(x_0)$ to remind ourselves that the first d components satisfy Neumann-type boundary conditions with a B-field, while we impose Dirichlet boundary conditions along the other $q = D - d$ directions. Let us first discuss the construction of the second factor $\| D(x_0) \rangle\!\rangle$ for components satisfying Dirichlet boundary conditions. In order to build the boundary state, we introduce the following family of coherent states,

$$| k, k \rangle\!\rangle_D = \exp\left(\sum_{n=1}^{\infty} \frac{g_{ab}}{n} \, \alpha_{-n}^a \, \overline{\alpha}_{-n}^b \right) | k \rangle \otimes \overline{| k \rangle} . \tag{1.35}$$

Let us stress that these coherent states are constructed from the ground states $| k \rangle \otimes \overline{| k \rangle}$ of the bulk theory. These are obtained as products of a ground state $| k \rangle$ and $\overline{| k \rangle}$ for the left- and right-moving current algebras, respectively. The latter are generated by two set of modes α_n^a and $\overline{\alpha}_n^a$ as usual. With the help of the standard commutation relations of α_n^a and $\overline{\alpha}_n^a$, it is not difficult to verify that

$$(\alpha_n^a - \overline{\alpha}_{-n}^a) \, | k, k \rangle\!\rangle_D = 0 . \tag{1.36}$$

Actually, it may be shown that the coherent states (1.35) form a basis for the space of solutions to the linear equations (1.36). The Dirichlet boundary state is a very particular linear combination, in which the coherent states are added up with coefficients in the numerator of the one-point coupling (1.30), i.e.

$$\| D(x_0) \rangle\!\rangle = \int \Pi_{a=d+1}^D (\sqrt{\alpha'/2} \, dk_a) \, e^{ik_a x_0^a} | k, k \rangle\!\rangle_D . \tag{1.37}$$

For the remaining components of the bosonic field X, the boundary state $\| N(B) \rangle\!\rangle$ is constructed similarly. To begin with, we construct solutions to the linear equations

$$(\alpha_n^j + \Omega_{B,i}^j \, \overline{\alpha}_{-n}^i) \, | k, k \rangle\!\rangle_B = 0 . \tag{1.38}$$

Note that the equations are altered compared to equations (1.36) to reflect the difference between the gluing conditions of currents. In this case, equation (1.38) has the solution

$$| 0, 0 \rangle\!\rangle_B = \exp\left(-\sum_{n=1}^{\infty} \frac{(\Omega_B)_{ij}}{n} \, \alpha_{-n}^i \, \overline{\alpha}_{-n}^j \right) | 0 \rangle \otimes \overline{| 0 \rangle} , \tag{1.39}$$

which is unique (up to a factor), and therefore the coherent state coincides with the Neumann boundary state, $\| B \rangle\!\rangle = | 0, 0 \rangle\!\rangle_B$.

In checking that equation (1.39) satisfies the condition (1.38), it must be kept in mind that the commutation relations of the oscillators α_μ^j for left-moving modes of the closed string involve the metric g^{ij}, and similarly for the right-moving modes $\overline{\alpha}_\mu^j$.

The full boundary state for our boundary theory is obtained by multiplying the boundary states (1.37) and (1.39) as in equation (1.34). By construction, the full boundary state has the property

$$\langle \phi(z, \bar{z}) \rangle_{(B,d);x_0} = \frac{\langle\!\langle B, d; x_0 \| \phi \rangle}{|z - \bar{z}|^{h_\phi + \bar{h}_\phi}}, \tag{1.40}$$

where $|\phi\rangle = \phi(0,0)|0\rangle \otimes |\bar{0}\rangle$ is the state of the bulk theory that is associated with the field ϕ. In other words, the boundary state is defined such that its overlap with $|\phi\rangle$ provides the coupling between the bulk field ϕ and the boundary. It is easy to check that the formula (1.40) reproduces the equations (1.30) and (1.32). For equation (1.30) we explicitly put the coupling into the boundary state. The conditions (1.36) and (1.38) guarantee that relation (1.40) is also true for descendants.

1.1.3 Boundary fields and their properties

After our discussion of bulk fields we now turn to a new set of fields which can be inserted at points on the boundary of the worldsheet. We will briefly discuss their spectrum before we compute correlators of boundary vertex operators. It is instructive to observe that the results for boundary correlators can be expressed with the help of the non-commutative Moyal–Weyl product. We will review the latter to make the presentation self-contained.

Spectrum of boundary fields

Boundary fields are in one-to-one correspondence with states of the boundary theory. The space of these states was constructed explicitly when we solved the model in the Subsection 1.1.1; it is given by

$$\mathcal{H}_{(B,d)} = \int d^d k \, \mathcal{V}_k \otimes \mathcal{V}_0 .$$

The spaces $\mathcal{V}_0$ and $\mathcal{V}_k$ were introduced around equations (1.7) and (1.14), respectively. Momenta for the directions along the brane are integrated over while the factor $\mathcal{V}_0$ is associated with the transverse space.

The space $\mathcal{H}_{(B,d)}$ is that on which all our bulk and boundary fields act. In particular, through equation (1.21), it carries an action of the Virasoro algebra. Among the Virasoro modes, L_0 is distinguished because it agrees with the Hamiltonian of the worldsheet theory up to a simple shift, namely $H = L_0 - \frac{D}{24}$. Using the explicit formula (1.22) for L_0 it is rather easy to calculate the partition function of the theory,

$$Z_{(B,d)}(q) := \text{tr}_{\mathcal{H}} \left(q^H \right) = \frac{1}{\eta^D(q)} \int d^d k (2\alpha')^{d/2} \left(\det G \right)^{1/2} q^{\alpha' k \diamond k} , \tag{1.41}$$

where $\eta(q) = q^{1/24} \prod_{n=1}^\infty (1 - q^n)$ is Dedekind's eta function. The factor $\eta^{-D}(q)$ counts the states that are created by oscillators with negative mode index. In addition, there are d zero modes, which give rise to the integral over the d-dimensional momentum k. The term $\alpha' k \diamond k$ in the exponent – see equation (1.28) for the definition of the $\diamond$ product – is the conformal weight of primary boundary vertex operators. We see that the partition function $Z(q)$ is an important quantity containing quite detailed information about the boundary condition.

It is interesting to note that one can, in a fairly simple manner, compute the boundary partition function (1.41) from the boundary state we have introduced in the previous subsection. In fact, $Z(q)$ is obtained from the following overlap

$$\langle\!\langle B, d; x_0 \| \, \tilde{q}^{(L_0 + \overline{L}_0)/2 - \frac{D}{24}} \, \| B, d; x_0 \rangle\!\rangle = \frac{1}{\eta^D(\tilde{q})} \int d^{D-d}k \, (\alpha'/2)^{D-d} \tilde{q}^{\frac{\alpha'k^2}{4}}$$

$$= Z_{(B,d)}(q) \, . \tag{1.42}$$

Here, we use the variables $\tilde{q} = \exp(-2\pi i/\tau)$ and $q = \exp 2\pi i\tau$; the reason for the relation between q and $\tilde{q}$ will be explained in Section 4.3. The Virasoro modes on the left-hand side of equation (1.42) are those from the theory on the full complex plane, since they act on states from the bulk theory.

In the calculation, we first use the gluing conditions (1.36) and (1.38) along with the quadratic expression of the Sugawara form (1.24) for the generators L_n and $\overline{L}_n$ of the bulk Virasoro algebra to conclude that

$$(L_n - \overline{L}_{-n}) \| B, d; x_0 \rangle\!\rangle = 0 \, . \tag{1.43}$$

When applied to the case $n = 0$, this gluing condition for the Virasoro generators allows us to trade $\overline{L}_0$ for L_0. Next we exploit the fact that

$$_D\langle\!\langle k, k | \tilde{q}^{L_0 - \frac{D}{24}} | k, k \rangle\!\rangle_D = \frac{1}{\eta(\tilde{q})^{D-d}} \, \tilde{q}^{\frac{\alpha'k^2}{4}} \, ,$$

$$_B\langle\!\langle 0, 0 | \tilde{q}^{L_0 - \frac{D}{24}} | 0, 0 \rangle\!\rangle_B C = \frac{1}{\eta(\tilde{q})^D} \, . \tag{1.44}$$

In deriving the two formulas in equation (1.44) we have expanded the exponentials in the expressions (1.35) and (1.39) for the coherent states. Since there is no more $\overline{L}_0$ inserted, the anti-holomorphic sector is determined by the scalar products of states appearing in the expansion. The remaining sum over states in the holomorphic component may be written as a trace and then expressed through Dedekind's η-function and the k-dependent exponential with an exponent given by the left-moving conformal dimension (1.31) of the bulk vertex operator $\phi_{k,k}$. Putting all this together, we arrive at the first equality in equation (1.42). The integral arises from the integral over coherent states in equation (1.37). After a modular transformation (see the formulas in Appendix A) from $\tilde{q}$ to q, we end up with the expression (1.41) for the partition function.

The overlap we have computed is only a very special case of a much larger set of quantities that may be extracted from boundary states. In fact, we could have inserted two different boundary states in that computation, e.g. by admitting different values of the coordinates x_0, a different number d of Neumann boundary directions or even boundary conditions associated with two hyperplanes at a non-trivial angle. The resulting quantity counts the number of fields that can be inserted at points where the boundary condition jumps. For two boundary conditions $(B = 0, d, x_0)$ and $(B = 0, d, x_0')$ describing two parallel hyperplanes separated by a distance $\Delta x = x_0 - x_0'$ the overlap of boundary states is easy to compute,

$$Z_{(d,x_0)(d,x_0')}(q) = \frac{q^{\frac{1}{4\alpha'}\left(\frac{\Delta x}{2\pi}\right)^2}}{\eta^D(q)} \int d^d k (2\alpha')^{d/2} q^{\alpha' k^2} . \tag{1.45}$$

In comparison to equation (1.41) we have set $B = 0$ so that $k \cdot k = k^2$ and $\det G = 1$. The additional factor involving the distance Δx is associated with the non-trivial vertex operator that must be inserted when the Dirichlet parameter jumps from x_0 to x_0'. In most other cases, the formulas for partition functions look much more complicated when different boundary conditions are imposed at the two ends of the interval. We shall see a few examples in the additional material section at the end of this chapter.

Correlation functions

Following the standard wisdom of CFT, there is a boundary field associated with each state in the space $\mathcal{H}_{(B,d)}$. For the ground states $|k\rangle$, the corresponding fields are the "open string tachyon vertex operators",

$$\psi_k(x) :- {}_\circ^\circ e^{i k_i X^i(x)} {}_\circ^\circ = e^{i k_i X_<^i(x)} e^{i k_i X_>^i(x)} , \tag{1.46}$$

where $k \in \mathbb{R}^d$ and where we used operator normal ordering of the oscillators acting on the state space of the boundary theory, i.e.

$$X_>^\mu(x) = -i\sqrt{2\alpha'}\,\alpha_0^\mu \ln x + i\sqrt{2\alpha'} \sum_{n>0} \frac{\alpha_n^\mu}{n} x^{-n} ,$$

$$X_<^\mu(x) = \hat{x}^\mu + i\sqrt{2\alpha'} \sum_{n<0} \frac{\alpha_n^\mu}{n} x^{-n} .$$

From our exact construction of the theory it is rather straightforward now to compute all the correlation functions of these tachyonic vertex operators. They can be evaluated easily with the help of the Baker–Campbell–Hausdorff formula,

$$\langle \psi_{k_1}(x_1) \cdots \psi_{k_n}(x_n) \rangle = \prod_{r<s} e^{-\frac{i}{2} k_r \times k_s}\, \delta\left(\sum_r k_r\right) |x_r - x_s|^{2\alpha' k_r \diamond k_s} . \tag{1.47}$$

Here, $r, s = 1, \dots, n$, and we have used the notations $k \times k' = k_i\, \Theta^{ij}\, k_j'$ and $k \diamond k' = k_i\, G^{ij}\, k_j'$ as before. Note that boundary correlation functions depend in several ways on the choice of boundary conditions. To begin with, the set of boundary fields it labelled by d-dimensional momentum vectors k, in other words it depends on the number of bosonic fields satisfying the Neumann boundary conditions. Next, the exponents $k_r \diamond k_s$ of $|x_r - x_s|$, and hence the conformal weights

$$h(\psi_k) = \alpha'\, k \diamond k \tag{1.48}$$

of the boundary fields, contain the deformed metric G^{ij} of rank d rather than our original matrix $g^{\mu\nu}$; in particular, G^{ij} depends on the antisymmetric matrix $B_{\mu\nu}$. Finally, the couplings possess a non-trivial dependence on the momenta k_r through the factor $\exp(-\frac{i}{2}k_r \times k_s)$. From equation (1.47) we can read off the short distance singularities as two boundary fields approach each other,

$$\psi_k(x)\,\psi_{k'}(x') = e^{-\frac{i}{2}k\times k'}\,|x - x'|^{2\alpha' k\diamond k'}\,\psi_{k+k'}(x') + \cdots, \qquad (1.49)$$

where the $\cdots$ stand for terms of lower order. The power of the leading singularity is determined by the conformal weights of our boundary fields. One may recall that the phase factor involving the crossed product $k \times k'$ also appears when a linear space with constant Poisson bracket is quantised, and we shall briefly review the relation before concluding this section.

The Weyl product

Let us digress for a moment and recall a few facts about the quantisation of a d-dimensional linear space V. We assume that V comes equipped with a constant antisymmetric $d \times d$ matrix Θ^{ij}, which defines a Poisson bracket for functions on V. More concretely, when evaluated on the coordinate functions $x^j, j = 1, \ldots, d$, the Poisson structure reads,

$$\{\,x^j\,,\,x^l\,\} = \Theta^{jl}\,. \qquad (1.50)$$

Quantisation means to associate a self-adjoint operator $\hat{x}^j : \mathcal{H} \to \mathcal{H}$ on some state space $\mathcal{H}$ to each coordinate function such that the commutators are

$$[\,\hat{x}^j\,,\,\hat{x}^l\,] = i\,\Theta^{jl}\,. \qquad (1.51)$$

More generally, one would like to associate a self-adjoint operator $Q(f)$ to any complex-valued function f on V such the commutator $[Q(f_1), Q(f_2)]$ is approximated by the Poisson bracket $\{f_1, f_2\}$ in a sense that we shall make more precise below. An appropriate mapping $f \mapsto Q(f)$ was suggested by Weyl [446],

$$Q(f) = \int d^d k\, \hat{f}(k)\, \exp(ik_j \hat{x}^j)$$

where $\hat{f}(k)$ denotes the Fourier transform of f. The operator $Q(f)$ is trace class if f is smooth and decreases, together with all its derivatives, faster than the reciprocal of any polynomial at infinity, but $Q(f)$ is defined for more general functions, in particular $Q(x^j) = \hat{x}^j$. A detailed discussion of appropriate spaces of functions can be found e.g. in [320].

We want to compute the product of any two operators $Q(f)$ and $Q(g)$ and compare this to the operator which Weyl's formula assigns to the Poisson bracket of the two functions f and g. Using the Baker–Campbell–Hausdorff formula it can be derived that

$$Q(f)\,Q(g) = Q(f * g),$$

$$\text{where } (f * g)\,(x) = \exp\left(\frac{i}{2}\Theta^{jl}\partial_j\tilde{\partial}_l\right)f(x)g(\tilde{x})|_{\tilde{x}=x}. \qquad (1.52)$$

The multiplication $*$ defined in the second line is known as the *Moyal product* [351] associated with the constant antisymmetric matrix Θ. It is an associative and non-commutative product for functions on the d-dimensional space V. Moreover, to leading order in the number of derivatives, it is found that

$$[f \overset{*}{,} g] := f * g - g * f = i\,\Theta^{jl}\,\partial_j f \partial_l g + \cdots = i\,\{f, g\} + \cdots .$$

Hence, the Moyal-commutator of the functions f and g is approximated by the Poisson bracket of these functions. In the same sense, the commutator of the operators $Q(f)$ and $Q(g)$ is approximated by the operator $Q(\{f, g\})$. If we choose f, g to be exponentials, the formula (1.52) specialises to the following formula

$$\exp(ik_j\hat{x}^j)\,\exp(ik'_l\hat{x}^l) = \exp\left(-\frac{i}{2}k_j\Theta^{jl}k'_l\right)\exp(i(k+k')_j\,\hat{x}^j). \qquad (1.53)$$

On the right-hand side, it involves the same phase factor as in our boundary operator product expansion (1.49). To convert this similarity into a precise statement, we introduce the "smeared" boundary vertex operators

$$\psi[f](x) = \int d^d k\,\hat{f}(k)\,\psi_k(x).$$

In the limit $\alpha' \to 0$, in which the singular factor disappears from the expansion (1.49), we obtain

$$\psi[f](x)\,\psi[g](x') \sim \psi[f * g](x'),$$

i.e. in this limit, the product of boundary vertex operators is determined by the non-commutative Weyl product of functions in a d-dimensional space. We will encounter similar phenomena in connection with boundary Wess–Zumino–Witten (WZW) models in Chapter 6. The above derivation of the Moyal product from the free boson boundary CFT with non-vanishing B-field is a special case of Kontsevich's deformation quantisation [322], see also [109].

1.2 Free fermionic field theory

We are now going to extend our previous analysis of free bosonic field theory on the upper half-plane to free two-dimensional fermions. After a brief introduction to the description of massless fermions and their boundary conditions, we shall solve the model and compute a few basic quantities. In the end we shall also comment on how bosonic and fermionic fields can be combined into a model that possesses two-dimensional worldsheet supersymmetry.

1.2.1 Solution of free fermionic field theory

Our objective is to analyse a purely fermionic two-dimensional free field theory involving D fermions $\psi^\mu, \mu = 1, \ldots, D$. The action of the model is given by

$$S[\psi] = \frac{1}{4\pi}\int d^2z\,g_{\mu\nu}\left(\psi^\mu\bar{\partial}\psi^\nu + \bar{\psi}^\mu\partial\bar{\psi}^\nu\right). \qquad (1.54)$$

Here, we integrate again over the upper half-plane with coordinates $z, \bar{z}$. By the bulk equations of motion, the fields $\psi^\mu = \psi^\mu(z)$ are holomorphic while $\bar{\psi}^\mu = \bar{\psi}^\mu(\bar{z})$ depend on $\bar{z}$ only.

Before we discuss boundary conditions, let us comment on a subtlety that also exists when the fermions live on the entire complex plane. In that case, fermionic fields are allowed to pick up a sign upon rotation around the origin at $z = 0$, i.e.

$$\psi^\mu(e^{2\pi i}z) = \epsilon\, \psi^\mu(z) \text{ where } \epsilon = \pm 1. \tag{1.55}$$

Whenever $\epsilon = 1$ the fermion ψ^μ is said to be in the Neveu–Schwarz (NS) sector. Fermions in the Ramond (R) sector satisfy relation (1.55) with a non-trivial sign $\epsilon = -1$. The same distinction exists when the fermions are defined on the upper half-plane.

After these comments we can spell out the boundary conditions we want to consider in this section,

$$\begin{aligned}
\bar{\psi}^j(\bar{z}) &= \psi^j(z)\big|_{z=\bar{z}>0} & \bar{\psi}^j(\bar{z}) &= \epsilon\psi^j(z)\big|_{z=\bar{z}<0} & \text{for} \quad j = 1,\ldots,d, \\
\bar{\psi}^a(\bar{z}) &= -\psi^a(z)\big|_{z=\bar{z}>0} & \bar{\psi}^a(\bar{z}) &= -\epsilon\psi^a(z)\big|_{z=\bar{z}<0} & \text{for} \quad a = d+1,\ldots,D.
\end{aligned}$$
$$\tag{1.56}$$

The gluing conditions in the left column should be compared with the analogous relations (1.9) and (1.38) for chiral currents in the bosonic model. While the conditions (1.9) and (1.56) for the directions $a = d+1,\ldots,D$ are identical (except for the appearance of ϵ), our gluing condition (1.56) for the first d components ψ^j corresponds to Neumann boundary conditions (1.38) for vanishing B-field $B = 0$. We could have used the more general matrix Ω_B defined in equation (1.38) for fermions as well, but – for simplicity – decided to set $B = 0$ in most of our discussion of the fermionic sector. The extension to non-vanishing B-field is straightforward.

As in the case of free bosons, see equation (1.33), we can restate the gluing conditions (1.56) in a single line,

$$\begin{aligned}
\bar{\psi}^\mu(\bar{z}) &= \Omega^\mu_{(D,d);\nu}\, \psi^\nu(z)\big|_{z=\bar{z}>0} \\
\bar{\psi}^\mu(\bar{z}) &= \epsilon\Omega^\mu_{(D,d);\nu}\, \psi^\nu(z)\big|_{z=\bar{z}<0}
\end{aligned}
\qquad \text{where } (\Omega^\mu_{\ \nu}) = \left(\Omega^\mu_{(D,d);\nu}\right) = \begin{pmatrix} \mathbf{1}_d & 0 \\ 0 & -\mathbf{1}_{D-d} \end{pmatrix},$$
$$\tag{1.57}$$

with $D - d$ being the number of components satisfying the gluing conditions in the second line of equations (1.56). If we want to allow for a non-vanishing B-field in the fermionic theory, we simply replace $\Omega_{(D,d)}$ by the more general $\Omega_{(B,d)}$ that was defined in equation (1.33).

In order to implement the distinction between Neveu–Schwarz and Ramond sectors, we had to split the real line into its positive and negative parts. The boundary conditions we impose along the two half-lines differ by a sign ϵ. Consequently, when a fermionic field $\psi^\mu(z)$ is moved around the origin, it picks up an overall sign ϵ, just as given in equation (1.55).

Solution of the boundary problem

Since the fermionic model is very easy to solve, we shall give explicit formulas for both types of boundary conditions at the same time. In the Ramond sector, the fields take the form

$$\psi^j(z) = \sum_{n\in\mathbb{Z}} \psi^j_n \, z^{-n-\frac{1}{2}}, \qquad \overline{\psi}^j(\bar{z}) = \sum_{n\in\mathbb{Z}} \psi^j_n \, \bar{z}^{-n-\frac{1}{2}},$$

$$\psi^a(z) = \sum_{n\in\mathbb{Z}} \psi^a_n \, z^{-n-\frac{1}{2}}, \qquad \overline{\psi}^a(\bar{z}) = -\sum_{n\in\mathbb{Z}} \psi^a_n \, \bar{z}^{-n-\frac{1}{2}}$$

for $j = 1, \ldots, d$ and $a = d+1, \ldots, D$, as in the previous section. Similarly, for the Neveu–Schwarz sector we may solve the theory through

$$\psi^j(z) = \sum_{r\in\frac{1}{2}+\mathbb{Z}} \psi^j_r \, z^{-r-\frac{1}{2}}, \qquad \overline{\psi}^j(\bar{z}) = \sum_{r\in\frac{1}{2}+\mathbb{Z}} \psi^j_r \, \bar{z}^{-r-\frac{1}{2}},$$

$$\psi^a(z) = \sum_{r\in\frac{1}{2}+\mathbb{Z}} \psi^a_r \, z^{-r-\frac{1}{2}}, \qquad \overline{\psi}^a(\bar{z}) = -\sum_{r\in\frac{1}{2}+\mathbb{Z}} \psi^a_r \, \bar{z}^{-r-\frac{1}{2}}.$$

Since the fermionic fields ψ^μ possess conformal weight $h_\psi = \frac{1}{2}$, their mode expansion contains a factor $z^{-1/2}$. The latter is compensated in the Neveu–Schwarz sector by summing over the half-integer r. Thereby the fermionic fields satisfy the periodicity condition (1.55). The fermionic modes ψ^μ_n and ψ^μ_r obey anti-commutation relations of the form

$$\{\psi^\mu_n, \psi^\nu_m\} = g^{\mu\nu}\delta_{m+n,0} \quad (\mathrm{R}),$$
$$\{\psi^\mu_r, \psi^\nu_s\} = g^{\mu\nu}\delta_{r+s,0} \quad (\mathrm{NS}).$$

Note that, in contrast to the boundary theory, the left- and right-moving fermion fields $\psi(z)$ and $\overline{\psi}(\bar{z})$ have independent (anti-commuting) modes if the theory is defined on the full complex plane; those bulk field modes will appear when we work with boundary states below.

Let us now construct the state space of the boundary theory. This is particularly simple for the Neveu–Schwarz sector, which has a unique ground state $|0\rangle$ such that

$$\psi^\mu_r |0\rangle = 0 \quad \text{for} \quad r > 0. \tag{1.58}$$

Consequently, the space $\mathcal{H}^D_{\mathrm{NS}}$ is spanned by states of the form

$$\psi^{\mu_1}_{r_1} \ldots \psi^{\mu_i}_{r_i} |0\rangle \quad \text{where} \quad r_1 < r_2 < \cdots < r_i \leq -\frac{1}{2}. \tag{1.59}$$

Since the fermionic modes ψ^μ_r obey $\psi^\mu_r \psi^\mu_r = 0$, a creation operator ψ^μ_r can be applied at most once. This explains why, on the above basis, we can restrict to $r_i < r_{i+1}$. Note that the state space in the Neveu–Schwarz sector does not depend on d.

The Ramond sector is a bit more difficult to deal with since this time there are zero modes ψ_0^μ. These form the D-dimensional Clifford algebra. For simplicity, let us assume that D is even (for applications to string theory we will set $D = 8$ later). In this case, the zero modes possess a unique representation on a $2^{D/2}$-dimensional space of ground states. To see this, let us combine the zero modes into the following pairs of operators

$$\Psi_{0\varrho} := \frac{1}{\sqrt{2}}\left(\psi_0^{2\varrho} - i\psi_0^{2\varrho-1}\right), \quad \Psi_{0\varrho}^\dagger := \frac{1}{\sqrt{2}}\left(\psi_0^{2\varrho} + i\psi_0^{2\varrho-1}\right).$$

The new fermionic operators $\Psi_{0\varrho}$ and $\Psi_{0\varrho}^\dagger$ provide $D/2$ pairs of fermionic creation and annihilation operators, i.e.

$$\{\Psi_{0\varrho}, \Psi_{0\varrho}^\dagger\} = 1$$

for $\varrho = 1, \ldots, D/2$. (For definiteness, we have assumed that the target carries the standard Euclidean metric here.) The space of ground states can therefore be generated with $\Psi_{0\varrho}^\dagger$ from a state $|R\rangle$ with the properties

$$\Psi_{0\varrho}^\dagger|R\rangle = 0 \text{ for } \varrho = 1, \ldots, D/2, \quad \psi_n^\mu|R\rangle = 0 \text{ for } n > 0. \tag{1.60}$$

Since there are $D/2$ fermionic creation operators, the space of ground states has dimension $2^{D/2}$. From this space we can now generate the entire Ramond sector $\mathcal{H}_R$ by application of the raising operators ψ_n^μ with $n < 0$,

$$\psi_{n_1}^{\mu_1} \ldots \psi_{n_i}^{\mu_i} \prod_{\varrho=0}^{D/2} \Psi_{0\varrho}^{M_\varrho}|R\rangle \quad \text{with } n_1 < n_2 < \cdots < 0, \quad M_\varrho \in \{0, 1\}. \tag{1.61}$$

Equations (1.58,1.59) define the state space of the Neveu–Schwarz sector $\mathcal{H}_{NS}$ and (1.60,1.61) the state space of the Ramond sector $\mathcal{H}_R$. The total state space of our fermionic boundary theory is given by the sum $\mathcal{H}_{NS} \oplus \mathcal{H}_R$. It does not depend on the choice of boundary condition, i.e. on the parameter d.

1.2.2 Some properties of the fermionic theory

Since the zero modes of fermionic fields square to zero, their bulk and boundary theories are not as rich as in the bosonic case. In the following paragraphs, we shall make a few comments on the Virasoro field, the boundary state and the boundary partition function.

The Virasoro field

As in the case of bosonic fields we can construct two chiral Virasoro fields out of the fermionic constituents ψ and $\overline{\psi}$. We have

$$T(z) = -\frac{1}{2}\, g_{\mu\nu} :\psi^\mu(z)\partial\psi^\nu(z):$$

in the holomorphic sector, and an analogous formula involving $\overline{\psi}^\mu$ for $\overline{T}$. All state spaces for our fermionic field theory therefore carry a representation of the Virasoro algebra. Since the Virasoro fields T and $\overline{T}$ are quadratic in the fermions, they are glued trivially along the boundary, $T(z) = \overline{T}(\bar{z})$, as long as the fermions satisfy the gluing condition (1.57); compare this with equation (1.20) for the bosonic model. Furthermore, we can employ equation (1.21) to recover the following Virasoro generators,

$$L_m^R = \frac{1}{2} \sum_{n \in \mathbb{Z}} \left(n + \frac{m}{2} \right) g_{\mu\nu} :\psi_{-n}^\mu \psi_{m+n}^\nu: + \frac{D}{16} \delta_{m,0},$$

$$L_m^{NS} = \frac{1}{2} \sum_{r \in \frac{1}{2} + \mathbb{Z}} \left(r + \frac{m}{2} \right) g_{\mu\nu} :\psi_{-r}^\mu \psi_{m+r}^\nu: .$$

The shift of L_0^R by a constant is necessary for the L_m^R to satisfy the usual Virasoro commutation relations. With a bit of extra work we find that the central charge of both L_m^R and L_m^{NS} is given by $c = D/2$, as opposed to $c = D$ in the case of free bosons.

Note that we have used the symbol $: \cdot :$ in the equations for the Virasoro modes, even though they obviously involve operator normal ordering. Since it is usually clear from the context whether we mean this type of normal ordering or the conformal normal ordering involving subtractions of free field correlators, we employ the simpler notation $: \cdot :$ in all cases, from now on.

Construction of the boundary state

Once more, we would like to construct a boundary state to store information about the one-point functions of our boundary theory. In our discussion we will now switch to the form (1.57) of the gluing conditions so that all fermionic components can be considered in one go. On the other hand, we still have to distinguish between the Ramond and Neveu–Schwarz sector of the model. Once again, the boundary state for the Neveu–Schwarz sector is a bit simpler,

$$\| D, d \rangle\!\rangle^{(NS)} \equiv | NS, NS \rangle\!\rangle_\Omega := \exp \left\{ i \sum_{r \in \mathbb{Z} + \frac{1}{2}} \Omega_{\mu\nu} \psi_{-r}^\mu \overline{\psi}_{-r}^\nu \right\} | NS \rangle \otimes \overline{| NS \rangle}. \tag{1.62}$$

Here, the ψ and $\overline{\psi}$ are left- and right-moving fermion modes acting on the bulk state space, and the whole exponential acts on the unique ground state of the NS–NS sector of the bulk theory. It involves the matrix elements $\Omega_{\mu\nu} = g_{\mu\rho} \Omega^\rho{}_\nu$ of the gluing matrix which we introduced in equation (1.57). The latter depends only on the integer d that is used to label different fermionic boundary states. As in the case of Neumann boundary conditions for the free bosons, the boundary state is a coherent state. It is the unique solution of the following set of linear equations

$$\left(\psi_r^\mu - i\Omega^\mu{}_\nu \overline{\psi}_{-r}^\nu \right) | NS, NS \rangle\!\rangle_\Omega = 0 \tag{1.63}$$

for all $r \in \frac{1}{2} + \mathbb{Z}$ and $\mu = 1, \ldots, D$. The general results to be explained in Section 4.3 below will show that the boundary state $\| D, d \rangle\!\rangle^{(\mathrm{NS})}$ encodes the one-point couplings, as we found in equation (1.40) for the free boson case.

In the Ramond sector, things work analogously. Once more, the entire boundary state can be written as a single coherent state,

$$\| D, d \rangle\!\rangle^{(\mathrm{R})} := |R, R\rangle\!\rangle_\Omega := \exp\left\{ i \sum_{n=1}^{\infty} \Omega_{\mu\nu}\, \psi^{\mu}_{-n} \overline{\psi}^{\nu}_{-n} \right\} |R, R\rangle_{(D,d)}. \tag{1.64}$$

The matrix Ω is the same as in equation (1.62). The only object that requires some further attention is the specific ground state $|R, R\rangle_{(D,d)}$ on which we act with our exponential. Since we want our entire boundary state to solve the following linear constraints

$$\left(\psi^{\mu}_{n} - i\Omega^{\mu}{}_{\nu}\overline{\psi}^{\nu}_{-n} \right) \| D, d \rangle\!\rangle^{(\mathrm{R})} = 0 \tag{1.65}$$

for all $n \in \mathbb{Z}$, the object $|R, R\rangle$ must obey equation (1.65) for $n = 0$. In order to construct such a state among all the 2^D ground states of the Ramond sector in the bulk theory, it is useful to introduce the operators

$$\psi^{\mu}_{\pm} := \frac{1}{\sqrt{2}} \left(\psi^{\mu}_{0} \pm i\overline{\psi}^{\mu}_{0} \right). \tag{1.66}$$

The only non-vanishing anti-commutators among the $\psi^{\mu}_{\pm}$ are given by

$$\{ \psi^{\mu}_{+}, \psi^{\nu}_{-} \} = g^{\mu\nu}.$$

Let us now define the state $|R, R\rangle$ to be the unique Ramond ground state that is annihilated by all the operators ψ^{μ}_{+}. From this we generate the $|R, R\rangle_{(D,d)}$ through application of some ψ^{α}_{-}, $a = d+1, \ldots, D$,

$$|R, R\rangle_{(D,d)} = \prod_{a=d+1}^{D} \psi^{a}_{-} |R, R\rangle$$

$$\text{where} \quad \psi^{\mu}_{+} |R, R\rangle = 0 \text{ for all } \mu = 1, \ldots, D \tag{1.67}$$

Since the definition of the operators $\psi^{\mu}_{\pm}$ mixes left and right movers from the bulk theory, the ground states $|R, R\rangle_{(D,d)}$ cannot be written as a simple product of a left- and a right-moving ground state, in contrast to the Neveu–Schwarz sector.

The boundary spectrum

We have described the construction of the state space in some detail above. Now we would like to compute the partition function of the theory by tracing over all states. There are actually two different quantities that we are interested in, namely the trace itself and the super-trace. The latter contains an additional insertion of $(-1)^F$ where F is the fermion number operator, counting how many ψ appear in a state (1.59) or (1.61).

Let us begin by discussing the ordinary trace over the state space $\mathcal{H} = \mathcal{H}_{\mathrm{NS}} \oplus \mathcal{H}_{\mathrm{R}}$ of our fermionic boundary theories

$$Z^+(q) = \mathrm{tr}_{\mathcal{H}}\left(q^H\right) = \chi_{\mathrm{NS}}(q) + \chi_{\mathrm{R}}(q)$$

$$= q^{-\frac{D}{48}} \prod_{n=0}^{\infty}(1+q^{n+\frac{1}{2}})^D + q^{\frac{D}{24}}\prod_{n=0}^{\infty}(1+q^n)^D = f_3(q^{\frac{1}{2}})^D + f_2(q^{\frac{1}{2}})^D$$

$$(1.68)$$

where $H = L_0 - D/48$ and where the functions f_i are defined through ϑ-functions as in Appendix A. If we insert $(-1)^F$, the corresponding quantity turns out to be

$$Z^-(q) = \mathrm{tr}_{\mathcal{H}}\left((-1)^F q^H\right) = \chi_{\widetilde{NS}}(q) + \chi_{\tilde{R}}(q)$$

$$= q^{-\frac{D}{48}}\prod_{n=0}^{\infty}(1-q^{n+\frac{1}{2}})^D + q^{\frac{D}{24}}\prod_{n=0}^{\infty}(1-q^n)^D = f_4(q^{\frac{1}{2}})^D. \quad (1.69)$$

Note that the contribution $\chi_{\tilde{R}}(q)$ from the Ramond sector vanishes since even and odd states come in equal numbers on each energy level.

In the string theory context, a particularly important role is played by the sum of the functions $Z^\pm(q)$,

$$Z^F(q) := \frac{1}{2}\left(Z^+(q) + Z^-(q)\right) = \frac{1}{2}\left(f_3(q^{\frac{1}{2}})^D + f_2(q^{\frac{1}{2}})^D + f_4(q^{\frac{1}{2}})^D\right). \quad (1.70)$$

This is the partition function of a "GSO-projected" fermionic boundary theory; its relevance will become clearer in later chapters. For the time being let us just check that Z^F can be recovered from our boundary states through a computation that is very similar to the one we performed in the bosonic case:

$$\langle\!\langle D,d\| \, \tilde{q}^{(L_0+\bar{L}_0)/2-\frac{D}{48}} \frac{1}{2}\left(1 + (-1)^F\right) \| D,d\rangle\!\rangle$$

$$= \frac{1}{2}\left(f_3(\tilde{q}^{\frac{1}{2}})^D + f_2(q^{\frac{1}{2}})^D + f_4(\tilde{q}^{\frac{1}{2}})^D\right) = Z^F(q).$$

In deriving this result, we followed the same steps as in equation (1.42) and we used the fact that the function f_2 is sent to f_4 under a modular transformation, i.e. $f_2(\tilde{q}^{\frac{1}{2}}) = f_4(q^{\frac{1}{2}})$. Under the same operation, the function f_3 is invariant, $f_3(\tilde{q}^{\frac{1}{2}}) = f_3(q^{\frac{1}{2}})$. Let us stress once more that the partition function of the fermionic boundary theory is independent of the boundary condition, i.e. of the parameter d.

1.2.3 Supersymmetric free field theory

Let us briefly explain how models can be built with worldsheet supersymmetry from fermionic and bosonic free field theories. By adding the two actions (1.1) and (1.54) we obtain a model with $N = 1$ superconformal symmetry. If the number of components D is even, the superconformal algebra may be extended to $N = 2$.

Boundary theories for these models are obtained in a straightforward manner from the previous findings. We shall see that under very simple conditions our boundary conditions preserve worldsheet supersymmetry.

Free field theory and $N = 1$ superconformal algebra

The simplest free action depending on the fermionic fields ψ^μ and the familiar bosonic fields X^μ is given by

$$S[\psi, X] = \frac{1}{4\pi} \int d^2z \, g_{\mu\nu} \left(\frac{2}{\sqrt{\alpha'}} \partial X^\mu \bar{\partial} X^\nu + \psi^\mu \bar{\partial} \psi^\nu + \bar{\psi}^\mu \partial \bar{\psi}^\nu \right). \qquad (1.71)$$

Being just a sum of the two models we studied before, the analysis of the new model and its quantisation is trivially achieved by combining the results from above. But there is one new aspect that is not present in the building blocks. In fact, it can easily be shown that the action (1.71) possesses a global supersymmetry. This implies that every field in the theory comes with a supersymmetric partner. For the traceless conserved stress tensor $T = T^F + T^B$ of our model the superpartner is

$$G = i\sqrt{\frac{2}{\alpha'}} \, \psi^\mu \partial X_\mu = \sqrt{\frac{2}{\alpha'}} \, \psi^\mu J_\mu \,, \qquad (1.72)$$

and similarly for its anti-holomorphic partner. Note that G is fermionic with conformal weight $h = \frac{3}{2}$. Furthermore, it possesses a Neveu–Schwarz and a Ramond sector, just as its fermionic constituent field ψ^μ. In the bulk and boundary theory, the mode expansion for the field G depends on the sector:

$$G(z) = \begin{cases} \sum_{n \in \mathbb{Z}} G_n z^{-n-\frac{3}{2}} & \text{(R)} \\ \sum_{r \in \mathbb{Z}+\frac{1}{2}} G_r z^{-r-\frac{3}{2}} & \text{(NS)} \end{cases}$$

The modes G_n and G_r can be expressed in terms of the modes α_n^μ and ψ_n^μ or ψ_r^μ of the fields ∂X^μ and ψ as follows:

$$G_n = \sum_m g_{\mu\nu} \, \alpha_m^\mu \psi_{n-m}^\nu \quad \text{(R)}$$

$$G_r = \sum_m g_{\mu\nu} \, \alpha_m^\mu \psi_{r-m}^\nu \quad \text{(NS)} \qquad (1.73)$$

From these formulas one can now determine the commutation relations of G_r and G_n with L_n and among the modes of G. The results read

$$[L_m, G_n] = \left(\frac{m}{2} - n \right) G_{m+n}$$

$$\{G_m, G_n\} = 2L_{m+n} + \frac{D}{2} \left(m^2 - \frac{1}{4} \right) \delta_{m+n,0}$$

in the Ramond sector, and the same for the Neveu–Schwarz sector with $m \to r, n \to s$. Obviously, the modes L_n obey the defining relations (1.23) of a Virasoro algebra with central charge $c = D + \frac{D}{2} = \frac{3D}{2}$.

After these comments on the bulk sector of the model (1.71), let us now turn to the boundary theory. We want to impose boundary conditions on the fermions and bosons such that supersymmetry is preserved. Recall that all our boundary conditions ensure that the Virasoro fields T are trivially glued, see equation (1.20) and the beginning of Subsection 1.2.2. Wordsheet supersymmetry demands that the same is true for the superpartner G of T, i.e.

$$G(z) = \overline{G}(\bar{z})\big|_{z=\bar{z}>0}, \qquad G(z) = \epsilon\, \overline{G}(\bar{z})\big|_{z=\bar{z}<0}. \tag{1.74}$$

The case $\epsilon = 1$ corresponds to the Neveu–Schwarz sector, $\epsilon = -1$ to the Ramond sector. Looking back at how G is composed from fermions and bosons, we conclude that we should use the same gluing map Ω in equations (1.33) for bosons and (1.57) for fermions in order for equation (1.74) to hold. The boundary states are then constructed as a tensor product of the bosonic part (1.34) and the fermionic part (1.62,1.64), namely

$$\| D, d; x_0 \rangle\!\rangle^{(NS)} = \| D, d; x_0 \rangle\!\rangle \otimes \| D, d \rangle\!\rangle^{(NS)},$$
$$\| D, d; x_0 \rangle\!\rangle^{(R)} = \| D, d; x_0 \rangle\!\rangle \otimes \| D, d \rangle\!\rangle^{(R)}.$$

More specifically, $\| D, d; x_0 \rangle\!\rangle$ is the boundary state of the bosonic system with $B = 0$. A non-zero B-field may be admitted, but must then be accompanied by the corresponding gluing condition in the fermionic sector in order to preserve supersymmetry. Similarly, we obtain the boundary partition function

$$Z^{N=1}_{(D,d)}(q) = Z^{B}_{(D,d)}(q) Z^{F}(q)$$

as a product of the bosonic result (1.41) for $Z = Z^B$ and its fermionic counterpart (1.70).

Free field theory and $N = 2$ superconformal algebra

In case the number D of component fields is even, the field theory (1.71) has more structure. In order to describe this, we combine real bosonic and fermionic fields into complex ones,

$$Z^\varrho = X^{2\varrho} + iX^{2\varrho-1}, \qquad \eta^\varrho = \psi^{2\varrho} + i\psi^{2\varrho-1}.$$

In terms of these, the Virasoro field T can then be written as

$$T = -\frac{1}{\alpha'} : \partial Z^\varrho \partial Z^*_\varrho : -\frac{1}{4}\left(:\eta^*_\varrho \partial \eta^\varrho : + :\eta^\varrho \partial \eta^*_\varrho : \right).$$

Its superpartner G admits a very natural split $G = G^+ + G^-$ into two fields $G^\pm$ which are given by

$$G^-(z) = \frac{i}{\sqrt{\alpha'}}\eta^*_\varrho(z)\, \partial Z^\varrho(z), \qquad G^+(z) = \frac{i}{\sqrt{\alpha'}}\eta^\varrho(z)\, \partial Z^*_\varrho(z). \tag{1.75}$$

In addition to the new fields $G^\pm$, we also define another current J through

$$J(z) = -\frac{1}{2} : \eta^*_\varrho(z)\eta^\varrho(z) : . \tag{1.76}$$

The modes of $T, G^{\pm}$ and J generate a $N = 2$ superconformal algebra, meaning that

$$[L_n, G_m^{\pm}] = \left(\frac{n}{2} - m\right) G_{n+m}^{\pm}, \quad [L_n, J_m] = -mJ_{n+m}, \quad [J_n, G_m^{\pm}] = \pm G_{n+m}^{\pm},$$

$$\{G_n^+, G_m^-\} = 2L_{n+m} + (n-m)J_{n+m} + \frac{c}{3}\left(n^2 - \frac{1}{4}\right)\delta_{n,-m}, \quad (1.77)$$

and similarly for the modes in the Neveu–Schwarz sector.

There are two interesting choices of boundary conditions. To begin with, let us assume that we have glued left and right movers through the gluing map $\Omega = \Omega_{(D,d)}$ defined in equation (1.57). Recall that we use the same gluing conditions for bosons and fermions in order to preserve the $N = 1$ superconformal algebra. In our present context it is natural to look for boundary conditions preserving the extended $N = 2$ superconformal symmetry. This imposes additional constraints on our gluing map $\Omega = \Omega_{(D,d)}$. One possibility is to demand that the number d of components satisfying Neumann boundary conditions should be even. Thereby we can ensure that the complex fields are glued as $\partial Z^{\varrho} = \pm \bar{\partial} Z^{\varrho}$ and $\eta^{\varrho} = \pm \bar{\eta}^{\varrho}$ all along the boundary. Note that the same conditions hold for the complex conjugates. Consequently, we find

$$J(z) = \bar{J}(\bar{z}), \quad G^{\pm}(z) = \bar{G}^{\pm}(\bar{z}) \quad \text{for } z = \bar{z} > 0 \qquad (1.78)$$

along with the usual trivial gluing conditions (1.20) for the Virasoro field. For reasons we shall elaborate on in later chapters, the gluing conditions (1.78) of the $N = 2$ superconformal algebra are referred to as "B-type".

Naturally, there are also "A-type" gluing conditions. These arise from a different choice of the gluing map Ω. It corresponds to imposing Neumann boundary conditions on half the component fields and Dirichlet boundary conditions on the others. For definiteness, let us impose Neumann boundary conditions on all the fields X^{μ} and ψ^{μ} with even label μ, then our gluing map reads

$$\Omega = \mathrm{diag}(-1, 1, -1, 1, \ldots, -1, 1).$$

Thereby, we glue the complex fields as $\partial Z_{\varrho} = \bar{\partial} Z_{\rho}^*$ and $\eta_{\varrho} = \bar{\eta}_{\varrho}^*$ to the antiholomorphic partner of their complex conjugate. Since η_{ϱ} are fermionic, we find

$$J(z) = -\bar{J}(\bar{z}), \quad G^{\pm}(z) = \bar{G}^{\mp}(\bar{z}) \quad \text{for} \quad z = \bar{z} > 0. \qquad (1.79)$$

Gluing the fields of an $N = 2$ superconformal algebra according to equations (1.79) is referred to as imposing an "A-type" boundary condition. We will study $N = 2$ supersymmetric boundary theories in more detail in Chapter 7.

1.A Additional material

Throughout our text we shall collect material that is slightly more technical and not indispensable for understanding the gist of later chapters in extra sections at the end of the chapter. The following few pages deal with torus compactifi-cations and their boundary conditions. In the first subsection we analyse a one-dimensional circle with both Dirichlet and Neumann boundary conditions. For a compactified bosonic field these two choices are related by T-duality. Then we put two circles together into a two-dimensional torus, which gives us more freedom in choosing boundary conditions, and in particular allows us to study boundary conditions that correspond to a one-dimensional brane, possibly rotated with respect to the coordinate axes. Through T-duality, such boundary conditions admit various other interpretations, in particular as a two-dimensional space-filling brane with a non-vanishing B-field. Finally, we compute the partition function for a system with two branes that intersect each other at some angle ϕ.

1.A.1 Boundary states for a circle compactification

The compactification of a free bosonic field X to a circle of radius r is described by a free field theory with the additional identification $X \sim X + 2\pi r$. The bulk and boundary spectra of this system get slightly more involved due to the appearance of winding modes. We did not incorporate the case of finite radius r from the very beginning so as to avoid cluttering the exposition in the main sections. But since torus compactifications play an important role, we should not discard them completely. To simplify our presentation, we shall keep the number D of component fields small, and in the present subsection we only consider $D = 1$. Since there is no non-trivial antisymmetric tensor B in one dimension, we only need to distinguish between ordinary Neumann and Dirichlet boundary conditions – as long as we restrict ourselves to boundary conditions that preserve the U(1) current algebra; see Subsection 4.A.5 for a wider class.

Let us then consider a boson compactified on a circle of radius r. The winding number of the map $X : \Sigma \longrightarrow S^1_r$ appears as a new degree of freedom, and at the quantum level the mode expansion of $X(z, \bar{z}) = X_L(z) + X_R(\bar{z})$ is given by

$$X_L(z) = \hat{x}_L - \frac{i}{4} p \ln z - \frac{i}{2} r \hat{w} \ln z + \frac{i}{2} \sum_{n \neq 0} \frac{\alpha_n}{n} z^{-n} \,,$$

$$X_R(\bar{z}) = \hat{x}_R - \frac{i}{4} p \ln \bar{z} + \frac{i}{2} r \hat{w} \ln \bar{z} + \frac{i}{2} \sum_{n \neq 0} \frac{\bar{\alpha}_n}{n} \bar{z}^{-n} \,, \tag{1.80}$$

where we have put $\alpha' = \frac{1}{2}$. We are considering the theory on the full plane for the moment, therefore α_n and $\bar{\alpha}_n$ are independent sets of oscillators.

With $\alpha_0 = \frac{p}{2} + r\hat{w}$, $\bar{\alpha}_0 = \frac{p}{2} - r\hat{w}$ and $x = \hat{x}_L + \hat{x}_R$, $\tilde{x} = \hat{x}_L - \hat{x}_R$, the non-trivial commutation relations of the zero-mode operators are

$$[x, p] = i \,, \qquad [\tilde{x}, \hat{w}] = \frac{i}{2r} \,.$$

The bulk spectrum of our theory contains ground states $|(k, w)\rangle$ labelled by an integer wave number $k \in \mathbb{Z}$ and an integer winding number $w \in \mathbb{Z}$

and satisfying

$$\alpha_0|(k,w)\rangle = \left(\frac{k}{2r} - wr\right)|(k,w)\rangle\,, \qquad \overline{\alpha}_0|(k,w)\rangle = \left(\frac{k}{2r} + wr\right)|(k,w)\rangle\,. \quad (1.81)$$

All other states of the theory on the full complex plane are obtained by acting with the creation operators α_n and $\overline{\alpha}_n$ for $n < 0$, just as in the uncompactified case. The bulk partition function is given by the expression

$$Z(q,\overline{q}) = \frac{1}{|\eta(q)|^2} \sum_{k,w\in\mathbb{Z}} q^{\frac{1}{2}\left(\frac{k}{2r}-wr\right)^2} \overline{q}^{\frac{1}{2}\left(\frac{k}{2r}+wr\right)^2}\,. \quad (1.82)$$

Let us recall that this family of free field theories possesses an important symmetry. In fact, if we replace the holomorphic current J by $-J$, then the commutation relations (1.6) for its modes and the Sugawara construction for L_n remain unchanged. Furthermore, the spectrum of the operators $(-\alpha_0, \overline{\alpha}_0)$ for radius r is identical to the spectrum of $(\alpha_0, \overline{\alpha}_0)$ for radius $\tilde{r} = 1/2r$. This symmetry is known as *T-duality*. Here, the letter T indicates that it is a transformation of the target space which leads to an equivalent model, in contrast to other types of dualities which appear in string theory.

In discussing possible boundary conditions we shall not go back and construct the entire theory from scratch. Instead, let us try to implement our description of boundary theories in terms of boundary states, as outlined in Subsection 1.1.2. This means that we have to solve the gluing conditions (1.38) for $B = 0$ and (1.36) for the boundary states in the state space of the compactified model

$$\left(\alpha_n \pm \overline{\alpha}_{-n}\right)|(k,w)\rangle\!\rangle = 0\,. \quad (1.83)$$

The plus (minus) sign is for Neumann (Dirichlet) boundary conditions, the integers k and w label the ground states in the bulk as above. For $n = 0$, we obtain $\hat{k}\,|(k,w)\rangle\!\rangle_N = 0$ (Neumann) and $\hat{w}\,|(k,w)\rangle\!\rangle_D = 0$ (Dirichlet), respectively. Using the canonical commutation relations, it is straightforward to show that the expressions

$$|(0,w)\rangle\!\rangle_N = \exp\left(-\sum_{n=1}^{\infty} \frac{1}{n}\alpha_{-n}\overline{\alpha}_{-n}\right)|(0,w)\rangle\,, \quad (1.84)$$

$$|(k,0)\rangle\!\rangle_D = \exp\left(\sum_{n=1}^{\infty} \frac{1}{n}\alpha_{-n}\overline{\alpha}_{-n}\right)|(k,0)\rangle \quad (1.85)$$

solve equation (1.83), where the states on the right-hand side are oscillator ground states with $k = 0$ and $w = 0$, respectively. Equations (1.84) and (1.85) extend the formulas (1.35) and (1.39) to the case of a single compactified free boson. Let us point out that we obtain an infinite number of solutions to equation (1.83) for both Dirichlet and Neumann boundary conditions.

We will see in Chapter 4 that, in general, only certain linear combinations of the coherent states (1.84) and (1.85) lead to acceptable boundary states of the theory. The main restriction arises from demanding that overlaps of boundary states produce partition functions of the boundary theory, cf. the computation (1.42) and the analogous one in Subsection 1.2.2.

For Dirichlet boundary conditions of the compactified boson, the previous formula (1.37) can be carried over to our new framework, by restricting the wave number in the coherent states $|(0,k)\rangle\!\rangle_D$ to integer values of k. The boundary state is given by

$$\|D(x_0)\,\rangle\!\rangle = \frac{1}{\sqrt{2r}} \sum_{k\in\mathbb{Z}} e^{ikx_0/r} |(k,0)\rangle\!\rangle_D \,. \tag{1.86}$$

The right-hand side is invariant under shifts of the parameter x_0 by integer multiples of $2\pi r$. Hence, we may still interpret x_0 as the value of the compactified boson X at the boundary $z = \bar{z}$. The normalisation with the prefactor $1/\sqrt{2r}$ is a new feature not present in the uncompactified case and will be justified below.

But first let us turn to the case of Neumann boundary conditions. In flat space, there was a unique coherent state (1.39), which could be identified with the boundary state. Now we have an infinite number of states $|(0,w)\rangle\!\rangle_N$ and we propose to sum them, just as we did for Dirichlet boundary conditions,

$$\|N(\tilde{x}_0)\,\rangle\!\rangle = \sqrt{r} \sum_{w\in\mathbb{Z}} e^{2irw\tilde{x}_0} |(0,w)\rangle\!\rangle_N \,. \tag{1.87}$$

The right-hand side is invariant under shifts of the new parameter $\tilde{x}_0$ by integer multiples of $\pi/r = 2\pi\tilde{r}$. A geometric and physical interpretation of $\tilde{x}_0$ will be discussed in later chapters. For the moment let us explain our expression (1.87) by noting that it is related to the formula (1.86) through T-duality. In fact, if we replace r by $\tilde{r} = 1/2r$ and rename the summation parameter, the coefficients of equations (1.87) and (1.86) are mapped onto each other. Furthermore, T-duality sends α_n to $-\alpha_n$ and $\bar{\alpha}_n$ to itself, hence the generalised coherent state $|(0,w)\rangle\!\rangle_N$ to $|(w,0)\rangle\!\rangle_D$. In other words, Dirichlet and Neumann boundary conditions switch roles upon T-duality, and the form of our boundary states (1.86) and (1.87) is consistent with this field theoretic symmetry. Thus, T-duality already provides one possible geometric meaning of the parameter $\tilde{x}_0$, which we introduced in the boundary state for Neumann boundary conditions: it describes the transverse position parameter for Dirichlet boundary conditions on the T-dual circle.

As promised above, let us explain the normalisation of our boundary states, by computing their overlaps in analogy to equation (1.42). To do this, we first determine the overlaps between pairs of coherent states,

$$_D\langle\!\langle (k,0)|\, \tilde{q}^{L_0-\frac{c}{24}}\, |(k',0)\rangle\!\rangle_D = \delta_{k,k'}\, \tilde{q}^{\frac{1}{2}\left(\frac{k}{2r}\right)^2} \eta(\tilde{q})^{-1} \,,$$

$$_N\langle\!\langle (0,w)|\, \tilde{q}^{L_0-\frac{c}{24}}\, |(0,w')\rangle\!\rangle_N = \delta_{w,w'}\, \tilde{q}^{\frac{1}{2}(2rw)^2} \eta(\tilde{q})^{-1} \,,$$

$$_D\langle\!\langle (k,0)|\, \tilde{q}^{L_0-\frac{c}{24}}\, |(0,w)\rangle\!\rangle_N = \delta_{k,0}\, \delta_{w,0}\, \tilde{q}^{-\frac{1}{24}} \prod_{n=1}^{\infty}(1+\tilde{q}^n)^{-1} \,.$$

The first two lines are completely standard, the third follows because the relative signs in the exponentials of equations (1.84) and (1.85) imply that inner products of states having an odd number of oscillator excitations contribute to the trace with a minus sign.

In order to obtain the partition functions as overlaps of full boundary states, we sum over k and w (with the coefficients in equations (1.86) and (1.87)) and apply Poisson re-summation or the formulas in Appendix A. This yields

$$Z_{N\tilde{x}_0, N\tilde{x}_0'}(q) = \eta(q)^{-1} \sum_{w \in \mathbb{Z}} q^{\frac{1}{2}[w/r + (\tilde{x}_0 - \tilde{x}_0')/\pi]^2} , \tag{1.88}$$

$$Z_{Dx_0, Dx_0'}(q) = \eta(q)^{-1} \sum_{k \in \mathbb{Z}} q^{\frac{1}{2}[2rk + (x_0 - x_0')/\pi]^2} , \tag{1.89}$$

$$Z_{Dx_0, N\tilde{x}_0'}(q) = q^{\frac{1}{48}} \prod_{n=1}^{\infty} (1 - q^{n-\frac{1}{2}})^{-1} . \tag{1.90}$$

If, in the first two lines, we set the $x_0 = x_0'$ and $\tilde{x}_0 = \tilde{x}_0'$, the partition functions describe the spectrum of a single Dirichlet or Neumann boundary condition. In both cases, the vacuum occurs precisely once, as it should: it is this requirement of having a unique vacuum state which poses constraints on the overall normalisation of the boundary state. The third line gives the spectrum of boundary fields which mediate a change from Dirichlet to Neumann boundary conditions at the point $z = \bar{z} = 0$ on the real line. The boundary conditions are therefore similar to those we imposed on fermions in the Ramond sector (see, e.g., equation (1.57)). The change from Dirichlet to Neumann conditions is induced by twist fields, conformal fields of dimension $1/16 + \frac{(2n+1)}{2}$ whose OPEs with the current have square-root branch cuts; see [195] for more details. The term $Z_{ND}(q)$ in equation (1.90) also provides the simplest example of a partition function involving "twisted characters". We observe that it is independent of the parameters $x_0, \tilde{x}_0$. The free boson boundary states and partition functions will enable us to make contact with string theory D-branes in the next chapter. The supersymmetric extension of this simple example is of great importance for brane physics. Note that the fermionic sector of strings on a compactified flat background is the same as in the uncompactified case.

1.A.2 Boundary states for rotated branes

The considerations of the previous subsection can of course be easily generalised from the circle to higher-dimensional toroidal targets, in particular to a rectangular two-torus with radii R_1 and R_2. By forming tensor products of the boundary states from the previous subsection, we can describe boundary conditions where the value of the boson over the boundary is fixed (using a tensor product of two Dirichlet boundary states), or free (using two Neumann boundary states), or confined to a one-dimensional circle within the torus (using a tensor product of a Dirichlet and a Neumann boundary state). In slight abuse of the string terminology (see the next chapter for the common jargon), we will in this subsection refer to the latter situation as a "D1-brane".

Depending on whether we choose Neumann boundary conditions for the first or the second coordinate, the "worldvolume" of this D1-brane extends along the first coordinate cycle or the second. In the following, we want to discuss more general D1-branes whose worldvolume intersects both coordinate axes of the two-torus at a non-zero angle.

To start with, consider a one-dimensional brane in the torus target such that Neumann gluing conditions are prescribed in the X^1-direction and Dirichlet conditions along the second coordinate. Writing $\boldsymbol{a}_n$ for the column vector $(\alpha_n^1, \alpha_n^2)^t$, the gluing conditions for this "unrotated" boundary state ($\phi = 0$) are

$$\left(\boldsymbol{a}_n + \Omega\, \bar{\boldsymbol{a}}_{-n} \right) \|\phi = 0; \tilde{x}_0, x_0\rangle\rangle_0 = 0 \qquad \text{with } \Omega = \Omega_{(1,1)} = \begin{pmatrix} 1 & 0 \\ 0 & -1 \end{pmatrix}, \quad (1.91)$$

where $\tilde{x}_0$ and x_0 are parameters from the boundary states in a circle target, as above. The solutions to these equations are then linear combinations of coherent states

$$|\boldsymbol{k}, \boldsymbol{w}\rangle\rangle_0 = \exp \left\{ \sum_{n>0} \frac{1}{n}\, \boldsymbol{a}_{-n}^t\, \Omega\, \bar{\boldsymbol{a}}_{-n} \right\} |\boldsymbol{k}, \boldsymbol{w}\rangle \qquad (1.92)$$

associated to $U(1)^2$ highest-weight states with certain momentum and winding quantum numbers $\boldsymbol{k} = (k_1, k_2)$ and $\boldsymbol{w} = (w_1, w_2)$, respectively. More precisely, using the zero-mode action

$$\alpha_0^\mu |\boldsymbol{k}, \boldsymbol{w}\rangle = \left(\frac{k_\mu}{2R_\mu} + w_\mu R_\mu \right) |\boldsymbol{k}, \boldsymbol{w}\rangle \quad \text{and} \quad \bar{\alpha}_0^\mu |\boldsymbol{k}, \boldsymbol{w}\rangle = \left(\frac{k_\mu}{2R_\mu} - w_\mu R_\mu \right) |\boldsymbol{k}, \boldsymbol{w}\rangle,$$

equation (1.91) for $n = 0$ shows that only the highest weight states with $k_1 = 0$ and $w_2 = 0$ contribute to boundary states for D1-branes extending in the X^1-direction.

If we rotate this D1-brane by an angle ϕ, it extends along the rotated $^R X^1$-direction and has a fixed locus with respect to the rotated $^R X^2$-coordinate. In order to describe the corresponding boundary state, we have to solve the gluing conditions (1.91) with the old oscillators $\boldsymbol{a}_n, \bar{\boldsymbol{a}}_n$ replaced by

$$^R\boldsymbol{a}_n := R(\phi)\,\boldsymbol{a}_n , \qquad ^R\bar{\boldsymbol{a}}_n := R(\phi)\,\bar{\boldsymbol{a}}_n \qquad \text{with } R(\phi) = \begin{pmatrix} \cos\phi & \sin\phi \\ -\sin\phi & \cos\phi \end{pmatrix}.$$
$$(1.93)$$

This transformation is an outer automorphism of the $U(1)^2$ algebra, leaving the canonical commutation relations of the oscillators, as well as the Virasoro generators L_n, invariant.

The rotated coherent states could be written in the same form (1.92) as for $\phi = 0$ with $^R\boldsymbol{a}_n$ instead of $\boldsymbol{a}_n$. Alternatively, we have

$$\left(\boldsymbol{a}_n + \Omega_\phi\, \bar{\boldsymbol{a}}_{-n} \right) |\boldsymbol{k}, \boldsymbol{w}\rangle\rangle_\phi = 0, \qquad (1.94)$$

where the matrix Ω_ϕ is given by

$$\Omega_\phi := R(\phi)^t \Omega R(\phi) = \begin{pmatrix} \cos 2\phi & \sin 2\phi \\ \sin 2\phi & -\cos 2\phi \end{pmatrix}. \tag{1.95}$$

After the rotation, a different selection of highest-weight states $|k, w\rangle$ contributes to the rotated boundary state. Evaluating equation (1.94) for $n = 0$, it can be seen that the quantum numbers k and w are subject to

$$\begin{pmatrix} \frac{k_1}{R_1} \\ \frac{k_2}{R_2} \end{pmatrix} = -\Omega_\phi \begin{pmatrix} \frac{k_1}{R_1} \\ \frac{k_2}{R_2} \end{pmatrix}, \qquad \begin{pmatrix} w_1 R_1 \\ w_2 R_2 \end{pmatrix} = \Omega_\phi \begin{pmatrix} w_1 R_1 \\ w_2 R_2 \end{pmatrix}. \tag{1.96}$$

Let us assume $0 < \phi < \frac{\pi}{2}$ so that we can write $\tan\phi =: Q\frac{R_2}{R_1}$ for some positive Q. Then we obtain the conditions $k_1 = -Qk_2$, $w_2 = Qw_1$.

If Q is irrational, $k = w = 0$ is the only solution of equations (1.96), and only the coherent state for the vacuum exists. Strictly speaking, the corresponding boundary state

$$|\Xi\rangle := |0, 0\rangle\!\rangle_\phi = \exp\left\{ -\sum_{n>0} \frac{1}{n} {}^R a_{-n}^t \, \Omega \, {}^R \overline{a}_{-n} \right\} |0, 0; 0, 0\rangle, \tag{1.97}$$

should be dismissed since it violates some important consistency conditions to be discussed below. To some extent it could still be interpreted as a "delocalised", rotation invariant superposition of branes. At the angles ϕ in question, the one-dimensional brane wraps around the torus without closing again and forms a dense foliation of the target. In agreement with this spacetime picture, $|\Xi\rangle$ is invariant under translations in coordinate space. Moreover, the corresponding partition function

$$Z_{\Xi\Xi}(q) = \int_{-\infty}^{\infty} dg_1 dg_2 \, \chi_{g_1}(q)\chi_{g_2}(q) \quad \text{with} \quad \chi_g(q) = q^{\frac{g^2}{2}} \eta^{-1}(q) \tag{1.98}$$

is independent of ϕ; it is identical to the partition function of Neumann boundary conditions in uncompactified $\mathbb{R}^2$. Although the target is compactified (which is, e.g., manifest in the set of coherent states that contribute to the boundary state) the boundary operators possess a continuous spectrum, as if the brane was located in $\mathbb{R}^2$.

Let us now turn towards the more interesting case of "rational rotations" where ϕ satisfies the condition

$$\tan\phi = \frac{N}{M}\frac{R_2}{R_1} \quad \text{with} \quad N, M \in \mathbb{N} \text{ coprime}. \tag{1.99}$$

At the angles from this series, the worldvolume of the D1-brane wraps M or N times around the two fundamental cycles of length R_1 and R_2, respectively; see also Figure 1.1. Now, the constraint (1.96) is solved by the highest-weight states

$$|k, w\rangle \equiv |k_1, k_2; w_1, w_2\rangle = |kN, -kM; wM, wN\rangle \quad \text{with} \quad k, w \in \mathbb{Z}.$$

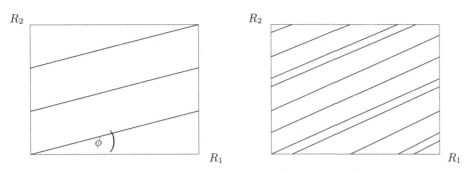

Figure 1.1 A "rational" rotated brane at an angle $\tan\phi = \frac{1}{3}\frac{R_2}{R_1}$ and an unfinished sketch of a "foliating brane" as described by the delocalised boundary state $|\Xi\rangle$.

Note that all the k_μ and w_μ are integers, which is why momenta and windings must be multiples of N or M. Thus, boundary states for a one-dimensional brane on the torus $S^1_{R_1} \times S^1_{R_2}$ rotated by an angle ϕ such that equation (1.99) holds are linear combinations of the coherent states

$$|k,w\rangle\!\rangle_\phi = \exp\left\{-\sum_{n>0}\frac{1}{n}\boldsymbol{a}^{\mathrm{t}}{}_n\,\Omega_\phi\,\overline{\boldsymbol{a}}_{-n}\right\}|kN,-kM;wM,wN\rangle. \qquad (1.100)$$

Here, we have used the matrix Ω_ϕ from equation (1.95) in the exponent in order to have the oscillators $\boldsymbol{a}_n, \overline{\boldsymbol{a}}_n$ acting on the ground states rather than ${}^R\boldsymbol{a}_n$ and ${}^R\overline{\boldsymbol{a}}_n$. The complete boundary states for rotated one-dimensional branes are given by

$$\|\phi;\tilde{x}_0,x_0\rangle\!\rangle = \kappa_\phi \sum_{k,w\in\mathbb{Z}} e^{\frac{ix_0 k}{R'_2}}\, e^{2iR'_1\tilde{x}_0 w}\,|k,w\rangle\!\rangle_\phi\,, \qquad (1.101)$$

with $R'_1 := \frac{MR_1}{\cos\phi}$ and $R'_2 := \frac{R_2\cos\phi}{M}$. Our phase factors follow the formulas (1.86) and (1.87) for a single compactified boson. In fact, as is evident from Figure 1.1, the coordinate for which we impose Neumann boundary conditions parametrises a circle of radius R'_1. The orthogonal circle possesses radius $\frac{R_1 R_2}{R'_1} = R'_2$. Since the corresponding field obeys Dirichlet boundary conditions, we can copy the k-dependent term in the phase factor from equation (1.86). This reasoning can also be used to determine the normalisation κ_ϕ of the boundary state,

$$\kappa_\phi = \sqrt{\frac{R'_1}{2R'_2}} = \frac{M}{\cos\phi}\sqrt{\frac{R_1}{2R_2}} = \frac{N}{\sin\phi}\sqrt{\frac{R_2}{2R_1}}. \qquad (1.102)$$

This choice may be confirmed easily through a direct computation of the overlap. The expressions (1.101) and (1.100) allow us to compute the excitation spectrum of a brane at a "rational" angle in the usual manner,

$$Z_{\phi\,\phi}(q) = \langle\!\langle\phi;\tilde{x}_0,x_0\|\,\tilde{q}^{L_0-\frac{c}{24}}\,\|\phi;\tilde{x}_0,x_0\rangle\!\rangle = \kappa_\phi^2 \sum_{k,w}\chi_{g_1(k,w)}(\tilde{q})\chi_{g_2(k,w)}(\tilde{q})$$

where we used the U(1) characters $\chi_g(\tilde{q}) = \tilde{q}^{\frac{g^2}{2}} \eta^{-1}(\tilde{q})$ that come with the charges

$$g_1(k,w) = \frac{kN}{2R_1} + R_1 wM \,, \qquad g_2(k,w) = -\frac{kM}{2R_2} + R_2 wN \,.$$

Apart from the restriction on the charges that occur in the boundary partition function, the characters are as in the unrotated case since Ω_ϕ is symmetric and orthogonal. Introducing $r := \frac{N^2}{4R_1^2} + \frac{M^2}{4R_2^2}$ and $s := M^2 R_1^2 + N^2 R_2^2$, inserting the normalisation (1.102) of the boundary state, and performing a standard modular transformation yields

$$Z_{\phi\phi}(q) = \eta^{-2}(q) \sum_{k,w} q^{\frac{k^2}{2r}} q^{\frac{w^2}{2s}} \,. \tag{1.103}$$

Let us list some important variants of the above situation. First of all, the boundary state for a rotated brane in an *uncompactified target* satisfies the same gluing conditions (1.94), only the representation content available from the bulk theory is different. The boundary state is

$$\| \phi; x_0 \rangle\!\rangle = \int dk \, e^{ix_0 k} \, |k\rangle\!\rangle_\phi \,, \tag{1.104}$$

with coherent states as in equations (1.97) or (1.100), but built over the highest weights $|k\rangle \equiv |k_1, k_2\rangle = |k \sin\phi, k \cos\phi\rangle$. In the uncompactified case, the brane excitation spectrum does not change under rotations, i.e. $Z_{\phi\phi}(q) = Z_{00}(q)$. The rotation is a symmetry of the target $\mathbb{R}^2$ – in contrast to the two-torus, where a rotation does influence the boundary field content. In our formula (1.104), the parameter x_0 gives the minimal distance of the rotated brane from the origin. It is sometimes preferable to describe the same state in terms of the coordinate $\beta = \beta(x_0, \phi) = x_0 / \cos\phi$ at which the brane intersects the 2-axis. This is achieved by a simple substitution $\tilde{k} = k \cos\phi$ after which the boundary state takes the form

$$\| \phi; x_0 \rangle\!\rangle = \frac{1}{\cos\phi} \int d\tilde{k} \, e^{i\beta\tilde{k}} \, |\tilde{k}\rangle\!\rangle_\phi \,, \tag{1.105}$$

where $|k\rangle\!\rangle_\phi$ is built on the highest weight states $|\tilde{k} \tan\phi, \tilde{k}\rangle$. In terms of the new parameters β and ϕ, the brane is localised along the line $x^2 = \beta + x^1 \tan\phi$.

This second way of writing the state is particularly common if the target is endowed with a *Minkowski signature* and the gluing conditions describe a Lorentz boost with the boost parameter v. This means that we impose the usual gluing conditions (1.91) on the boosted modes

$$^B a_n := B(u) \, a_n \,, \qquad ^B \bar{a}_n := B(u) \, \bar{a}_n \qquad \text{with } B(u) = \begin{pmatrix} \cosh u & \sinh u \\ \sinh u & \cosh u \end{pmatrix} . \tag{1.106}$$

and $u = \text{arctanh}\, v$. To avoid confusion we stress that the letter B stands for "boost" and is not to be confused with the B-field. The matrix Ω_ϕ in equation (1.95) is to be replaced by $\Omega_u := B(-u)\Omega B(u) = \Omega B(2u)$. The associated boundary states have been discussed in many works, e.g. in [57, 87, 123, 261, 292]. They are obtained from equation (1.105) by the replacements $i\phi = u$ and $\sqrt{1 - v^2} := \cosh^{-1} u = \cos^{-1} \phi$, which leads to the boundary state

$$\| v; b \rangle\rangle \;=\; \sqrt{1 - v^2} \int dk \; e^{ibk} \, | \boldsymbol{k} \rangle\rangle_v \; ; \tag{1.107}$$

the coherent states are built over the ground states $|kv, k\rangle$. The boosted brane we have described here is located along a one-dimensional line $x^1 = b + vx^0$ where x^0 is the time-like coordinate and x^1 is space-like.

Next, we want to show that a rotated D1-brane in a two-dimensional target is closely related to the boundary theory of two compactified bosons with Neumann boundary conditions and a B-field switched on. To see this, we start from the gluing conditions (1.94) for a rotated D1-brane and perform a T-duality along one of the two circles, say along the second coordinate X_2. Then we obtain a boundary theory for a torus with radii R_1 and $\tilde{R}_2 = 1/2R_2$. The new gluing conditions for the currents J^1 and $\tilde{J}^2 = -J^2$ are obtained from the gluing map

$$\tilde{\Omega} = \begin{pmatrix} 1 & 0 \\ 0 & -1 \end{pmatrix} \Omega_\phi \begin{pmatrix} 1 & 0 \\ 0 & -1 \end{pmatrix} = \begin{pmatrix} \cos 2\phi & -\sin 2\phi \\ \sin 2\phi & \cos 2\phi \end{pmatrix} .$$

In contrast to Ω_ϕ itself, this new matrix is orthogonal, and can in fact be written in the form

$$\tilde{\Omega} = \frac{1}{1 + b^2} \begin{pmatrix} 1 - b^2 & 2b \\ -2b & 1 - b^2 \end{pmatrix} = \frac{1 + B}{1 - B} = \Omega_B \quad \text{with} \quad B = \begin{pmatrix} 0 & b \\ -b & 0 \end{pmatrix} .$$

In other words, the gluing condition in the T-dual model is precisely of the form (1.38) describing Neumann boundary conditions with a non-vanishing B-field. The angle ϕ and the parameter b of the B-field are related by

$$\frac{1}{\cos \phi} = \sqrt{1 + b^2} = \sqrt{\det(1 + B)} .$$

In the case of Minkowski signature, b gets replaced by $b \to ie$ and the parameter e is interpreted as an electric rather than a magnetic field, see also [1, 89, 123, 248, 249, 261].

1.A.3 Partition functions for branes at relative angles

Here we derive an expression for the partition function $Z_{0,\phi}(q)$ of a system of two D1-branes in a two-torus, which are rotated relative to each other by an angle $0 < \phi < \pi/2$. After an analytic continuation to imaginary ϕ, the quantity $Z_{0,\phi}(q)$ serves as point of departure for the discussion of brane–brane scattering in Section 2.A.2.

The boundary states for unrotated and rotated branes on a torus were given above. If we consider the overlap between an unrotated brane and one rotated by an angle ϕ, only the coherent state associated with the vacuum representation contributes to the partition function

$$Z_{0,\phi}(q) = \kappa \cdot {}_0\langle\!\langle 0| \, \tilde{q}^{L_0 - \frac{c}{24}} \, |0\rangle\!\rangle_\phi \, . \tag{1.108}$$

Here, $\kappa = \kappa_0 \kappa_\phi$ denotes the product of the boundary state normalisations given in equation (1.102). Formula (1.108) also holds for a non-compact, two-dimensional target with an appropriately chosen factor κ, see below. Computing the sandwich of coherent states (1.84,1.85) and (1.100) is a lengthy combinatorial exercise. Other methods to deal with closely related expressions, using functional integrals, may be found in the literature, e.g. in [89]. To start our pedestrian derivation, we decompose the quantity (1.108) into "elementary pieces":

$$
\begin{aligned}
{}_0\langle\!\langle 0| \, \tilde{q}^{L_0 - \frac{c}{24}} \, |0\rangle\!\rangle_\phi &= \langle 0,0| \prod_{n>0} e^{-\frac{1}{n}\alpha_n^t \Omega \bar\alpha_n} \, \tilde{q}^{L_0 - \frac{c}{24}} \prod_{m>0} e^{-\frac{1}{m}\alpha_{-m}^t \Omega_\phi \bar\alpha_{-m}} \, |0,0\rangle \\
&= \tilde{q}^{\frac{c}{24}} \langle 0,0| \prod_{n>0} e^{-\tilde{q}^n \frac{1}{n}\alpha_n^t \Omega \bar\alpha_n} \prod_{m>0} e^{-\frac{1}{m}\alpha_{-m}^t \Omega_\phi \bar\alpha_{-m}} \, |0,0\rangle \\
&= \tilde{q}^{\frac{c}{24}} \langle 0,0| \prod_{n>0} e^{-\tilde{q}^n \frac{1}{n}\alpha_n^t \Omega \bar\alpha_n} \, e^{-\frac{1}{n}\alpha_{-n}^t \Omega_\phi \bar\alpha_{-n}} \, |0,0\rangle \\
&= \tilde{q}^{\frac{c}{24}} \prod_{n>0} \langle 0,0| \, e^{-\tilde{q}^n \frac{1}{n}\alpha_n^t \Omega \bar\alpha_n} \, e^{-\frac{1}{n}\alpha_{-n}^t \Omega_\phi \bar\alpha_{-n}} \, |0,0\rangle \, .
\end{aligned}
$$

In the second line, we have used $q^{-L_0} e^{\alpha_n} q^{L_0} = e^{q^n \alpha_n}$, in the third line that α_n commutes with α_{-m} for $m \neq n$. The last equality follows from the same fact, defining the vacuum expectation value of the infinite product of exponentials as the limit of the vev of a finite product. Thus the basic objects we have to compute are

$$
\begin{aligned}
V_n &:= \langle 0,0| \, e^{-\tilde{q}^n \frac{1}{n}\alpha_n^t \Omega \bar\alpha_n} \, e^{-\frac{1}{n}\alpha_{-n}^t \Omega_\phi \bar\alpha_{-n}} \, |0,0\rangle \\
&= \sum_{l,m=0}^{\infty} \frac{(-1)^{l+m}}{l!m!} \langle 0,0| \left(\tilde{q}^n \frac{1}{n}\alpha_n^t \Omega \bar\alpha_n \right)^l \left(\frac{1}{n}\alpha_{-n}^t \Omega_\phi \bar\alpha_{-n} \right)^m |0,0\rangle \\
&= \sum_{l=0}^{\infty} \frac{1}{(l!)^2 n^{2l}} \langle 0,0| \left(\tilde{q}^n \alpha_n^t \Omega \bar\alpha_n \right)^l \left(\alpha_{-n}^t \Omega_\phi \bar\alpha_{-n} \right)^l |0,0\rangle \, .
\end{aligned}
$$

It is not difficult to prove by induction that

$$
\begin{aligned}
&\frac{1}{n^{2l}} \langle 0,0| \left(\alpha_n^t \Omega \bar\alpha_n \right)^l \left(\alpha_{-n}^t \Omega_\phi \bar\alpha_{-n} \right)^l |0,0\rangle \\
&= \left[\prod_{i=1}^{l} \Omega_{\mu_i \nu_i} \Omega_{\phi; \rho_i \sigma_i} \right] \cdot \left[\sum_{\pi' \in S_l} \delta_{\mu_1 \rho_{\pi'(1)}} \cdots \delta_{\mu_l \rho_{\pi'(l)}} \right] \left[\sum_{\pi'' \in S_l} \delta_{\nu_1 \sigma_{\pi''(1)}} \cdots \delta_{\nu_l \sigma_{\pi''(l)}} \right] .
\end{aligned}
$$

The contractions will give sums of products of traces of powers of $M := \Omega\Omega_\phi \equiv R(2\phi)$. For example, for $l = 1$ we obtain $\operatorname{tr} M$, for $l = 2$ the above expression is equal to $2(\operatorname{tr} M)^2 + 2\operatorname{tr}(M^2)$, and so on.

In order to find a closed formula, let us first write the second permutation π' in the form $\pi'' = \pi' \circ \pi$. Then we can pull out the sum over π' above. Inspection shows that terms with $\pi = \mathrm{id}$ all contribute $(\mathrm{tr}\, M)^l$, and that it is anyway only π that matters for the structure of the "product of traces of powers of M" – more precisely, it is the cycle structure of π that matters. Thus, we get an overall prefactor $l!$ from the summation over π', and if π can be decomposed into m cycles of lengths $p_1, \ldots, p_m$,

$$\pi = (\, 1\ \pi(1)\ \pi^2(1)\ \cdots\ \pi^{p_1-1}(1)\,) \cdots (\, k\ \pi(k)\ \pi^2(k)\ \cdots\ \pi^{p_m-1}(k)\,)$$

(with some k which does not occur in the previous cycles), then the corresponding term contributes $\mathrm{tr}\,(M^{p_1}) \cdots \mathrm{tr}\,(M^{p_m})$.

Therefore, we have to count how many permutations $\pi \in S_l$ can be written as products of m cycles of lengths $p_1, \ldots, p_m$; call that number $c_{\{p_i\}}$. We will "fill" the m cycles one after the other: for the first one, we can choose p_1 out of l numbers and order them in $(p_1 - 1)!$ different ways (cyclic re-orderings within the cycles do not change π). For the next cycle, we have to pick p_2 elements from the $l - p_1$ remaining ones, and we can form $(p_2 - 1)!$ different cycles from these. The elements of the last cycle are fixed but can still be ordered in $(p_m - 1)!$ inequivalent ways.

But not all of these "inequivalent fillings" of the numbers from 1 to l into m cycles give different permutations: as soon as two (or more) cycles have the same length, they can be interchanged (as entire cycles) without affecting π. Therefore, we write the partition of l in the form

$$l = k_1 \cdot p_1 + \cdots + k_m p_m \quad \text{with}\ \ k_i, p_i \in \mathbb{N}\ \ \text{and}\ \ p_i \neq p_j\ \ \text{for}\ \ i \neq j;$$

then we have – in an obvious notation to display the multiplicities k_i –

$$c_{\{p_i, k_i\}} = \frac{((p_1 - 1)!)^{k_1}}{k_1!} \binom{l}{p_1} \binom{l - p_1}{p_1} \cdots \binom{l - (k_1 - 1)p_1}{p_1}$$

$$\cdot\, \frac{((p_2 - 1)!)^{k_2}}{k_2!} \binom{l - k_1 p_1}{p_2} \cdots \binom{l - k_1 p_1 - (k_2 - 1)p_2}{p_2}\ \cdots$$

$$\cdot\, \frac{((p_m - 1)!)^{k_m}}{k_m!} \binom{l - \sum_{i=1}^{m-1} k_i p_i}{p_m} \cdots$$

$$\binom{l - \sum_{i=1}^{m-1} k_i p_i - (k_m - 1)p_m}{p_m}$$

$$= \frac{l!}{k_1! \cdots k_m!}\, \frac{1}{p_1^{k_1} \cdots p_m^{k_m}}\,.$$

Plugging this in, and remembering the overall prefactor $l!$ from the π'-summation, we arrive at the following intermediate result:

$$V_n = \sum_{l=0}^{\infty} \tilde{q}^{nl} \sum_{\{p_i, k_i\}} \frac{p_1^{-k_1}}{k_1!} \cdots \frac{p_m^{-k_m}}{k_m!} \left(\mathrm{tr}\,(M^{p_1})\right)^{k_1} \cdots \left(\mathrm{tr}\,(M^{p_m})\right)^{k_m}.$$

The second sum runs over all partitions of l with p_i appearing k_i times.

Finally, we have to carry out the summations – which is, in fact, very easy. We first note that the sum over l is completely superfluous, we just need to sum over *all* partitions $k_1 p_1 + \cdots + k_m p_m$ of *any* non-negative integer. If we further allow the k_i to take the value 0, we simply have to sum over all numbers $k_1 \cdot 1 + k_2 \cdot 2 + k_3 \cdot 3 + \cdots$, with the k_i running from 0 to ∞ *independently*. (The "1", "2" and "3" replace the pairwise different p_i from above.)

To proceed, we distribute the prefactor $\tilde{q}^{nl}$ over the traces and introduce the notation $M_n := \tilde{q}^n M$. Then we obtain

$$V_n = \sum_{m=1}^{\infty} \sum_{k_1,\ldots,k_m=0}^{\infty} \frac{1}{k_1!} \left(\operatorname{tr}\left(M_n\right)\right)^{k_1} \frac{1}{k_2!} \left(\operatorname{tr}\frac{1}{2}\left(M_n^2\right)\right)^{k_2} \cdots \frac{1}{k_m!} \left(\operatorname{tr}\frac{1}{m}\left(M_n^m\right)\right)^{k_m}$$

$$= \prod_{m=1}^{\infty} \exp\left\{ \operatorname{tr}\frac{M_n^m}{m} \right\} = \exp \operatorname{tr}\left[\sum_{m=1}^{\infty} \frac{M_n^m}{m}\right] = \exp \operatorname{tr}\left[-\log\left(1 - M_n\right)\right]$$

$$= \det\left(1 - M_n\right)^{-1}.$$

Since M is just the rotation by 2ϕ, the determinant is given by $\det\left(1 - M_n\right) = 1 - 2\tilde{q}^n \cos 2\phi + \tilde{q}^{2n} = (1 - \theta\,\tilde{q}^n)(1 - \theta^{-1}\,\tilde{q}^n)$ with the phase $\theta := e^{2i\phi}$. Thus we have a closed expression for the partition function of two branes intersecting at an angle ϕ, although still given in terms of $\tilde{q}$ instead of q:

$$Z_{0,\phi}(q) = \kappa\,\tilde{q}^{-\frac{1}{12}} \prod_{n=1}^{\infty} \left((1 - e^{2i\phi}\tilde{q}^n)(1 - e^{-2i\phi}\tilde{q}^n)\right)^{-1}. \qquad (1.109)$$

We can apply the Jacobi triple product identity to convert equation (1.109) into a quotient of an eta and a theta function: putting $w = -\tilde{q}^{-\frac{1}{2}}\theta$ in equation (A.10) from Appendix A, and using the theta function $\vartheta_1(z,\tau)$ for $z = \phi\tau/\pi$, see equation (A.15), a short calculation shows that equation (1.109) takes the form

$$Z_{0,\phi}(q) = 2\kappa \sin\phi\, e^{-i\pi\tau\frac{\phi^2}{\pi^2}}\, \eta(q)\, \vartheta_1\left(\frac{\phi\tau}{\pi},\tau\right)^{-1}. \qquad (1.110)$$

With the product formula (A.16) for the theta function, we obtain an alternative expression for the partition function of two branes at an angle ϕ,

$$Z_{0,\phi}(q) = 2\kappa \sin\phi\, q^{-\frac{1}{12}}\, q^{\frac{\phi}{2\pi}(1-\frac{\phi}{\pi})} \prod_{n=0}^{\infty} \left(1 - q^{n+\frac{\phi}{\pi}}\right)^{-1}\left(1 - q^{n+1-\frac{\phi}{\pi}}\right)^{-1}. \qquad (1.111)$$

Finally, we plug in the expressions (1.102) into the factor $\kappa = \kappa_0 \kappa_\phi$ and obtain

$$Z_{0,\phi}(q) = NM\, q^{-\frac{1}{12}}\, q^{\frac{\phi}{2\pi}(1-\frac{\phi}{\pi})} \prod_{n=0}^{\infty} \left(1 - q^{n+\frac{\phi}{\pi}}\right)^{-1}\left(1 - q^{n+1-\frac{\phi}{\pi}}\right)^{-1}. \qquad (1.112)$$

This answer has a simple interpretation. Note that the q-dependent factors form the character of a twisted representation for two free bosonic fields. More precisely, let us consider a complex boson Z with boundary condition $Z(ze^{2\pi i}) = e^{2i\phi}Z(z)$. The modes of such a field raise the L_0 eigenvalue by an amount of $n \pm \frac{\phi}{\pi}$. They act on a ground state of weight $h_\phi = (1 - \frac{\phi}{\pi})\frac{\phi}{2\pi}$. The exponent $-\frac{1}{12} = -\frac{c}{24}$ contains the central charge of a complex boson (two real bosons). The twisted character is multiplied by the integer NM which counts the number of times our branes intersect. In the non-compact case, the final answer is identical, except that the factor NM is missing since the branes intersect in a single point.

For an application to brane–brane scattering in superstring theory, we shall also need the partition functions of rotated fermionic boundary states. As we have pointed out before, free fermions admit the same gluing maps as free bosons. Just as for the simpler cases considered in Section 1.2.3, the gluing conditions for bosons and fermions must be coupled so as to preserve an $N = 1$ worldsheet symmetry algebra. Working out the details of rotated fermion boundary states is, in principle, straightforward and the computation of the analogue of $Z_{0,\phi}(q)$ is easier due to the nilpotency of fermionic modes. Here we only spell out the resulting expression for the partition function. The relevant building block is given by

$$Z_{0,\phi}^{\pm}(q) = q^{-\frac{1}{24}} \, q^{\frac{\phi}{2\pi}\left(\frac{\phi}{\pi}-1\right)} \prod_{n=0}^{\infty}\left(1 \pm q^{n+\frac{\phi}{\pi}}\right)\left(1 \pm q^{n+1-\frac{\phi}{\pi}}\right) = \frac{q^{\frac{\phi^2}{2\pi^2}-\frac{1}{8}}}{\eta(q)}\,\vartheta\genfrac{}{}{0pt}{}{}{1}_{2}\left(\frac{\phi\tau}{\pi},\tau\right).$$

$$(1.113)$$

In the notation we use here, the index "1,2" of the ϑ-function on the right is correlated with the superscript "$+,-$" of the partition function on the left. The expression for Z^+ describes the character of a twisted representation for a complex fermion $\eta(z)$ with boundary condition $\eta(ze^{2\pi i}) = e^{2i\phi}\eta(z)$, and can be expressed in terms of ϑ_2. The partition function Z^- is obtained by inserting $(-1)^F$ into the trace, and can be expressed in terms of ϑ_1.

To sum over Neveu–Schwarz and Ramond sectors we must add $Z^{\pm}$ for ϕ and $\phi + \frac{\pi}{2}$. The additional contributions are (with the same shorthand notation concerning ϑ-indices)

$$Z_{0,\phi+\pi/2}^{\pm}(q) = q^{-\frac{1}{24}} \, q^{\frac{\phi^2}{2\pi^2}-\frac{1}{8}} \prod_{n=0}^{\infty}\left(1 \pm q^{n+\frac{1}{2}+\frac{\phi}{\pi}}\right)\left(1 \pm q^{n+\frac{1}{2}-\frac{\phi}{\pi}}\right)$$

$$= \frac{q^{\frac{\phi^2}{2\pi^2}-\frac{1}{8}}}{\eta(q)}\,\vartheta\genfrac{}{}{0pt}{}{}{3}_{4}\left(\frac{\phi\tau}{\pi},\tau\right). \qquad (1.114)$$

We will discuss this in some more detail in Subsection 2.A.1, in connection with brane–brane scattering.

2

Superstrings and branes

Our main goal in the opening chapters is twofold: namely, to introduce some very basic concepts of boundary conformal field theory (CFT); and to link them to developments in string theory. We have now dispensed with the first task, as far as is possible without the more advanced technology we will provide in Chapters 3 and 4.

Here, we will link the main concepts of the previous chapter, such as boundary conditions, boundary states and boundary excitations, to target space concepts in string theory. Before that, we will sketch some of the very basics of string theory, mostly dating back to the 1970s and 1980s. Obviously, our exposition cannot serve as a textbook introduction to string theory, but we hope that it is useful for those readers who have been exposed to string theory only sporadically.

After sketching various approaches to first-quantised closed strings and the route from CFT correlators to string scattering amplitudes and effective actions, including supergravity, we turn to D-branes. We start with solitonic p-branes and their low-energy effective action, then try to set up a dictionary between notions from target space physics and various worldsheet quantities that occurred in the first chapter.

Some more technical background material needed to build models with target space supersymmetry, and an analysis of brane-brane scattering based on CFT partition functions, are collected at the end of this chapter.

2.1 A very brief reminder of closed string theory

In the following few subsections, we shall describe the Polyakov action of bosonic string theory and discuss various approaches to its quantisation. Along the way the role of two-dimensional (conformal) field theory will become apparent. In particular, we shall review the relation between vibrational modes of the string and states/fields in the worldsheet field theory. Next, we shall discuss the construction and computation of string scattering amplitudes, mostly focusing on tree level. The generalisation to closed superstrings and the emergence of type IIA/B supergravity concludes this section.

2.1.1 Polyakov action and quantisation of closed bosonic strings

String theory starts from the assumption that the fundamental objects extend in one dimension rather than being point-like. The action of a relativistic point particle is the length of the worldline it sweeps out in spacetime. Analogously, the classical action of a relativistic closed string can be defined as the area of its worldsheet Σ embedded into the target spacetime M. The mathematical implementation of this geometric idea leads to what is known as the Nambu–Goto action. Since the Nambu–Goto action involves a square-root, it is usually replaced by the more convenient (but classically equivalent) Polyakov action

$$S_P[X, h] = -\frac{1}{4\pi\alpha'} \int_\Sigma d^2\sigma \sqrt{h}\, h^{\alpha\beta}\, g_{\mu\nu}(X)\, \partial_\alpha X^\mu\, \partial_\beta X^\nu. \tag{2.1}$$

This describes a σ-model of maps $X : \Sigma \longrightarrow M$ from the worldsheet Σ – viewed as a two-dimensional parameter space with metric $h_{\alpha\beta}$ – to the D-dimensional target manifold M with metric $g_{\mu\nu}$. The prefactor contains the string tension $T = (2\pi\alpha')^{-1}$ of dimension [mass]^2. The action (2.1) describes the motion of a string in the presence of a possibly non-trivial metric g. There are a number of further terms that are often considered,

$$S[X, h] = S_P + \frac{1}{4} \int_\Sigma d^2\sigma \left(\Lambda \sqrt{h} + \sqrt{h}\, r_h\, \phi(X) + \frac{1}{\alpha'} \epsilon^{\alpha\beta}\, B_{\mu\nu}(X)\, \partial_\alpha X^\mu\, \partial_\beta X^\nu \right), \tag{2.2}$$

where Λ is a (worldsheet) cosmological constant, ϕ is a dilaton coupling to the scalar curvature r_h of Σ, and $B_{\mu\nu}dx^\mu dx^\nu$ is a two-form on M.

For the moment, let us assume that the additional terms in equation (2.2) are not present. Deviating from the conventions in the first chapter we will also assume that $M = \mathbb{R}^D$ is equipped with a Minkowski metric. We will label the time direction by $\mu = 0$ and use $\mu = 1, \dots, D-1$ for spatial components.

The worldsheet is a parameter space rather than an object of physical significance. If we focus on the freely propagating closed string, we can take Σ to be a cylinder (with light-cone coordinates $\sigma_\pm = \sigma_0 \pm \sigma_1$), or map it to a sphere with two punctures and use complex coordinates, as in the first chapter. In the Euclidean setting, it is conceptually straightforward to generalise to arbitrary Riemann surfaces Σ.

At the classical level, the action (2.1) enjoys invariance under Weyl transformations (local rescalings of $h_{\alpha\beta}$), and under orientation-preserving diffeomorphisms of the worldsheet. One can employ these symmetries to trivialise the worldsheet metric $h_{\alpha\beta}$ at least locally. If such a gauge is assumed the classical action (2.1) reads

$$S[X] = -\frac{1}{4\pi\alpha'} \int_\Sigma d^2\sigma\, g_{\mu\nu}(X)\, \partial_\alpha X^\mu\, \partial^\alpha X^\nu \tag{2.3}$$

and the equations of motion for the worldsheet metric,

$$T_{\alpha\beta} \equiv -4\pi\alpha' \frac{1}{\sqrt{h}} \frac{\delta S_P}{\delta h^{\alpha\beta}} = 0, \tag{2.4}$$

are to be interpreted as constraints. As we have indicated in the previous equation, the variation of the Polyakov action with respect to the worldsheet metric gives the energy-momentum tensor of our classical two-dimensional field theory. In light-cone coordinates on a worldsheet cylinder, the constraints (2.4) become

$$T_{+-}(\sigma_+, \sigma_-) = 0, \quad T_{++}(\sigma_+) = 0 = T_{--}(\sigma_-). \quad (2.5)$$

The component T_{+-} is the trace of the energy-momentum tensor. It vanishes if the two-dimensional theory (2.3) is conformally invariant. This is the case for flat backgrounds with $g = \eta$ and a special class of curved geometries; see Subsection 2.1.3 for a few remarks. Whenever $T_{+-} = 0$, the components $T_{\pm\pm}$ may be shown to be chiral, i.e. to depend on one of the light-cone coordinates only. Setting these two chiral fields to zero provides a strong constraint on the allowed configurations of our classical field theory.

There are different approaches to the quantisation of the bosonic string. In each of them, demanding the absence of anomalies which could destroy classical symmetries leads to certain conditions on the background M and on parameters of the quantisation procedure.

Canonical covariant quantisation

In canonical covariant quantisation, the unconstrained system (2.3) is quantised first. The constraints (2.5) are afterwards imposed in the quantum theory. The unconstrained system to be quantised is obtained from the action (2.1) or (2.2) upon insertion of a trivial worldsheet metric $h_{\alpha\beta} = \delta_{\alpha\beta}$. Thereby, the Polyakov action reduces to an ordinary bosonic field theory. The number of bosonic component fields is given by the spacetime dimension D, and the metric $g_{\mu\nu}$ possesses one time-like direction. Quantisation of this field theory on the complex plane, i.e. on the worldsheet of a closed string, is straightforward. Classical solutions of the equation (2.1) can be parametrised by the centre-of-mass coordinates x^μ and momenta p^μ along with a set of bosonic oscillators α_n^μ and $\overline{\alpha}_n^\mu$. In contrast to the case of open string boundary conditions, left- and right-moving oscillations of a bosonic field on a cylinder are independent of each other. This is why we need two infinite sets of oscillators. After quantisation of the classical system, we can build up a Fock space $\mathcal{F} = \mathcal{F}^{\mathrm{bos}}$ in the same way as we did in Chapter 1. Due to the Lorentzian signature of the metric $g = \eta$, however, the space $\mathcal{F}$ contains negative norm states. In addition, we need to implement the classical constraints (2.4) in the quantum system. As with Gupta–Bleuler quantisation of electrodynamics, both problems are tackled in one go, by projecting onto a subspace of *physical states*. They are defined as those elements in $\mathcal{F}$ which satisfy

$$\left(L_n - a\,\delta_{n,0}\right)|\mathrm{phys}\rangle = \left(\overline{L}_n - a\,\delta_{n,0}\right)|\mathrm{phys}\rangle = 0 \qquad \text{for } n \geq 0, \quad (2.6)$$

with some real parameter a from normal-ordering ambiguities. The Virasoro modes L_n of the energy-momentum tensor in the quantum theory are constructed from the basic oscillators through the Sugawara formula, i.e. an expression of the form (1.22). The conditions (2.6) are as close as we can get to imposing the two equations on the right-hand side of equation (2.5) in the quantum theory, since the algebra of the modes L_n of the energy-momentum tensor turn out to possess an anomaly. In fact, the generators L_n of the chiral diffeomorphisms obey the defining relations of a Virasoro algebra with central charge $c = D$. Because of the central extension, i.e. the term containing the central charge c, it is not possible to set all the L_n to zero in the strong sense. The equations (2.6) ensure that the constraints (2.5) hold for all *matrix elements* that are evaluated with physical states.

It turns out that the space of solutions to the equations (2.6) is free of negative norm states if $a = 1$ and if the target has dimension $D = 26$. Note, however, that there are still zero-norm states. They signal for example gauge symmetries in target space physics and get removed through a simple quotient; see [252]. What we end up with is the space $\mathcal{H}$ of (one-particle) states of the closed bosonic string.

Each physical state in the space $\mathcal{H}$ corresponds to an excitation of the string. On this space of physical states, the operators L_0 and $\bar{L}_0$ are constant. A glance back at the Sugawara-like formula (1.24) reminds us that L_0 has the form

$$L_0 = \frac{1}{2}\alpha_0^\mu \alpha_0^\nu \eta_{\mu\nu} + N = \frac{\alpha'}{4}p^2 + N = -\frac{\alpha'}{4}m^2 + N\,,$$

where we inserted the relation $\alpha_0^\mu = p^\mu \sqrt{\alpha'/2}$ between the modes α_0 and the centre-of-mass momentum operator p of the closed string. The definition of the mass operator $m^2 = -p^2$ was used in the second step. The operator N, finally, is defined as

$$N = \sum_{n \geq 1} \alpha_{-n}^\mu \alpha_{\mu n}\,.$$

It counts the total number of left-moving oscillators excited in the physical state, each weighted with its mode number n. From this observation, along with the physical state condition $L_0 = 1$, we can now determine the form of the mass operator on physical states,

$$m^2 \equiv -p^2 = \frac{4}{\alpha'}(N - 1)\,. \tag{2.7}$$

Since the same reasoning applies to the right-moving oscillators, there exists a second expression for m^2 which takes the same form with $\bar{N}$ instead of N. Together, the equations (2.7) and its right counterpart imply a "level matching condition" $N = \bar{N}$.

It is quite straightforward to count the physical states with a given mass. The lowest-lying level ($N = \bar{N} = 0$) consists of the states

$$|v\rangle = |k\rangle \otimes |\bar{k}\rangle \in \mathcal{H}\,,$$

with $k^2 = k^\mu k_\mu = -4/\alpha'$. From the spacetime point of view, these states correspond to a tachyonic scalar particle (with respect to the Lorentz group in D dimensions). The associated "vertex operator" is defined as in equation (1.25).

At the next mass level ($N = \overline{N} = 1$), the physical states are obtained by acting with the oscillators α_{-1}^μ and $\overline{\alpha}_{-1}^\mu$,

$$|v\rangle = |\epsilon(k)\rangle = \epsilon_{\mu\nu}(k)\,\alpha_{-1}^\mu \overline{\alpha}_{-1}^\nu\,|k\rangle \otimes |\overline{k}\rangle \tag{2.8}$$

where $k^2 = 0$ and where the set of functions $\epsilon_{\mu\nu}(k)$ must satisfy $k^\mu \epsilon_{\mu\nu} = k^\nu \epsilon_{\mu\nu} = 0$ in order to satisfy the physical state conditions. Once more, there is a particular field associated with massless states (2.8). Again, we can associate a vertex operator to our states with $N = 1$. It is given by

$$\Phi(|\epsilon(k)\rangle; z, \bar{z}) = \phi_{\epsilon(k)}(z, \bar{z}) = \epsilon_{\mu\nu}(k)\,{:}J^\mu(z)\overline{J}^\nu(\bar{z})e^{ikX(z,\bar{z})}{:}. \tag{2.9}$$

Decomposing the matrix $\epsilon_{\mu\nu}$ into its trace, its traceless symmetric and its antisymmetric part, we obtain three linearly independent $\epsilon_{\mu\nu}^{\mathbf{m}}$, $\mathbf{m} = \phi, g, B$. The massless states of a closed string have the spacetime interpretation of a scalar (or "dilaton") ϕ, of a graviton $g_{\mu\nu}$, and of an antisymmetric two-form $B_{\mu\nu}$ (also called "Kalb–Ramond tensor"). Obviously, the appearance of a massless spin 2 particle is most welcome.

Proceeding in this way, we can isolate linearly independent physical states

$$|v\rangle = \phi_{\text{eff}}^{\mathbf{m}}(k)\,|k; \mathbf{m}\rangle \ \in \ \mathcal{H} \tag{2.10}$$

at each mass level. The states $|k; \mathbf{m}\rangle$ denote certain descendants of the ground states $|k\rangle \otimes |\overline{k}\rangle$, i.e. states that are obtained by action of creation operators α_n^μ and $\overline{\alpha}_n^\mu$. Our sketchy notation in equation (2.10) hides Lorentz indices as well as conditions, which ensure that we are actually dealing with physical states (like $k^\mu \epsilon_{\mu\nu} = 0$, etc.). Altogether, the excitations of the string give rise to an infinite tower of "effective" spacetime fields supported on $\mathbb{R}^D$, whose Fourier components we have denoted by $\phi_{\text{eff}}^{\mathbf{m}}(k)$. To be precise, string theory provides only "on-shell" spacetime fields, since the physical state condition fixes the value of k^2.

Other quantisation procedures

Light-cone gauge quantisation offers an alternative way to pass from the classical to the quantised string. Here, (most of) the constraints in the classical theory are implemented and then the reduced theory is quantised. But the elimination of constraints comes at a price: in the reduced system the target-space Lorentz symmetry is no longer manifest. To be a bit more concrete, we introduce light-cone coordinates in our target space $\mathbb{R}^D$ by $X^\pm = (X^0 \pm X^{D-1})/\sqrt{2}$, and recall that the gauge is now fixed by setting $X^+(\sigma_0, \sigma_1) = x^+ + p^+ \sigma_0$, where x^+ and p^+ are constants and σ_0, σ_1 are worldsheet coordinates on the cylinder. The "Virasoro constraint" (2.4) can then be used to express X^- as a bilinear in the oscillators α_n^i for $i = 1, \dots, D - 2$, up to an integration constant x^-. Upon quantisation, the normal-ordering constant a shows up as in the covariant approach. The Fock space formed out of the transverse oscillators α_n^i is a Hilbert space $\mathcal{H}$ of physical states only. Note that, once the light-cone coordinates are eliminated, the index i runs over $i = 1, \dots, D - 2$ only and that the remaining $D - 2$ bosonic fields have a Euclidean metric now. It can be shown that Lorentz invariance can be reinstalled only for a 26-dimensional target, and only if the normal-ordering constant is fixed to $a = 1$. At the end of the day, the same mass formula (2.7) and the same multiplets arise as before.

Let us point out that the field theory of the transverse coordinates $X^i, i = 1, \ldots, D - 2$ is a free field theory of strings moving in a flat target. For more general backgrounds, the light-cone gauge quantisation may lead to non-conformal and even non-relativistic two-dimensional field theories. On the other hand, there are many cases for which the $D - 2$ transverse degrees of freedom are still described by a CFT. These provide an important class to which the two-dimensional CFT can be applied.

To wrap up our brief sketch of quantisation schemes, we would like to mention *BRST quantisation* (where the BRST refers to Becchi, Rouet, Stora and Tyutin), which is perhaps the most elegant procedure, and the formulation of which is least dependent on which string background CFT is considered. Here, the total Fock (or more general CFT state) space is enlarged by further unphysical objects, namely ghost fields (anti-commuting operators of integer conformal dimension). In this bigger space, the BRST charges Q and $\overline{Q}$ are formed from the Virasoro and the ghost modes (involving a normal-ordering constant a as before). If $D = 26$ (or more generally: if the central charge of the CFT describing the string background is 26) and $a = 1$, these operators Q and $\overline{Q}$ are nilpotent, and their cohomology can be considered. Physical states are identified with representatives of a certain fixed ghost number; spelling out the condition explicitly yields equation (2.6); see e.g. [153, 252, 337, 377] for more details.

The BRST approach emerges from the functional integral quantisation of strings. There, the ghosts arise as Faddeev–Popov ghosts from fixing diffeomorphism and Weyl invariance on the worldsheet. The details can be found in the reviews mentioned above, but see also [194] for a short summary. Due to the inclusion of ghost contributions, CFTs with vanishing Virasoro central charge show up in BRST quantisation. Such models are interesting on their own, but will not play a particular role throughout this text.

2.1.2 String scattering amplitudes and two-dimensional correlators

We have identified string excitations with effective spacetime fields, and we can ask what their effective physics is. String theory allows the computation of scattering processes of the $\phi^{\mathbf{m}}_{\text{eff}}$. We then determine the action $S_{\text{eff}}[\phi^{\mathbf{m}}_{\text{eff}}]$ of an effective field theory in such a way that the same amplitudes are produced. This idea can be formulated (somewhat schematically) in terms of functional integrals,

$$\int_{\phi_i, \phi_f} \prod_{\mathbf{m}} D\phi^{\mathbf{m}}_{\text{eff}} e^{iS_{\text{eff}}[\phi^{\mathbf{m}}_{\text{eff}}]} = \int_{X_i, X_f} DX \, d\mu_{\Sigma} \, e^{iS_P[X]} . \tag{2.11}$$

The integral on the right-hand side describes a string scattering process with X_i and X_f as initial and final string states. They determine the ϕ_i, ϕ_f on the left-hand side. By $d\mu_\Sigma$, we denote integration over all "moduli" (in particular all possible metrics) of the worldsheet Σ – the unphysical characteristics of the parameter space are averaged out. The "volume" of the symmetry group (diffeomorphisms $\times$ Weyl transformations) is divided out in the measure. On the left-hand side, we integrate over the infinite family of effective spacetime fields $\phi_{\text{eff}}^{\mathbf{m}}$. Altogether, the effective action $S_{\text{eff}}[\phi_{\text{eff}}^{\mathbf{m}}]$, which is an integral over spacetime and which implicitly depends on the topology of Σ, is defined by the "change of variables" that trades the string coordinate X for the effective spacetime fields $\phi_{\text{eff}}^{\mathbf{m}}$.

Conformal field theory helps to obtain concrete expressions for the right-hand side of the prescription (2.11), using correlators of conformal fields (often called "vertex operators" in the string literature). After an appropriate conformal map, we may assume that Σ is a closed Riemann surface with punctures $(z_j, \bar{z}_j)$ that represent the asymptotic in- and out-states v_j. As we recalled in Section 2.1.1, such states are created by inserting local field operators $\varphi_{v_j}(z_j, \bar{z}_j) = \Phi(v_j; z_j, \bar{z}_j)$ – such as the fields (1.25) for the free boson CFT, but cf. the general state-field correspondence explained in Chapter 3 below. The locations of the punctures are part of the worldsheet moduli and have to be integrated over in the string theory calculation of scattering amplitudes

$$\mathcal{A}_\Sigma\left(\phi_{\text{eff}}^{\mathbf{m}_1}(k_1), \dots, \phi_{\text{eff}}^{\mathbf{m}_n}(k_n)\right)$$

$$\sim \phi_{\text{eff}}^{\mathbf{m}_1}(k_1) \cdots \phi_{\text{eff}}^{\mathbf{m}_n}(k_n) \int d\mu_\Sigma \, \langle \varphi_{k_1, \mathbf{m}_1}(z_1, \bar{z}_1) \cdots \varphi_{k_n, \mathbf{m}_n}(z_n, \bar{z}_n) \rangle_\Sigma \, .$$

$$(2.12)$$

Here, $\varphi_{k_l, \mathbf{m}_l}(z_l, \bar{z}_l)$ is the vertex operator corresponding to the state $|k_l; \mathbf{m}_l\rangle$ that appears in equation (2.10), and $\langle \cdot \rangle_\Sigma$ denotes the CFT correlation function on the worldsheet Σ – realising the DX-integral in equation (2.11). If Σ is a punctured sphere, the correlators can, in the flat-target case, be computed from repeated application of Wick's theorem. The integration requires a few more comments. In the case of a punctured sphere, we only integrate over insertion points. But the correlation functions enjoy a number of symmetries that would make these integrations divergent. Translation symmetry, for example, implies that the integration over simultaneous shifts of all insertion points is infinite. Similar arguments apply to all the conformal (or Möbius) transformations of the complex plane. Since these symmetries may be used to send the first three insertion points z_1, z_2, z_3 to 0, 1 and ∞, the integration should only involve insertion points z_j with $j \geq 4$. Hence, no z-integrations have to be carried out for scattering processes with less than four asymptotic states. The remaining z-integral in a four-point function yields, for strings in a flat target, the familiar Virasoro–Shapiro amplitude.

Higher genus worldsheets arise from interactions of strings, which can be pictured as splitting and joining – as "handle formation" on the worldsheet. The details of the interaction are already dictated by the properties of the free string: the states of the latter are given by field operators of a CFT, and it is their operator product expansion (OPE) which fixes the decomposition of the states of two interacting strings into those of the third. Since the worldsheet is two-dimensional and since the whole theory is diffeomorphism invariant, no independent higher-order string "vertices" can occur.

A dimensionless coupling constant g_S is assigned to each three-string vertex, and each amplitude (with a prescribed collection of inserted vertex operators) becomes a weighted sum over closed oriented Riemann surfaces Σ_g of genus g,

$$\mathcal{A} = \sum_g g_S^{\chi(\Sigma_g)} \mathcal{A}_{\Sigma_g} . \tag{2.13}$$

The exponent of g_S is determined by the number N_3 of three-vertices in a pant decomposition of the underlying punctured Riemann surface, i.e. by the number of splitting and joining processes that occur in the scattering process. Even though such a pant decomposition is not unique, the number N_3 is. As can easily be seen, it is given by $N_3 = 2g + n - 2$ where n denotes the number of external legs. After suitable normalisation, the exponent $\chi(\Sigma_g)$ of g_S is given by the Euler characteristics of Σ_g, i.e. $\chi(\Sigma_g) = 2g - 2$; see e.g. [252]. The g_S-prefactor can also be understood from the coupling (2.2) of the dilaton to the scalar curvature of Σ.

We see that perturbative string theory naturally includes the notion of putting "one and the same" CFT – which describes the perturbative spectrum and the interaction of strings – on all possible Riemann surfaces. The effective action defined in equation (2.11) also receives contributions from every worldsheet topology. Terms from surfaces with $g \geq 1$ represent quantum (loop-) corrections to classical (tree-level) string theory. For most of this text, we shall content ourselves with the lowest-order terms in g_S.

Having sketched how to compute string scattering amplitudes, at least in principle, it would be tempting to discuss a few examples. We shall, however, postpone such computations to Section 2.2 to include branes as well. Let us mention here that in most situations, we are only interested in the effective physics that string theory induces at low-energy scales – low enough that no massive string excitations can occur as asymptotic states. To obtain the relevant *low-energy effective action* (LEEA), all massive fields on the left-hand side of equation (2.11) are integrated. Relation (2.12) remains valid: only φ_{k_l, j_l} that correspond to massless spacetime fields occur as "external legs", but there is no truncation of the OPEs among them (the "internal lines" can carry arbitrary string states). It is one of the more encouraging features of string theory that familiar and physically relevant theories originate from string theory in the low-energy limit, at least in the supersymmetric case.

Gravity and beta functions

Incorporating curved target spaces is a straightforward extension at the conceptual level, and establishes the famous link between string theory and general relativity. If $\mathbb{R}^D$ is replaced by a more general (curved) manifold M with metric $g_{\mu\nu}$, the corresponding σ-model (2.3) for string propagation in such a target becomes an interacting theory which can be approached via perturbative expansion in α' (or around the small curvature limit). Because of the non-trivial couplings in the worldsheet action (2.3), the quantised σ-model will usually have non-trivial beta functions, signalling a breakdown of conformal symmetry due to quantum effects. As we have explained earlier, string theory demands conformal invariance. In particular, the generators L_n and $\overline{L}_n$ of conformal transformations feature in the selection (2.6) of physical states. Since the couplings of the sigma model depend on geometrical data of M, vanishing of the beta functions imposes restrictions on the target manifolds which are admitted; see e.g. [194, 252, 365, 377]. To first order in α' we find

$$\beta_{ij}^g = \alpha' R_{ij} + 2\alpha' \nabla_i \nabla_j \Phi - \frac{\alpha'}{4} H_{imn} H_j^{mn} + O(\alpha'^2), \qquad (2.14)$$

$$\beta_{ij}^B = -\frac{\alpha'}{2} \nabla^m H_{mij} + \alpha' \nabla^m \phi H_{mij} + O(\alpha'^2), \qquad (2.15)$$

$$\beta^\phi = \frac{D-26}{6} - \frac{\alpha'}{2} \nabla^2 \phi + \alpha' \nabla_m \phi \nabla^m \phi - \frac{\alpha'}{24} H_{mij} H^{mij} + O(\alpha'^2). \qquad (2.16)$$

In particular, the vanishing of the beta function for the metric implies that a target manifold M (with $\Phi = 0$ and $H = 0$) has to be Ricci-flat: in this way, the Einstein equations can be derived from string theory. Moreover, "stringy corrections" to classical relativity can also be obtained if higher orders in α' are taken into account. Note that the beta functions are often taken as the starting point to define an LEEA: the action is chosen in such a way that $\beta_n = 0$ is the equation of motion for the nth effective field.

2.1.3 The closed superstring and supergravity

The main flaws of bosonic string theories are the presence of tachyons and the lack of spacetime fermions among the physical states. Both problems can be cured by introducing superstrings. After specialising to the gauge $h_{\alpha\beta} = \delta_{\alpha\beta}$, the worldsheet action contains left- and right-moving free fermions along with the bosons, as in the action (1.71). As we have discussed before, this theory of free bosons and fermions enjoys worldsheet supersymmetry, and thus the Virasoro constraints (2.4) are supplemented with corresponding ones for the superpartner G of the energy-momentum tensor T. In the quantum theory, the modes of G generate an $N = 1$ superconformal algebra (1.73) with central charge $c = \frac{3D}{2}$.

The Fock space is now built up from bosonic oscillators and from NS- and R-sectors of the free fermions ψ^μ and $\overline{\psi}^\mu$. Note that all D left-moving fermions must be either in the NS- or in the R-sector because of D-dimensional Lorentz covariance; analogously for the right-movers. The physical states in this space have to satisfy equations (2.6) for the Virasoro modes together with

$$G_r |\text{phys}\rangle = \overline{G}_r |\text{phys}\rangle = 0 \qquad \text{for } r \geq \tfrac{1}{2} \text{ (NS) or } r \geq 0 \text{ (R)}. \qquad (2.17)$$

The normal-ordering constant a from equation (2.6) is $a = \tfrac{1}{2}$ for NS-states and $a = 0$ for R-states. Consistency furthermore requires that we have $D = 10$ bosons and fermions, i.e. that the total central charge of the $N = 1$ superconformal field theory is $c = 15$. As in the case of bosonic strings, the mass formula follows from the physical state conditions.

The effective spacetime fields now include bosons from physical states that have both left- and right-moving components in the same sector. Such states are denoted by NS–NS or R–R. But since the Ramond ground states form a representation of the D-dimensional Clifford algebra, see Subsection 1.2.2, the superstring also contains fermionic spacetime excitations. These are associated with physical states in the NS–R and R–NS sectors of the model. In order to remove the NS–NS tachyon, and in order to match fermionic and bosonic degrees of freedom in each mass level, the GSO projection is performed: all physical states with odd fermion number are removed from the NS-sector, with the convention that $(-1)^F |k\rangle = -|k\rangle$ for the tachyonic ground state. In the R-sector, the GSO projector P^{GSO} is a combination of fermion number and chirality operator for the ten-dimensional spinors, see Subsection 2.A.1 for more explicit formulas. The chiralities in P_L^{GSO} and P_R^{GSO} can be chosen independently, leading to type IIA (opposite chiralities) and type IIB (equal chiralities) superstring theory.

After the GSO projection, the superstring spectrum is tachyon-free and spacetime supersymmetric. As a result, the total partition function vanishes, since spacetime fermions contribute to the loop diagram with a sign. Spacetime supersymmetry charges, which map spacetime bosons into fermions, can be formed from the worldsheet fields which map the Neveu-Schwarz to the Ramond sector, i.e. from the spin fields; see [186] and our more detailed remarks in Section 2.A.

The massless states from the NS–NS sector have the same spacetime interpretation as for the bosonic string, although the worldsheet expressions are different. Since now $a = \tfrac{1}{2}$, the massless physical states are obtained from

$$|\epsilon(k)\rangle = \epsilon_{\mu\nu}(k)\,\psi^\mu_{-\frac{1}{2}}\,\overline{\psi}^\nu_{-\frac{1}{2}}\,|k\rangle\,,$$

with the same restrictions on $\epsilon_{\mu\nu}(k)$ and k^μ as before. The massless NS–R and R–NS states contribute spacetime gravitinos λ_L and λ_R to the spectrum; they satisfy the massless Dirac equation in ten dimensions due to equation (2.17). After the GSO projection, we are left with Majorana–Weyl spinors, the relative chirality of which depends on that of P_L^{GSO} and P_R^{GSO}. The bosons from the R–R sector come as products of two spacetime spinors. With the help of ten-dimensional gamma matrix identities, they can be rewritten as n-forms $F_{(n)}$ on the spacetime manifold, with n even for type IIA and n odd for type IIB, and with the relation $F_{(n)} \sim *F_{(10-n)}$. The physical state conditions translate into equation of motion and Bianchi identity,

$$d * F_{(n)} = 0, \quad d F_{(n)} = 0,$$

so there are local potentials $C_{(n-1)}$ such that $F_{(n)} = d C_{(n-1)}$.

In summary, the massless spectrum of the closed superstring coincides with that of type IIA or type IIB $N = 2$ supergravity in ten dimensions. The type IIB string theory contains chiral spacetime fermions and n-form fields $F_{(n)}$ with n odd, including a self-dual 5-form. The massless spectrum of type IIA superstrings, on the other hand, is non-chiral and contains form fields $F_{(n)}$ of even degree n. Type IIA supergravity can be obtained from 11-dimensional $N = 1$ supergravity by dimensional reduction.

As in the case of bosonic strings, scattering amplitudes of massless closed superstrings can be determined from correlation functions of the associated vertex operators on the worldsheet Σ and encode the results in an LEEA. The bosonic part of this LEEA is given by (up to Chern–Simons terms which couple NS–NS to R–R potentials)

$$S_{\mathrm{II}} = \frac{1}{2\kappa_0^2} \int d^{10}x \, \sqrt{-G} \, e^{-2\phi} \left(R_G - \frac{1}{12} H^2 + 4 \, \partial_\mu \phi \, \partial^\mu \phi \right)$$
$$+ \int d^{10}x \, \hat{F}_{(n)} \wedge * \hat{F}_{(n)} , \tag{2.18}$$

with $H = dB$ and with rescaled R–R fields $\hat{F}_{(n)} = e^{-\phi} F_{(n)}$. Only even and odd n contribute for type IIA and IIB supergravity, respectively. Half of the forms are Hodge duals of the other ones; in particular, for type IIB an additional self-duality constraint $F_{(5)} = * F_{(5)}$ must be imposed. Of course, higher-order corrections in α' are omitted in this low-energy limit. The term κ_0 plays the role of the ten-dimensional gravitational constant in this supergravity theory.

Superstrings in curved backgrounds

The beta-function equations (2.14–2.16) of bosonic string theory impose restrictions on a curved background manifold, to satisfy the requirement of (bosonic) conformal symmetry. If we extend this to superstrings, further conditions on the admissible targets arise.

One way to see this is to work directly with supergravity as the LEEA of superstring theory, and analyse the conditions guaranteeing the existence of covariantly constant spinors necessary for spacetime supersymmetry; see e.g. [252]. We arrive at a restriction on the holonomy group of the target: the latter must be a Calabi–Yau manifold, i.e. Ricci-flat and Kähler.

Working from the worldsheet, we first find that in order to formulate the physical state condition of superstring theories the Virasoro field must possess a supersymmetric partner G; moreover, it can be shown [35, 377] that spacetime supersymmetry actually requires that the $N = 1$ superconformal symmetry generated by T and G gets enhanced to an $N = 2$ super-Virasoro algebra, the relations of which were given in Subsection 1.2.3, in equation (1.77). In particular, the U(1) current J appearing there can be used to define the GSO projection in a convenient way (see Chapter 7 for an example).

The two points of view fit together in the sense that it may be shown that such an $N = 2$ superconformal symmetry exists at least at the classical level for sigma models with Calabi–Yau targets [17, 252]. As with Einstein equations arising from the lowest-order beta functions, the Calabi–Yau condition is modified by higher-order corrections [256] so that, strictly speaking, classical Calabi–Yau manifolds can appear as string targets only in the large-volume (or small curvature) limit. It turns out, however, that various interesting quantities do not depend on the Kähler moduli of the target (the parameters that control changes of the metric), and that they do not change qualitatively. This justifies the use of classical large-volume pictures in the stringy regime to some extent.

Taking $M = \mathbb{R}^4 \times K^6$ with K^6 a compact Calabi-Yau three-fold "solves" the problem of how to fit our four-dimensional real world into string theory: the internal manifold K is supposed to be small, thus invisible at low energies. Its geometric structure merely provides internal degrees of freedom for effective spacetime fields supported on the flat external spacetime $\mathbb{R}^4$. In this way, algebraic geometry enters string phenomenology.

Since the only restrictions on acceptable string targets come from (super-) conformal invariance, and since all the indispensable consistency conditions of string theory, in particular the physical state condition, may be formulated in abstract CFT terms, we can immediately pass to the following, more radical generalisation, namely replace the σ-model with classical target M by any (super-) conformal field theory on the worldsheet Σ, with total central charge 26 (or 15). Again, we usually stick to tensor product theories which involve external free bosons (and free fermions), in order to have the familiar notions of flat four-dimensional spacetime available from the start. But it should be noted, for conceptual clarity, that target geometry can be defined even in the absence of any such "classical input" into the closed string background. Methods on how to extract (quantised) targets from CFT were, e.g., developed in [110, 193, 396], using the formalism of non-commutative geometry.

For the flat-target case, we have formulated the computations giving excitation spectra and effective physics in "CFT-style", and they carry over without change to the general case. For a standard tensor product background, the vertex operators are products of an external free boson part with fields from a (unitary) internal CFT, which again determine internal degrees of freedom. For example, affine Lie algebras are used as internal CFTs in the context of the heterotic string to produce low-energy effective gauge theories.

Gepner models are perhaps the most famous class of superstring compactifications formulated in terms of abstract CFTs, and they will be the focus of

Chapter 7. Their low-energy spectra come relatively close to that of the standard model. It should be pointed out that Gepner models can be regarded as special points in the moduli space of geometric compactifications on Calabi–Yau manifolds, see [253] for a review, so that geometric and algebraic string compactifications may in fact be closely related.

2.2 Branes and boundary conformal field theory

Our crash course on string theory has brought us to a point now at which we can begin to link it up to the study of boundary conditions in two-dimensional CFT. After a brief review of solitonic solutions in supergravity we shall explain how these objects are modelled in string theory. Thereby, we are able to embed the results of the first chapter into perturbative string theory. Among other things, we shall explain how to interpret, e.g., one-point functions and boundary spectra in terms of target-space physics. In addition, we sketch how (non-commutative) gauge theories emerge from boundary excitations and their correlation functions.

2.2.1 Introductory remarks on branes

Before we plunge into the technical details of the worldsheet description of D-branes, let us briefly outline some historical developments in the discovery of branes and state some formulas that are important ingredients in the spacetime picture.

Solitonic p-branes

Within string theory, non-perturbative objects were first encountered in a "low-energy disguise". As we reviewed in the first section, the classical action which governs the effective physics of massless modes in type II string theory is ten-dimensional supergravity,

$$S = \frac{1}{2\kappa_0} \int d^{10}x \sqrt{-g} \left(R(g) - \frac{1}{8}(\nabla\phi)^2 - \frac{1}{2(p+2)!} \, e^{\frac{p-3}{4}\phi} \, (dC_{(p+1)})^2 \right). \quad (2.19)$$

In contrast to equation (2.18) we have passed to the so-called Einstein frame, ignored the B-field and used the R–R potential $C_{(p+1)}$ rather than the associated field strength $F_{(p+2)}$. The restrictions on the degrees of the R–R potentials were described in Section 2.1. We can ask whether the equations of motion for the action (2.19) admit non-trivial classical solutions with properties similar to solitons or instantons in gauge theories. This is indeed the case, and the field configurations (metric, dilaton and antisymmetric R–R potentials) of these "*solitonic*" *p-branes* have the following form,

$$ds^2 = A^{-\frac{7-p}{8}}(y) \, \eta_{ij}dx^i dx^j + A^{\frac{p+1}{8}}(y) \left(dy^2 + y^2 d\Omega_{8-p}^2 \right),$$
$$e^{\phi} = A(y)^{-\frac{p-3}{2}}, \quad (2.20)$$
$$C_{(p+1)} = \pm A^{-1}(y) \, dx_0 \wedge dx_1 \wedge \cdots \wedge dx_p.$$

The x^i for $i = 0, \ldots, p$ are coordinates tangential to what is called the brane's worldvolume, while the function $A(y)$ depends only on the transverse distance $y^2 = x_{p+1}^2 + \cdots + x_9^2$ from the brane. The function $A(y)$ is harmonic for $y \neq 0$ and given by

$$A(y) = 1 + \frac{2\kappa_0 T_p}{(7-p)\Omega_{8-p}} \frac{1}{y^{7-p}} \qquad (2.21)$$

for $-1 \leq p \leq 6$. Because of the structure of the action, only solitonic branes with p even or odd exist for IIA and IIB supergravity, respectively. Here, T_p is a positive constant related to tension and charge of the brane (see below), while Ω_d is the volume of the d-dimensional unit sphere. Far away from the brane, i.e. for large y, the function $A(y)$ approaches $A(y = \infty) = 1$ and spacetime is asymptotically flat. It can be shown that charge and mass densities of these solitonic configurations are concentrated around the $(p+1)$-dimensional hyperplane spanned by the coordinates x^i, $i = 0, 1, \ldots, p$. The so-called "D-instanton" $(p = -1)$ can be regarded as a point-like defect in spacetime. For $p = 7$ we have

$$A(y) = 1 + \frac{\kappa_0 T_p}{\pi} \ln(y/y_0) \, .$$

We refer to [152, 432] and also to [123] for a more detailed discussion and for further references.

Apart from their characteristic shape, the configurations (2.20, 2.21) have several interesting general properties. First of all, they are constructed as BPS states of the spacetime supersymmetry algebra and preserve half of the spacetime supersymmetry. BPS states saturate the Bogomolny bound which relates masses (or tensions) and central charges (charge densities) of the supersymmetry algebra. Here, the central charges are just the charge densities of the field configuration (2.20) with respect to the R–R fields $F_{(p+2)} = dC_{(p+1)}$, and the BPS property simply means that mass and charge have the same modulus (with suitable normalisations). The most important feature of p-branes is their nonperturbative nature: it turns out that the mass and charge density of a p-brane asymptotically scale as $e^{-\phi}$. But in string theory the vacuum expectation value of e^{ϕ} is (proportional to) the value of the string coupling constant g_S. Hence, just like gauge theory monopoles, branes become heavy at weak coupling. The g_S^{-1} dependence of the mass and charge density is characteristic of the above "solitonic" p-branes, and leads to $e^{-\frac{1}{g_S}}$-corrections in the low-energy effective action – cf. Shenker's observations in [426].

There is another non-perturbative string theory object, the so-called Neveu–Schwarz five-brane (see e.g. [56, 86, 172, 394]), whose mass and charges scale as g_S^{-2}. Despite the name, the NS five-brane is very different from the D-branes to be discussed here, at least in a perturbative expansion. It requires a nonperturbative formulation, such as M-theory, to treat both types of defects on the same footing, invoking U-duality. At the worldsheet level, NS five-branes correspond to a certain closed string background CFT, not to boundary conditions.

In connection with the BPS property, the non-perturbative character of p-branes allows us to make statements about the strong coupling behaviour of string theory: as long as supersymmetry is not destroyed by quantum effects, the algebraic constraints behind the Bogomolny bound protect BPS states against "mixing" with other states, so their perturbative properties can be extrapolated to arbitrarily large string coupling g_S. This is why solitonic p-branes prove to be a useful device in the discussion of non-perturbative effects in string theory, of S-dualities and of the tentative properties of M-theory. The derivation of the black hole entropy from branes [284, 285, 342, 434] was particularly celebrated as providing the first concrete application of strings to gravity.

Dirichlet p-branes

Solitonic p-branes would not, however, have become so important had it not been for a truly string theoretic description as *Dirichlet branes*, or *D-branes* for short. These were discovered in several steps, the first of which was taken at about the same time as, but independently of, the discovery of solitonic p-branes: Polchinski and co-workers encountered hyperplanes in spacetime when they studied open string theories in flat targets [119]. More specifically, they investigated the effect of T-duality on a compactified target direction X^{μ_0} (with radius R), and asked themselves what happens if R is sent to zero or infinity so as to recover flat extended targets.

In a system consisting of open and closed strings, the effect of T-duality on the latter is simply an interchange of winding and momentum modes. In particular, the limits $R \to 0$ and $R \to \infty$ of the compactification radius are equivalent; in both cases, closed string theory in an uncompactified target is approached. For $R \to 0$, the momentum modes become infinitely heavy, while the spacings in the winding spectrum become arbitrarily narrow. The open strings – which in [119] were of course supposed to obey Neumann conditions at the worldsheet boundary – have no winding modes. Letting R tend to zero therefore simply removes the momentum modes (and all higher excitations) from the spectrum.

Polchinski *et al.* nevertheless found certain "left-overs" of open strings even in the limit $R = 0$: over the string endpoints, the dual coordinate $X^{\mu_0\prime}(z,\bar{z}) := X_L^{\mu_0}(z) - X_R^{\mu_0}(\bar{z})$ must take a definite value (it satisfies Dirichlet conditions). When the same procedure is applied to $D - 1 - p$ directions, it results in confining the endpoints of the "dual" open strings to a $(p+1)$-dimensional hyperplane in the D-dimensional target. Applying worldsheet duality to a one-loop open string diagram, this also means that the closed strings couple to this hyperplane. Altogether, a sequence of T-dualities and limits takes a theory of open and closed strings into a system containing closed strings and a new extended object of higher dimension – called a Dirichlet p-brane.

In a companion paper to [119], Leigh found a classical action to describe the low-energy physics of the new objects, including both its coupling to the massless closed string states and its internal physics. The action written in [328] is the *(Dirac)–Born–Infeld action*, previously encountered in open string theory in [174],

$$S_{\text{BI}} = -T_p \int d^{p+1}\xi \, e^{-\tilde{\phi}} \sqrt{-\det(\tilde{g} + \tilde{B} + 2\pi\alpha' F)}, \qquad (2.22)$$

where T_p is the tension of the p-brane. The functions $\tilde{\phi}$ and $\tilde{B}$ are the closed string dilaton field and Kalb–Ramond tensor restricted (pulled back) to the $(p+1)$-dimensional worldvolume of the D-brane, respectively, F is the field strength of a $U(1)$ gauge field on the brane, and $\tilde{g}$ is the metric on the brane induced by the closed string bulk metric,

$$\tilde{g}_{ij} = \frac{\partial X^\mu}{\partial \xi^i} \frac{\partial X^\nu}{\partial \xi^j} \, g_{\mu\nu}(X(\xi)); \qquad (2.23)$$

again, we use indices i, j for coordinates tangential to the brane, while a, b will denote normal directions.

The action (2.22) constitutes a generally covariant extension of abelian gauge theory with non-polynomial interactions, originally intended to describe electromagnetism at short distances. The integral over $\det^{\frac{1}{2}}(\tilde{g})$ is simply the worldvolume of an embedded p-brane. That the F- and B-fields enter the determinant seems natural considering, e.g., that g and B enter the Polyakov action (2.1) and (2.2) only through their sum $g + B$; see also equation (1.1), and our comments after equation (1.2).

To see the connection to gauge theory, let us set $B = 0$ and expand the Born–Infeld action (2.22) for small field strength $2\pi\alpha' F$ and for small fluctuations of the embedding coordinates $X^\alpha = 2\pi\alpha'\phi^\alpha$ around a flat background. This implies that the induced metric takes the form

$$\tilde{g}_{ij} \approx \eta_{ij} + (2\pi\alpha')^2 \, \partial_i\phi^a \, \partial_j\phi^b \, \delta_{ab} + \cdots .$$

Inserting this into equation (2.22), we obtain an action with a $(p+1)$-dimensional abelian gauge field and with $D - p - 1$ scalar fields, one for each direction transverse to the brane:

$$S_{\text{BI}} \approx -(2\pi\alpha')^2 T_p \int d^{p+1}\xi \left(\tfrac{1}{2}\partial_j\phi^a\partial^j\phi_a + \tfrac{1}{4}F_{ij}F^{ij}\right) + \cdots . \qquad (2.24)$$

Since the kinetic term for the scalar fields involves ordinary derivatives rather than covariant ones, the scalars carry no charge with respect to the gauge field.

In the supersymmetric case, the full low-energy brane action contains additional *Wess–Zumino terms*, which describe the coupling of closed string fields from the R–R sector – the antisymmetric forms $C_{(n+1)}$ – to the brane. The Wess–Zumino terms have the simple form (see [143, 251, 335])

$$S_{\text{WZ}} = i \, Q_p \int d^{p+1}\xi \, e^{\tilde{B} + 2\pi\alpha' F} \sum_n \tilde{C}_{(n+1)}. \qquad (2.25)$$

The exponential is nothing but the Chern character of the vector bundle over the brane worldvolume, which carries a connection with curvature $\mathcal{F} = \tilde{B} + 2\pi\alpha' F$.

The constant Q_p is the charge of the D-brane with respect to the $(p+1)$-form part of the integrand. In the simplest case of vanishing fields F and $\tilde{B}$, only the R–R gauge potential $C_{(p+1)}$ couples (minimally) to the worldvolume of the p-brane. As is customary, we have only given the bosonic terms of the D-brane action; for the supersymmetric completion, see, e.g., [439].

The method that was used in [328] to arrive at action (2.22) is based on a σ-model description of the hyperplane propagating in the ambient space-time. The beta functions for the σ-model fields can be computed perturbatively, and the vanishing of the beta functions – i.e. perturbative conformal invariance of the σ-model – leads to a system of coupled differential equations, which can be rewritten as classical equations of motion of a low-energy effective action. There is some freedom in this last step, which is one reason to perform string calculations to test that the LEEA is given by

$$S_{\text{D-brane}} = S_{\text{BI}} + S_{\text{WZ}}.$$

2.2.2 Dirichlet branes and boundary conditions

Let us now try to establish explicit connections between boundary CFT data and the standard example of flat D-branes for strings in a flat target. We will first relate gluing conditions and one-point functions to spacetime quantities, then discuss how to extract interaction terms in D-brane effective actions from CFT correlators.

Gluing conditions and worldvolume dimension

The bulk CFT describing strings in $\mathbb{R}^D$ is given by a D-fold tensor product of free bosons. For the moment we do not specify the signature of the flat-target metric. We shall impose the boundary conditions (1.33) along the boundary of the upper half-plane. This implies Neumann-type boundary conditions with possibly non-vanishing B-field for the first d coordinates $i = 1, \ldots, d$ and Dirichlet boundary conditions for the remaining ones. The associated boundary state

$$\|\alpha\rangle\!\rangle_\Omega = \|N(B)\rangle\!\rangle \otimes \|D(x_0)\rangle\!\rangle \tag{2.26}$$

is a product of the two boundary states (1.39) and (1.37) defined in Subsection 1.1.2. This means that, over the boundary of the worldsheet, the bosonic string coordinates take values in a d-dimensional hyperplane in $\mathbb{R}^D$ which is parallel to the first d-coordinate axes and confined to x_0^a for $a = d+1, \ldots, D$.

If the target metric has Minkowski signature, the natural gluing condition for the time coordinate is Neumann, and the convention is to use the name "Dirichlet p-brane" for a hyperplane with $(p+1)$-dimensional *worldvolume*, i.e. $d = p+1$. The case $p = -1$ does occur and is interpreted as a D-instanton (after analytic continuation to an Euclidean target). Let us also mention that, in the light-cone gauge, the natural gluing conditions for $X_\pm$ are Dirichlet – see [249]. Then also the branes of higher dimension should be interpreted as $(p+1)$-instantons, because of the Euclidean signature of their worldvolumes.

One-point functions and source terms in the LEEA

Beyond gluing conditions, the next simplest pieces of data specifying a boundary condition are one-point functions such as (1.30) or (1.32). We will discuss their string theoretic meaning presently, but let us first make one further general remark on string amplitudes.

Once D-branes are present, we need to extend the perturbation expansion (2.13) so as to include summations over open worldsheets Σ^{op} with boundaries,

$$\mathcal{A} = \sum_{g,b} g_S^{\chi(\Sigma_{g,b})}\, \mathcal{A}_{\Sigma_{g,b}}\,. \tag{2.27}$$

The relative weight g_S^{χ} of a diagram in the perturbation series is now determined from the Euler characteristic for a surface with genus g and with b boundary components, $\chi(\Sigma_{g,b}) = 2g - 2 + b$. In particular, the lowest-order contributions (to which we will restrict ourselves in the following) arise from disk diagrams which carry a weight g_S^{-1}. This is another way to see the non-perturbative nature of D-branes.

The amplitudes $\mathcal{A}_{\Sigma_{g,b}}$ are obtained from boundary CFT correlators, by integrating over the moduli of the relevant worldsheet diagrams. To calculate scattering of asymptotic in- and out-states $\phi_{\mathrm{eff}}^{\mathbf{m}}$ of perturbative string theory, we have to start from worldsheet correlation functions of the corresponding Virasoro primary fields $\varphi_{\mathbf{m}}$ in the associated CFT, as was reviewed briefly in Section 2.1.

As before, when restricting to massless asymptotic states we obtain amplitudes which are to be reproduced by the low-energy effective action. But since we are starting from a worldsheet with boundary, there are new types of correlators which lead new terms in the effective action, when compared to the closed string scenario.

The simplest new terms arise from one-point functions of non-trivial bulk fields $\varphi_{\mathbf{m},k}(z, \bar{z})$, which must vanish on the sphere but need not on the disk. Since the one-point functions depend on the boundary conditions, they introduce closed string *source terms* into the LEEA, which characterise the D-brane,

$$S_\alpha[\phi_{\mathrm{eff}}^{\mathbf{m}}] \sim g_S^{-1} \int d^d k\; \phi_{\mathrm{eff}}^{\mathbf{m}}(k)\, \langle \varphi_{\mathbf{m},k}(\tfrac{i}{2}, -\tfrac{i}{2}) \rangle_\alpha\,. \tag{2.28}$$

The disk diagram has no moduli to be integrated over, and invariance under $\mathrm{SL}(2, \mathbb{R})$ permits us to fix the insertion point, e.g. as $z = \tfrac{i}{2} = -\bar{z}$. As before, the index α stands for the boundary condition we impose.

If the field $\phi_{\mathrm{eff}}^{\mathbf{m}}$ has a specific spacetime interpretation, e.g. as a graviton or as some gauge potential, the coupling acquires a special meaning, too, e.g. that of a *tension* or a *charge* of the brane described by the boundary condition α. In this way, certain "physical quantum numbers" can be assigned to the brane; the necessary information is encoded in the set of bulk-field one-point functions.

In the simplest case of massless closed string modes, fields of momentum k correspond to operators of the form (2.9) with $k^2 = 0$ and a transverse polarisation tensor ϵ, see the comments after equation (2.8). The one-point functions of these massless states for a brane of dimension d were computed in equation (1.32) so that we obtain

$$
\begin{aligned}
S_\alpha &= g_S^{-1} \gamma_d^{\text{flat}} \int d^D k \, \epsilon_{\mu\nu}(k) \langle :J^\mu(z) \overline{J}^\nu(\bar{z}) e^{ikX} : \rangle_\alpha \\
&= g_S^{-1} \gamma_d^{\text{flat}} \int d^D k \, \epsilon_{\mu\nu}(k) \, \delta^{(d)}(k_i) \, e^{ik_a x_0^a} \, \Omega_{(B,d)}^{\mu\nu} \\
&= g_S^{-1} \gamma_d^{\text{flat}} \int d^D x \, \delta^{(D-d)}(x^a - x_0^a) \, \epsilon_{\mu\nu}(x) \, \Omega_{(B,d)}^{\mu\nu} \\
&= g_S^{-1} \gamma_d^{\text{flat}} \int d^d x \, \epsilon_{\mu\nu}(x)|_{x^a = x_0^a} \, \Omega_{(B,d)}^{\mu\nu}.
\end{aligned}
\tag{2.29}
$$

Here, k_a and k_i denote components of the momentum orthogonal and longitudinal to the flat worldvolume, respectively, and analogously for the target (or centre-of-mass) coordinates x that were introduced upon Fourier transformation. The gluing matrix $\Omega_{(B,d)}$ was defined in equation (1.33) and encodes both the worldvolume dimension and the B-field. Let us just quote the normalisation factor

$$
\gamma_d^{\text{flat}} = \frac{1}{2} \frac{\sqrt{\pi}}{2^{(D-10)/4}} (4\pi^2 \alpha')^{(D-2-2d)/4} \sqrt{\det(1+B)}
\tag{2.30}
$$

from [123, 175] and our discussion in Section 1.A.2. Equation (2.29) applies, strictly speaking, only to on-shell "effective fields" $\epsilon_{\mu\nu}(x)$. As long as there is no definition of string theory beyond first-quantised level, the ad-hoc (but commonly used) extension of effective actions to off-shell fields remains, to some extent, guesswork.

Up to this subtlety, we see that the specific momentum dependence of Dirichlet and Neumann one-point functions implies that the source term in the low-energy effective action is an integral over the worldvolume of the flat D-brane – just as in the Born–Infeld action (2.22). Put differently, only longitudinal momenta are conserved in interactions of closed strings with D-branes. In order to perform a more detailed comparison between the answer obtained from boundary conformal and the Born–Infeld action, we expand the latter up to first order in the bulk fluctuation fields. Assuming $F = 0$ and vanishing fluctuations of the brane's transverse position, i.e. $\xi^i = X^i$, $X^a = x_0^a$, we use the expansions

$$
\tilde{g}_{ij} \approx \eta_{ij} + h_{ij}, \quad \tilde{B}_{ij} \approx B_{ij} + b_{ij}, \quad \tilde{\phi} \approx \phi + \varphi.
$$

Substituting this into the Born–Infeld action and keeping only linear terms in the fluctuations h, b and φ, we obtain

$$\delta S_{\text{BI}} \approx T_p \int d^d\xi e^{-\phi} \sqrt{\det_\shortparallel(1+B)} \left[\varphi + \frac{1}{2}\text{tr}_\shortparallel \left(\frac{1}{1+B}(h+b) \right) \right]. \tag{2.31}$$

The subscripts $\shortparallel$ remind us to take the trace and determinant only over the d directions along the brane.

The outcome (2.31) still looks a bit different from our result (2.29) from the boundary state. However, we can match the expressions by rewriting the integrand as

$$\epsilon_{\mu\nu}\Omega^{\mu\nu}_{(B,d)} = \text{tr}_\shortparallel \left(\frac{1-B}{1+B}\epsilon \right) - \text{tr}_\perp(\epsilon) = \text{tr}(\Omega_{(B,d)}\epsilon^\phi)$$

$$+ \text{tr}_\shortparallel \left(\frac{2}{1+B}(\epsilon^g + \epsilon^B) \right). \tag{2.32}$$

In the first equality we expressed the left-hand side as a trace over the D directions of the target space using the equation $\Omega^{\mu\nu}_{(B,d)} = \Omega^{\nu\mu}_{(-B,d)}$ that follows from the antisymmetry of B. At the same time we split the trace into a part involving the directions along the brane and the remaining perpendicular ones. Furthermore, we inserted formula (1.33) for Ω. In passing to the second expression, we decomposed

$$\epsilon = \epsilon^g + \epsilon^B + \epsilon^\phi$$

into a symmetric traceless part ϵ^g, the antisymmetric tensor ϵ^B and the trace ϵ^ϕ. Since $\epsilon^B + \epsilon^g$ is traceless, we were able to replace $\frac{(1-B)}{(1+B)}$ by $\frac{2}{(1+B)}$.

If we identify the tensors ϵ^m with the fluctuations h, b and φ, respectively, the expressions (2.29) and (2.32) coincide with equation (2.31) provided that

$$T_p \sqrt{\det(1+B)} = 2\gamma^{\text{flat}}_{p+1}. \tag{2.33}$$

For the comparison we must recall the relation $g_S \sim \exp(-\phi)$. The coefficient of the fluctuation h of the metric measures the mass of the D-brane. Because of the overall factor g_S^{-1}, the mass of a D-brane is large for weak string coupling.

Evaluating the dilaton term requires some more care. Let us recall that the polarisation tensor $\epsilon_{\mu\nu}$ must satisfy the on-shell conditions, i.e. $\epsilon_{\mu\nu}k^\nu = 0 = k^\mu \epsilon_{\mu\nu}$. For ϵ^ϕ this is ensured through

$$\epsilon^\phi_{\mu\nu} = \frac{\epsilon^\phi}{\sqrt{d-2}} \left(\eta_{\nu\mu} - k_\mu l_\nu - k_\nu l_\mu \right),$$

where l is a light-like vector satisfying $kl = 1$; see [123]. Inserting this expression into the first term of equation (2.32) gives

$$\text{tr}(\Omega_{(B,d)}\epsilon^\phi) \sim \frac{1}{\sqrt{d-2}} \left(2\text{tr}_\shortparallel(1-B^2)^{-1} - D + 2 \right) \epsilon^\phi.$$

In the short derivation we used the fact that components of the closed string momentum k pointing along the brane may be set to zero. Note that the matrix $(1 - B^2)^{-1}$ in the argument of the trace is the open string metric G that was introduced in equation (1.13). For a vanishing B-field, the coefficient of ε^ϕ is given by $(2p + 4 - d)/\sqrt{d - 2}$; see [123] for a detailed comparison with the Born–Infeld action. The B-field modifies the coefficient. This was to be expected since the B-field interpolates between Neumann and Dirichlet boundary conditions. In particular, sending (some components of) the B-field to infinity reduces the spatial dimension p of the brane that appears in the prefactor of ε^ϕ for $B = 0$.

If the background B-field vanishes, both equations (2.29) and (2.31) show that our bosonic D-branes do not couple to the fluctuations b of the Kalb–Ramond tensor $B_{\mu\nu}$. For this massless field, the polarisation $\epsilon^B_{\mu\nu}$ is anti-symmetric so that the contraction with $\Omega_{\mu\nu}$ in equation (2.29) vanishes. The same is actually true for the supersymmetric extension of our simple setup. In this sense "elementary" BPS D-branes carry no NS–NS charge. On the other hand, turning on a background B-field results in a non-vanishing coupling between the D-brane and the fluctuations b of the NS–NS two-form potential B. Such branes are therefore often interpreted as bound states of a fundamental string (which couples to B almost by definition).

Let us take the opportunity to add a few informal remarks on *D-instantons* here. These special branes have no massless internal excitations and thus the LEEA for a system of strings and D-instantons depends on effective fields from the closed string sector only – in contrast to branes with extended worldvolumes, where massless internal degrees of freedoms lead to interesting new low-energy effects, see Subsection 2.2.3.

Due to this simplicity, some general statements can be made on string amplitudes (2.27) where all boundaries satisfy Dirichlet gluing conditions in all D directions. First of all, the mere presence of a D-instanton introduces a non-perturbative g_S-dependence into the couplings of the LEEA. To see this, first note that every bulk field correlator, hence every closed string amplitude, will be proportional to the one-point function $A_0^\alpha = \langle 1 \rangle_\alpha$ of the identity operator for the D-instanton boundary condition α_0. Next, it can be shown that the leading contribution to the series (2.27) comes from disjoint disks [247, 374]. Taking care of symmetry factors and Wick rotations, we can perform the summation and obtain a field-independent factor

$$S_{\text{one-inst}} = e^{-\frac{A_0^\alpha}{g_S}} \qquad (2.34)$$

modifying the closed string couplings. The relation (2.34) shows the characteristic g_S-dependence of non-perturbative effects in string theory predicted by Shenker [426] even before the advent of D-branes.

In addition, disk diagrams with D(-1)-boundary conditions and bulk field insertions result in new instanton-induced interactions for the closed string effective fields, which have to vanish for symmetry reasons in theories with Neumann boundary conditions. One example is the R^4-terms in the LEEA, which provide a rich playground for tests of S- and U-dualities. We will briefly return to R^4-interactions in Chapter 7 in the context of Gepner models, but see, e.g., [250, 262, 362] for general reviews and further references.

2.2.3 Correlators and LEEA interaction terms

We have seen in Chapter 1 that imposing a boundary condition on the fields of a CFT on the plane inevitably introduces new operators confined to the boundary, and we would now like to fit these boundary fields into the string picture of D-branes.

With bulk and boundary fields at our disposal, we can compute string amplitudes of the form

$$\mathcal{A} \sim \int d\mu \, \langle \phi_1(z_1, \bar{z}_1) \dots \psi_1(x_1) \dots \rangle_\alpha$$

as integrals of correlators of a number of bulk and boundary fields ϕ_i, ψ_j on the disk or on the upper half-plane (we restrict ourselves to the lowest order in the string coupling). Summation over inequivalent orderings of the boundary field arguments is understood to be included in the measure $d\mu$ on the worldsheet moduli space. Of course, fields corresponding to massless string states play the most important role for low-energy effective physics.

Before going into any details, we can make a few qualitative observations on the amplitudes above:

If only bulk fields are included, $\mathcal{A}$ describes the *elastic scattering* of closed string states off the brane α – which here is always considered as a fixed target that can absorb any amount of transversal momentum $k_\perp$. This is justified at weak coupling where branes are very heavy and where recoil effects are suppressed. In particular, note that the amplitude for elastic scattering of two closed string states in a flat target can be analysed for poles in the Mandelstam variables s and t, which offers an alternative method to extract LEEA source terms; see, e.g., [221, 259, 271].

Mixed bulk–boundary correlators lead to amplitudes for *inelastic scattering* of closed strings at a brane. A simple example is an amplitude with one closed string and two open string states, which is sometimes interpreted as a contribution to "Hawking radiation" of the brane.

Amplitudes involving only boundary fields belong to *transitions in the internal state* of the brane via open string excitation, and can be used to test expectations on the form of the LEEA for the (weak-coupling) brane dynamics; they will be the one we focus on below.

Boundary fields and internal brane excitations

First we need to "count" the boundary fields that are available. In the special case of a "spacetime-filling" brane, i.e. a boundary state (2.26) where Neumann conditions apply to all coordinates, boundary fields ψ_k are of course well known as open string vertex operators. The massless excitations $|\epsilon(k)\rangle = \epsilon_\mu(k)\alpha^\mu_{-1}|k\rangle$ of open strings produce a vector of gauge fields in spacetime.

In order to see what happens for D-brane boundary conditions, we compute the partition function $Z_{\alpha\alpha}(q)$ for the brane boundary condition α in equation (2.26), which provides the required counting of open string states. Multiplying the contributions from Neumann and Dirichlet directions, the result was obtained in equation (1.41)

$$Z_{\alpha\alpha}(q) = \int d^d k \, (2\alpha')^{\frac{d}{2}} (\det G)^{\frac{1}{2}} \frac{q^{\alpha' k^2}}{\eta(q)^D}. \tag{2.35}$$

Here we have assumed $B = 0$. There are D eta functions from the Fock spaces generated by the currents J^μ for $\mu = 0, \ldots, D-1$, but the integration runs only over momenta parallel to the brane worldvolume, which we will sometimes denote by $k^\parallel$ in the following.

The primary boundary fields are products of vertex operators as in equation (1.46) for the Neumann directions with an identity operator for the Dirichlet directions. In addition, we can multiply with the D currents $J^\mu(x)$ and their derivatives. The partition function (2.35) describes the excitation spectrum of a D-brane α. It contains fields with spacetime mass zero, which belong to states $\alpha^\mu_{-1}|k_\parallel\rangle$. Because of the momentum dependence, the first d components transform like a d-vector under the worldvolume Lorentz (or Euclidean) group, while the last $D-d$ components are scalars. Vertex operators for massless modes take the usual form

$$\psi(|\epsilon(k^\parallel)\rangle; x) = \epsilon_\mu(k^\parallel) :J^\mu(x)e^{ik_j X^j(x)}: . \tag{2.36}$$

The summation over j and the functional dependence of ϵ involves directions along the brane only. After a Fourier transformation in k, we find that the corresponding effective spacetime fields depend only on coordinates $x^\parallel$ *within the brane worldvolume*: these fields live on the D-brane and can be interpreted as its internal excitations. The associated effective vector fields $A^i(x^\parallel)$, $i = 0, \ldots, p$, make up a U(1) gauge potential on the brane, while the remaining scalars $\phi^a(x^\parallel)$ for $a = p+1, \ldots, D-1$ may be viewed as Higgs fields.

Gauge theories on the brane

It is well known that interacting open strings are inconsistent without introducing closed strings – the most direct and general argument will be given in Chapter 4 in the context of Cardy's condition, which is nothing but the re-interpretation of a one-loop open string diagram as a closed string propagator via worldsheet duality. Similarly, a theory of closed strings interacting with D-branes cannot be formulated without inducing internal excitations on the latter. The spacetime intuition for this is that the closed string with its non-vanishing tension pulls at the brane, so that the seemingly static hyperplane acquires non-trivial dynamics.

We can ask what the low-energy effective action for the internal physics of a brane is. The terms bilinear or trilinear in the effective fields on the brane follow immediately from the boundary CFT data: as before, string amplitudes and terms from the effective action are obtained from CFT correlators by integrating over worldsheet moduli of disk diagrams (we consider the lowest order in g_S only). Since $\mathrm{SL}(2, \mathbb{R})$-invariance allows the fixing up to three insertion points (here of the boundary fields), no integrations have to be carried out for two- and three-point functions. For n-point functions with $n \geq 3$, we must also sum over non-cyclic permutations of the boundary fields: the ordering of boundary field arguments constitutes an additional "discrete worldsheet modulus".

In this way, it can be shown that, for a vanishing B-field, the LEEA of internal excitations of a flat brane is given by dimensionally reduced $\mathrm{U}(1)$ gauge theory, with the same action (2.24) that is obtained from expanding the Born–Infeld action around flat space. The computation of the relevant string amplitudes can be found, e.g., in [377]; we will spell out a few details of the derivation from CFT correlators below, in a slightly more general situation where $B \neq 0$ is allowed.

We can of course also consider open strings with different boundary conditions at the two ends, and we have in fact already computed partition functions for such situations in Subsection 1.A.1. A particularly simple, yet physically highly interesting special case arises when we consider multiple copies, say N, of the same brane. The open string states on such a "stack of branes" are described by boundary fields $M \otimes \psi$, where ψ is a boundary field associated with a single such brane and M is an $N \times N$ matrix, the so-called *Chan–Paton factor*. It is clear that the OPE of two such boundary fields can be formed only if the "end-brane" of the first coincides with the "start-brane" of the second open string state – in other words, the OPE of boundary fields on a single brane is overlayed with matrix multiplication of the Chan–Paton factor, and, in particular, summation over boundary field orderings becomes non-trivial. All in all, it is not surprising that non-abelian gauge theories arise as LEEAs of open strings on stacks of branes.

The fact that dimensionally reduced Yang–Mills–Higgs theories describe the low-energy physics of flat branes paves the way towards various interesting applications of D-branes to problems of gauge theory. Reviewing these developments is beyond the scope of the present work, and we refer to the original literature and to reviews such as [238].

Non-commutative gauge theory

We have seen that a single brane gives rise to a $\mathrm{U}(1)$ gauge theory, which does not possess any non-trivial self-interaction. In order to make the discussion of interaction terms more interesting, we will now include a non-vanishing background B-field. The resulting LEEA theory for a single brane has many features in common with the non-abelian gauge theories obtained for $B = 0$ by introducing Chan–Paton factors, and has at the same time a natural interpretation as a gauge theory on a non-commutative space.

Before we sketch how to compute the interaction terms, we need to specify how we want to take the low-energy limit. The relevant limiting procedure for functionals $F(\alpha'; g, B)$ is the *decoupling limit* introduced by [415],

$$F^{\mathrm{DL}}(g, B) = \lim_{\epsilon \to 0} F(\epsilon \alpha'; \epsilon^2 g, \epsilon B). \qquad (2.37)$$

This includes sending the string scale α' to zero, simply because we are interested in the low-energy physics; at the same time, the magnetic field B and the closed string metric g (more precisely its components g_{ij} along the brane) are scaled in such a way that the geometry the brane "sees", i.e. the open string metric G and the antisymmetric tensor Θ defined in equation (1.13), remain finite.

Let us digress briefly to give a more physical motivation behind this limit. It is well known that the coordinates of a charged particle in a magnetic background have a non-vanishing Poisson bracket and hence they do not commute after quantisation. The above limit produces this quantum-mechanical situation from an open string: recall that the action functional (1.1) for open strings has two terms. To begin with, there is a boundary term which describes the motion of the charged open string ends in a magnetic field. To reach a point particle limit, we have to suppress the string oscillations, i.e. we have to send α' to zero. The B-field should be scaled down at the same rate so that $\frac{B}{\alpha'}$ remains constant. But even in this limiting regime the resulting theory for the string endpoints does not approach the theory for charged particles in a magnetic field because of the bulk term in the action (1.1) for open strings. This term makes the open string ends remember that they are attached to a string which becomes very stiff as we try to turn off the oscillations. Consequently, the open string ends dissipate energy into these tails and the strength of this dissipation is given by g_{ij}/α', see, e.g., [91]; obviously, only the components along the brane matter. If we want to suppress this dissipation, these components of the closed string metric g have to vanish at a faster rate than α'. All this is achieved by the decoupling limit we defined above.

In order to determine interaction terms for massless open string fields, we have to evaluate correlation functions of the vertex operators (2.36). They are built from tachyonic vertex operators ψ_k, whose correlators have been computed already in Subsection 1.1.3, see equation (1.47). In taking the decoupling limit of this result, the coordinate dependence disappears (since $\alpha' G^{ij}$ vanishes) and we find

$$\langle \psi_{k_1}(x_1) \; \cdots \; \psi_{k_n}(x_n) \rangle \xrightarrow{\mathrm{DL}} \prod_{r<s} e^{-\frac{i}{2} k_r \times k_s} \, \delta(\textstyle\sum_r k_r). \qquad (2.38)$$

Here, $r, s = 1, \ldots, n$, and we have used the notation $k \times k' = k_i \Theta^{ij} k'_j$, as in Chapter 1. The δ-function enforces momentum conservation, and the phase factors are identical to the ones we encountered in multiplying two exponential functions using the non-commutative Moyal–Weyl $*$ product associated with the antisymmetric tensor Θ; see equation (1.53). Generalising this to fields $\psi[f](x) := \int d^d k \, \hat{f}(k) \, \psi_k(x)$, we can restate the result in a concise form as follows

$$\langle \psi[f_1](x_1) \cdots \psi[f_n](x_n) \rangle^{\mathrm{DL}} = \int_V d^d x \, f_1 * \cdots * f_n \,. \qquad (2.39)$$

The formulation we have followed here was found in [410], but related observations were made by several authors [23, 111, 114, 148].

We still need to compute string amplitudes of the massless vector fields associated with the worldsheet boundary fields (2.36), so we need to compute correlation functions of the latter. The additional CFT ingredients we need concern normal ordering and Ward identities arising from simple OPEs.

The normal ordering procedure in (2.36) means that annihilation operators should be moved to the right of the creation operators, explicitly

$$:J^\mu(x)\psi_k(x): = J^\mu_<(x)\,\psi_k(x) + \psi_k(x)\,J^\mu_>(x)\,, \qquad (2.40)$$

where the summands $J^\mu_>(x)$ and $J^\mu_<(x)$ in $J(x) = J^\mu_>(x) + J^\mu_<(x)$ are defined by

$$J^\mu_>(x) = \sum_{n>-1} \alpha^\mu_n x^{-n-1} \quad \text{and} \quad J^\mu_<(x) = \sum_{n\le-1} \alpha^\mu_n x^{-n-1}\,.$$

To compute correlation functions of the above fields in the decoupling limit, we proceed in two steps. The first one is to remove all the currents from the correlators with the help of Ward identities; see below. Once we are left with correlation functions of the exponential fields ψ_k we can then use the Moyal product formulas resulting from the discussion above. Since we are only interested in the decoupling limit of the correlation function, we can drop sub-leading terms whenever they arise in the computation.

Let us discuss this in some more detail. Using a bit of algebra, it is not difficult to compute the singular parts of the following operator product expansions:

$$(J^\mu(x_1)J^\nu(x_2))_{\mathrm{sing}} = [\,J^\mu_>(x_1), J^\nu(x_2)\,] = \frac{\alpha'}{2}\frac{G^{\mu\nu}}{(x_1-x_2)^2} \qquad (2.41)$$

$$(J^\mu(x_1)\psi_k(x_2))_{\mathrm{sing}} = [\,J^\mu_>(x_1), \psi_k(x_2)\,] = \frac{\alpha'}{2}\frac{G^{\mu i}k_i}{x_1-x_2}\psi_k(x_2) \qquad (2.42)$$

These commutators along with the properties $J^\mu_>(x)|0\rangle = 0$ and $\langle 0|J^\mu_<(x) = 0$ of the vacuum can be used to remove all currents from an arbitrary correlator. In order to do so, we commute the lowering term $J^\mu_>$ of each current insertion to the right until it meets the vacuum, and similarly the raising terms are all pushed to the left. The commutation relations give rise to two different contributions. If one current hits another, both of them disappear from the correlator and we obtain a factor of $\alpha' G^{\mu\nu}$. Commuting a current through an exponential, on the other hand, removes only one current insertion and furnishes a factor $\alpha' G^{\mu i}k_i$ instead. Since both factors are of the same order in α', the leading contribution to an n-point function is obtained when we contract as many pairs of currents as possible, i.e. $\frac{n}{2}$ for even n and $\frac{(n-1)}{2}$ for odd n. In the latter case, the last current will necessarily lead to a linear dependence on the momentum k.

It follows that three- and four-point functions both contain terms which are second order in α'. All n-point correlators with $n > 4$ are sub-leading. While the dominant contributions to the three-point function contain a factor linear in the momenta k, the corresponding factor for the four-point function is independent of k. Both these factors are finally multiplied with correlators (2.38) of the exponential fields. The latter certainly introduce a strong k dependence whenever there is a non-vanishing B-field.

Apart from these momentum-dependent phase factors in the Moyal product, the computation and the outcome are the same as in the derivation of the low-energy string amplitudes for ordinary open strings with Neumann boundary conditions [377, 412], which can be reproduced by a Yang–Mills theory. For instance, the three-gluon vertex is linear in the external momenta, just as the leading α'-contribution from the three-point function is.

Inclusion of Chan–Paton factors is straightforward, and we can conclude that the low-energy effective action for a stack of M branes is given by the Yang–Mills action for gauge fields $A_\mu \in \mathrm{Mat}_M(\mathit{Fun}\,(\mathbb{R}^D))$ on a non-commutative $\mathbb{R}^D$ [415],

$$S_N(A) = \frac{1}{4} \int d^D x \; \mathrm{tr} \; (F_{\mu\nu} * F^{\mu\nu}) \,.$$

The integration extends over the worldvolume of the brane, the field strength is defined by

$$F_{\mu\nu}(A) = \partial_\mu A_\nu - \partial_\nu A_\mu + i\,[A_\mu \stackrel{*}{,} A_\nu]$$

and $*$ denotes the Moyal product as before. It can easily be verified that this non-commutative Yang–Mills theory is invariant under the following gauge transformations,

$$A_\mu \longrightarrow A_\mu + \partial_\mu \lambda + i\,[A^\mu \stackrel{*}{,} \lambda],$$

for $\lambda \in \mathrm{Mat}_M(\mathit{Fun}\,(\mathbb{R}^D))$. In writing down $S_N(A)$, we have assumed that $D = d$ for simplicity; the action for lower-dimensional branes is obtained by dimensional reduction. It is also worth noting that the same type of non-commutative gauge theories arise from superstring theory [415].

The relation between branes in flat space and non-commutative geometry has triggered a lot of interest in non-commutative field theories among string theorists. In particular, this has led to significant progress in constructing classical solutions (see, e.g., [9, 243, 257, 269, 361, 381]), which allow for an interpretation as tachyon condensates on branes. We shall come back to related issues in Chapter 6 when we analyse branes on a three-sphere. For an overview of many of the developments in this field and other aspects that we have not touched, we recommend, e.g., [151, 268, 319, 320, 415] and references therein.

2.2.4 Concluding remarks

This completes the basic part of our "dictionary" between string-theory Dirichlet branes in a flat target and boundary CFT. We have seen how gluing conditions and one-point functions for a free field theory reproduce the more conventional target picture of these non-perturbative objects. We have sketched how scattering amplitudes and new terms in the low-energy effective action can be derived from CFT quantities. Boundary CFT partition functions have so far only been used in open string state counting, but we will extract further string theoretic information from them below in the additional material section.

The translation between string theory D-branes and boundary conditions makes it possible to address the construction of "D-branes" for more general string backgrounds and in non-geometric compactifications given in terms of some internal CFT. The most important class of examples (from the point of view of string phenomenology), the so-called Gepner models, are dealt with in Chapter 7. An intermediate situation, between algebraic CFT backgrounds and flat targets, is provided by toroidal orbifolds, which were studied extensively in the string literature, see, e.g., [73, 147, 234, 292, 382]. In contrast to Gepner models, much of the structure of torus orbifolds can be understood in the spacetime picture, so they are tractable without too much worldsheet machinery, at least as long as the orbifold group acts symmetrically on left- and right-moving modes. We will briefly discuss the orbifold construction in Subsection 4.A.2 below, but see also [59, 207, 367] for rather detailed and more general investigations of (symmetric) orbifold theories within the boundary CFT framework.

Let us remark that there are a number of physical issues we will not discuss, in spite of their importance for string theory and beyond. Among the most notable, and in fact earliest, applications of branes are calculations to derive the Bekenstein–Hawking black-hole entropy from string theory [284, 285, 434]. At first, only the language of solitonic p-branes was used in this context, but later also CFT methods, in particular applied to systems of D1–D5-branes, proved useful. A review of these ideas can be found in [90, 342].

Branes can also be used for a new type of string compactification, so-called "brane world models": here, our universe is regarded as the worldvolume of a Dirichlet three-brane, and worldsheet computations are in fact useful in determining the effective low-energy corrections induced by gravitons and other excitations in the perpendicular dimensions; see, e.g., [22] and references therein, and also [368] for M-theory variations of this, based on Hořava's and Witten's work [279].

It is probably gauge theory where branes had the strongest impact, partly due to the fact that such models arise as low-energy actions as sketched above, partly due to Maldacena's discovery of the gauge–gravity correspondence [343, 456]. It turns out that various constructions in gauge theory have simple descriptions in terms of stacks of branes, viewed within string or M-theory, see, e.g.,

[238, 267, 455]. Some of those ideas have been cast in the worldsheet language; for other ideas, e.g. branes ending on branes [433], such a formulation is missing.

2.A Additional material

In this section we will first explain how to treat Dirichlet branes in superstring theory from the worldsheet point of view, which in particular requires implementation of the GSO projection. One important physical consequence of the cancellation of Neveu–Schwarz and Ramond sector contributions is the vanishing of the force between two static parallel Dp-branes in superstring theory. In Subsection 2.A.2, we study the potential between two branes at slow relative motion, i.e. the non-relativistic scattering of two branes. This provides a nice application of the formulas collected in connection with rotated branes, and also suggests some interesting spacetime interpretations.

2.A.1 *GSO projection and target-space supersymmetry*

In our expository dictionary between string theoretic aspects of D-branes and notions from boundary CFT, we have deliberately left supersymmetry aside, except for some marginal remarks, partly because of the heavier notation to be carried through supersymmetric computations, which might have blurred the salient points in our translation from boundary CFT to brane physics.

But even though supersymmetry is not an indispensable ingredient of the boundary CFT approach to D-branes, the supersymmetric extension of the above considerations is of great importance for applications to string theory. First, only superstring theory seems to allow for a stable spacetime theory since the tachyon can consistently be removed from the spectrum by means of the GSO projection. Secondly, it is the supersymmetric phenomenon of *BPS states* that allows us to make statements on the strong coupling behaviour of string theories from perturbative (weak coupling) calculations in the new D-brane sectors: as long as the brane configuration preserves some supersymmetry, certain quantities receive no loop corrections, as can be made plausible by algebraic methods ("short" multiplets of the supersymmetry algebra) or with the help of path integral formulas (integration over fermionic zero modes).

In the following, we want to deduce the most important new properties of supersymmetric D-branes, still restricting ourselves to free fields. During the construction of the fermionic boundary state, we will recover the relation between "allowed" D-brane dimensions and chirality type (IIA or IIB) of closed superstrings. Then, Polchinski's calculation of the one-loop potential between two parallel identical branes will be reviewed, which he used to demonstrate that D-branes have the BPS property.

In order to keep formulas relatively simple and closer to those from the bosonic case, we work in the Ramond–Neveu–Schwarz (RNS) formalism (i.e. with worldsheet fermions) and in the light-cone gauge as in [47]. The Green–Schwarz formulation, which has been presented comprehensively in [249], seems to lack one major advantage of the worldsheet approach: it relies on spacetime supersymmetry and is therefore hardly suitable for the construction of non-BPS branes.

In the light-cone gauge, the longitudinal degrees of freedom $X_{\pm} = (X^0 \pm X^9)/\sqrt{2}$ in a spacetime of dimension $D = 10$ satisfy Dirichlet conditions and will be ignored in the following. So we should actually speak of $(p+1)$-instantons with $-1 \le p \le 7$ rather than of Dirichlet p-branes. We could also employ a Lorentz covariant RNS-formulation with BRST-charge at the expense of tensoring with ghosts and superghosts. This would not pose any fundamental problems since our general framework does not require unitarity. Moreover, boundary conditions and boundary states for the ghosts do not depend on those for the free bosons and fermions (or for more general internal CFTs like in the Gepner models): the same ghost boundary states as for ordinary open strings (Neumann in all directions) can be used. They link right- and left-moving BRST-charges by $(Q + \overline{Q}) \| B_{\text{gh}} \rangle\rangle = 0$. We refer to [89, 123, 335, 378] and in particular [460] for a detailed discussion.

Recall from Section 1.2 that (in the RNS-formulation of superstring theory) there is one bosonic coordinate X^j together with its fermionic superpartner ψ^j for each direction j of the flat target. If we demand that the $N = 1$ supercurrent $G = i \psi_j \partial X^j$ is preserved by the boundary conditions, i.e. that the boundary theory enjoys superconformal invariance, then the gluing conditions for the individual currents $\partial X^i(z) = \Omega^i_j \bar\partial X^j(\bar z)$ enforce gluing conditions for the individual fermions

$$\psi^i(z) = \eta \, \Omega^i_j \, \overline{\psi}^j(\bar z) \,. \tag{2.43}$$

In comparison to our earlier treatment in Chapter 1, we have admitted an additional sign $\eta = \pm 1$. Since it is independent of the label j, the gluing conditions (2.43) respect all target-space symmetries of the configuration. The relevance of the sign η will become clear in a moment. It also appears in the gluing conditions of the chiral field G that was introduced in equation (1.72) above,

$$G(z) = \eta \, \overline{G}(\bar z) \quad \text{for } z = \bar z \,.$$

For simplicity, let us again restrict ourselves to the case that describes a Dp-brane without a magnetic field B. In the light-cone approach where j runs through $j = 1, \ldots, D - 2$, this means that

$$\Omega^i_j \equiv \Omega^i_{p;j} = \varepsilon^{(j)} \delta_{i,j} \quad \text{where} \quad \varepsilon^{(j)} = \begin{cases} +1 & \text{for } j = 1, \ldots, p+1 \\ -1 & \text{for } j = p+2, \ldots, D-2 \,, \end{cases}$$

with $D = 10$. Based on the constructions and notations of Subsection 1.2.2, in particular equations (1.62)–(1.67), we introduce the boundary states

$$\| p, \eta \rangle\rangle^{(\text{NS})} = \exp \left\{ i \eta \sum_{r>0} \Omega_{ij} \psi^i_{-r} \overline{\psi}^j_{-r} \right\} |NS, NS\rangle \,, \tag{2.44}$$

$$\| p, \eta \rangle\rangle^{(\text{R})} = \exp \left\{ i \eta \sum_{n>0} \Omega_{ij} \psi^i_{-n} \overline{\psi}^j_{-n} \right\} |R, R\rangle_{p;\eta} \tag{2.45}$$

where the Ramond ground states are related by the action of linear combinations (1.66) of left- and right-moving fermion zero modes

$$|R, R\rangle_{p,+} = \psi_-^{p+2} \cdots \psi_-^8 |R, R\rangle \,, \quad |R, R\rangle_{p,-} = \psi_-^1 \cdots \psi_-^{p+1} |R, R\rangle$$

as in equation (1.67). The ground state $|R, R\rangle$ is annihilated by all ψ_+^j.

Physical boundary states for closed superstrings are GSO-projected. To formulate the conditions, we use two operators F and Γ_9 together with their right-moving analogues $\overline{F}$ and $\overline{\Gamma}_9$. The fermion number operator F is given by

$$F = \sum_{i=1}^8 \sum_{r>0}^\infty \psi_{-r}^i \psi_r^i$$

with $r \in \mathbb{Z}$ for the Ramond and $r \in \mathbb{Z} + \frac{1}{2}$ for the Neveu–Schwarz sector. We define its action on the Neveu–Schwarz ground state by

$$(-1)^F |NS, NS\rangle = -|NS, NS\rangle \,.$$

In addition we introduced the ten-dimensional light-cone gauge chirality operator in the Ramond sector by

$$\Gamma_9 = 16 \, \psi_0^1 \cdots \psi_0^8 \,.$$

The numerical factor 16 ensures that $\Gamma_9^2 = 1$.

The GSO conditions in the *NS-sector* say that physical closed string states $|\xi\rangle$ must satisfy

$$(-1)^F |\xi\rangle = |\xi\rangle = (-1)^{\overline{F}} |\xi\rangle \,. \tag{2.46}$$

Since by construction of F we have $(-1)^F |NS, NS\rangle = -|NS, NS\rangle$, the condition eliminates the tachyon from the superstring spectrum and shows that

$$(-1)^F \| p, \eta \rangle\!\rangle^{(\mathrm{NS})} = (-1)^{\overline{F}} \| p, \eta \rangle\!\rangle^{(\mathrm{NS})} = - \| p, -\eta \rangle\!\rangle^{(\mathrm{NS})}$$

because $(-1)^F$ anti-commutes with ψ_r^j but commutes with $\overline{\psi}_{-r}^j$. A GSO invariant fermionic boundary state in the NS-sector can be formed as a linear combination of the two possible choices of η (the two possible spin structures),

$$\| p \rangle\!\rangle_{\mathrm{ferm}}^{(\mathrm{NS})} = \kappa_{\mathrm{NS}} \left(\| p, + \rangle\!\rangle^{(\mathrm{NS})} - \| p, - \rangle\!\rangle^{(\mathrm{NS})} \right) , \tag{2.47}$$

with an up to now free normalisation factor κ_{NS}. On the right-hand side, we have introduced the GSO projectors $P_{\mathrm{NS}}^{\mathrm{GSO}} := \frac{1}{2}(1 + (-1)^F)$, etc. We have placed a subscript "ferm" on the boundary state to remind us that the state is still to be multiplied with a factor in the bosonic sector of the model, i.e. with a boundary state of the form (2.26).

In the *R-sector*, physical states must be invariant under $\Gamma_{\mathrm{R}} = \Gamma_9(-1)^F$ and $\overline{\Gamma}_{\mathrm{R}} = \overline{\Gamma}_9(-1)^{\overline{F}}$. As above, application of these operators to $\| p, + \rangle\!\rangle^{(\mathrm{R})}$ switches the overall sign in the gluing conditions from $\eta = +1$ to $\eta = -1$, and vice versa. Our normalisation ensures that

$$\Gamma_9 \, |R, R\rangle_{p,+} = |R, R\rangle_{p,-} \quad \text{and} \quad \overline{\Gamma}_9 \, |R, R\rangle_{p,+} = (-1)^{7-p} \, |R, R\rangle_{p,-} \, . \quad (2.48)$$

In the Ramond sector the physical state conditions differ for type IIA and type IIB superstrings: $|\xi\rangle$ remains in the spectrum if

$$\begin{aligned} \Gamma_{\mathrm{R}} \, |\xi\rangle = |\xi\rangle = -\overline{\Gamma}_{\mathrm{R}} \, |\xi\rangle \quad &\text{for type IIA} \, , \\ \Gamma_{\mathrm{R}} \, |\xi\rangle = |\xi\rangle = +\overline{\Gamma}_{\mathrm{R}} \, |\xi\rangle \quad &\text{for type IIB} \, . \end{aligned} \quad (2.49)$$

The eigenvalue equation for the right-moving $\overline{\Gamma}_{\mathrm{R}}$ dictates that the physical boundary state in the R-sector is

$$\| p \rangle\!\rangle_{\mathrm{ferm}}^{(\mathrm{R})} = \kappa_{\mathrm{R}} \left(\| p, + \rangle\!\rangle^{(\mathrm{R})} + \| p, - \rangle\!\rangle^{(\mathrm{R})} \right) . \quad (2.50)$$

On the other hand, we have $\Gamma_{\mathrm{R}} \| p \rangle\!\rangle_{\mathrm{ferm}}^{(\mathrm{R})} = (-1)^{7-p} \| p \rangle\!\rangle_{\mathrm{ferm}}^{(\mathrm{R})}$ because of equation (2.48). This fits with the GSO conditions (2.49) for type IIA only if p is *even*, and only if p is *odd* for the type IIB theory.

The restriction carries over to the full fermionic boundary state, since it must receive contributions from both the Neveu–Schwarz and the Ramond sector if we require the brane to preserve some spacetime supersymmetry; see below. In this way, we recover the conditions encountered in the classical solutions (2.20) of the LEEA. The constraints we used within the boundary CFT framework were linear ones, namely the closed string GSO projection and (worldsheet supersymmetry preserving) gluing conditions.

For toroidal compactifications, the considerations above apply without changes. Almost by definition of Dirichlet boundary conditions, we know that T-duality along a circular direction of the target changes the dimension of a brane by one; see Subsection 1.A.1. This is in agreement with the "even/odd classification" because, at the same time, a type IIA string theory is mapped to type IIB: under T-duality in the 1-direction, say, the left-moving component of the bosonic coordinate is reflected, $X^1_R(\bar{z}) \longmapsto -X^1_R(\bar{z})$. Worldsheet supersymmetry also requires the transformation of the fermionic coordinate $\overline{\psi}^1(\bar{z}) \longmapsto -\overline{\psi}^1(\bar{z})$. Therefore, the right-moving GSO projector picks up a sign, since $\overline{\Gamma}_9$ does, thus the chirality of the states in the R-sector is reversed.

The link between dimensions of BPS-branes and the type of gluing conditions also persists in more complicated string compactifications to lower dimensions (e.g. using Calabi–Yau manifolds), albeit with slight modifications; see Chapter 7.

Let us note that for non-BPS branes, the rule that only Dirichlet $(2p)$-branes exist in type IIA theory, etc., maybe be violated. For example, a D-particle in IIB string theory can be constructed as a bound state in a brane–anti-brane system via tachyon condensation; see [421] and related observations in Section 5.A.2. In fact, the GSO invariant NS-boundary state (2.47) is perfectly sensible on its own, without adding some term in the Ramond sector – apart from questions of supersymmetry and stability.

Next, we want to compute the excitation spectrum of a (fermionic) p–p'-system, i.e. the quantity

$$Z_{pp'}^{\text{ferm}}(q) = \langle\!\langle \Theta(p') \| \, \tilde{q}^{L_0 - \frac{c}{24}} \, \| p \rangle\!\rangle \tag{2.51}$$

with $\| p \rangle\!\rangle \equiv \| p \rangle\!\rangle_{\text{ferm}} = \| p \rangle\!\rangle_{\text{ferm}}^{(\text{NS})} + \| p \rangle\!\rangle_{\text{ferm}}^{(\text{R})}$ as above, and analogously for $\| p' \rangle\!\rangle$. In this special configuration, both branes extend along the given coordinate axes, and each Neumann direction of the p'-brane is also a Neumann direction of the p-brane (the worldvolumes are parallel, but we admit $p \geq p'$). The full excitation spectrum of a two-brane configuration in superstring theory factorises into the fermionic part and the excitation spectrum of the bosonic partners. Furthermore, the system contains p–p'-open strings along with p'–p-, p–p- and p'–p'-strings; all these contributions have to be added up.

The left-hand side of equation (2.51) contains the CPT operator Θ that we included because the two boundaries of the annulus have opposite orientation and orientation reversal is a non-trivial operation in fermionic systems; see also Chapter 4 for further remarks on the CPT operator. By definition, Θ is an anti-linear operator on the state space of the bulk theory, which satisfies the intertwining property

$$\Theta \psi_r^j = i \psi_r^j \Theta, \quad \Theta \overline{\psi}_r^j = i \overline{\psi}_r^j \Theta$$

and acts on the ground states according to

$$\Theta |NS, NS\rangle = |NS, NS\rangle, \quad \Theta |R, R\rangle = \psi_-^1 \cdots \psi_-^8 |R, R\rangle.$$

These properties are sufficient to compute the action of Θ on any state in the bulk theory, and in particular on the boundary states.

The partition function (2.51) splits into contributions from NS- and R-parts of the boundary states, which in turn are combined from $\eta = +1$ and $\eta = -1$ coherent states (2.44) and (2.45). In the NS case, the CPT operator acts on coherent states as

$$\Theta \| p, \eta \rangle\!\rangle^{(\text{NS})} = \| p, -\eta \rangle\!\rangle^{(\text{NS})}. \tag{2.52}$$

Note that the exponential appearing in the coherent states (2.44) can be rewritten as a product of terms $(1 + i\varepsilon^{(j)} \psi_{-r}^j \overline{\psi}_{-r}^j)$ with $\varepsilon^{(j)} = \pm 1$. On these factors, the operator $\tilde{q}^{L_0}$ is easily evaluated, and we obtain the following basic building block

$$^{(\text{NS})}\langle\!\langle \Theta(p', \eta') \| \, \tilde{q}^{L_0 - \frac{c}{24}} \, \| p, \eta \rangle\!\rangle^{(\text{NS})} = \kappa_{\text{NS}}^2 \, \tilde{q}^{-\frac{8}{48}} \prod_{j=1}^{8} \prod_{n=0}^{\infty} \left(1 - \eta' \eta \varepsilon^{(j)'} \varepsilon^{(j)} \, \tilde{q}^{n+\frac{1}{2}} \right).$$

$$\tag{2.53}$$

The sign factors $\varepsilon^{(j)}$ for a Dp-brane were introduced after equation (2.43) and we use the same prescription to determine the signs $\varepsilon^{(j)'}$ for the Dp'-brane. Without restriction we assume that $p \geq p'$ so that there are $(p' + 1) + (7 - p)$ directions with $\varepsilon^{(j)} \varepsilon^{(j)\prime} = +1$ and $p - p'$ directions with $\varepsilon^{(j)} \varepsilon^{(j)\prime} = -1$. Summation over the extra signs η, η' ensures GSO invariance as in equation (2.47), and the NS-part of the p–p'-partition function is obtained,

$$
Z_{pp'}^{(\mathrm{NS})}(q) = 2\kappa_{\mathrm{NS}}^2 \, \tilde{q}^{-\frac{8}{48}} \left\{ \left[\prod_{n=0}^{\infty} (1 - \tilde{q}^{n+\frac{1}{2}}) \right]^{8-p+p'} \left[\prod_{n=0}^{\infty} (1 + \tilde{q}^{n+\frac{1}{2}}) \right]^{p-p'} \right.
$$
$$
\left. - \left[\prod_{n=0}^{\infty} (1 + \tilde{q}^{n+\frac{1}{2}}) \right]^{8-p+p'} \left[\prod_{n=0}^{\infty} (1 - \tilde{q}^{n+\frac{1}{2}}) \right]^{p-p'} \right\}
$$
$$
= 2\kappa_{\mathrm{NS}}^2 \left\{ f_2(q^{\frac{1}{2}})^8 \left(\frac{f_3(q^{\frac{1}{2}})}{f_2(q^{\frac{1}{2}})} \right)^{p-p'} - f_3(q^{\frac{1}{2}})^8 \left(\frac{f_2(q^{\frac{1}{2}})}{f_3(q^{\frac{1}{2}})} \right)^{p-p'} \right\}. \quad (2.54)
$$

Here, $p - p'$ is the (T-duality invariant) number of coordinates where different gluing conditions (DN or ND) are imposed at the open string ends. To reach the last line, we have applied a modular transformation to the theta functions – see Appendix A – and inserted the q-series $f_i(q)$ defined in equations (A.6).

We still need to compute the contribution $Z_{pp'}^{\mathrm{R}}$ from the component of the boundary state that lies in the R–R sector. The oscillator parts of $\| p \rangle\rangle_{\mathrm{ferm}}^{\mathrm{R}}$ and of $Z_{pp'}^{\mathrm{R}}(q)$ are treated as in the NS-sector. The main new element concerns the treatment of Ramond ground states. It is fairly easy to see that – for $p \geq p'$ – the only non-vanishing overlaps of ground states are given by

$$
{}_{p',\pm}\langle \Theta(R, R) | R, R \rangle_{p,\pm} = \delta_{p',-1} \delta_{p,7}, \quad {}_{p',\mp}\langle \Theta(R, R) | R, R \rangle_{p,\pm} = (-i)^{\frac{1+(-1)^p}{2}} \delta_{p',p}.
$$

Consequently, $Z_{pp'}^{\mathrm{R}}(q)$ *vanishes* unless $p = p'$ or $p - p' = 8$. In the first case $p = p'$, the fermionic parts of the two branes are in fact completely identical and we find

$$
Z_{pp}^{(\mathrm{R})}(q) = 2\kappa_{\mathrm{R}}^2 \, (-i)^{\frac{1+(-1)^p}{2}} \, \tilde{q}^{\frac{8}{24}} \prod_{n=1}^{\infty} (1 + \tilde{q}^n)^8
$$
$$
= 2\kappa_{\mathrm{R}}^2 \, (-i)^{\frac{1+(-1)^p}{2}} \, 2^{-4} f_4(q^{1/2})^8 . \quad (2.55)
$$

We have used formulas from Appendix A. Note in particular that the factor 2^{-4} arises from the definition of $f_4(q^{\frac{1}{2}})$ in (A.6). The R excitations of an open string stretched between a D-instanton and a seven-brane are counted by

$$
Z_{7,-1}^{(\mathrm{R})}(q) = 2\kappa_{\mathrm{R}}^2 \, \eta(\tilde{q})^8 . \quad (2.56)
$$

Let us see which conditions on the normalisation factors κ_{NS}, κ_{R} we can infer from $Z_{pp'}(q)$. The ratio of κ_{NS} and κ_{R} follows from requiring *spacetime supersymmetry*. We investigate the excitation spectrum of a single p-brane first (always ignoring the multiplicative contribution from worldsheet bosons X^j). Spacetime supersymmetry requires that boson and fermion loops cancel, i.e. that $Z_{pp}(q) = Z_{pp}^{\mathrm{NS}}(q) + Z_{pp}^{\mathrm{R}}(q) = 0$. This is satisfied thanks to the "abstruse identity"

$$
f_3(q)^8 - f_2(q)^8 - f_4(q)^8 = 0
$$

if κ_R is appropriately tuned: comparing equations (2.54) and (2.55), the condition for $Z_{pp}(q) = 0$ turns out to be

$$\kappa_R^2 = \begin{cases} 16\kappa_{NS}^2 & p \text{ odd}, \\ 16i\kappa_{NS}^2 & p \text{ even}. \end{cases} \tag{2.57}$$

As we will see in Chapter 4, overall normalisations of boundary states can, in a bosonic worldsheet CFT, be determined from Cardy's constraints (overlaps of boundary states have to produce a partition function *counting* open string states). In theories with spacetime supersymmetry, however, they cannot hold in their usual form: the partition function involves loops of spacetime fermions, and the corresponding characters must contribute with a minus sign. At the same time, the GSO projection of superstring theory removes all tachyons from the spectrum – including the CFT vacuum state. We can still derive κ_{NS} if we realise that the GSO projection is an operation that produces superstring amplitudes from expressions that have been computed in a consistent CFT before (cf. the functional integral picture of the GSO projection, where it amounts to a sum over spin structures at the end of the calculations). Thus, we demand that the vacuum occurs precisely once in the full partition function $Z_{pp}^{\mathrm{ferm}}(q)$ with the minus sign in front of the f_3-term in equation (2.54) reversed. This yields

$$\kappa_{NS} = \frac{i}{2},$$

which indeed is also the normalisation ensuring that there are eight massless bosons on a supersymmetric p-brane.

The freedom to choose the relative signs of κ_R and κ_{NS} finds its explanation in the existence of *anti-branes*. Indeed, if $\| p \rangle\rangle = \| p \rangle\rangle_{\mathrm{ferm}}^{(NS)} + \| p \rangle\rangle_{\mathrm{ferm}}^{(R)}$ describes a BPS brane with $Z_{pp}(q) = 0$, then so does the boundary state

$$\| \bar{p} \rangle\rangle := \Theta \| p \rangle\rangle = \| p \rangle\rangle_{\mathrm{ferm}}^{(NS)} - \| p \rangle\rangle_{\mathrm{ferm}}^{(R)} \tag{2.58}$$

obtained by applying the CPT operator. On the level of effective spacetime physics, all the R–R charges of $\| \bar{p} \rangle\rangle$ – i.e. the one-point functions $\langle \varphi^{RR} \| \bar{p} \rangle\rangle$ of closed string states that correspond to R–R potentials in spacetime – are reversed compared to $\| p \rangle\rangle$. This is the conventional way to obtain anti-branes. The simple formula (2.58) can be derived using the properties of the CPT operator and relation (2.57) for the relative normalisations, but it only holds for GSO invariant boundary states.

Any combination $\| \tilde{p} \rangle\rangle$ of NS- and R-boundary states which violates equation (2.57) will lead to a non-vanishing partition $Z_{\tilde{p}\tilde{p}}(q)$. Usually, this entails an "instability" of the brane in the sense that a tachyon occurs in the spectrum. This can also happen if we combine two branes that are supersymmetric by themselves but "incompatible" with each other: The simplest non-supersymmetric configuration of this kind is provided by a brane–antibrane system:

$$Z_{p\bar{p}}(q) \sim 2\kappa_{NS}^2 \left(f_3(q^{1/2})^8 - f_2(q^{1/2})^8 + f_4(q^{1/2})^8 \right). \tag{2.59}$$

Here, the GSO projection in the open string NS-sector is done "in the wrong way": The tachyon remains in the theory while the massless NS-states are projected out. This example is the starting point of brane–anti-brane annihilation processes which lead – in suitably modified settings – to interesting "descent relations" between branes and BPS or non-BPS bound states thereof, see [421]. These annihilation or tachyon condensation processes also form the basis for the K-theory picture of branes proposed in [457].

Let us briefly look at p–p'-systems with $p' \neq p, \bar{p}$. To check supersymmetry of the spectrum, it is still enough to consider the fermionic partition functions from above, since the X^j spectrum just provides an overall factor (if a more interesting one because now DN-strings are present). We have seen that the R–R contribution to $Z_{pp'}^{\mathrm{ferm}}(q)$ vanishes whenever the difference of dimensions $p - p' \not\equiv 0 \, (\mathrm{mod} \, 8)$. Then, $Z_{pp'}(q) = 0$ is possible only if the NS–NS contribution vanishes on its own. Equation (2.54) shows that this happens precisely for $|p - p'| = 4$. The case $p = 5$, $p' = 1$ turned out to be of particular interest for D-brane applications to (near-extremal) black holes [284, 285, 434].

The fact that the BPS p-brane boundary states preserve half of the spacetime supersymmetry can of course be established more directly and more rigorously than only from the partition function $Z_{\mu\mu}(q)$. It must be shown that they are annihilated by linear combinations of left- and right-moving (spacetime) supercharges Q^α and $\overline{Q}^\alpha$, which in turn are obtained as integrals of worldsheet spin fields $S^\alpha(z)$, $\overline{S}^\alpha(\bar{z})$, see [186]. Recall that we are working with the RNS formulation, but in the light-cone gauge. The familiar covariant expression for the spacetime supercharges is

$$Q^\alpha(z) \sim \int dz \, e^{-\frac{\phi}{2}} S^\alpha(z) \,,$$

where now S^α is a ten-dimensional spin field and where ϕ bosonises the superghost. Both here and in our light-cone gauge, the integrand is a worldsheet fermion of conformal dimension 1.

The light-cone gauge spin fields take a concise form after having bosonised the fermions ψ^k as $\psi^{2j-1}\psi^{2j} = e^{iH_j}$ for $j = 1, \ldots, 4$. Introducing four-vectors α with all entries equal to $\pm\frac{1}{2}$, the spin fields can be written as

$$S^\alpha(z) = e^{i\alpha \cdot H(z)} \,,$$

and analogously for the right-movers. The $S^\alpha(z)$ map the NS- to the R-sector and vice versa, i.e. their OPEs with the worldsheet fermions have branch cuts,

$$\psi^k(z)S^\alpha(w) \sim (z - w)^{-\frac{1}{2}} \, \Gamma_{\alpha\beta}^k \, S^\beta(w) \,, \tag{2.60}$$

where Γ^k is a 16×16 gamma matrix, i.e. the spin fields transform as spacetime spinors. In particular, the NS-vacuum is mapped into a Ramond ground state $S^\alpha(0) \, |NS\rangle$ by a spin field. Using this fact and the OPEs among the spin fields, which follow from the bosonisation formula, the action of the supercharges on the boundary states can be computed. This is an exercise in not getting lost in notations and keeping track of normalisations, and we will not repeat it here but rather refer to the literature, e.g. to [34, 123, 335], where this question has been studied in various frameworks. A few remarks on the analogous problem in the case of Gepner models will be made in Chapter 7.

The outcome of the computation is easy to guess. The correct relative normal-isation of NS- and R-parts of the boundary state, which ensures that a linear combination $Q^\alpha + M^p_{\alpha\beta}\, \overline{Q}^\beta$ of supersymmetry charges is conserved, is just the one in equation (2.57). The matrix $M^p_{\alpha\beta}$, on the other hand, follows from con-sistency of the fermionic gluing conditions with the OPE (2.60): The product

$$M^p = \Gamma_9 \Gamma^{p+2}\, \Gamma_9 \Gamma^{p+3} \cdots \Gamma_9 \Gamma^8$$

commutes with all Γ^i, $i = 0, \ldots, p+1$, and anti-commutes with Γ^a for the transverse directions $a = p+2, \ldots, 8$. Using this, it can be shown that the $\eta = +1$ fermion boundary state $\| p, + \rangle\rangle = \kappa_{\mathrm{NS}} \| p, + \rangle\rangle^{(\mathrm{NS})} + \kappa_{\mathrm{R}} \| p, + \rangle\rangle^{(\mathrm{R})}$ with normalisations as in equation (2.57) is annihilated by the linear combinations $Q^\alpha_{p,1} := Q^\alpha + M^p_{\alpha\beta}\overline{Q}^\beta$. In the papers [375, 376, 379], this result was derived from the usual (Neumann) open string supercharges using T-duality.

We have seen how tensions and charges of D-branes can be extracted from boundary states or from closed string scattering amplitudes. Brane–brane par-tition functions provide another derivation, which is the one originally used by Polchinski [375]. He compared effective field-theory results to the static brane–brane interaction arising from closed string exchange.

The field theory amplitude is calculated as a tree diagram involving bulk field propagators from ten-dimensional supergravity (2.19) and vertices from the Born–Infeld action (2.22) and from the Wess–Zumino term (2.25) – which contain the tension and the R–R charge, respectively. The relevant amplitude on the string theory side is just the partition function of two parallel BPS p-branes, integrated over the worldsheet modulus of the cylinder diagram,

$$\mathcal{A} = \int d\mu_q\; Z^{\Delta x}_{pp}(q)\,.$$

The integrand is the product of the fermionic partition function (2.51) with the bosonic part (1.45), which alone depends on the distance Δx of the branes. The measure is given by $\int d\mu_q \sim \int_0^\infty \frac{dt}{2t}$ with $q = e^{-2\pi t}$. In view of the relation between string scattering amplitudes and terms in the effective action, the amplitude $\mathcal{A}$ can be interpreted as (proportional to) the potential between two static branes; see also Subsection 2.A.2.

For the comparison with low-energy effective field theory, only the contribution of massless closed string states to the partition function counts, NS–NS must be separated from R–R states because the full partition function vanishes. In field theory terms, this means that the NS–NS field attraction, which includes the con-tributions from the graviton, exactly compensates the repulsion from R–R field exchange. Projection onto massless (long-range) fields is achieved in the limit of an infinitely long worldsheet cylinder, i.e. for $q \to 1$ or, after a modular transformation, for $\tilde{q} \to 0$. Performing the Gaussian integration over $k_\parallel$ in equation (1.45) and using the simple behaviour of $Z_{pp}^{\Delta x}(q)$ under $q \mapsto \tilde{q}$, we obtain

$$\mathcal{A}_{\text{NSNS}} \sim \int_0^\infty \frac{dt}{2t}\, t^{-\frac{p-1}{2}}\, e^{-\frac{(\Delta x)^2}{4\pi}t}\, t^4 \left(8 + \mathcal{O}(e^{-\frac{\pi}{t}})\right) \simeq \text{const} \cdot G_{9-p}((\Delta x)^2).$$
$$(2.61)$$

On the right-hand side, the scalar field propagator in $9 - p$ dimensions appears, just as in the field theory computation (recall that momentum is conserved only in the $p + 1$ directions parallel to the branes). We will not repeat the exercise of getting symmetry factors and α'-dependence right in equation (2.61) and in the field theory computation – see the reviews [28, 376]. The results for tension and charges coincide with those obtained from the boundary state,

$$T_p^2 = \frac{1}{2} Q_p^2 = \pi \left(4\pi^2 \alpha'\right)^{3-p}.$$

Note that the spacetime picture tells us that a $(6 - p)$-brane should be the dual object to a p-brane, since it couples to the $(7 - p)$-form R–R vector potential, whose "magnetic" field strength form is Hodge dual (in ten dimensions) to an "electric" $(p + 2)$-form coupling to the worldvolume of a p-brane. In accordance with electromagnetic duality, the R–R charges of such D-branes indeed satisfy a Dirac quantisation condition $Q_p Q_{6-p} = \pm 2\pi$.

2.A.2 Brane–brane scattering

Let us now go beyond the case of static branes and try to analyse the scattering of two D-branes. To this end, we study two branes in relative motion (with constant velocity) and the effective spacetime potential between them. We restrict ourselves to the case of two parallel p-branes (although the generalisation to a p-and a p'-brane can be discussed with little additional effort). Many more details, in particular concerning the interpretation of the results, can be found in [27, 149].

The boundary states for the two branes we scatter coincide in eight of the spatial dimensions of the ten-dimensional flat Minkowski target, and we take these parts to be given by the expressions discussed in Subsections 2.2.2 and 2.A.1. In the remaining two directions, the first brane is static (extending along X^0, fixed in X^1), while the second is boosted to move with velocity v in the (X^0, X^1)-plane, i.e. it is localised along the line $x^1 = b + x^0 v$, where b is the impact parameter of the scattering process. We have discussed the corresponding contribution to the boundary state at the end of Subsection 1.A.2. Due to the non-trivial X^0-dependence, the scattering amplitude is more conveniently computed using BRST quantisation rather than the light-cone gauge.

The basic worldsheet quantity to consider is, once more, the partition function $Z_{\alpha\beta}(q)$ of open strings between the boundary conditions $\alpha = (p; 0, 0)$ for a Dp-brane as before and $\beta = (p; v, b)$ for the "boosted" p-brane. This has to be integrated over the worldsheet modulus (the annulus width or cylinder length) so as to obtain the lowest-order contribution (in g_S) to the brane–brane scattering amplitude in string theory, the phase shift

$$\mathcal{A}_{\mathrm{bb}} = \int_0^\infty \frac{dt}{t} \, Z_{\alpha\beta}(q) = \int_0^\infty \frac{dt}{t} \, \langle\!\langle \Theta(p; 0, 0) \| \, \tilde{q}^{L_0 - \frac{c}{24}} \, \| (p; v, b) \rangle\!\rangle \qquad (2.62)$$

where $\tilde{q} = e^{-\frac{2\pi}{t}}$ or $\tau = it$; cf. Polchinski's calculation of the potential between two static branes which was sketched briefly in the previous subsection. Obviously, equation (2.62) can only give a non-relativistic approximation to the full scattering amplitude since processes like creation or annihilation of branes are not taken into account. Thus, we assume that the relative velocity v of the branes is small.

The integrand in equation (2.62) can be put together from the various partition functions we have seen in this and the previous chapter. While the contribution from the (X^0, X^1)-plane is described by equations (1.110, 1.113) and (1.114), we can employ formulas (2.35), (2.54) and (2.55) to build the factors arising from the remaining eight directions. After the substitution $\phi = iu := i\,\mathrm{artanh}\,v$ we obtain

$$Z_{\alpha\beta}(q) \sim \frac{1}{t^{\frac{p}{2}}\eta(it)^6} \, \frac{e^{-b^2/2\pi\alpha'}\eta(it)}{\vartheta_1\left(\frac{ut}{\pi}, it\right)} \cdot \frac{1}{\eta(it)^4} \sum_{a=2,3,4} (-1)^a \, \vartheta_a^3(0, it) \, \vartheta_a\left(\frac{ut}{\pi}, it\right) .$$

$$(2.63)$$

Let us quickly go through the individual factors. All the terms before the · are associated with bosonic modes. As we know from equation (2.35) eight bosonic spacetime directions contribute a factor η^{-8} along with an integral $\int d^p k \, q^{\frac{k^2}{2}} = t^{-\frac{p}{2}}$, where p is the number of spatial directions the branes extend into and $q = \exp(-2\pi t)$, as before.

The (X^0, X^1)-plane provides an additional term $\frac{\eta}{\vartheta_1}$ from equation (1.110) along with some factors (namely powers of q) that cancel against similar fermionic contributions. A factor η^2 is removed by the physical state conditions.

The fermionic terms appear after the · and take the form of equations (2.54) and (2.55), except that one factor of f_a gets replaced by $\vartheta_a(\frac{ut}{\pi}, it)/\eta(it)$ as seen from equations (1.113) and (1.114). In addition, we have inserted the definition $f_a = \frac{\vartheta_a}{\eta}$ for the remaining fermionic directions. Before we begin discussing the physics implications of the formula (2.63) it is useful to rewrite the expression using the following special case of Jacobi's fundamental formulas, see e.g. [447],

$$\sum_{a=2,3,4} (-1)^a \, \vartheta_a^3(0, it) \, \vartheta_a\left(\frac{ut}{\pi}, it\right) = 2\vartheta_1^4\left(\frac{ut}{2\pi}, it\right) .$$

In order to compare with the standard form of Jacobi's fundamental formulas we note that the sum over a on the left-hand side can be extended to the full range $a = 1, \ldots, 4$ since $\vartheta_1(0, it) = 0$. Insertion into equation (2.63) gives

$$Z_{\alpha\beta}(q) \sim \frac{2 e^{\frac{-b^2}{2\pi\alpha'}}}{t^{\frac{1}{2}} \eta(it)^9} \frac{\vartheta_1^4\left(\frac{ut}{2\pi}, it\right)}{\vartheta_1\left(\frac{ut}{\pi}, it\right)} =: \frac{e^{\frac{-b^2}{2\pi\alpha'}}}{t^{\frac{p}{2}}} f(ut, t). \qquad (2.64)$$

We can now try to rephrase the scattering process as the motion of one brane in an effective potential generated by the other one, by writing $\mathcal{A}_{bb}$ as an integral

$$\mathcal{A}_{bb} = \int_{-\infty}^{\infty} dT \, V(v, r^2 = b^2 + v^2 T^2) \qquad (2.65)$$

over the time direction T in the target. With the help of the usual Gaussian integral it is easy to determine the potential from the function f that was introduced in equation (2.64),

$$V(v, r^2) = \frac{1}{\sqrt{\pi}} \int_0^{\infty} \frac{dt}{t} \frac{e^{-r^2 t}}{t^{\frac{p-1}{2}}} v \, f(t \operatorname{artanh} v, t). \qquad (2.66)$$

A physical interpretation is most easily obtained after expanding the potential in powers of the velocity, $V = \sum_{n \geq 0} v^n V_n(r)$. The constant term $V_0(r)$ is the static inter-brane potential considered before and vanishes for our supersymmetric configuration. Next, observe that V is an even function of v (using $\vartheta_1(-z, \tau) = -\vartheta_1(z, \tau)$), so that the next term is quadratic in the velocity. Ideas initiated by Manton [233, 346] say that scattering of slow, heavy solitons is nothing but geodesic motion in their moduli space. Accordingly, the $\mathcal{O}(v^2)$ term in the effective potential (2.66) is a correction to the simple flat metric on the p-brane moduli space, and it generates a kinetic term

$$T_{\text{KIN}} = \frac{m_{\text{D}p}}{2} v^2 + v_i v_j \frac{\partial^2 V(0, r^2)}{\partial v_i \partial v_j}$$

for this "quantum mechanics on the moduli space". However, as first shown by Bachas, the moduli space stays flat in our case [27]: the contribution $V_2(r)$ vanishes because each of the theta functions in equation (2.63) satisfies $\vartheta_1(z, \tau) \sim -2\pi z \eta^3(\tau) + \mathcal{O}(z^2)$ which implies

$$V(v, r^2) \sim \int_0^{\infty} \frac{dt}{t} \frac{e^{-r^2 t}}{t^{\frac{p-7}{2}}} v^4 + \mathcal{O}(v^6)$$

for small velocity. The fact that V behaves like $v^4 V_4(r) + \mathcal{O}(v^6)$ can be traced back to the large amount of supersymmetry preserved by two parallel D-branes of the same dimension. Had we started with a p-p' brane system with $p' = p - 4$, a non-trivial correction to the moduli space metric would have emerged, see [149].

We would like to note that this method of computing a moduli space metric rests on very peculiar properties of the flat space example: there is a parameter v in the gluing conditions which happens to have an interpretation as the time derivative of a modulus x_0^1. Such a relation of parameters will not be available in general – in contrast to the Zamolodchikov metric on the moduli space, which solely rests on worldsheet concepts.

The formulas used above in the analysis of brane–brane scattering at small velocities also suggest [149] that branes can resolve spacetime at *sub-stringy distances*. To arrive at this interpretation, first note that the integral (2.66) receives its most important contributions from the region where $r^2 t \approx 1$, i.e. from large t for small r and from small t for large r. This links spacetime and worldsheet parameters, and allows the argument [149] that open string diagrams dominate the short-distance physics, while they become exponentially suppressed compared to closed string exchange at larger distances. Here, we confine ourselves to a few more qualitative remarks. Since the low-velocity expansion leads to singularities in the limit $r \to 0$, we should instead approximate $f(ut, t)$ in equation (2.66) by its large t asymptotics

$$f(ut, t) \approx -16 \, \frac{\sin^4\left(\frac{ut}{2}\right)}{\sin ut} \,,$$

which is easily obtained from the sum representation of the theta function ϑ_1 and is valid for very small spacetime separations r. The integral (2.66) is still dominated by $t \approx r^{-2}$, thus $ut \approx \frac{v}{r^2}$ and the sines oscillate rapidly for $r \to 0$ even if the velocity v is small. This means that the singularity is smoothed out, and that our effective picture for D-brane scattering is appropriate down to distances

$$r \gtrsim v^{\frac{1}{2}} \alpha'^{\frac{1}{2}} \,. \tag{2.67}$$

As long as the D-brane probe moves slowly, this length scale is indeed *smaller than the string scale*: closed strings cannot achieve finer resolutions than $\alpha'^{\frac{1}{2}}$ due to T-duality.

If we specialise to D0-branes and imagine them as quantum mechanical particles (see e.g. [120, 300]), we obtain further interesting bounds: viewing the brane as a "wave packet", its position is at best as sharp as its momentum $m_{D0} v = g_S^{-1} \alpha'^{\frac{-1}{2}} v$ allows, thus

$$r \gtrsim g_S \, \alpha'^{\frac{1}{2}} \, v^{-1} \,. \tag{2.68}$$

Both bounds are optimised for a velocity $v = g_S^{\frac{2}{3}}$ – which should be small, in keeping with a weakly coupled string theory. We conclude that the minimal effective distance that can be resolved by scattering of quantum-mechanical D0-branes is

$$r_{\min} \approx g_S^{\frac{1}{3}} \alpha'^{\frac{1}{2}} \,. \tag{2.69}$$

This is nothing but the Planck length $l_p^{(11)}$ in 11 dimensions – related to quantities for supergravity in ten dimensions by means of dimensional reduction. The fact that the minimal length that can be resolved through D0-brane scattering in superstring theory coincides with the minimal length we would expect to resolve in 11-dimensional supergravity is one of the observations that suggest the existence of M-theory underlying both [453].

Finally, let us note that the bounds on the spacetime resolution by brane–brane scattering induce another uncertainty relation, namely

$$\Delta x \, \Delta t \gtrless \alpha'$$

if we use equation (2.67) and $\Delta t = r/v$ for the scattering time, see [377]. Such uncertainty relations involving only coordinates and no conjugate momenta are expected in any theory that is to unify quantum physics and gravitation, as has been argued on very general grounds, e.g. in [137, 190, 194]. Obviously, they are also bound to occur in non-commutative spaces.

3
Conformal field theory on the plane

In this chapter, we give a brief overview of conformal field theory (CFT) on the plane, in an attempt to make the whole text more or less self-contained as far as worldsheet aspects are concerned. Some of the concepts to be discussed have been used in Chapter 1 already, but they will be put in a more general, model-independent context in the following. Many of the notions and tools to be reviewed will re-appear in the subsequent investigation of boundary CFT.

There are various alternative formulations of CFT, but our exposition will mainly follow the spirit of Belavin, Polyakov and Zamolodchikov [45]. One reason is that, up to now, no other framework has been adapted to boundary theories with equal success. However, each formulation has its virtues, and a great deal of the fascination of CFT comes from the interplay of very different mathematical structures.

The functional integral framework has the advantage that theories on surfaces of higher genus can be treated essentially on the same footing as CFTs on the complex plane. In addition, many abstract results find natural geometrical interpretations. We refer to [166, 223–225, 227] and the literature quoted there for more details. The disadvantage of functional integrals is that, by definition, we have to start from a description in terms of a classical action. Although a large class of models can be related to the rather small number of theories that have classical counterparts (including free fields and Wess–Zumino–Witten (WZW) models), there are interesting examples for which no Lagrangian is available. In fact, it is one of the greatest achievements of CFT that we can work directly at the quantum level without the need to borrow ideas from classical pictures.

In this sense, CFT models come close to the axiomatic framework of algebraic quantum field theory (QFT) [138, 265]. This approach is the most elegant one for investigations of statistical properties [189] and provides various insights into the general structure of CFT; see, e.g., [176, 210, 392]. It does not, on the other hand, lend itself easily to the description of specific models.

In the mathematical literature, there are two predominant formulations of CFT. Segal's definition [414] is motivated by string theory and functional integrals and can be seen as an extension of Atiyah's axioms for topological field theories, encoding field-theoretical properties into a bi-functor from a category of spaces (e.g. Riemann surfaces) to infinite-dimensional vector spaces, both with extra structure. Vertex operator algebras, on the other hand, are relatively close to the BPZ framework. There are some mathematical problems that could be solved using tools from vertex operator algebras, most notably the construction of the moonshine module for the monster; see [68, 212] for details and references.

Since we aim at giving some "practical recipes" to deal with the main examples of CFTs, we will not present complete proofs in the following, although most aspects of CFTs can be analysed rigorously. Some statements might require certain implicit assumptions which are satisfied in the models from everyday life, but perhaps not in each and every CFT that can be constructed. On the other hand, we will sometimes point out different ways to formulate and deduce certain statements rather than taking the shortest route only. In this way, we hope to cover all the notions that will be needed later, as well as to make accessible most of the original literature.

Let us emphasise that this chapter is by no means intended as a supplement to CFT textbooks, reviews and lecture notes. To fill in gaps and develop a more complete picture of this subject, the reviews and textbooks [19, 63, 121, 209, 212, 214, 224, 225, 235, 239, 296, 357, 411] are useful, among others.

It is natural, at least within the BPZ approach, to break up the construction of a CFT into three steps: analysis of the symmetry algebra; study of its representation theory, in particular in field space; and formulation and solution of non-linear constraints to build a physical theory from these building blocks. Our exposition will follow this sequence.

3.1 Symmetry algebras and highest weight representations

In this first section, we discuss the properties which lend CFTs their name, namely conformal transformations and their realisation in QFT by way of the Virasoro algebra and extensions thereof. Since these chiral algebras have infinite dimension, their representation theory provides much more severe constraints on the structure of the QFT than, e.g., in four-dimensional Poincaré covariant models. To start with, we look at the intrinsic structure of the symmetry algebras themselves, while the second subsection lists some general remarks concerning the representations that occur in CFT. Now and then, we will refer back to free field models discussed in Chapter 1, and also to the so-called Virasoro minimal models, as concrete examples.

3.1.1 Conformal symmetry and chiral algebras

In a CFT on a spacetime manifold Σ, covariance is demanded under all coordinate transformations $y \mapsto y'$ which leave the conformal class of the metric $h_{ab}(y)$ on Σ invariant, i.e. which satisfy

$$h_{ab}(y') = \Omega(y)\, h_{ab}(y) \tag{3.1}$$

for some positive function $\Omega(y)$. If Σ is flat and of dimension three or more, the conformal group extends the Poincaré group (transformations with $\Omega(y) = 1$), but is still finite-dimensional. In contrast, an infinite-dimensional symmetry algebra acts in a conformally invariant field theory on a two-dimensional spacetime manifold Σ. We will usually refer to Σ as the "worldsheet" in order to avoid confusion with (ten-dimensional) spacetime target manifolds of string theory. We will restrict ourselves to flat worldsheets here, like the cylinder $\Sigma = \mathbb{R} \times S^1$ with coordinates (σ_0, σ_1) and metric $\eta = \mathrm{diag}(1, -1)$. After Wick-rotating and passing to Euclidean light-cone coordinates, the cylinder can be mapped to the complex plane by

$$z = e^{\sigma_0 + i\sigma_1}, \qquad \bar{z} = e^{\sigma_0 - i\sigma_1}. \tag{3.2}$$

The origin of the plane corresponds to the distant past on the cylinder. In the new coordinates, it is easy to see that all (invertible) holomorphic and anti-holomorphic transformations

$$z \longmapsto f(z), \qquad \bar{z} \longmapsto \bar{f}(\bar{z}) \tag{3.3}$$

of z and $\bar{z}$, respectively, are allowed as conformal transformations (3.1) of the complex plane. More precisely, equation (3.3) solves the infinitesimal version of equation (3.1). The transformations that are globally well-defined and invertible generate two copies of the Möbius group $\mathrm{PSL}(2, \mathbb{R})$, under which

$$z \longmapsto \frac{az + b}{cz + d} \qquad \text{with} \quad \begin{pmatrix} a & b \\ c & d \end{pmatrix} \in \mathrm{SL}(2, \mathbb{R}),$$

analogously for $\bar{z}$. Due to the factorisation or "chiral splitting" which arises in equation (3.3), we can treat z and $\bar{z}$ as independent complex variables for most of the time. Because of their relation to the light-cones on the cylinder, quantities depending only on z or $\bar{z}$ will also be called left- and right-moving, respectively.

The infinitesimal generators of $z \mapsto f(z)$ are given by the vector fields $l_n := -z^{n+1}\partial_z$, which satisfy the so-called Witt algebra

$$[l_n, l_m] = (n - m)\, l_{n+m}.$$

This can be identified with the Lie algebra of $\mathrm{Diff}_+(S^1)$, the orientation-preserving diffeomorphisms of S^1; again, the circle shows up due to the range of the left-moving light-cone coordinate $\sigma_0 + \sigma_1$ on the cylinder.

In classical field theory, the Noether current for general coordinate transformations is the energy-momentum tensor T_{ab}. It is conserved and symmetric and – for a conformally invariant theory – traceless, $T_a{}^a = 0$. In two dimensions, these relations imply that $T_{ab}(\sigma_0, \sigma_1)$ splits into the parts $T_{++}(\sigma_0 + \sigma_1)$ and $T_{--}(\sigma_0 - \sigma_1)$, each depending on one light-cone coordinate only.

Passing to Euclidean signature and complex coordinates, these two components are denoted by T_{zz} and $T_{\bar{z}\bar{z}}$, and they only depend on the holomorphic and anti-holomorphic variables, respectively: we have

$$\partial_{\bar{z}} T_{zz} = \partial_z T_{\bar{z}\bar{z}} = T_{z\bar{z}} = 0,$$

the last relation being the condition of vanishing trace, and we introduce the notations

$$T(z) := T_{zz} \quad \text{and} \quad \overline{T}(\bar{z}) := T_{\bar{z}\bar{z}} .$$

The infinitesimal generators of the coordinate transformations (3.3) become the Laurent modes of $T(z)$ and $\overline{T}(\bar{z})$, respectively.

In QFT, it is projective representations of the covariance group that matter – or, equivalently, representations of the central extension $\widehat{\text{Diff}}_+(S^1)$. The associated Lie algebra is known as the *Virasoro algebra, Vir*, with generators L_n, $n \in \mathbb{Z}$, and commutation relations

$$[L_n, L_m] = (n - m) L_{n+m} + \frac{c}{12} (n^3 - n) \delta_{n+m,0} , \tag{3.4}$$

where c is a central generator which commutes with all the L_n. The elements L_0, $L_{\pm 1}$ generate a sub-Lie algebra of *Vir* (without central extension), which is isomorphic to $\text{sl}(2, \mathbb{R})$. Commutation relations analogous to equation (3.4) hold for the generators $\overline{L}_n$ of the right-moving Virasoro algebra, which furthermore commute with the left-movers:

$$[L_n, \overline{L}_m] = 0 ,$$

in agreement with the chiral splitting into independent transformations of z and $\bar{z}$ that we encountered in equation (3.3). In the following, we will treat the central elements c and $\bar{c}$ as numerical constants and consider only situations with $c = \bar{c}$.

Conformal QFTs in two dimensions are distinguished by the requirement that they are covariant under global $\text{PSL}(2, \mathbb{R}) \times \text{PSL}(2, \mathbb{R})$ coordinate transformations and by demanding that the Laurent series

$$T(z) = \sum_{n \in \mathbb{Z}} z^{-n-2} L_n , \qquad \overline{T}(\bar{z}) = \sum_{n \in \mathbb{Z}} \bar{z}^{-n-2} \overline{L}_n ,$$

where L_n, $\overline{L}_n$ are Virasoro generators as above, define *local field operators* of the QFT.

Consequently, the whole infinite-dimensional symmetry algebra *Vir* × *Vir* (and its universal enveloping algebra) acts on the space of states and forms part of the observable algebra of a CFT. (Here we permit ourselves to use terminology from algebraic QFT without worrying about unbounded observables and point-field limits – issues that can be settled in a satisfactory way for CFTs in two dimensions.) In analogy to the charges of the classical energy-momentum tensor on the cylinder, the special combinations $L_0 + \overline{L}_0 - \frac{c}{12}$ and $L_0 - \overline{L}_0$ provide Hamiltonian and momentum operators of a CFT, respectively. In a slight abuse of terminology, the chiral operators L_0 and $\overline{L}_0$ are often called (conformal) "energy operators" as well.

In many examples, the symmetry algebra is larger than just *Vir* × *Vir*. The chiral left–right splitting persists in all non-pathological cases so that the full symmetry algebra has the form $W \times \overline{W}$. Note that while both W and $\overline{W}$ must contain *Vir*, they need not be isomorphic, in general. The chiral observable algebras W, $\overline{W}$ go under the name *W-algebras*. They contain additional generators $W_m^{(l)}$ which satisfy the commutation relations

$$[\, L_n, W_m^{(l)}\,] = (\,(h_l - 1)n - m\,)\; W_{n+m}^{(l)} \tag{3.5}$$

for all m and all $n \in \mathbb{Z}$. This implements conformal symmetry on the extra generators. The objects $W^{(l)}(z) = \sum_m W_m^{(l)} z^{-m-h_l}$ are called *primary chiral fields* of conformal dimension h_l. As we did for the energy-momentum tensor, we also demand that the $W^{(l)}(z)$ are (semi-)local field operators, thus h_l and the modes m are (half-)integer.

The algebra W also contains fields that obey relation (3.5) only for $n = 0, \pm 1$; they are called *quasi-primary*. The energy-momentum tensor itself is an example, cf. equation (3.4). As long as equation (3.5) is valid for $n = 0$, we can still assign a definite conformal dimension to the field in question. Examples are derivatives of quasi-primary fields. We will say more about the structure of W-algebras below but, for most purposes, one can simply regard W as the universal enveloping algebra spanned by the L_n and the $W_m^{(l)}$, modulo commutation and null-field relations. Note also that we will make no distinction between the algebra generated by the modes or by the fields.

In general, the additional generators $W(z)$ in W cannot be integrated to a group of symmetry transformations. This is still possible for (the Virasoro fields and for) the symmetry generators that appear in *WZW models* [317, 448]. These have a classical description as sigma-models with Lie group target; the basic field occurring in the Lagrangian is a map $g : \Sigma \longrightarrow G$. From this group-valued field $g = g(z, \bar{z})$, we form

$$J := -k\partial g g^{-1} \quad \text{and} \quad \overline{J} := kg^{-1}\bar{\partial}g\,,$$

which satisfy the equations of motion $\bar{\partial}J = \partial\overline{J} = 0$. This means that the currents are actually holomorphic and anti-holomorphic fields, $J = J(z)$ and $\overline{J} = \overline{J}(\bar{z})$, respectively, and that there are infinitely many symmetry generators.

A closer study of the symmetry algebras $W \times \overline{W}$ of the quantum theory shows that it consists of two copies of the Lie algebra of the centrally extended loop group, i.e. W is an affine Lie algebra (sometimes called Kac–Moody algebra) $\hat{g}$ associated with the finite-dimensional (and, say, semi-simple, simply connected, compact) Lie algebra g of the target Lie group G; see e.g. [121, 198, 241, 385] for detailed descriptions.

We will discuss WZW models in more detail in Chapter 6, but let us record some basic facts on their symmetry algebras now. The generating fields of W are primary fields $J^a(z)$ of dimension one ("currents" for short) with $a = 1, \ldots, \dim g$, arising from an expansion $J(z) = \sum_a t^a J^a(z)$ into a Lie algebra basis t^a. If this basis is suitably chosen, the modes of the currents obey the commutation relations

$$[J_n^a, J_m^b] = i\, f^{abc}\, J_{n+m}^c + k\, n\, \delta^{a,b}\, \delta_{n+m,0} \,. \tag{3.6}$$

Here, f^{abc} are the structure constants of g with basis t^a, and the so-called *level* k is the value that the central element of $\widehat{g}$ takes in the model under consideration. Unitarity implies $k \in \mathbb{Z}_{>0}$.

It turns out that an affine Lie algebra already contains Virasoro generators given by the "Sugawara formula" (for simplicity we restrict to simply-laced algebras)

$$L_n = \frac{1}{2} \frac{1}{k + h^\vee} \sum_a \sum_{m \in \mathbb{Z}} : J_m^a J_{n-m}^a : \tag{3.7}$$

where $h^\vee$ is the dual Coxeter number of g and where the normal-ordering prescription simply means placing the generator with the higher mode index to the right. These L_n satisfy the Virasoro algebra (3.4) with central charge $c = \frac{k \dim g}{k+h^\vee}$, and the J_m^a are indeed primary of dimension one with respect to L_n. In other words, affine Lie algebras are examples of W-algebras with additional symmetry generators of conformal dimension one.

In fact, we have already encountered the very simplest case of an affine Lie algebra in Chapter 1: the chiral symmetry algebra of a free boson is generated by an abelian current $J(z) = \sum_{m \in \mathbb{Z}} a_m\, z^{-m-1}$ with Heisenberg commutation relations

$$[a_n, a_m] = n\, \delta_{n+m,0} \,; \tag{3.8}$$

we could introduce a "level" here by changing the normalisation of the current.

The other type of free field we encountered, the free fermion $\psi(z) = \sum_{r \in \mathbb{Z}+\frac{1}{2}} \psi_r\, z^{-r-\frac{1}{2}}$, is a primary field of dimension $h = \frac{1}{2}$ whose modes satisfy canonical anti-commutation relations

$$\{ \psi_r, \psi_s \} = \delta_{r+s,0} \,. \tag{3.9}$$

In both cases, the energy-momentum tensor can be expressed in terms of the basic fields just as in the Sugawara construction,

$$L_n^{\text{fb}} = \frac{1}{2} \sum_{m \in \mathbb{Z}} : a_{n-m}\, a_m : \quad \text{and} \quad L_n^{\text{ff}} = \frac{1}{2} \sum_{s \in \mathbb{Z}+\frac{1}{2}} (s - \tfrac{1}{2}) : \psi_{n-s} \psi_s : , \tag{3.10}$$

respectively, yielding conformal algebras of central charge $c^{\text{fb}} = 1$ and $c^{\text{ff}} = \frac{1}{2}$, respectively.

Note that there are similar expressions which allow, e.g., the expression of the generators of current algebras (at special levels) through free fermions; see [235] and references therein. Obviously, additional computational tools become available whenever a model admits such a "free field construction".

The examples given so far share the feature that the commutators of basic generators (L_n, J_n^a, etc.) close among the same generating modes. For more general W-algebras, the commutators of W_m are linear combinations of L_n, W_m and modes of further composite fields, called *normal-ordered products*. This happens in particular when W has generators $W(z)$ of conformal dimension $h_W > 2$, which is the situation the term W-algebra is usually reserved for.

We have already encountered simple examples of normal-ordered products in the Sugawara formulas (3.7) and (3.10) expressing Virasoro modes in terms of currents or fermions. In general, the modes of two W-fields $W^{(i)}(z)$ and $W^{(j)}(z)$ allow the formation of modes of a normal-ordered product by

$$N\big(W^{(i)}, W^{(j)}\big)_m = \sum_{n \leq -h_i} W_n^{(i)} W_{m-n}^{(j)} + \sum_{n > -h_i} W_{m-n}^{(j)} W_n^{(i)} . \qquad (3.11)$$

Here, we can also take $W^{(i)}(z) = T(z)$, or in fact any quasi-primary field in W. The field defined by the modes (3.11) has a definite conformal dimension $h_i + h_j$, but in general fails to be quasi-primary. However, it can be shown [354] that a W-algebra admits a linear basis of quasi-primary fields, which is extremely useful information for structural investigations of W-algebras. Indeed correction terms can be added to equation (3.11) so as to obtain a quasi-primary normal-ordered product $\mathcal{N}(W^{(i)}, W^{(j)})(z)$ of dimension $h_i + h_j$. The explicit formula is due to Nahm and can be found, e.g., in [62]:

$$\mathcal{N}\big(W^{(i)}, W^{(j)}\big) = N\big(W^{(i)}, W^{(j)}\big) - \sum_k C_{ij}^k \binom{2h_j - 1}{h_{ijk}} \frac{(2h_k+1)!}{(h_i+h_j+h_k-1)!} \, \partial^{h_{ijk}} \phi^{(k)} .$$

$$\qquad (3.12)$$

The sum runs over all quasi-primary fields $\phi^{(k)}$ in W (including normal-ordered products) with the restriction $h_{ijk} \equiv h_i + h_j - h_k \geq 1$ on their dimension h_k. The structure constants $C_{ij}^k = C_{ijl}(d^{-1})^{kl}$ can be computed from the modes via

$$d_{ij} = \langle 0 | \phi_{h_i}^{(i)} \phi_{-h_j}^{(j)} | 0 \rangle \quad \text{and} \quad C_{ijk} = \langle 0 | \phi_{h_k}^{(k)} \phi_{h_j - h_k}^{(i)} \phi_{-h_j}^{(j)} | 0 \rangle ,$$

which relates them to basic parameters like the central charge. $|0\rangle$ denotes the vacuum state, see below. Formula (3.12) still holds if the $W^{(i)}$ are replaced by arbitrary quasi-primary fields; for a generalisation of normal-ordered products of derivative fields see [62].

All fields in a W-algebra are linear combinations of the generators $T(z)$ and $W^{(i)}(z)$ – also called "simple fields" as distinct from normal-ordered products – of their derivatives and of (derivatives of) normal-ordered products thereof. In particular, the commutators of the modes of quasi-primary fields in a W-algebra form an ordinary (if "very" infinite-dimensional) Lie algebra, in spite of somewhat misleading statements that can be found in the literature. Because of conformal invariance, the commutators depend only on the conformal dimensions and the structure constants above; see [62] for a universal formula.

The quasi-primary normal-ordered product (3.12) can be taken as the starting point for an algorithmic construction of W-algebras; see, e.g., [62]. Given the conformal dimensions h_i of the primary generators $W^{(i)}(z)$ as input data, the structure constants appearing in the commutator of simple fields have to be determined. These parameters are constrained by the Jacobi identity and, as was shown in [354, 357], only a finite number of commutators have to be checked in order to guarantee that the whole W-algebra $\mathcal{W}(2, h_1, \ldots, h_n)$ is consistent. We refer to [61, 62, 70, 169, 309] for examples, and also to [461] where conformal W-algebras were introduced and where some simple cases like $\mathcal{W}(2,3)$ are discussed explicitly. Here, the additional simple generator $W(z)$ has conformal dimension three, and the basic commutation relations of $W(z)$ with itself read

$$
[W_n, W_m] = \frac{1}{3}(n - m)\Lambda_{n+m} + \frac{1}{30}\frac{22 + 5c}{48}(n - m)(2n^2 - nm + 2m^2 - 8)\, L_{n+m}
$$
$$
+ \frac{c}{360}\frac{22 + 5c}{48} n(n^2 - 4)(n^2 - 1)\, \delta_{n+m,0} \tag{3.13}
$$

where $\Lambda := \mathcal{N}(T, T) = N(T, T) - \frac{3}{10}\partial^2 T$ is the unique quasi-primary normal-ordered product of dimension four that can be formed from the energy-momentum tensor alone.

The algorithm used in [62] provides a systematic construction of all W-algebras with given dimensions of the generators. It is, however, restricted to small h_i because of limited computing power. There are some methods to derive W-algebras with generators of arbitrary high dimension from simpler models, e.g. one can obtain $\mathcal{W}(2, 3, \ldots, N)$ as "Casimir algebra" from an SU(N) WZW model [33], but it remains doubtful whether such constructions are exhaustive and can also uncover spurious W-algebras which exist only at isolated points of the parameter space.

In string theory, supersymmetric extensions of the Virasoro algebra are of central importance. In the $N = 1$ case, there is one additional fermionic W-algebra generator $G(z)$ of spin $\frac{3}{2}$, so that we could also write $\mathcal{W}(\frac{3}{2}, 2)$ for this algebra – even though no normal-ordered products occur in the basic (anti-)commutators. The field G is the worldsheet superpartner of the energy-momentum tensor. The relations were already stated in Subsection 1.2.3:

$$
[\, L_n, G_r \,] = \left(\frac{n}{2} - r\right) G_{n+r}\, ,
$$
$$
\{\, G_r, G_s \,\} = 2\, L_{r+s} + \frac{c}{3}\left(r^2 - \frac{1}{4}\right)\delta_{r+s,0}\, . \tag{3.14}
$$

The moding of G is half-integer ($r \in \mathbb{Z} + \frac{1}{2}$) in the Neveu–Schwarz and integer in the Ramond sector. The first relation tells us that $G(z)$ is a Virasoro primary field, while the second in particular allows us to write the conformal Hamiltonian as an anti-commutator, $L_0 = \frac{1}{2}\{Q, Q^*\}$ with $Q = G_{-\frac{1}{2}}$; in the Ramond sector, $G_0^2 = L_0 - \frac{c}{24}$.

We can also consider Virasoro extensions with higher supersymmetry content, and in fact the $N = 2$ *super Virasoro algebra* plays the most prominent role for string compactifications. Here, the supermultiplets arise from the action of two fermionic generators $G^\pm(z)$ of spin $\frac{3}{2}$ and, as a consequence, the full W-algebra contains a $\widehat{U(1)}$-current $J(z)$, too. The new non-vanishing (anti-)commutators of this $\mathcal{W}(1, \frac{3}{2}, \frac{3}{2}, 2)$ read

$$[\,J_n, J_m\,] = \frac{c}{3}\, n\, \delta_{n+m,0}\,,$$

$$[\,J_n, G^\pm_{r\pm a}\,] = \pm G^\pm_{n+r\pm a}\,, \tag{3.15}$$

$$\{\,G^+_{r+a}, G^-_{s-a}\,\} = 2\,L_{r+s} + (r - s + 2a)\,J_{r+s} + \frac{c}{3}\left((r+a)^2 - \frac{1}{4}\right)\delta_{r+s,0}$$

cf. Subsection 1.2.3. In these relations, we take $n, m, r, s \in \mathbb{Z}$, but the offset is $a = \frac{1}{2}$ in the Neveu–Schwarz sector and $a = 0$ in the Ramond sector. The reason for this notation is that, in fact, there is a whole family of $N = 2$ superconformal algebras, parametrised by a real number η and related to each other by the so-called *spectral flow*: The linear map α_η defined by

$$\alpha_\eta(L_n) = L_n + \eta J_n + \frac{c}{6}\eta^2\delta_{n,0}\,,\quad \alpha_\eta(J_n) = J_n + \frac{c}{3}\eta\delta_{n,0}\,,\quad \alpha_\eta(G^\pm_{r\pm a}) = G^\pm_{r\pm a\pm\eta}$$

$$\tag{3.16}$$

preserves the Lie algebra relations (3.16). For $\eta \in \mathbb{Z}$, it can in fact be viewed as an automorphism of the $N = 2$ algebra (then labelled by a parameter in $\mathbb{R}/\mathbb{Z}$). The existence of this group of Lie algebra isomorphisms is due to the $U(1)$ gauge symmetry of the defining relations (3.15), see [413]. As a consequence, representations of $N = 2$ superconformal algebras at different η are identical as far as purely algebraic properties are concerned. For example, there are as many irreducible representations (or "irreps") in the Neveu-Schwarz as in the Ramond sector. Observe, however, that highest weight states in a representation at η may be mapped to descendants at η', as follows, e.g., from the action of the spectral flow on L_0.

Upon bosonisation of the $\widehat{U(1)}$-current $J(z) = i\sqrt{\frac{c}{3}}\,\partial\phi(z)$, a realisation of the spectral flow is obtained (see, e.g., [253]) in terms of operator products with

$$U_\eta(z) = e^{i\sqrt{\frac{c}{3}}\eta\phi(z)}\,. \tag{3.17}$$

The special importance of the spectral flow becomes clear when we specialise to $\eta = \frac{1}{2}$: The corresponding operator induces an isomorphism between the Neveu–Schwarz and the Ramond sector of the $N = 2$ superconformal model. In string theory, these sectors are related to spacetime bosons and spacetime fermions, respectively. Therefore, a *spacetime supersymmetry generator* can be constructed from the spectral flow operator $U_{\frac{1}{2}}$ as long as this is (semi-)local with respect to all other worldsheet fields. From the bosonisation formula (3.17), we see that the last condition requires that all U(1)-charges occurring in the CFT are (half-)integer, a condition intimately related to the generalised GSO projection to be used in the context of Gepner models; see Chapter 7. Together with this extra requirement, $N = 2$ supersymmetry on the worldsheet guarantees $N = 1$ spacetime supersymmetry [35] – which explains the attention that $N = 2$ super-conformal models have received in string theory.

These theories have many other interesting properties, which will be discussed in more depth later, in particular in Chapter 7.

3.1.2 Structure of the state spaces

Covariance of a QFT under a symmetry algebra $\mathcal{W} \times \overline{\mathcal{W}}$ implies that its *state space* carries an action of this algebra. In this text, we restrict ourselves to theories whose state space decomposes into a direct sum (or sometimes a direct integral)

$$\mathcal{H} = \bigoplus_{(i,\bar{\imath})\in I} \mathcal{H}_i \otimes \mathcal{H}_{\bar{\imath}} \tag{3.18}$$

where $\mathcal{H}_i$ and $\mathcal{H}_{\bar{\imath}}$ are irreducible representations or (superselection) "sectors" of the chiral symmetry algebras $\mathcal{W}$ and $\overline{\mathcal{W}}$, respectively. We demand that the energy operator is bounded from below, which means that the $\mathcal{H}_i$ and $\mathcal{H}_{\bar{\imath}}$ have to be highest weight representations: in each sector $\mathcal{H}_i$, a spanning system of states is generated by the action of L_n, $W_n^{(l)}$ on a (finite) set of highest weight states $|h^{(\iota)}\rangle$ which satisfy

$$L_0 |h^{(\iota)}\rangle = h_i |h^{(\iota)}\rangle, \quad L_n |h^{(\iota)}\rangle = W_n^{(l)} |h^{(\iota)}\rangle = 0 \quad \text{for } n > 0. \tag{3.19}$$

The eigenvalue h_i is called the conformal dimension of the representation $\mathcal{H}_i$. Analogous conditions hold in $\mathcal{H}_{\bar{\imath}}$ for the right-movers. Furthermore, we are only interested in CFTs where L_0 and $\overline{L}_0$ can be diagonalised on all of $\mathcal{H}$ – although there are "logarithmic CFTs" where L_0 can be brought in Jordan normal form only, see, e.g., [212] and references therein. Sometimes, maximal abelian subalgebras of $\mathcal{W}$ with more elements than just L_0 exist; then we assume that a basis of common eigenstates in $\mathcal{H}$ can be found.

While the two sets of all possible irreducible representations are determined by $\mathcal{W}$ and $\overline{\mathcal{W}}$, the coupling of left- and right-moving modules is not. It turns out that there are constraints on the choice of the index set I, which come from the fusion rules and from modular invariance; see below. However, solutions to these constraints need not be unique, and there may be different CFTs with the same symmetry algebra. It is always assumed that the vacuum representation $\mathcal{H}_0 \otimes \mathcal{H}_0$ is contained in $\mathcal{H}$, with $\mathcal{H}_0$ being built up over a unique highest weight vector $|0\rangle$ that is $SL(2,\mathbb{R})$-invariant, i.e. satisfies $L_n|0\rangle = 0$ for $n = 0, \pm 1$.

We call a theory *rational* if I is finite; these theories are the easiest ones to deal with since $\mathcal{W} \times \overline{\mathcal{W}}$-covariance leaves only a finite set of parameters to determine, as will be discussed below. A CFT is called unitary if all the $\mathcal{H}_i$ and $\mathcal{H}_{\bar{\imath}}$ in equation (3.18) admit scalar products such that

$$(W_n)^* = W_{-n}, \quad (\overline{W}_n)^* = \overline{W}_{-n} .$$

Here, W_n stands for both Virasoro and other generators, and we use the same symbol for the abstract $\mathcal{W}$-element and its implementation on the state space. It is easy to see that unitarity implies $c \geq 0$ and h_i, $h_{\bar{\imath}} \geq 0$. Unitary CFTs are automatically "non-logarithmic", i.e. L_0 is diagonalisable. However, unitarity does not seem to play as important a role for CFT in two dimensions as it does in general QFT; having finite-dimensional eigenspaces of L_0 appears to make non-unitary CFTs just as well-behaved.

The simplest highest weight representations of *Vir* are given by the so-called Verma modules $V_{c,h}$ with basis vectors

$$L_{-n_k} \cdots L_{-n_1} |h\rangle \qquad \text{with} \ \ n_k \geq \cdots \geq n_1 > 0 \ \ \text{and} \ \ k \in \mathbb{Z}_+ , \qquad (3.20)$$

built over a single highest weight vector $|h\rangle$ of conformal dimension h as in definition (3.19). Note that the states in equation (3.20) have L_0 eigenvalue ("energy") $h + \sum_i n_i$.

Conformal characters $\chi_{\mathcal{H}_i}(q)$ are introduced to count the degeneracy of the energy levels in a representation $\mathcal{H}_i$,

$$\chi_{\mathcal{H}_i}(q) := \mathrm{tr}_{\mathcal{H}_i} q^{L_0 - \frac{c}{24}} , \qquad (3.21)$$

where q is a formal variable for the time being. The character of a Verma module $V_{c,h}$ takes the simple form:

$$\chi_{V_{c,h}}(q) = q^{h - \frac{c}{24}} \sum_{N=0}^{\infty} p(N) q^N = q^{h - \frac{c-1}{24}} \eta(q)^{-1} , \qquad (3.22)$$

where $p(N)$ denotes the number of partitions of the integer N, and $\eta(q) = q^{\frac{1}{24}} \prod_{n=1}^{\infty}(1 - q^n)$ is the Dedekind eta function, which we already encountered in free boson partition functions in Chapter 1.

The irreducible highest weight representations of $\mathcal{W} = Vir$ were completely classified in the seminal work of Feigin and Fuchs [159]. The module $V_{c,h}$ is irreducible in general, but this is no longer true for special choices of the parameters c and h. However, all irreducible highest weight representations of the Virasoro algebra arise as factors of Verma modules [159].

For $h = 0$, for example, the vector $L_{-1}|0\rangle$ is a *singular vector* or *null-vector*, i.e. it is annihilated by all L_n with $n > 0$ without being the highest weight state. To obtain an irreducible representation from this "degenerate" Verma module, we have to divide out (at least) the sub-Verma module built up over $L_{-1}|0\rangle$ – in other words, we set $L_{-1}|0\rangle = 0$ – cf. the SL$(2, \mathbb{R})$-invariance of the ground state in the irreducible vacuum sector. At the level of characters, we have to subtract the character of the sub-module from equation (3.22). As another example, consider the state

$$|v_2\rangle = \left(L_{-2} - \frac{3}{2(2h+1)} L_{-1}^2 \right)|h\rangle ; \tag{3.23}$$

reshuffling modes using the Virasoro relations, it is an easy exercise to show that this state is a singular vector at the second energy level if $h = \frac{1}{16}(5 - c \pm \sqrt{(c-1)(25-c)})$.

This relation between central charge and conformal dimension in particular shows up in the so-called *Virasoro minimal models*, which were the main examples of CFTs considered by BPZ in [45]. Minimal models are the only rational CFTs with symmetry algebra $Vir \times Vir$; their central charges have the form

$$c(p, p') = 1 - 6\frac{(p - p')^2}{pp'} \tag{3.24}$$

with $p, p' \in \mathbb{Z}_{\geq 2}$ coprime. Unitarity requires $p' = p + 1$ [187]. The Virasoro modules of the minimal models are labelled by integers r, s with $1 \leq r < p'$ and $1 \leq s < p$, modulo identifying pairs (r, s) and $(p' - r, p - s)$ in this rectangle (this is the so-called "conformal grid symmetry"). The conformal dimensions of the associated highest weight vectors are listed in the finite "Kac table"

$$h_{r,s}(p, p') = \frac{(rp - sp')^2 - (p - p')^2}{4pp'} . \tag{3.25}$$

For $(r, s) = (1, 2)$ and $(r, s) = (2, 1)$ we have the relation between h and c mentioned before, i.e. each of these Verma modules contains a singular vector at the second energy level. But as was shown by Feigin and Fuchs, the Verma modules for $c(p, p')$ and $h_{r,s}(p, p')$ in fact contain two independent singular vectors, at levels rs and $(p' - r)(p - s)$ – in agreement with conformal grid symmetry.

This implies that, in order to obtain the conformal characters $\chi_{r,s}(q)$ of the irreducible representation with highest weight $h_{r,s}(p, p')$, we start from the Verma module character, subtract the Verma module characters over the singular vectors, then add contributions from the intersection of the two, and so on. The final formula can be written as

$$\chi_{r,s}(q) = q^{-\frac{c-1}{24}} \eta(q)^{-1} \sum_{k \in \mathbb{Z}} \left(q^{h_{r,s+2pk}} - q^{h_{-r,s+2pk}} \right) ; \tag{3.26}$$

see [395] and also [19, 121] for derivations and a detailed discussion.

We will see later that relations derived from the existence of null vectors are of great importance in solving a CFT. For very detailed expositions of the representation theory of Vir and of singular vectors, we refer to [159] and [38, 46], respectively.

From the physical point of view, it should be mentioned that, while most of the minimal models have no direct Lagrangian formulation, the series of minimal models contains descriptions of various well-known second-order phase transitions in two dimensions: for $p = p' - 1 = 3$ we obtain the Ising model, $p = p' - 1 = 5$ yields the tetra-critical Ising model and the three-states Potts model, while the non-unitary case $p = 2$, $p' = 5$ corresponds to the Lee–Yang edge singularity. We will not pursue the very interesting connections between CFT and critical phenomena here but refer to textbooks such as [121] and literature quoted there.

But let us note here that conformal characters are useful in establishing relations between CFTs that look very different at first sight. For example, the characters of the minimal model with central charge $c(3, 4) = \frac{1}{2}$ coincide with GSO projected characters of the free fermion (also with $c = \frac{1}{2}$),

$$\chi_0(q) = \frac{1}{2} \left(\chi_{\mathrm{NS}}(q) + \chi_{\widetilde{\mathrm{NS}}}(q) \right), \quad \chi_{\frac{1}{2}}(q) = \frac{1}{2} \left(\chi_{\mathrm{NS}}(q) - \chi_{\widetilde{\mathrm{NS}}}(q) \right),$$

$$\chi_{\frac{1}{16}}(q) = \frac{1}{2} \chi_{\mathrm{R}}(q),$$

where we have labelled the irreducible Virasoro characters by their conformal dimensions $h_{r,s}(3, 4)$, and where the tilde denotes free fermion characters in the Neveu-Schwarz or Ramond sector with $(-1)^F$ inserted, F being the fermion number; see also Subsection 1.2.2 and equations (1.68) and (1.69).

In their work [159], Feigin and Fuchs also clarified the structure of Virasoro modules for $c = 1$, and fully-fledged physical models with this central charge were classified later in [132, 236]. The *Vir* modules for $c > 1$ are "trivial" in the sense that all the irreducible representations are Verma modules, i.e. no singular vectors have to be divided out. To construct consistent physical theories in this range is, however, more difficult, one reason being that they are all non-rational unless the symmetry algebra is enlarged.

If, on the other hand, the symmetry algebras $\mathcal{W}$ and $\overline{\mathcal{W}}$ are extensions of *Vir*, the state-space decomposition (3.18) is one into tensor products of irreducible highest weight representations of those larger algebras. Typically, each $\mathcal{H}_i$ then contains infinitely many irreducible Virasoro modules.

3.2 Fields and representation theoretic constraints on correlation functions

Since the chiral symmetry algebra determines a large part of the structure of a CFT, it is important to have a handle on the construction of W-algebras. But there is more to a CFT than just its observable algebra $\mathcal{W} \times \overline{\mathcal{W}}$, namely *field operators* and their *correlation functions*. In the next two subsections, we consider field operators and how they are acted on by the chiral algebra, namely via differential operators – which will immediately lead to differential equations on correlators, the so-called Ward identities. In Subsection 3.2.3, we present some ingredients of a more abstract description of the representation categories of W-algebras, often referred to as the "duality data" of the CFT.

3.2.1 Fields

Because of the state-field correspondence, which holds in local QFTs under very general assumptions, the local fields of a CFT are "indexed" (one-to-one) by the vectors of the state space:

$$\varphi_v(z, \bar{z}) = \Phi(v; z, \bar{z}) \quad \text{for} \quad v \in \mathcal{H} \quad \text{and} \quad z, \bar{z} \in \mathbb{C} . \tag{3.27}$$

Thus fields can be viewed as state vectors inserted at the worldsheet coordinate $(z, \bar{z})$. On the other hand they act on $\mathcal{H}$ as (unbounded) linear operators and generate all states in $\mathcal{H}$ from the vacuum $|0, 0\rangle := |0\rangle \otimes |0\rangle$,

$$|v\rangle = \lim_{z, \bar{z} \to 0} \varphi_v(z, \bar{z}) |0, 0\rangle . \tag{3.28}$$

These properties are sometimes summarised by saying the vacuum is a cyclic and separating state.

The (maximally extended) symmetry algebra $\mathcal{W} \times \overline{\mathcal{W}}$ of the CFT can simply be defined as the chiral fields, i.e. as those local fields φ_v that depend only on z or only on $\bar{z}$. They correspond to vectors in the vacuum representation $\mathcal{H}_0 \otimes \mathcal{H}_0$. The identity operator is the field associated to the vacuum, $\mathbf{1} = \Phi(|0, 0\rangle; z, \bar{z})$, and for the energy-momentum tensor we can write

$$T(z) = \Phi\big((L_{-2}|0\rangle) \otimes |0\rangle; z, \bar{z} \big) \quad \text{and} \quad \overline{T}(\bar{z}) = \Phi\big(|0\rangle \otimes \overline{L}_{-2}|0\rangle; z, \bar{z} \big) .$$

As a consequence of the state-field correspondence, the whole space of fields is acted on by $\mathcal{W} \times \overline{\mathcal{W}}$ and decomposes into the same irreducible representations as $\mathcal{H}$. Fields associated with a highest weight vector in $\mathcal{H}_i \otimes \mathcal{H}_{\bar{i}}$ are called *primary fields* and will usually be denoted by $\varphi_{i,\bar{i}}(z, \bar{z})$. They can be viewed as "top components" of the infinite-dimensional $\mathcal{W} \times \overline{\mathcal{W}}$ "tensor multiplet", and one can use covariance under $\mathcal{W} \times \overline{\mathcal{W}}$ to reduce (infinitely many) quantities involving other fields to ones involving primaries (and a few descendants) only, see below.

Virasoro primary fields – i.e. those associated with states annihilated by all $L_n, \overline{L}_n$ with $n > 0$ – enjoy particularly simple transformation rules under analytic reparametrisations:

$$U_f \, \varphi_{i,\bar{i}}(z, \bar{z}) \, U_f^{-1} = \left(\frac{\partial f}{\partial z} \right)^{h_i} \left(\frac{\partial \bar{f}}{\partial \bar{z}} \right)^{h_{\bar{i}}} \varphi_{i,\bar{i}}(f(z), \bar{f}(\bar{z})) , \tag{3.29}$$

where h_i and $h_{\bar{i}}$ are the conformal dimensions of the Virasoro highest weight state, thus of $\varphi_{i,\bar{i}}(z, \bar{z})$.

In a CFT, infinitesimal coordinate transformations $f(z) = z + \epsilon(z)$ with $\epsilon(z)$ holomorphic and "small", are implemented by the energy-momentum tensor. More precisely, in our complex coordinates with "time" running in radial direction, the integral

$$T(\epsilon) = \oint_{C_0} \frac{dz}{2\pi i} \, \epsilon(z) T(z) \tag{3.30}$$

over a contour C_0 around the origin is the generator of the coordinate transformation. Expanding equation (3.29) to first order in ϵ, we obtain

$$[\,T(\epsilon), \varphi_{i,\bar{\imath}}(w,\bar{w})\,] = \left(\epsilon(w)\partial_w + \epsilon'(w)h_i\right)\varphi_{i,\bar{\imath}}(w,\bar{w})\,. \tag{3.31}$$

Formulating the theory on the complex plane allows for a very concise rewriting of the transformation properties of primary fields: The equal time commutator on the left-hand side of equation (3.31) can be expressed as a contour integral $\oint_{C_w} \frac{dz}{2\pi i}\,\epsilon(z)T(z)$ around the insertion point w of $\varphi_{i,\bar{\imath}}$. Using Cauchy's formula and the fact that $\epsilon(z)$ is an arbitrary holomorphic function, we infer

$$T(z)\varphi_{i,\bar{\imath}}(w,\bar{w}) = \frac{h_i}{(z-w)^2}\,\varphi_{i,\bar{\imath}}(w,\bar{w}) + \frac{1}{z-w}\,\partial_w\varphi_{i,\bar{\imath}}(w,\bar{w}) \;+\; \mathrm{reg}_{z\to w}\,. \tag{3.32}$$

This is the *operator product expansion* (OPE) of the left-moving energy-momentum tensor with a Virasoro primary field. The left-hand side is tacitly assumed to involve radial ordering (the field closer to the origin is placed to the right), and the right-hand side of the OPE describes the short-distance singularities (of correlation functions) when the two fields approach each other; contributions which are regular as z approaches w are not recorded explicitly.

Operator product expansions are a general concept in local QFT, but for conformally invariant theories they become particularly useful because covariance under analytic coordinate transformations fixes their form to a large extent: the OPE of two field operators with left- and right-moving conformal dimensions $h_i, h_{\bar{\imath}}$ and $h_j, h_{\bar{\jmath}}$, respectively reads

$$\varphi_{v_i}(z,\bar{z})\,\varphi_{v_j}(w,\bar{w}) = \sum_{v_k} C_{v_i v_j v_k}\,(z-w)^{h_k - h_i - h_j}(\bar{z}-\bar{w})^{h_{\bar{k}} - h_{\bar{\imath}} - h_{\bar{\jmath}}}\,\varphi_{v_k}(w,\bar{w}) \tag{3.33}$$

where the sum runs over a basis of field operators φ_{v_k} with fixed conformal dimensions (i.e. quasi-primaries or derivatives thereof). The $C_{v_i v_j v_k}$ are structure *constants*, and the coordinate dependence is determined by the conformal dimensions $h_l, h_{\bar{l}}$ of the states v_l. For non-logarithmic CFTs, only powers of $z-w$ and $\bar{z}-\bar{w}$ occur. We will discuss OPEs, in particular non-linear constraints on the structure constants, in some more detail in Section 3.3.

Since modes W_n of $\mathcal{W}$-generators are obtained from $W(z)$ upon contour integration around the origin, the OPE (3.32) immediately yields the commutators

$$[\,L_n, \varphi_{i,\bar{\imath}}(w,\bar{w})\,] = w^n\left(w\partial_w + h_i(n+1)\right)\varphi_{i,\bar{\imath}}(w,\bar{w}) \tag{3.34}$$

for Virasoro primary fields. Formulas analogous to equations (3.32) and (3.34) hold of course for the right-moving energy-momentum tensor $\bar{T}(\bar{z})$ and its components $\bar{L}_n$. Note that relation (3.34) shows that $\varphi_{i,\bar{\imath}}(0,0)|0,0\rangle$ is indeed a Virasoro highest weight state.

The contour integral technique also enables us to express the commutation relations of $\mathcal{W}$-modes through the singular part of OPEs of generating fields. In particular, the Virasoro algebra becomes

$$T(z)T(w) = \frac{\frac{c}{2}}{(z-w)^4} + \frac{2}{(z-w)^2}T(w) + \frac{1}{z-w}\partial T(w) + \text{reg}_{z \to w}. \quad (3.35)$$

An analogous relation holds for $\overline{T}(\bar{z})\overline{T}(\bar{w})$, while the chiral splitting of the conformal symmetry $Vir \times Vir$ translates into the fact that the OPE $T(z)\overline{T}(\bar{w})$ is regular.

Whenever $c \neq 0$, the OPE (3.35) already shows that $T(z)$ is not a primary field. Instead, the behaviour under finite coordinate transformations is given by

$$T(z) = (f'(z))^2 T(f(z)) + \frac{c}{12}\{f(z), z\} \quad \text{where} \quad \{f(z), z\} := \frac{f'f''' - \frac{3}{2}(f'')^2}{(f')^2}. \quad (3.36)$$

The last term is the *Schwarz derivative*; it vanishes for rational maps $f(z)$ – another way of saying that the energy-momentum tensor is still quasi-primary.

From a primary field, we can form so-called *descendants* by acting with $\mathcal{W}$- or $\overline{\mathcal{W}}$-modes:

$$\big(\widehat{W}_{-n}\varphi_{i,\bar{\imath}}\big)(w, \bar{w}) = \Phi\big(W_{-n}|v_{i,\bar{\imath}}\big); w, \bar{w}\big) \quad (3.37)$$

if $\varphi_{i,\bar{\imath}} - \Phi(|v_{i,\bar{\imath}}\rangle)$. The conformal dimensions of the descendant field $\widehat{W}_{-n}\varphi_{i,\bar{\imath}}$ are $(h_i + n, h_{\bar{\imath}})$. Covariance and invertibility of the state-field correspondence may be used to show that descendants can equivalently be defined in terms of the OPE:

$$W(z)\varphi_{i,\bar{\imath}}(w, \bar{w}) = \sum_{k \geq 0}(z-w)^{k-h_W}\big(\widehat{W}_{-k}\varphi_{i,\bar{\imath}}\big)(w, \bar{w})$$

with

$$\big(\widehat{W}_{-k}\varphi_{i,\bar{\imath}}\big)(w, \bar{w}) = \oint_{C_w}\frac{dz}{2\pi i}(z-w)^{h_W - 1 - k}W(z)\varphi_{i,\bar{\imath}}(w, \bar{w}). \quad (3.38)$$

As simple special cases, we read off the Virasoro descendants

$$\widehat{L}_0\varphi_{i,\bar{\imath}} = h_i\varphi_{i,\bar{\imath}} \quad \text{and} \quad \widehat{L}_{-1}\varphi_{i,\bar{\imath}} = \partial\varphi_{i,\bar{\imath}}. \quad (3.39)$$

The space formed by the descendants of a primary, i.e. its "$\mathcal{W}$-tensor multiplet", is often called its conformal family and denoted $[\varphi_{i,\bar{\imath}}]$.

3.2.2 Correlation functions and Ward identities

The most important quantities that can be formed from field operators are *correlation functions*. In many texts, "solving a CFT" is used as a synonym for being able to compute all correlation functions, at least in principle. *N-point correlation functions* can be defined as vacuum expectation values

$$\mathcal{F}_{v_1\ldots v_N}(z_1, \bar{z}_1, \ldots, z_N, \bar{z}_N) = \langle\,\varphi_{v_1}(z_1, \bar{z}_1)\cdots\varphi_{v_N}(z_N, \bar{z}_N)\,\rangle \quad (3.40)$$

of products of linear operators on the space of states (leaving domain questions aside). In this notation, it is understood that φ_{v_N} acts on a vacuum state on the right and that a dual vacuum is placed on the left. Dual "out-states" at infinity (which corresponds to the distant future on the cylinder) can be defined with the help of field operators as

$$\langle v| = \lim_{z,\bar{z} \to \infty} \langle 0,0| \, \varphi_v(z,\bar{z}) \, z^{2h_v} \bar{z}^{2\bar{h}_v} \,. \tag{3.41}$$

The functions (3.40) are analytic everywhere except when arguments coincide. These short-distance singularities are determined by the OPE (3.33). Indeed, if all OPEs are known, we could in principle compute arbitrary correlators by reducing them to one-point functions, but in view of the infinite sums occurring in OPEs this is of course not very useful in practice. Therefore, let us concentrate on the consequences of conformal symmetry for correlation functions – or more generally on the question to what extent representation theory determines the correlators of a CFT.

Covariance under $Vir \times Vir$ allows any correlation function to be reduced to that of Virasoro primary fields (and similar statements hold for general W-algebra symmetries): conformal covariance imposes linear differential equations on the correlators, the so-called (conformal) *Ward identities*, which can be derived easily from T-ϕ-OPEs using the calculus of residues: consider an N-point function of primary fields $\varphi_{v_k}(z_k, \bar{z}_k)$ of dimensions $(h_k, h_{\bar{k}})$, and insert the symmetry generator $T(\epsilon)$ from equation (3.30) with the contour C_0 encircling all insertion points z_k. Because of analyticity, this integral splits into a sum of N integrals around the z_k, each of which can be expressed through the commutator (3.31). Cauchy's formula then yields a concise form for the "generating" conformal Ward identity:

$$\langle T(z)\varphi_{v_1}(z_1,\bar{z}_1)\cdots\varphi_{v_N}(z_N,\bar{z}_N)\rangle$$
$$= \sum_{k=1}^{N} \left(\frac{h_k}{(z-z_k)^2} + \frac{1}{z-z_k}\partial_{z_k} \right) \langle \varphi_{v_1}(z_1,\bar{z}_1)\cdots\varphi_{v_N}(z_N,\bar{z}_N)\rangle. \tag{3.42}$$

Next, we can use the expansion (3.38) of the OPE $T(z)\varphi_{v_N}(z_N,\bar{z}_N)$ into Virasoro descendants on the left-hand side – recall that $T(z)$ is local with respect to all other fields in the theory, so that it can be moved freely within correlation functions. Then expand the prefactors on the right-hand side of equation (3.42) around z_N, and compare coefficients. This allows correlators of descendants to be related to those of primary fields by means of certain differential operators: for $m \geq 1$, we have

$$\langle \varphi_{v_1}(z_1,\bar{z}_1)\cdots\varphi_{v_{N-1}}(z_{N-1},\bar{z}_{N-1})\big(\widehat{L}_{-m}\varphi_{v_N}\big)(z_N,\bar{z}_N)\rangle$$
$$= \sum_{k=1}^{N-1} \left(\frac{(m-1)h_k}{(z_k-z_N)^m} - \frac{1}{(z_k-z_N)^{m-1}}\partial_{z_k} \right) \langle \varphi_{v_1}(z_1,\bar{z}_1)\cdots\varphi_{v_N}(z_N,\bar{z}_N)\rangle.$$
$$\tag{3.43}$$

For theories with more general symmetry algebras $\mathcal{W} \times \overline{\mathcal{W}}$, we proceed in the analogous way: starting from the OPE of a generator $W(z)$ with a primary $\varphi(w, \bar{w})$ and using the same contour integral techniques, we can express correlators containing $\widehat{W}_{-m}\varphi(z_N, \bar{z}_N)$ in terms of powers of $(z_k - z_N)$ and correlators containing $\widehat{W}_{-n}\varphi_{v_k}(z_k, \bar{z}_k)$ for $0 \leq n \leq h_W - 1$ only.

The spaces generated by the action of these "small" modes $\widehat{W}_{-n}$ on $\mathcal{W}$ primary fields are finite-dimensional for so-called quasi-rational CFTs [356]. However, even for those theories the action of the $\widehat{W}_{-n}$ need not be given by simple differential operators such as equation (3.39) from the Virasoro case and, at present, there seems to be no general representation theory of the relevant small modes for arbitrary W-algebras.

In the most important case of extended symmetries, namely WZW models with non-abelian affine Lie algebra symmetry, this is still under control: since the currents all have conformal dimension $h_J = 1$, the only "small" modes are the zero modes J_0^a. By construction, the spaces generated with the help of zero modes carry irreducible representations of the underlying finite-dimensional Lie algebra **g**. Thus, the basic Ward identity in WZW models reads

$$\langle J_a(z)\varphi_{v_1}(z_1, \bar{z}_1) \cdots \varphi_{v_N}(z_N, \bar{z}_N) \rangle = \sum_{j=1}^{N} \frac{t_a^{(j)}}{z - z_j} \langle \varphi_{v_1}(z_1, \bar{z}_1) \cdots \varphi_{v_N}(z_N, \bar{z}_N) \rangle$$

(3.44)

where $t_a^{(j)}$ denotes a Lie algebra generator acting on the vector v_j, for $j = 1, \ldots, N$ and $a = 1, \ldots, \dim$ **g**.

We note in passing that covariance under $\mathcal{W} \times \overline{\mathcal{W}}$ also restricts the OPEs of field operators. To keep notations simpler, let us focus on the Virasoro case, $\mathcal{W} = Vir$, and use $\varphi_{k,\bar{k}}^{n,\bar{n}}(w, \bar{w})$ to denote an orthogonal basis of descendants in the $\mathcal{W} \times \overline{\mathcal{W}}$ representation built over the primary field $\varphi_k(w, \bar{w}) := \phi_{k,\bar{k}}(w, \bar{w})$, and let $(h_k + N(\boldsymbol{n}), h_{\bar{k}} + N(\overline{\boldsymbol{n}}))$ denote their dimensions. (Here, $\boldsymbol{n}$ and $\overline{\boldsymbol{n}}$ are multi-indices labelling $\mathcal{W}$-modes acting on the primary.) Then we can write the OPE as

$$\varphi_{i,\bar{i}}(z, \bar{z})\, \varphi_{j,\bar{j}}(w, \bar{w}) = \sum_{(k,\bar{k})} C_{ijk}\, (z - w)^{h_k - h_i - h_j}(\bar{z} - \bar{w})^{h_{\bar{k}} - h_{\bar{i}} - h_{\bar{j}}}$$
$$\times \sum_{\boldsymbol{n},\overline{\boldsymbol{n}}} \beta_{ijk}^{\boldsymbol{n}} \beta_{\bar{i}\bar{j}\bar{k}}^{\overline{\boldsymbol{n}}}\, (z - w)^{N(\boldsymbol{n})}(\bar{z} - \bar{w})^{N(\overline{\boldsymbol{n}})}\, \varphi_{k,\bar{k}}^{n,\bar{n}}(w, \bar{w})$$

(3.45)

and the coefficients $\beta_{ijk}^{\boldsymbol{n}}$ and $\beta_{\bar{i}\bar{j}\bar{k}}^{\overline{\boldsymbol{n}}}$ in front of the descendants, i.e. the relative couplings within the conformal family $[\varphi_{k,\bar{k}}]$, are universal and follow from representation theory, while only the "reduced matrix elements" C_{ijk} for three primaries – finitely many in a rational CFT – are left undetermined. Explicit expressions for the $\beta_{ijk}^{\boldsymbol{n}}$ in the Virasoro case can be derived from the Ward identities (3.43); see [45].

Let us continue our discussion of correlation functions. We have seen after equation (3.43) that, thanks to the Ward identities, we can express arbitrary correlators by those containing only primaries and descendants built with modes $\widehat{W}_{-n}$ for $0 \leq n \leq h_W - 1$. Moreover, N-point functions of Virasoro primary fields are themselves constrained by conformal symmetry: from the commutators (3.34) and $SL(2,\mathbb{R})$-invariance of the vacuum, we conclude that if all φ_{v_i} in the correlator (3.40) are (quasi-)primary, then

$$\sum_{i=1}^{N} \partial_{z_i} \mathcal{F}_{v_1 \dots v_N} = \sum_{i=1}^{N} (z_i \partial_{z_i} + h_i) \mathcal{F}_{v_1 \dots v_N} = \sum_{i=1}^{N} (z_i^2 \partial_{z_i} + 2 z_i h_i) \mathcal{F}_{v_1 \dots v_N} = 0 \,.$$

This behaviour under Möbius transformation fixes one-, two- and three-point functions of (quasi-)primary fields up to normalisations: $\langle \varphi_v(z, \bar{z}) \rangle$ vanishes unless φ_v is the identity operator. The two-point functions read

$$\langle \varphi_{i,\bar{i}}(z_1, \bar{z}_1) \varphi_{j,\bar{j}}(z_2, \bar{z}_2) \rangle = \frac{C_{i,\bar{i}} \, \delta_{j,i^+} \, \delta_{\bar{j},\bar{i}^+}}{(z_1 - z_2)^{2h_i} (\bar{z}_1 - \bar{z}_2)^{2h_{\bar{i}}}} \tag{3.46}$$

where i^+ labels the conjugate sector of i (such that $h_{i^+} = h_i$ because the conformal Hamiltonian is CPT-invariant). We normalise primary bulk fields such that $C_{i,\bar{i}} = 1$. The three-point function of primary fields is given by

$$\langle \varphi_{i,\bar{i}}(z_1, \bar{z}_2) \varphi_{j,\bar{j}}(z_2, \bar{z}_2) \varphi_{k,\bar{k}}(z_3, \bar{z}_3) \rangle = \frac{C_{\text{ijk}}}{|z_1 - z_2|^{2h_{ijk}} |z_1 - z_3|^{2h_{ikj}} |z_2 - z_3|^{2h_{jki}}}$$
$$\tag{3.47}$$

with $h_{ijk} := h_i + h_j - h_k$, etc., and with structure constants from the OPE (3.33) for three Virasoro highest weight vectors. In writing equation (3.47), we have assumed that $h_l = h_{\bar{l}}$, just to keep the denominator simpler, and we have again used shorthands $\mathrm{i} = (i, \bar{i})$ for the indices. But it should be emphasised that the OPE coefficients are non-chiral data, and not determined by representation theory.

The functional dependence of four-point functions of primaries is only partially constrained by conformal invariance, essentially because the $SL(2,\mathbb{R})$-invariant cross-ratio

$$\eta = \frac{(z_1 - z_2)(z_3 - z_4)}{(z_1 - z_3)(z_2 - z_4)} \tag{3.48}$$

can be formed from four complex variables. A right-moving $\bar{\eta}$ is defined analogously. It can be shown that

$$\mathcal{F}_{(i_1, \bar{i}_1), \dots, (i_4, \bar{i}_4)}(z_1, \bar{z}_1, \dots, z_4, \bar{z}_4)$$
$$= \prod_{1 \leq l < m \leq 4} (z_l - z_m)^{H - h_l - h_m} (\bar{z}_l - \bar{z}_m)^{\bar{H} - h_{\bar{l}} - h_{\bar{m}}} \, G^{(4)}(\eta, \bar{\eta}) \,, \tag{3.49}$$

with $H := \frac{1}{3} \sum_{l=1}^{4} h_{i_l}$, $\bar{H} := \frac{1}{3} \sum_{l=1}^{4} h_{\bar{i}_l}$; the function $G^{(4)}(\eta, \bar{\eta})$ is model-dependent and needs to be determined from additional conditions.

N-point functions for $N > 4$ are similar products of a universal prefactor and a model-dependent function of $N - 3$ cross-ratios, but it turns out that the study of four-point functions furnishes virtually all interesting constraints on the CFT.

The relations stated so far hold in every conformal theory, but typically there are additional and much more restrictive differential equations which depend on the respective model and which may allow its correlation functions to be computed explicitly. These additional differential equations arise from null-vectors.

In the case of degenerate representations of the Virasoro algebra, we have seen that certain combinations of Virasoro modes create singular vectors from the highest weight states, which are then set to zero in an irreducible representation. Through the state-field correspondence, such null-vectors $|n\rangle$ become null-fields $n(z, \bar{z})$, and all correlation functions involving $n(z, \bar{z})$ vanish.

In particular, null-states in the vacuum module yield chiral null-fields, i.e. certain normal-ordered products of $\mathcal{W}$ generators vanish. All other irreducible representations of $\mathcal{W}$ occurring in a model with this vacuum module are representations of $\mathcal{W}/\mathcal{N}$, where $\mathcal{N}$ is the sub-W-algebra generated by $n(z)$, its derivatives and normal-ordered products with $\mathcal{W}$-fields; $\mathcal{N}$ is called the "annihilating ideal" in [160].

As an example, consider a Virasoro highest weight state $|h\rangle$ with conformal dimension h and central charge c related by $h = \frac{1}{16}\left(5 - c \pm \sqrt{(c-1)(25-c)}\right)$, and denote by $\varphi_{|h\rangle}(z, \bar{z})$ the primary field corresponding to $|h\rangle$. The level two descendant from equation (3.23) is a null-vector, and all correlation functions which contain the corresponding field

$$n_{|h\rangle;2}(z, \bar{z}) := \left(\widehat{L}_{-2} - \frac{3}{2(2h+1)}\widehat{L}_{-1}^2\right)\varphi_{|h\rangle}(z, \bar{z})$$

vanish. Using the translation rule (3.43), these correlators can be expressed through a second-order differential operator acting on correlators containing the primary field itself:

$$0 = \left\langle \varphi_{v_1}(z_1, \bar{z}_1) \cdots \varphi_{v_{N-1}}(z_{N-1}, \bar{z}_{N-1})\, n_{|h\rangle;2}(z_N, \bar{z}_N)\right\rangle$$
$$= \mathbf{D}\left\langle \varphi_{v_1}(z_1, \bar{z}_1) \cdots \varphi_{v_{N-1}}(z_{N-1}, \bar{z}_{N-1})\varphi_{|h\rangle}(z_N, \bar{z}_N)\right\rangle$$

with the differential operator

$$\mathbf{D} := -\frac{3}{2(2h+1)}\partial_{z_N}^2 + \sum_{k=1}^{N-1}\left(\frac{h_k}{(z_k - z_N)^2} - \frac{1}{z_k - z_N}\partial_{z_k}\right).$$

Specialising to $N = 4$, and stripping off universal factors from the four-point function as in equation (3.49), an ordinary differential equation is obtained for $G^{(4)}(\eta, \bar{\eta})$ of the hyper-geometric type. This was the main tool used in [45] and [142] to construct exact correlation functions of Virasoro minimal models; see also [19] for a review.

Another famous example for a differential equation arising from a null-field is the *Knizhnik–Zamolodchikov equation* for correlation functions of WZW models [317]. The null-field in question,

$$n(z) = T(z) - \frac{1}{2} \frac{1}{k + h^\vee} : J^a(z) J^a(z) :$$

(summation over a understood), comes from the Sugawara formula for the energy-momentum tensor and leads to the differential equation

$$\partial_{z_i} \mathcal{F}_{v_1 \dots v_N}(\mathbf{z}, \bar{\mathbf{z}}) = \frac{1}{2} \frac{1}{k + h^\vee} \sum_{j; j \neq i} \frac{t_a^{(i)} t_a^{(j)}}{z_i - z_j} \mathcal{F}_{v_1 \dots v_N}(\mathbf{z}, \bar{\mathbf{z}}) \qquad (3.50)$$

with the notations of equation (3.44) and $\mathbf{z} = (z_1, \dots, z_N)$. The Knizhnik–Zamolodchikov equation (3.50) holds in every WZW model, but model-specific differential equations arise from further null-vectors: in the spin j representation of the SU(2) WZW model with level k, for example, we have that the highest weight state $|j; j\rangle$ satisfies [231]

$$(J_{-1}^-)^{k - 2j + 1} |j; j\rangle = 0,$$

which in particular leads to the conclusion that only highest weights $j = 0, \frac{1}{2}, \dots, \frac{k}{2}$ can occur in the conformal theory. Their conformal dimensions are

$$h_j = j(j + 1)/(k + 2) . \qquad (3.51)$$

We can introduce generalisations of the Knizhnik–Zamolodchikov equation in a large class of CFTs without any relation to affine Lie algebras. Then the subspaces of lowest dimensions are replaced by other finite-dimensional special subspaces of the $\mathcal{H}_i$; see [11, 356] for details.

3.2.3 Duality data

It is worthwhile going beyond concrete examples and differential equations and identifying some general properties of W-algebra representations. This more abstract description is, e.g., useful for establishing links with mathematical areas (in particular those with a category-theoretic component), but is also useful in stating the non-linear constraints to be discussed in Section 3.3 clearly. Moreover, the data we will extract will also play an important role later for boundary CFT.

The simplest of these data are the *fusion rules* N_{ij}^k of the chiral symmetry algebra W. They have the same conceptual meaning as for finite-dimensional Lie algebras and describe the decomposition into irreducibles of a tensor product of representations, but defining the latter is more involved for W-algebras: we use the so-called *fusion product*.

In equation (3.38), the action of W-modes on fields $\varphi_v(z_1, \bar{z}_1)$ inserted somewhere in the complex plane was defined in terms of a contour integral over

$W(z)\omega(z)$ around z_1, where ω denotes some meromorphic one-form. Left-moving $\mathcal{W}$-modes acting on a product of field operators $\varphi_{v_1}(z_1, \bar{z}_1)\,\varphi_{v_2}(z_2, \bar{z}_2)$ can be defined analogously by integrating $W(z)\omega(z)$ along a contour C_{12} which encircles both z_1 and z_2. Splitting this integral into a sum of contour integrals around z_1 and z_2, respectively, allows us to relate this $\mathcal{W}$-representation Δ_{z_1,z_2} on the (tensor) product of two fields to the known representations. Explicit formulas are obtained upon expanding around the z_i: with $h \in \mathbb{Z}$ being the conformal dimension of the (bosonic) field $W(z)$, we have

$$\Delta_{z,0}(\widehat{W}_n) = \mathbf{1} \otimes \widehat{W}_n \; + \; \sum_{k=0}^{\infty} \binom{n+h-1}{k} z^{n+h-1-k} \widehat{W}_{k+1-h} \otimes \mathbf{1}\,; \qquad (3.52)$$

see [350], and also [211], for a more detailed discussion. The terms $\widehat{W}_m \otimes \mathbf{1}$ act non-trivially on $\varphi_{v_1}(z, \bar{z})$ while leaving the second field $\varphi_{v_2}(0,0)$ unchanged. An analogous fusion product exists for $\overline{\mathcal{W}}$. Observe that, as required, central charges do not add up under the fusion product; they would if we were using the usual tensor product that is familiar from the representation theory of finite-dimensional Lie algebras.

Denoting the original $\mathcal{W}$-representations by π_1 and π_2, the new one is given by $(\pi_1 \otimes \pi_2) \circ \Delta$. In general, the fusion product of two irreducible $\mathcal{W}$-representations π_i and π_j is not irreducible again. We restrict overselves to cases where the new representation can be decomposed into irreducibles (see [212] and references therein for other situations). The fusion rules N_{ij}^k count how often the subrepresentation π_k occurs in the fusion product $(\pi_i \otimes \pi_j) \circ \Delta$. Whenever $N_{ij}^k = 0$, no field $\varphi_{k,\bar{k}}$ appears in the OPE of $\varphi_{i,\bar{i}}$ and $\varphi_{j,\bar{j}}$ (and analogously for the right-moving labels). Let us list some abstract properties of the fusion rules and of the *fusion matrices* defined by

$$\left(N_i\right)_{jk} = N_{ij}^k\,.$$

We label the vacuum representation of $\mathcal{W}$ by 0; it behaves like a unit under fusion, i.e. $N_{0j}^k = \delta_{j,k}$. The conjugate sector i^+ of a representation i is distinguished by the property

$$N_{ij}^0 = \delta_{j,i^+}\,,$$

and we have the relations $N_{i^+j}^k = N_{ik}^j$ and $N_{i^+j^+}^{k^+} = N_{ij}^k$. The representation content of the OPE is independent of the arguments z_i in the fusion product, so that in particular

$$N_{ji}^k = N_{ij}^k\,.$$

Furthermore, the result of iterated fusion does not depend on the ordering: fusion is an associative procedure,

$$\sum_m N_{im}^l\,N_{jk}^m = \sum_n N_{ij}^n\,N_{nk}^l\,;$$

the sums run over all representations of the chiral algebra, a set we will denote I_W. This is finite for a rational CFT.

Altogether, the fusion rules N_{ij}^k define an associative commutative ring with generators $[\phi_i]$, $i \in I_W$, with an involution $[\phi_i] \mapsto [\phi_{i^+}]$, and with a distinguished unit element N_0, such that

$$[\phi_i] \times [\phi_j] = \sum_k N_{ij}^k \, [\phi_k],$$

where the structure constants N_{ij}^k are non-negative integers. The fusion matrices themselves form the regular representation of this ring. Because of $N_{i^+} = N_i^T$, they can be simultaneously diagonalised, and their eigenvalues provide the one-dimensional representations of the fusion rules. Fusion rings can be studied independently of conformal models, and various results concerning general structure and a classification for small $|I_W|$ have been obtained; see, e.g., [108, 116, 199].

To give an example, we list the fusion rules of the Virasoro minimal models with central charge $c(p, p')$ as in formula (3.24) and (left-moving) conformal dimensions $h_{r,s}(p, p')$ of the primary fields given in the Kac table (3.25). The fusion rules read

$$N_{(r_1,s_1)\,(r_2,s_2)}^{(r_3,s_3)} = \begin{cases} 1 & \text{if } |r_1 - r_2| < r_3 < \min(r_1 + r_2, 2p' - r_1 - r_2) \\ & \text{and } |s_1 - s_2| < s_3 < \min(s_1 + s_2, 2p - s_1 - s_2) \quad (3.53) \\ 0 & \text{otherwise}. \end{cases}$$

A very detailed derivation of this relation is given in [121]. Focusing on the r_i labels or on the s_i labels, we recognise a resemblance to decomposition rules for tensor products of SU(2) representations. This is due to the fact that Virasoro minimal models arise from SU(2) WZW theories via the coset construction; see also Chapter 6.

The representation labelled (r, s) with $r = p' - 1$ and $s = p - 1$ is an example of a so-called *simple current*: these are primary fields $\phi_{\text{s.c.}}$ which are invertible at the level of fusion rules; they satisfy

$$[\phi_{\text{s.c.}}] \times [\phi_{\text{s.c.}}^c] = [0] \qquad (3.54)$$

with their conjugate field $\phi_{\text{s.c.}}^c$. This implies that the fusion of $\phi_{\text{s.c.}}$ with any other primary contains only a single irreducible representation of W.

The fusion rules provide an important piece of partial information on the OPE, namely which conformal families $[\varphi_{k,\bar{k}}]$ may appear at all on the right-hand side of the OPE (3.45), which has to respect the W-symmetry. Note that fusion rules can also be determined in a less algebraic manner, from null-vector differential equations on correlation functions. In the example of Virasoro minimal models mentioned before, the hyper-geometric differential equations provide characteristic equations for the singularities as $z_1 \approx z_2$. They yield conditions on the conformal dimensions h_k which can appear in the OPE of $\varphi_{|h_{1,2}}$ with any other primary field; again, see [45] for a very explicit description of the procedure.

If the fusion rules N_{ij}^k do not vanish, then by definition there exist intertwiners

$$\left(\begin{smallmatrix} i \\ k\,j \end{smallmatrix}\right)_{z;\,\alpha} : \mathcal{H}_i \otimes \mathcal{H}_j \longrightarrow \mathcal{H}_k \tag{3.55}$$

between the $\mathcal{W}$-actions $(\pi_i \otimes \pi_j) \circ \Delta_{z,0}$ and π_k. The label $\alpha = 1, \dots, N_{ij}^k$ enumerates different "fusion channels" between the representations. We can use the intertwiners to define *chiral vertex operators* (CVOs)

$$\phi_{kj}^{i;\,\alpha}(\xi; z) := \left(\begin{smallmatrix} i \\ k\,j \end{smallmatrix}\right)_{z;\,\alpha}(\xi \otimes \cdot) : \mathcal{H}_j \longrightarrow \mathcal{H}_k$$

associated to each state $\xi \in \mathcal{H}_i$ from the chiral representation space $\mathcal{H}_i$. Right-moving CVOs for $\overline{\mathcal{W}}$ are introduced analogously. These operators do not directly correspond to physical fields in the theory, but we will soon see that they are very useful intermediate objects. Their covariance properties parallel those of fully fledged field operators $\varphi_v(z, \bar{z})$ from before, with appropriate adjustments: the left-moving $\phi_{kj}^{i;\,\alpha}(\xi; z)$ are insensitive to the right-moving symmetry algebra $\overline{\mathcal{W}}$, and relations like the commutator (3.34) with Virasoro modes have to be projected onto the chiral highest weight representations involved: a primary CVO corresponding to a highest weight vector $\xi \in \mathcal{H}_i$ satisfies

$$\pi_k(L_n)\,\phi_{kj}^{i;\,\alpha}(\xi; z) - \phi_{kj}^{i;\,\alpha}(\xi; z)\,\pi_j(L_n) = z^n \left(z\partial_z + h_\xi(n+1)\right) \phi_{kj}^{i;\,\alpha}(\xi; z).$$

In formal analogy to physical fields $\varphi_v(z, \bar{z})$, we can introduce correlation functions of CVOs, the so-called *conformal blocks*, as chiral expectation values of suitable concatenations,

$$F_{\boldsymbol{i}}^{\boldsymbol{p},\alpha}(\mathbf{z}) = \langle \xi_{i_1} | \phi_{i_1,p_1}^{i_2;\,\alpha_1}(\xi_{i_2}; z_2)\, \phi_{p_1,p_2}^{i_3;\,\alpha_2}(\xi_{i_3}; z_3) \cdots \phi_{p_{N-3},i_N}^{i_{N-1};\,\alpha_{N-2}}(\xi_{i_{N-1}}; z_{N-1}) | \xi_{i_N} \rangle,$$
$$\tag{3.56}$$

with obvious abbreviations for multi-indices on the left-hand side. We have also used Möbius covariance to put $z_1 = \infty$ and $z_N = 0$, along with chiral analogues of the state-field correspondences (3.28) and (3.41). We will also use the following graphical notation [350]:

$$F_{\boldsymbol{i}}^{\boldsymbol{p},\alpha}(\mathbf{z}) = i_1 \quad \begin{array}{ccccccc} i_2 & i_3 & & i_{N-2} & i_{N-1} \\ | & | & & | & | \\ \cdots & & & & \cdots \end{array}$$

with labels $z_2\,|\,\alpha_1$, $z_3\,|\,\alpha_2$, $z_{N-2}\,|\,\alpha_{N-3}$, $z_{N-1}\,|\,\alpha_{N-2}$, i_N and p_1, p_2, p_{N-4}, p_{N-3}

The transformation properties under $\mathcal{W}$ ensure that these conformal blocks obey the same differential equations as the physical correlators. In fact, if we let p_l and α_m in equation (3.56) range over the values $p_l \in I_{\mathcal{W}}$ and $\alpha_m = 1, \dots, N_{i_{l+1}p_l}^{p_{l-1}}$ (where we set $p_0 := i_1$ and $p_{N-2} := i_N$), respectively, we obtain a fundamental system for the differential null-vector equations. The dimension of the space of conformal blocks is determined by the fusion rules as $\sum_{\boldsymbol{p}} N_{i_2 p_1}^{i_1} N_{i_3 p_2}^{p_1} \cdots N_{i_{N-1} i_N}^{p_{N-3}}$.

In contrast to physical fields, chiral vertex operators are not local with respect to each other, so the conformal blocks are multi-valued functions. If we define them for a fixed ordering $|z_2| > |z_3| > \cdots > |z_{N-1}| > 0$ of arguments, and use an analytic continuation along a path $\gamma_\pm$ to interchange two neighbouring points z_i and z_{i+1}, the result usually depends on the (class of the) path. (For definiteness, we denote the paths in such a way that $\pm\Im(z_i - z_{i+1}) > 0$ along $\gamma_\pm$.) The fields in $\mathcal{W}$, however, are local with respect to the CVOs, thus the conformal blocks for both regimes $|z_i| > |z_{i+1}|$ and $|z_{i+1}| > |z_i|$ provide fundamental systems for the Ward identities. The theory of linear differential equations then tells us that there is a constant invertible linear transformation between both systems. Such changes of basis relating conformal blocks evaluated in different regimes are sometimes referred to in CFT as *duality matrices*.

From now on, we focus on four-point blocks since they are sufficient to introduce the basic duality matrices. Using the graphical notation from above, the *braiding matrix* is defined by

$$
i\underset{p}{\underline{\quad\overset{\displaystyle j}{\overset{\Big|}{z_2\,\big|\,\alpha_1}}\quad\overset{\displaystyle k}{\overset{\Big|}{z_3\,\big|\,\alpha_2}}\quad}}l = \sum_{q,\,\beta_1\beta_2} B_{pq}\begin{bmatrix} j & k \\ i & l \end{bmatrix}^{\beta_1\beta_2}_{\alpha_1\alpha_2}(\varepsilon)\quad i\underset{q}{\underline{\quad\overset{\displaystyle k}{\overset{\Big|}{z_2\,\big|\,\beta_1}}\quad\overset{\displaystyle j}{\overset{\Big|}{z_3\,\big|\,\beta_2}}\quad}}l
$$

$$(3.57)$$

More explicitly, we first analytically continue the conformal block $F^p_{ijkl}(z_2, z_3)$ on the left-hand side along γ_+ or γ_- so as to interchange z_2 and z_3. Then the resulting function is expressed as a linear combination of conformal blocks with the original ordering of arguments, but with representation k inserted at z_2 and j at z_3. The sign ε in relation (3.57) indicates the dependence of the braiding matrix on the path γ_ε of analytic continuation – and implicitly accounts for the name: in 1+1 dimensions, the chiral conformal blocks carry representations of the braid group rather than the permutation group [189]. The braid group B_n on n strands can be presented using generators σ_i for $i = 1, \ldots, n$ and relations

$$\sigma_i\,\sigma_j = \sigma_j\,\sigma_i \quad \text{for } |i - j| > 1\,,$$

$$\sigma_i\,\sigma_{i+1}\sigma_i = \sigma_{i+1}\sigma_i\,\sigma_{i+1}\,.$$

Note that the second relation can be regarded as a Yang–Baxter equation without spectral parameters.

If the additional relations $\sigma_i^2 = 1$ are imposed, the σ_i become transpositions generating the permutation group; permutations arise from braids by dropping the distinction between over- and under-crossings of strands (i.e. between ε and $-\varepsilon$). On the other hand, in QFTs with braid-group statistics (such as the "chiral halves" of CFTs), we merely have the relation $\mathsf{B}(\varepsilon) = \mathsf{B}(-\varepsilon)^{-1}$.

The above considerations on solutions of differential equations and on locality of $\mathcal{W}$-fields show that B is not only independent of the insertion points, but also of the specific states ξ_i in the CVOs $\phi^i_{kj}(\xi_i; y)$. The braiding matrices can therefore be associated directly with the intertwiners (3.55), i.e. they are data from the representation category of the chiral algebra.

The same is true for the *fusing matrices*. They arise when we compare conformal blocks in the regime $|z_2 - z_3| \gg |z_3|$ to those with $|z_3| \gg |z_2 - z_3|$ without changing the order $|z_2| > |z_3|$. This means that we evaluate OPEs of CVOs at z_2, z_3 and 0 in two different ways, $\phi(z_2)(\phi(z_3)\phi(0))$ and $(\phi(z_2)\phi(z_3))\phi(0)$, respectively. The fusing matrix is simply the "re-associator" relating the different bracketings, and it connects the associated fundamental systems. Graphically,

$$
\sum_{q,\,\beta_1\beta_2} \mathsf{F}_{pq}\!\left[\begin{matrix} j & k \\ i & l \end{matrix}\right]^{\beta_1\beta_2}_{\alpha_1\alpha_2}
$$
(3.58)

This becomes a precise definition if we fix the normalisation of the CVOs by demanding that the conformal blocks $F_{ij,kl}$ and $F_{il,jk}$, respectively – in an obvious notation for the fusion structure on the left- and right-hand side of relation (3.58), respectively – behave asymptotically as

$$
F^p_{ij,kl}(x) = x^{h_p - h_k - h_l}\,(1 + P_1(x)) \quad \text{for } x \approx 0,
$$
$$
F^q_{il,jk}(1 - x) = (1 - x)^{h_q - h_j - h_k}\,(1 + P_2(1 - x)) \quad \text{for } x \approx 1, \tag{3.59}
$$

where we have chosen $z_2 = 1$ and $z_3 = x$, and where $P_i(x')$ denote power series which vanish at $x' = 0$.

Let us spell out fusion, fusing and braiding matrices for the case of the free boson, even though this machinery can be avoided here by computing all correlators directly from Wick's theorem. Chiral vertex operators are introduced in analogy to full local fields, $\phi_g(z) = :e^{ig X_L(z)}:$ where we have decomposed the bosonic coordinate as $X(z, \bar{z}) = X_L(z) + X_R(\bar{z})$. These CVOs have conformal dimensions $h_g = \alpha' \frac{g^2}{4}$, and their conformal blocks are computed as in Chapter 1, equation (1.47). In particular, the fusion rules $N^{g_3}_{g_1 g_2} = \delta_{g_3, g_1 + g_2}$ reflect conservation of the $U(1)$ charge, and the chiral four-point blocks have the form

$$
F_{g_1 g_2 g_3 g_4}(z_2, z_3) = (z_2 - z_3)^{\frac{\alpha'}{2} g_2 g_3}\, z_2^{\frac{\alpha'}{2} g_2 g_4}\, z_3^{\frac{\alpha'}{2} g_3 g_4} \tag{3.60}
$$

with $g_1 = -g_2 - g_3 - g_4$. It is easy to see that this function satisfies both expansions in equation (3.59) at the same time,

$$
F_{g_1 g_2 g_3 g_4}(z_2, z_3) = z_3^{h_{g_3 + g_4} - h_{g_3} - h_{g_4}}\,(1 + P_1(z_3))
$$
$$
= (z_2 - z_3)^{h_{g_2 + g_3} - h_{g_2} - h_{g_3}}\,(1 + P_2(z_2 - z_3)),
$$

thus the only non-vanishing element of the fusing matrix is equal to 1:

$$
\mathsf{F}_{pq}\!\left[\begin{matrix} g_2 & g_3 \\ g_1 & g_4 \end{matrix}\right] = \begin{cases} 1 & \text{if } g_1 = -g_2 - g_3 - g_4\,,\ p = g_3 + g_4\,, q = g_2 + g_3\,, \\ 0 & \text{otherwise} \,. \end{cases} \tag{3.61}
$$

To determine the braiding matrix, we interchange z_2 and z_3 in equation (3.60) by an analytic continuation along γ_ε and compare to $F_{g_1g_3g_2g_4}(z_2, z_3)$, with the result

$$B_{pq}\begin{bmatrix} g_2 & g_3 \\ g_1 & g_4 \end{bmatrix}(\varepsilon) = \begin{cases} e^{i\varepsilon\pi\frac{\alpha'}{2}g_2g_3} & \text{if } g_1 = -g_2 - g_3 - g_4\,, p = g_3 + g_4\,, q = g_2 + g_4\,, \\ 0 & \text{otherwise}\,. \end{cases}$$
(3.62)

There are many more chiral algebras for which fusion, fusing and braiding matrices are known explicitly, including Virasoro minimal models; see [19, 142, 165]; and also WZW models; see Chapter 6 for more details. The authors of [142] obtained these chiral data, as well as full correlation functions for physical fields, by employing the so-called Coulomb gas representation, which (forcibly) expresses the fields of the Virasoro minimal model in terms of a free boson with a background charge: this involves a shift in U(1) charge conservation and a shift in the energy-momentum tensor by a term proportional to $\partial^2 X$ compared to the usual definition $T \sim \partial X \partial X$ in the Sugawara formula (3.10). This ingenious method was put into the rigorous context of BRST cohomology in [161] and proved useful in many examples beyond minimal models.

The duality matrices introduced with the help of four-point conformal blocks also describe braiding and fusing on neighbouring points in arbitrary N-point blocks, since chiral OPEs can be used to fuse the remaining CVOs. Thus, more complicated "duality moves" can be written as sequences of elementary ones. It is found that certain chains of transformations lead back to the original basis of conformal blocks so that F and B must satisfy certain consistency conditions, called *polynomial equations* [350]. The simplest ones, the pentagon identity,

$$\sum_s F_{qs}\begin{bmatrix} j_2 & j_3 \\ p & j_6 \end{bmatrix} F_{pj_4}\begin{bmatrix} j_1 & s \\ j_5 & j_6 \end{bmatrix} F_{sr}\begin{bmatrix} j_1 & j_2 \\ j_4 & j_3 \end{bmatrix} = F_{pr}\begin{bmatrix} j_1 & j_2 \\ j_5 & q \end{bmatrix} F_{qj_4}\begin{bmatrix} r & j_3 \\ j_5 & j_6 \end{bmatrix},$$
(3.63)

and the hexagon identity (here given in the form of the Yang–Baxter equation),

$$\sum_p B_{j_6p}\begin{bmatrix} j_2 & j_3 \\ j_1 & j_7 \end{bmatrix}(\varepsilon)\, B_{j_7j_9}\begin{bmatrix} j_2 & j_4 \\ p & j_5 \end{bmatrix}(\varepsilon)\, B_{pj_8}\begin{bmatrix} j_3 & j_4 \\ j_1 & j_9 \end{bmatrix}(\varepsilon)$$

$$= \sum_q B_{j_7q}\begin{bmatrix} j_3 & j_4 \\ j_6 & j_5 \end{bmatrix}(\varepsilon)\, B_{j_6j_8}\begin{bmatrix} j_2 & j_4 \\ j_1 & q \end{bmatrix}(\varepsilon)\, B_{qj_9}\begin{bmatrix} j_2 & j_3 \\ j_8 & j_5 \end{bmatrix}(\varepsilon),$$
(3.64)

are obtained from a sequence of braidings or fusings on five-point blocks. It can be shown that (together with a few relations on the modular S-matrix) they already ensure that all higher-degree equations are satisfied. We refer to the literature for further details, e.g., to [19, 44, 350]. Apart from a number of symmetries among the entries of the fusing matrix, the polynomial equations also imply that the braiding matrix can be expressed in terms of F,

$$B_{pq}\begin{bmatrix} j & k \\ i & l \end{bmatrix}(\varepsilon) = e^{i\varepsilon\pi(h_i + h_l - h_p - h_q)}\, F_{pq}\begin{bmatrix} j & l \\ i & k \end{bmatrix}.$$
(3.65)

The fusing matrix together with the modular S-matrix (to be introduced below) and the values of central charge and conformal dimensions provide the data of what was called "duality groupoid" in [350]. More to the point, they specify a semi-simple rigid braided monoidal C^*-tensor category, which encodes all information on the behaviour of representations of the chiral algebra $\mathcal{W}$ under fusion etc. A detailed introduction to these categories was given by Fröhlich and Kerler in [196], where it was also shown how the duality data can be obtained from the algebraic approach to CFT in an elegant way; see also [210] and Rehren's works [392], which extends [176, 393]. In algebraic QFT, (localised) superselection sectors π_i are defined as unitary equivalence classes of (localised) endomorphisms ρ_i of the observable algebra, $\pi_i \simeq \pi_0 \circ \rho_i$, where π_0 is the vacuum representation. Composition of endomorphisms yields a natural definition of fusion, and there are so-called statistics operators $\varepsilon(\rho_i, \rho_j)$, which relate $\rho_i \circ \rho_j$ and $\rho_j \circ \rho_i$. From them, one can prepare braid-group representations (and also F and B from above) in a rather straightforward fashion [176, 192, 392].

Obviously, the emergence of braid-group statistics for chiral vertex operators is at the root of the interrelations between CFT in two dimensions and knot invariants [197, 450] or invariants of topological three-manifolds. It turned out that duality categories of various CFT models are isomorphic with representation categories of certain quantum groups or, more generally, of weak quasi-triangular quasi-Hopf algebras H, see e.g. [16, 18, 158, 196, 338, 349, 408]. These quantum symmetry algebras of chiral CFTs substitute for the ordinary Lie groups that furnish the global gauge symmetries of local QFT in four or more dimensions [139].

3.3 Non-linear constraints

Up to now, we have concentrated on linear, representation theoretic properties in CFTs. In the remainder of this chapter, we will discuss non-linear constraints (also called *"sewing relations"*) arising from certain physical principles, which impose restrictions on both the OPE coefficients and the selection of irreducible representations of $\mathcal{W} \times \overline{\mathcal{W}}$ that can occur in the state and field space. These underlying principles are the locality of physical fields, associativity of the OPE, and modular invariance of the torus partition function. With the help of these, it is, in principle, and for many important cases also in practise, possible to solve the CFT completely.

3.3.1 Locality, and associativity of the OPE

The chiral data introduced in the last section enter the sewing relations, since physical bulk fields can be written as bilinears of left- and right-moving chiral vertex operators,

$$\varphi_v^a(z, \bar{z}) = \sum_{j,k,\alpha,\bar{\alpha}} D_{i,\bar{i}}^a [j\bar{j}\,k\bar{k}]_{\alpha\bar{\alpha}} \; \phi_{kj}^{i;\,\alpha}(\xi_i, z) \, \overline{\phi}_{\bar{k}\bar{j}}^{\bar{i};\,\bar{\alpha}}(\bar{\xi}_{\bar{i}}, \bar{z}) \qquad (3.66)$$

for $v = \xi_i \otimes \bar{\xi}_{\bar{\imath}} \in \mathcal{H}$. Here, it is understood that CVOs ϕ^i_{kj} vanish on chiral representations $\mathcal{H}_l$ other than $\mathcal{H}_j$, and the index a indicates the possibility that the same left- and right-moving CVOs may be combined in more than one way.

We can now insert equation (3.66) into physical correlation functions, which thereby become bilinears of chiral conformal blocks. For example, the four-point function with $z_1 = \infty$, $z_4 = 0$ and with the normalisation from equation (3.59), reads

$$\mathcal{F}_{i_1,\ldots,i_4}(z_2, \bar{z}_2, z_3, \bar{z}_3) = \sum_{p,\bar{p}} C_{i_1 i_2 p}\, C_{i_3 i_4 p^+}\, F^p_{i_1 i_2, i_3 i_4}(z_2, z_3)\, F^{\bar{p}}_{\bar{\imath}_1 \bar{\imath}_2, \bar{\imath}_3 \bar{\imath}_4}(\bar{z}_2, \bar{z}_3)\,,$$

where C_{ijk} are OPE coefficients of primary fields. They can be expressed through the D coefficients in the expansion (3.66).

The full correlators $\mathcal{F}$ satisfy $W \times \overline{W}$ Ward identities by construction of the CVOs, but there are non-representation theoretic requirements: physical field operators must be (semi-)local, i.e. (anti-)commute with each other for spacelike separations, and their OPE must be an associative operation.

For the correlation functions $\mathcal{F}$, *locality* means that they are analytic functions in the "configuration space" $\mathbb{C}^N \setminus \{z_i \neq z_j, i \neq j\}$ – after having identified each right-moving coordinate $\bar{z}_i$ with the complex conjugate of z_i: while left- and right-moving degrees of freedom are completely independent at the level of Ward identities, within physical correlators their relation to real light-cone variables (3.2) must be accounted for. Since the conformal blocks are multi-valued functions in general, with cuts due to non-integer conformal dimensions, locality in particular prescribes that $h - \bar{h}$ must be (half-)integer for (semi-)local physical fields $\varphi_v(z, \bar{z})$. More abstractly, left- and right-chiral braid-group representations have to combine in such a way as to collapse to representations of the permutation group, which imposes relations between the (chiral) fusing and braiding matrices on the one hand and the D coefficients from equation (3.66) on the other.

That *associativity of the OPE* inside arbitrary correlation functions is an independent requirement will become obvious in the context of boundary CFT: boundary fields cannot be expected to be local, but associativity of the OPE still provides strong constraints. The basic non-linear sewing relation, which guarantees associativity, can be extracted by following the procedure used to define (chiral) fusing matrices, but now it has to be applied to non-chiral four-point functions $\mathcal{F}$ instead of chiral blocks. We evaluate $\mathcal{F}$ in the two regimes $|z_2 - z_3| \gg |z_3|$ and $|z_3| \gg |z_2 - z_3|$, which in particular involves different (sequences of) OPEs. With $z_2 = 1$, $z_3 = x$, we have

$$\mathcal{F}_{i_1,\ldots,i_4}(x, \bar{x}) = \sum_{p,\bar{p}} C_{i_1 i_2 p}\, C_{i_3 i_4 p^+}\, F^p_{i_1 i_2, i_3 i_4}(x)\, F^{\bar{p}}_{\bar{\imath}_1 \bar{\imath}_2, \bar{\imath}_3 \bar{\imath}_4}(\bar{x})$$

$$= \sum_{q,\bar{q}} C_{i_1 q i_4}\, C_{i_2 i_3 q^+}\, F^q_{i_1 i_4, i_2 i_3}(1 - x)\, F^{\bar{q}}_{\bar{\imath}_1 \bar{\imath}_4, \bar{\imath}_2 \bar{\imath}_3}(1 - \bar{x})\,.$$

This must hold because the physical correlators cannot depend on a choice of a fundamental system for the Ward identities. The conformal blocks for the different regimes are related by fusing matrices, and linear independence of the blocks allows the elimination all x-dependent parts. We arrive at non-linear equations for OPE coefficients, which involve fusing matrices for left- and right-chiral halves:

$$C_{i_1 i_2 p}\, C_{i_3 i_4 p^+} = \sum_{q,\bar{q}} \mathsf{F}_{pq}\left[\begin{smallmatrix} i_2 & i_3 \\ i_1 & i_4 \end{smallmatrix}\right] \mathsf{F}_{\bar{p}\bar{q}}\left[\begin{smallmatrix} \bar{i}_2 & \bar{i}_3 \\ \bar{i}_1 & \bar{i}_4 \end{smallmatrix}\right] C_{i_1 q i_4}\, C_{i_2 i_3 q^+} \,. \tag{3.67}$$

Note that here, and at times above, we have neglected indices α labelling different fusion channels if some $N_{ij}^k > 1$. They would lead to independent couplings $C_{ijk}^{(\alpha)}$ of the same conformal families in the OPE.

Explicit solutions for the OPE coefficients were obtained for several, fundamental classes of models, including Virasoro minimal models [142, 164] and the SU(2) WZW models [113, 370, 463], along with some non-rational CFTs like free bosons and Liouville theory [436].

3.3.2 Modular invariance

This condition provides further constraints, notably on the possible combinations of left- and right-moving CVOs that are allowed, and it is a very useful tool in attempting to classify conformal field theories with a given W-algebra. Usually, modular invariance is introduced as a condition on the partition function of the CFT on the torus – where it is understood that CFTs "living" on different surfaces arise from the same model if all local properties like OPEs are the same. As was argued, e.g. in [350], modular invariance together with the non-linear constraints reviewed above is sufficient to guarantee that a CFT can be defined on any closed Riemann surface of arbitrary genus. Roughly speaking, the reason is that any such surface can be "sewn together" from spheres with up to three punctures (hence the name "sewing" or "factorisation" relations for the non-linear constraints). Having extensions to higher Riemann surfaces at your disposal is important for applications to string theory. There, modular invariance was the main principle behind the construction of the heterotic string.

All the ingredients of the torus partition function are available in the plane CFT. A torus can be obtained from a cylinder if we identify the Euclidean times 0 and $2\pi\tau_2$, allowing for a simultaneous twist by $2\pi\tau_1$ in space direction. The terms τ_1 and τ_2 are real with $\tau_2 > 0$. The identifications imply that the Euclidean "vacuum functional", or *partition function*, of the theory is a trace of the corresponding translation operator,

$$Z(\tau,\bar{\tau}) = \mathrm{tr}_{\mathcal{H}}\, e^{-2\pi\tau_2 H + 2\pi i \tau_1 P} = \mathrm{tr}_{\mathcal{H}}\, q^{L_0 - \frac{c}{24}} \bar{q}^{\overline{L}_0 - \frac{c}{24}} \,,$$

where we have introduced $\tau = \tau_1 + i\tau_2$ and $q := e^{2\pi i \tau}$ along with their complex conjugates $\bar{\tau}$ and $\bar{q}$, and where we have expressed the cylinder Hamiltonian and momentum operator, respectively, in terms of Virasoro modes on the plane, $H = L_0 + \overline{L}_0 - \frac{c}{12}$ and $P = L_0 - \overline{L}_0$. The partition function can be expressed in terms of the previously mentioned conformal characters

$$\chi_i(q) := \mathrm{tr}_{\mathcal{H}_i} q^{L_0 - \frac{c}{24}} . \tag{3.68}$$

Recall that the conformal characters are generating functions for the dimensions of L_0 eigenspaces in a representation. In case the maximal abelian subalgebra of $\mathcal{W}$ contains further generators $W_0^{(i)}$, refined characters measuring the degeneracies of common eigenspaces can be introduced; we will encounter an example below. Right-moving analogues of $\chi_i(q)$ are defined using $\bar{q}$ and $\overline{L}_0$.

Exploiting the fact that the total state space $\mathcal{H}$ decomposes into tensor products of irreducibles, with $\mathcal{H}_i \otimes \mathcal{H}_{\bar{\imath}}$ occurring $M_{i,\bar{\imath}}$ times, we see that the partition function is bilinear in left- and right-moving characters:

$$Z(\tau, \bar{\tau}) = \sum_{(i,\bar{\imath}) \in I} M_{i,\bar{\imath}} \, \chi_i(q) \, \chi_{\bar{\imath}}(\bar{q}) . \tag{3.69}$$

The so-called "modular parameter" τ in the upper half-plane can be used to write the torus as $\mathbb{C}/\Gamma_\tau$ with the lattice $\Gamma_\tau = \mathbb{Z} + \tau\mathbb{Z}$. The torus stays the same if we pass to a different set of primitive periods in Γ_τ, i.e. if we act on τ with an element of the modular group $\mathrm{SL}(2, \mathbb{Z})$, $\tau \mapsto \frac{a\tau+b}{c\tau+d}$ with $a, b, c, d \in \mathbb{Z}$ and $ad - bd = 1$. (Note that both $\mathbf{1}_2$ and $-\mathbf{1}_2$ act trivially on τ, so that strictly speaking the modular group is the quotient of $\mathrm{SL}(2, \mathbb{Z})$ by $\mathbb{Z}_2$; we will ignore this subtlety here.)

Modular invariance simply states that the torus partition function must be independent of the specific choice of primitive periods used to parametrise that torus, i.e.

$$Z(\tau, \bar{\tau}) = Z\left(\frac{a\tau+b}{c\tau+d}, \frac{a\bar{\tau}+b}{c\bar{\tau}+d}\right) . \tag{3.70}$$

This condition becomes a very effective computational tool as soon as we have control over the behaviour of the conformal characters under modular transformations. It turns out that in rational CFTs, and also in many other interesting cases, the $\chi_i(q)$ transform linearly under $\mathrm{SL}(2, \mathbb{Z})$. We note that in [355] it was shown that such linear transformation laws even hold when arbitrary quasi-primary fields are inserted into the trace.

To formulate the constraints in a concise way, recall that the discrete group $\mathrm{SL}(2, \mathbb{Z})$ is generated by two elements S and T with the relations $S^2 = (ST)^3 = 1$. On τ, they act as $T : \tau \longmapsto \tau + 1$ and $S : \tau \longmapsto -\frac{1}{\tau}$. The T-action on the characters is diagonal,

$$\chi_i(q(\tau + 1)) = e^{2\pi i (h_i - \frac{c}{24})} \chi_i(q(\tau)) , \tag{3.71}$$

and modular invariance (3.70) implies that $M_{i,\bar{\imath}} = 0$ unless $h_i - h_{\bar{\imath}} \in \mathbb{Z}$: Modular invariance in its strict form would enforce that the total "spin" of any field in the CFT is an integer, thus it has to be slightly relaxed for fermionic systems, where only T^2 is to leave Z invariant.

The second generator S roughly corresponds to an interchange of the two coordinate axes of the torus (of worldsheet "time" and "space"), and its action on the characters is more complicated:

$$\chi_i\left(q\left(-\tfrac{1}{\tau}\right)\right) = \sum_j S_{ij} \, \chi_j(q(\tau)), \qquad (3.72)$$

with complex coefficients making up the so-called *modular S-matrix*. A rather general proof for transformation property (3.72) has been given by Zhu [464] in the context of vertex operator algebras. In unitary rational models, we find that

$$S^* = S^{-1}, \qquad S = S^{\mathrm{t}}, \qquad S^2 = C \qquad (3.73)$$

where $C_{ij} = \delta_{j,i^+}$ implements charge conjugation on the characters; see, e.g., [199].

Again, let us look at Virasoro minimal models to provide a concrete example (see, e.g., [99, 297]). With representations labelled by $i = (r,s)$ as before (such that $(1,1)$ is the vacuum sector), the S-matrix is

$$S_{(r_1,s_1),(r_2,s_2)} = \sqrt{\frac{8}{p'p}} \, (-1)^{r_1 s_2 + s_1 r_2 + 1} \, \sin\left(\frac{p r_1 r_2}{p'}\pi\right) \sin\left(\frac{p' s_1 s_2}{p}\pi\right).$$

The sine functions also occur in the modular S-matrix of SU(2) WZW models; see Chapter 6 for precise expressions.

Inserting the transformation (3.72) into the decomposition (3.69) of the partition function, and using the relations (3.73), we obtain the condition that the multiplicity matrix $M_{i,\bar{\imath}}$ has to commute with the modular S-matrix,

$$S\,M = M\,S\,,$$

which is a surprisingly powerful constraint in the classification of rational CFTs.

Two simple general solutions can be written down right away: given a left–right symmetric model whose chiral symmetry algebra $\mathcal{W} = \overline{\mathcal{W}}$ has only a finite list of irreducible representations $\mathcal{H}_i$, $i \in I_{\mathcal{W}}$, we can always work with the following two total spaces

$$\mathcal{H}^{\mathrm{diag}} = \bigoplus_{i \in I_{\mathcal{W}}} \mathcal{H}_i \otimes \mathcal{H}_i\,, \qquad \mathcal{H}^{\mathrm{ch.c.}} = \bigoplus_{i \in I_{\mathcal{W}}} \mathcal{H}_i \otimes \mathcal{H}_{i^+}\,, \qquad (3.74)$$

which lead to the so-called *diagonal* and *charge-conjugate* partition functions, respectively. To show that those are modular invariant, we use the properties (3.73) of S. Note that $\mathcal{H}^{\mathrm{diag}} = \mathcal{H}^{\mathrm{ch.c.}}$ if all representations of $\mathcal{W}$ are self-conjugate.

The relevance of (truly) non-diagonal left–right couplings was first advocated by Cardy [99]. The Virasoro algebra with, e.g., $c(5,6) = \frac{4}{5}$, admits a non-diagonal modular invariant which describes the critical three-state Potts model. Some of the irreducible representations from the Kac table (3.25) do not occur in Z_{Potts}, others contribute with multiplicity two. The existence of this modular invariant is connected with the fact that the symmetry algebra is not maximally extended: The spin three field from the list (3.25) for $c(5,6)$ generates a $\mathcal{W}(2,3)$-algebra as in equation (3.13).

If $\mathcal{W}$ and $\overline{\mathcal{W}}$ are maximal, then by definition the left- and right-moving vacuum sector can be paired only with the right- and left-moving vacuum sector, respectively. If the CFT is furthermore rational and unitary, it is rather easy to show that the possible modular invariant couplings all have the form [131]

$$\mathcal{H}^\tau = \bigoplus_{i \in I_\mathcal{W}} \mathcal{H}_i \otimes \mathcal{H}_{\tau(i)}, \tag{3.75}$$

where $\tau : I_\mathcal{W} \longrightarrow I_{\overline{\mathcal{W}}}$ is an isomorphism of the fusion rules of the maximal chiral algebras, $N_{ij}^k = N_{\tau(i)\tau(j)}^{\tau(k)}$, with $\tau(0) = 0$. This result holds for $\mathcal{W} \neq \overline{\mathcal{W}}$, too, and implies that $\overline{\mathcal{W}}$ must have as many sectors as $\mathcal{W}$. In particular, no left–right-coupling occurs more than once in equation (3.75).

In the maximally extended situation, the list of partition functions follows from the structure of the fusion ring. Typically, however, we are faced with the much more difficult task of finding the possible modular invariants for a whole series of CFTs like the unitary Virasoro minimal models with central charge $c(m, m+1)$, or the level k WZW models for some Lie algebra g, which admit non-trivial extensions only for certain values of m and k that may not be known a priori. In a classic work, Cappelli, Itzykson and Zuber gave a complete classification of modular invariants for SU(2) WZW and for Virasoro minimal models [97]. It turned out subsequently that the non-trivial invariants for $\widehat{SU(2)}_k$ can indeed be written as diagonal partition functions of extended W-algebras or in the form (3.75). An understanding of the somewhat surprising *ADE* pattern discovered in [97] was developed only later; see, e.g., [353] and also [66, 67], the latter works using operator algebraic techniques and attacking general WZW theories. The underlying structures are probably related to graph algebras [363], which, in physics, were first introduced in the context of two-dimensional integrable lattice models [369], before their potential for the classification of rational CFTs was uncovered in [122, 371, 465]. Meanwhile, breakthroughs towards the complete classification of modular invariants for SU(3) and higher WZW models have been achieved by Gannon [218], using more elaborate number-theoretic techniques.

We will encounter a few examples of non-diagonal modular invariants below, e.g. in Chapter 6 in the context of WZW models, and we will see that the orbifold construction (briefly reviewed in Section 4.A.2) provides a general method to obtain non-diagonal modular invariants from diagonal theories with a discrete group symmetry. But we should also refer back to the partition function (1.82) of a compactified free boson, which is perhaps the most natural theory with non-diagonal left–right coupling.

It turns out that there is a rather surprising connection between the modular S-matrix and the fusion rules, namely the so-called *Verlinde formula* [441]

$$N_{ij}^k = \sum_l \frac{S_{il}\,S_{jl}\,\overline{S}_{kl}}{S_{0l}}\,. \tag{3.76}$$

Summation runs over all sectors $l \in I_{\mathcal{W}}$ of the W-algebra, and as before "0" denotes the chiral vacuum sector.

Note that it is a relatively easy exercise in linear algebra to show that the matrix which diagonalises the fusion rules has the properties (3.73); see, e.g., [199, 210, 350, 392], and that N_{ij}^k can be expressed by an equation of the type (3.76). However, the fact that this diagonalisation matrix should coincide with the S-matrix describing the modular transformations of conformal characters is entirely non-trivial. Exploiting a pictorial presentation of the duality data, Verlinde devised ingenious arguments that equation (3.76) should hold for any rational CFT (with bosonic W-algebra) in [441]. A rigorous proof using some additional technical assumptions was given in the vertex operator algebra framework in [289].

Setting questions about the proof aside, let us remark that the Verlinde formula (3.76) will be of great practical use later, in the construction of boundary states. For this application, the relevant feature is that the specific combination of S-matrix elements on the right-hand side is a positive integer, thus providing an immediate solution to a basic non-linear constraint on boundary conditions – see Section 4.4.

The Verlinde formula states that the modular S-matrix diagonalises the fusion rules, so in particular all eigenvalues of the fusion matrices N_i can be expressed through S. The so-called *quantum dimension* D_i of a conformal family $\mathcal{H}_i$ is the maximal eigenvalue of N_i (which exists and is unique due to the Perron–Frobenius theorem). The dimension D_i measures the "asymptotic size" of the W-module $\mathcal{H}_i$ relative to the vacuum module, more precisely

$$D_i = \lim_{q \to 1^-} \frac{\chi_i(q)}{\chi_0(q)} = \frac{S_{i0}}{S_{00}}\,; \tag{3.77}$$

the last equality follows directly from equation (3.72) as long as the CFT is unitary. The D_i form a representation of the fusion rules.

Let us mention that (at least in a class of CFTs including WZW theories and Virasoro minimal models) the $q \to 1$ asymptotics and modular transformations of characters can also be used to derive unexpected expressions for the central charge and the highest weights in terms of the dilogarithm function – see, e.g., [358] and the subsequent generalisations in [311, 312]. These dilogarithm identities can also be obtained as critical limits of thermodynamic Bethe ansatz (TBA) relations which hold in (massive) integrable field theories; they also hint at a deeper connection between CFT in two dimensions and the geometry of hyperbolic three-manifolds, but the precise interplay remains to be understood.

Summarising, the constraints from locality and associativity of the OPE and from modular invariance provide non-linear constraints on the structure constants of a CFT. This set of equations is over-determined, but on the other hand there may exist several consistent theories for given left- and right-moving chiral algebras and fixed modular invariant partition function. Up to now, no general (model-independent) solution to the sewing constraints is known. In this sense, many of the CFT models dealt with in the literature are still "under construction" because not all constraints have been checked. Truly constructive approaches (e.g. based on rigorous functional integral methods which ensure consistency under sewing) exist only for a very limited number of cases. But even without a complete solution to all sewing constraints at hand, there are often good reasons for the existence of a certain model, e.g. due to relations to integrable field theories or lattice models. Moreover, new tools to study CFTs are still being discovered, and, for example, the investigation of boundary conditions seems to shed new light also on the structure of CFTs on the plane – see, e.g., the remarks in Section 4.4 in the next chapter.

4

Boundary conformal field theory

In this chapter, we will introduce conformal field theory (CFT) on surfaces with boundaries and lay the technical foundations for the remainder of this text. As with CFT on surfaces without boundary, we can in principle take different approaches to boundary CFT, but from our point of view, and for applications to string theory, the BPZ or "representation theoretic" picture involving W-algebras and operators on Hilbert spaces provides the most effective, explicit and flexible way to tackle boundary conditions in CFT.

The whole subject of boundary CFT was founded by Cardy, who almost single-handedly pushed its development very far. His works [98–100] still provide an excellent introduction, together with papers by Ishibashi [294] and [102, 334] by Cardy and Lewellen. In the late 1980s, Sagnotti realised the potential of boundary CFT for the description of open strings in arbitrary backgrounds [399], and later constructed many models, see, e.g., [53–55, 383, 384, 401], using Cardy's methods as well as additional constraints pertinent to open string theory. The latter are tadpole cancellation conditions involving non-orientable worldsheets – which have to be taken into account in type I string theory, but not necessarily when discussing D-branes. Since those will be our main focus in this book, we will almost everywhere restrict ourselves to orientable worldsheets – and mainly to the lowest genus case, at that: our boundary CFTs will be defined on the disk, on a strip or simply on the upper half-plane.

What was not realised right away was that it would have taken just a slight generalisation of Cardy's setup (namely studying more general gluing conditions) to discover D-branes as "non-perturbative sectors" of string theory, via the world-sheet approach. After Polchinski's ground-breaking paper [375], D-branes triggered a lot of work on conformal boundary conditions in string theory – although most of these works [34, 49, 51, 52, 57, 87, 123, 175, 248, 249, 292, 335, 425] stick to the coherent states that are special to free theories and that were used earlier by Callan, Polchinski and their collaborators [89, 378]. There are other articles where a more general point of view is taken, admitting also applications to "interacting" string backgrounds, notably [205, 364, 389–391, 427, 430]. In following years, new results on structural questions have been obtained, e.g. in [44, 59, 162, 163, 207, 390, 397, 398], and also on the connection of boundary CFT and integrable lattice models [7, 42–44, 206], a subject that was started already in [39].

The logic behind our exposition is as follows: if we try to restrict a ("bulk") CFT from the full plane to the upper half-plane, symmetries are lost (most obviously translations perpendicular to the boundary); trying to restore as much symmetry as possible, we find that left- and right-moving symmetry generators (most importantly $T(z)$ and $\overline{T}(\bar{z})$) can no longer be independent of each other along the boundary, but need to satisfy so-called gluing conditions. These will ensure that one copy of the Virasoro (and perhaps of an extended) symmetry still acts in the theory on the half-plane, but the Ward identities differ from the original ones. One consequence is that non-trivial primary fields $\varphi_{i,\bar{\imath}}(z,\bar{z})$ can have non-vanishing one-point functions in the boundary theory – see Section 4.1.

Another consequence of the modified Ward identities is that bulk field correlation functions have singularities not only when two arguments coincide, but also at the boundary. As is standard in quantum field theory (QFT), such singularities hint at the existence of extra fields, the boundary fields (introduced in Section 4.2). While the "bulk fields", i.e. the field operators from the original "bulk" CFT, depend on $(z,\bar{z})$ as before, the arguments of boundary fields are confined to the boundary. In the string theory context, boundary fields correspond to open string excitations, while the bulk fields are associated with closed string modes, as we have already seen in Chapters 1 and 2.

Section 4.3 will introduce the boundary state formalism, which provides very efficient tools for the construction of conformally invariant boundary conditions starting from quantities from the original bulk CFT on the full plane. In this formalism, gluing conditions are implemented in the form of so-called Ishibashi states, objects that generalise the coherent states we encountered in the special case of free bosons and fermions in Chapter 1. Boundary states are linear combinations of Ishibashi states; a choice of coefficients corresponds to a choice of bulk field one-point functions. We will give two (equivalent) definitions of boundary states in Section 4.3; one of them involves finite-temperature correlators and allows direct contact to be made to the partition function of the boundary CFT (and the computation of the spectrum of boundary fields).

Combining this interpretation of finite-temperature correlators as partition functions with ideas from representation theory of conformal symmetry algebras, in particular the modular properties of characters, leads to non-linear constraints on the one-point function coefficients: these constraints are known as Cardy conditions and will be discussed in Subsection 4.4.1. The remainder of Section 4.4 deals with other non-linear constraints on coefficients left undetermined by the Ward identities, namely sewing relations that arise from demanding associativity of operator product expansions – here meaning operator product expansions (OPEs) of bulk and of boundary fields. We will in particular note the following result: suppose we have an arbitrary rational bulk CFT with diagonal (or charge-conjugate) modular invariant bulk partition function, and that we chose so-called trivial gluing conditions (the simplest type, see below); then there exists a special class of boundary states (called Cardy boundary states) and a choice of OPE coefficients (often referred to as Runkel's solutions) which satisfy all the non-linear constraints and thus define a consistent boundary CFT.

However, we often need to consider more general situations, with different types of bulk partition functions or different gluing conditions. Then, we need to look "beyond the Cardy case" and try to modify the formulas for Cardy boundary states or even try to construct consistent boundary conditions from scratch. We collect some material on such cases in the additional material section, which considers permutation branes, orbifold theories and also how to construct the complete set of conformal boundary conditions for the free boson theory. Moreover, in Subsection 4.A.1 we briefly discuss orientifolds, which involve a different kind of gluing condition necessary for the study of non-oriented strings.

4.1 Defining data of a conformal boundary condition

Throughout most of this text, the definition of a boundary CFT will start from a given CFT on the complex plane, which we assume to be known in as much detail as necessary – and which is then *restricted to the upper half-plane* $\mathbb{H} = \{\, z \in \mathbb{C} \,|\, \Im z \geq 0 \,\}$. For reasons that will become obvious in Subsection 4.1.1, we assume that the left- and right-moving symmetry algebra of this "parent" CFT coincide, $\mathcal{W}_L = \mathcal{W}_R = \mathcal{W}$. Its state space $\mathcal{H}^{(P)}$ – the superscript $^{(P)}$ indicates the plane – then decomposes into a sum of $\mathcal{W} \times \mathcal{W}$ irreducible representations $\mathcal{H}_i \otimes \mathcal{H}_{\bar{\imath}}$, with fields $\Phi^{(P)}(v; z, \bar{z})$ associated with vectors $v \in \mathcal{H}^{(P)}$ as usual. Often, we will only display primary fields $\varphi_{i,\bar{\imath}}(z, \bar{z})$ for the sake of notational convenience, even if a statement holds for non-primary fields as well.

The local properties in the interior of the upper half-plane are inherited from the parent CFT. In particular, the singularity structure of correlation functions of $\varphi_{i,\bar{\imath}}(z, \bar{z})$ for $\Im z > 0$ is determined by the OPE of those bulk fields. What does change, on the other hand, are global features like states. In particular, in the boundary CFT the space $\mathcal{H}^{(P)}$ is nothing more than an "index set" for the set of bulk fields $\Phi^{(P)}(v; z, \bar{z})$; it is no longer the state space in the theory with-boundary.

The global properties of the "descendant" theory on the upper half-plane are encoded in the boundary conditions that are imposed along the real line. These are naturally divided into two kinds of data: first, we shall discuss "gluing conditions", linear constraints on which the symmetries of the boundary CFT depend. In Subsection 4.1.2, we turn towards the remaining freedom in bulk field correlators, namely coefficients of one-point functions. These can be non-trivial only in the presence of a boundary, and they are subject to certain constraints.

4.1.1 Gluing conditions and Ward identities

Restricting a CFT on the full plane to the upper half-plane $\mathbb{H}$ inevitably reduces the amount of symmetry realised in the theory. Translations orthogonal to the boundary, for example, are no longer a legitimate worldsheet transformation. Conformal maps $f : \mathbb{H} \longrightarrow \mathbb{H}$ which respect the boundary,

$$f(x) \in \mathbb{R} \quad \text{for} \quad x \in \mathbb{R}, \tag{4.1}$$

should, however, remain symmetries of the theory on $\mathbb{H}$, if it is to be called a boundary *conformal* field theory. These maps include the globally defined Möbius transformations $\mathrm{SL}(2, \mathbb{R})$ but, as we did for CFTs on the plane, we also admit transformations that have isolated singularities and generate an infinite-dimensional Virasoro symmetry.

Exploiting the fact that the energy-momentum tensor implements coordinate transformations, it is not too difficult to show that the reality condition (4.1) simply means that no energy flows across the boundary [98], i.e. $T_{xy}(x, 0) = 0$, where the Cartesian coordinates x and y on $\mathbb{H}$ are the real and imaginary parts of $z = x + iy$. Note that on higher genus "open" Riemann surfaces, analogous conditions on T have to hold along the boundaries, in appropriate local coordinates.

Switching to variables $z, \bar{z}$, the constraint $T_{xy}(x, 0) = 0$ becomes

$$T(z) = \overline{T}(\bar{z}) \quad \text{for} \quad z = \bar{z}, \tag{4.2}$$

which we will refer to as the *gluing condition* for the energy-momentum tensor because it links left- and right-moving components of $T(z, \bar{z})$ – which, on the full plane, were completely independent.

Relation (4.2) makes it possible to define the following set of modes (we will often use the superscript $^{(H)}$ to distinguish quantities in the boundary CFT on the upper half-plane):

$$L_n^{(H)} := \frac{1}{2\pi i} \int_C z^{n+1} T(z) \, dz - \frac{1}{2\pi i} \int_C \bar{z}^{n+1} \overline{T}(\bar{z}) \, d\bar{z}. \tag{4.3}$$

They span a *single* copy of the Virasoro algebra with the same central charge as in the parent CFT. Above, C is a semi-circle in $\mathbb{H}$ with ends on the real line. The circle around the origin, which occurred in the definition of L_n, $\overline{L}_n$ on the full plane, is no longer available, but can be replaced by the difference of two integrals over semi-circles as long as the integrands coincide along the real interval that would close the contours.

Obviously, equation (4.2) can be regarded as prescribing an analytic continuation of $T(z)$ from the upper into the lower half-plane. The "field"

$$\mathsf{T}(z) := \sum_{n \in \mathbb{Z}} L_n^{(H)} z^{-n-2} = \begin{cases} T(z) & \text{for} \quad \Im z \geq 0 \\ \overline{T}(\bar{z}) & \text{for} \quad \Im z < 0 \end{cases} \tag{4.4}$$

has the same algebraic characteristics as one chiral half of the energy-momentum tensor of a CFT on the full plane.

The commutation relations of the modes $L_n^{(H)}$ with primary bulk fields follow from (4.3) and the given OPE of $T(z)$ and $\overline{T}(\bar{z})$ with $\varphi_{i,\bar{i}}(z,\bar{z})$ in the interior of the upper half-plane:

$$[L_n^{(H)}, \varphi_{i,\bar{i}}(z,\bar{z})] = z^n(z\partial + h_i(n+1))\,\varphi_{i,\bar{i}}(z,\bar{z}) + \bar{z}^n(\bar{z}\overline{\partial} + h_{\bar{i}}(n+1))\,\varphi_{i,\bar{i}}(z,\bar{z})\,.$$

(4.5)

Observe that the new modes are sensitive to the dependence of bulk fields on both z and $\bar{z}$. In boundary CFT correlation functions, $\varphi_{i,\bar{i}}(z,\bar{z})$ behaves like a product of two chiral vertex operators with conformal weights h_i and $h_{\bar{i}}$ inserted at z and $\bar{z}$, respectively, as far as Ward identities associated with the Virasoro algebra are concerned.

Imposing the gluing condition (4.2) on the energy-momentum tensor is an indispensable ingredient of specifying a conformal boundary condition along the real line. If the symmetry algebra $\mathcal{W} \times \mathcal{W}$ of the bulk theory contains further chiral generators $W(z)$, $\overline{W}(\bar{z})$ besides the energy-momentum tensor, we may optionally wish to ensure that the boundary condition respects (all or part of) this extended symmetry, as well. This can be achieved by prescribing an analytic continuation as for $T(z)$. However, the symmetries associated with the additional $\mathcal{W}$-generators do not possess an equally fundamental meaning in terms of worldsheet geometry as those generated by $T(z)$. Indeed, it pays to work with a more general gluing condition, namely

$$W(z) = \Omega(\overline{W})(\bar{z}) \qquad \text{for } z = \bar{z}$$

(4.6)

where $\Omega : \mathcal{W} \longrightarrow \mathcal{W}$ is a local *automorphism* of the chiral symmetry algebra, which leaves the energy-momentum tensor fixed, $\Omega T = T$. By "local" we mean that Ω acts point-wise, $(\Omega W)(z) = \Omega(W(z))$, so that it commutes with the mode expansion. As a W-algebra automorphism, Ω must be compatible with all W-operations such as taking commutators or derivatives and forming normal-ordered products; therefore, it is sufficient to specify the action of Ω on simple fields. It is also convenient to assume that Ω commutes with the CPT operator. Gluing conditions with $\Omega = \mathrm{id}$ are called "trivial".

For later purposes, note that Ω allows $\mathcal{H}_i$ to carry another irreducible representation where elements $A \in \mathcal{W}$ act on states through

$$\pi_i^\Omega(A)|h\rangle := \pi_i(\Omega A)|h\rangle\,.$$

(4.7)

The space $\mathcal{H}_i$ equipped with this new action of $\mathcal{W}$ is isomorphic (as a $\mathcal{W}$-module) to some unique representation space $\mathcal{H}_{\omega(i)}$; schematically,

$$\pi_i \circ \Omega \simeq \pi_{\omega(i)}\,.$$

(4.8)

If Ω is outer, $\omega(i)$ is inequivalent to i.

Once a prescription (4.6) for an analytic continuation is fixed, we can proceed as for the Virasoro field and construct modes

$$W_n^{(H)} := \frac{1}{2\pi i} \int_C z^{n+h_W-1} W(z)\, dz - \frac{1}{2\pi i} \int_C \bar{z}^{n+h_W-1} \Omega \overline{W}(\bar{z})\, d\bar{z}, \qquad (4.9)$$

which act in the CFT on the upper half-plane. The properties of the automorphism Ω ensure that the $W_n^{(H)}$ generate a (single) copy of the symmetry algebra $\mathcal{W}$.

We can again combine the chiral fields $W(z)$ and $\Omega \overline{W}(\bar{z})$ into a single object $\mathsf{W}(z)$, which is defined on the whole complex plane by

$$\mathsf{W}(z) := \begin{cases} W(z) & \text{for } \Im z \geq 0, \\ \Omega \overline{W}(\bar{z}) & \text{for } \Im z < 0. \end{cases}$$

Because of the gluing condition along the boundary, this field is analytic and we can expand it in a Laurent series $\mathsf{W}(z) = \sum_n W_n^{(H)} z^{-n-h}$, thereby recovering the modes (4.9).

The action of $\mathcal{W}$ leads to Ward identities for correlation functions of the boundary theory. They follow directly from the singular parts of the operator product expansions of the field W with the bulk fields $\varphi_v(z,\bar{z}) = \Phi^{(P)}(v; z, \bar{z})$ inherited from the bulk theory, cf. the discussion in Section 1.1. With the help of the state-field correspondence, this OPE can be written in the form

$$\mathsf{W}(w)\varphi_v(z,\bar{z}) \sim \sum_{n > -h_W} \left(\frac{1}{(w-z)^{n+h_W}} \Phi^{(P)}\left(W_n^{(P)} v; z, \bar{z}\right) \right.$$

$$\left. + \frac{1}{(w-\bar{z})^{n+h_W}} \Phi^{(P)}\left(\Omega \overline{W}_n^{(P)} v; z, \bar{z}\right) \right). \qquad (4.10)$$

Here, the symbol $\sim$ means that regular terms are suppressed on the right-hand side; we have placed a superscript $^{(P)}$ on the modes $W_n, \overline{W}_n$ to display clearly that they act on the elements $v \in \mathcal{H}^{(P)}$ labelling the bulk fields in the theory.

The sum on the right-hand side of equation (4.10) is actually finite because v is annihilated by all modes $W_m, \overline{W}_m$ with sufficiently large m. For $\Im w > 0$, only the terms involving W_n can become singular and the singularities agree with the singular part of the OPE between $W(w)$ and $\varphi_v(z,\bar{z})$ in the bulk theory. Similarly, the singular part of the OPE between $\Omega \overline{W}(w)$ and $\varphi_v(z,\bar{z})$ in the bulk theory is reproduced by the terms which contain $\overline{W}_n$ and can become singular if $\Im w < 0$.

In complete analogy to CFTs on the plane, cf. Section 3.2, equation (4.10) yields Ward identities for arbitrary N-point functions of bulk fields $\varphi_{i_l,\bar{i}_l}(z_l, \bar{z}_l)$ for $l = 1, \ldots, N$. Notice, however, that the Ward identities in boundary CFTs depend on the gluing automorphism Ω since $\mathsf{W}(z)$ and the modes (4.9) do. Furthermore, equation (4.10) shows that the Ward identities for N-point functions in the presence of a boundary have the same form as those for *chiral conformal blocks in a bulk CFT with $2N$ insertions* of chiral vertex operators carrying charges $i_1, \ldots, i_N, \omega(\bar{i}_1), \ldots, \omega(\bar{i}_N)$, where $\omega(\bar{i})$ is the representation induced on

Figure 4.1 In the presence of a boundary with gluing automorphism Ω, a bulk field $\varphi_{i,\bar{\imath}}(z,\bar{z})$ "splits" into a charge i at the point z in the upper half-plane and an image charge $\omega(\bar{\imath})$ at $\bar{z}$ on the other side of the boundary.

$\mathcal{H}_{\bar{\imath}}$ by Ω, see equation (4.8). This phenomenon sometimes goes under the name "doubling trick"; it was first observed in [98] for the $\Omega = \mathrm{id}$ case and in [389] for general gluing conditions. In many concrete examples, there are rather explicit expressions for the chiral blocks in question. In particular, we see that objects familiar from the construction of bulk CFTs can be used as building blocks of correlators in the boundary theory.

The intuitive picture behind the boundary Ward identities is that each bulk field $\varphi_{i,\bar{\imath}}$, when placed near a boundary, splits into a chiral charge i in front of the boundary and a chiral *mirror charge* $\omega(\imath)$ behind the looking glass; see Figure 4.1.

In early investigations of boundary CFT and also in applications to string theory, the possibility of non-trivial gluing automorphisms Ω was overlooked or ignored. As we have seen before, it is precisely such gluing conditions that are connected with the existence of D-branes.

As was hinted at above, there are more general boundary conditions for a $\mathcal{W} \times \mathcal{W}$ bulk theory than those covered by equation (4.6). While the gluing conditions for the energy-momentum tensor are indispensable if the boundary theory is to be conformal, boundary conditions which break the extended $\mathcal{W}$-symmetry can be admitted. For example, gluing conditions can be required only for some proper sub-$\mathcal{W}$-algebra $\mathcal{W}^{\mathrm{open}} \subset \mathcal{W}$, which must contain the Virasoro algebra.

While this may seem artificial at first sight, boundary CFTs with reduced symmetry sometimes arise naturally as the result of certain manipulations of maximally symmetric boundary conditions, such as marginal perturbations, or superpositions of several boundary conditions.

The reduction of symmetry poses no enormous difficulties in the rather special case where the theory is still rational with respect to $\mathcal{W}^{\mathrm{open}}$; see [59, 207]. In general, however, the boundary CFT may be non-rational with respect to the smaller symmetry algebra $\mathcal{W}^{\mathrm{op}}$ even if the original bulk CFT is rational, and there is indeed no systematic and case-independent procedure for the construction of arbitrary (merely conformal) boundary conditions associated with a bulk theory with extended $\mathcal{W}$-symmetry.

Let us collect some *examples for gluing conditions* of W-algebras that will play a role later – or have already appeared in Chapter 1.

The free boson CFT is governed by the $\widehat{U(1)}$ current algebra, whether the target is compactified or not. The basic OPE of the currents or, equivalently, the canonical commutation relations (3.8) of the modes a_n, allow for one non-trivial automorphism $\Omega_D := -\operatorname{id}$, i.e.

$$\bar{J}(\bar{z}) \longmapsto \Omega_D \bar{J}(\bar{z}) = -\bar{J}(\bar{z}),$$

just as in Chapter 1, equation (1.9). Expressing currents through the boson, we obtain

$$J(z) = -\bar{J}(\bar{z}) \quad \Longleftrightarrow \quad \partial_\| X(z,\bar{z}) = 0 \tag{4.11}$$

for $z = \bar{z}$. This means that X satisfies Dirichlet conditions $X(z,\bar{z}) = x_0 = \text{const}$ along the real line. In contrast, the trivial automorphism $\Omega_N = \operatorname{id}$ leads to Neumann conditions for X,

$$J(z) = \bar{J}(\bar{z}) \quad \Longleftrightarrow \quad \partial_\perp X(z,\bar{z}) = 0, \tag{4.12}$$

which coincides with equation (1.3) from Chapter 1 for vanishing B-field. Since by the Sugawara construction, the energy-momentum tensor is quadratic in the currents, T is invariant under Ω_D as well as Ω_N.

This observation is rather trivial for the $\widehat{U(1)}$ current algebra, but the Sugawara construction and invariance of T prove to be very selective in the non-abelian case. It can be shown that any automorphism of an affine Lie algebra $\hat{g}_k$ which leaves T fixed comes from an automorphism of the underlying finite-dimensional Lie algebra g; in terms of modes, we have $\Omega J_n^a = \Omega_{ab} J_n^b$.

For $g = su(2)$, only inner automorphisms are available, i.e.

$$\Omega J_n^a = \operatorname{ad}_g(J^a) := g\, J_n^a\, g^{-1} \quad \text{for some} \quad g = e^{i\lambda_a J_0^a} \in SU(2). \tag{4.13}$$

Such twists will occur naturally in our study of marginal perturbations, Chapter 5. Other Lie algebras admit outer automorphisms as well, the simplest example being $\widehat{U(1)}^d$ and the rotated branes in a d-dimensional flat target as discussed in Section 1.A.2.

As in the case of the abelian current algebra, it is possible to find a geometric interpretation of the WZW gluing conditions in terms of sigma models with group targets [14]. This will be reviewed in Chapter 6.

In the worldsheet approach to D-branes in superstring theory, we will inevitably need to consider W-algebras with *fermionic generators* and their gluing conditions. The free fermion admits the same local automorphisms $\Omega = \pm\operatorname{id}$ as the free boson, cf. Chapter 1.

The $N = 2$ super Virasoro algebra (3.16) is more interesting. There are two types [364] of gluing conditions for the chiral fields $G^\pm, J, T$,

$$\text{A-type}: \quad J(z) = -\bar{J}(\bar{z}), \quad G^\pm(z) = \eta\, \bar{G}^\mp(\bar{z}), \tag{4.14}$$

$$\text{B-type}: \quad J(z) = \bar{J}(\bar{z}), \quad G^\pm(z) = \eta\, \bar{G}^\pm(\bar{z}), \tag{4.15}$$

supplemented by $T = \overline{T}$ in both cases. (The "A,B" nomenclature stems from similarities with topologically twisted $N = 2$ theories.)

All algebraic relations are preserved as long as the complex parameter η is a pure phase, but in string theory applications one requires that an $N = 1$ subalgebra with generating supercurrent $G(z) := \frac{1}{\sqrt{2}}(G^+(z) + G^-(z))$ or $G'(z) := \frac{i}{\sqrt{2}}(G^+(z) - G^-(z))$ is preserved by the boundary condition – so as to have unbroken spacetime supersymmetry. This leaves us with the choice $\eta = \pm 1$. If we admit the "twisted" gluing conditions $G(z) = \overline{G}'(\bar{z})$, then we can take η to be a fourth root of unity.

For $\eta = 1$, it is seen that B-type gluing conditions are trivial, i.e. $\Omega = \mathrm{id}$, while A-type conditions involve $\Omega = \Omega_M$, the so-called *mirror automorphism*

$$\Omega_M : J \longmapsto -J, \quad G^\pm \longmapsto G^\mp \tag{4.16}$$

of the $N = 2$ super Virasoro algebra. The existence of this map has far-reaching consequences in mathematics (triggering the mirror symmetry programme in various guises). In the present context, suffice it to remark that Ω_M is the obvious extension of the *T-duality map* $\Omega_D : J \mapsto -J$ of the free boson theory: both are discrete symmetries of the respective CFTs and are responsible for non-classical features of stringy geometry, namely for the "minimal length" a string can resolve, and for mirror symmetry with all its consequences such as topology changes, etc. (see, e.g., [253] for more details). Thus it turns out that non-classical symmetries are intimately related to the existence of non-trivial D-branes in string theory.

4.1.2 One-point functions

The choice of gluing conditions for the symmetry algebra does not, in general, specify the boundary CFT associated with a given bulk CFT uniquely. In terms of bulk field correlation functions, the Ward identities (determined by Ω) together with the coefficients of the OPE in the bulk allow us to reduce the computation of correlators involving N bulk fields to the evaluation of *one*-point functions $\langle \varphi_{i,\bar{i}}(z, \bar{z}) \rangle_\alpha$. On the plane, these vanish for $\varphi_{i,\bar{i}} \neq 1$ but, as translational invariance along the imaginary axis is broken, this is no longer true for a theory on the upper half-plane.

To control the amount of the additional freedom, we first note the transformation properties of a primary $\varphi_{i,\bar{i}}$ with respect to $L_n^{(H)}$, $n = 0, \pm 1$, and the zero-modes $W_0^{(H)}$,

$$[W_0^{(H)}, \varphi_{i,\bar{i}}(z, \bar{z})] = X_W^i \, \varphi_{i,\bar{i}}(z, \bar{z}) - \varphi_{i,\bar{i}}(z, \bar{z}) \, X_{\Omega \overline{W}}^{\bar{i}},$$
$$[L_n^{(H)}, \varphi_{i,\bar{i}}(z, \bar{z})] = z^n (z\partial + h_i(n+1))\varphi_{i,\bar{i}}(z, \bar{z}) + \bar{z}^n(\bar{z}\overline{\partial} + h_{\bar{i}}(n+1))\varphi_{i,\bar{i}}(z, \bar{z}),$$

where X_W^i denotes the linear map induced by W_0 in $\mathcal{H}_i^{(0)}$, the lowest-energy subspace of the irreducible representation $\mathcal{H}_i$ of $\mathcal{W}$. These commutator relations determine the one-point functions up to scalar factors in a rather elementary way: we insert the commutators into the vacuum expectation values and use the fact that the vacuum is annihilated by the modes in question; the commutator with $L_{-1}^{(H)}$ then yields a differential equation which implies that the one-point functions depend only on $z - \bar{z}$; the $L_0^{(H)}$ commutator determines a power law $(z - \bar{z})^{h_i + h_{\bar{i}}}$; the commutator with $L_1^{(H)}$ demands $h_i = h_{\bar{i}}$. All in all, we find that $\langle \varphi_{i,\bar{i}} \rangle_\alpha$ must be of the form

$$\langle \varphi_{i,\bar{i}}(z, \bar{z}) \rangle_\alpha = \frac{A_{i\bar{i}}^\alpha}{|z - \bar{z}|^{h_i + h_{\bar{i}}}} \, \delta_{\bar{i}, \omega^{-1}(i^+)} \tag{4.17}$$

where the linear map $A_{i\bar{i}}^\alpha : \mathcal{H}_{\bar{i}}^{(0)} \longrightarrow \mathcal{H}_i^{(0)}$ obeys

$$X_W^i \, A_{i\bar{i}}^\alpha = A_{i\bar{i}}^\alpha \, X_{\Omega\overline{W}}^{\bar{i}} . \tag{4.18}$$

The intertwining relation (4.18) also implies

$$\bar{i} = \omega^{-1}(i^+) \tag{4.19}$$

as a necessary condition for having a non-vanishing one-point function (i^+ is the representation conjugate to i). The image charge picture, together with the bulk–boundary OPE (see below) and translation invariance along the real axis, provides another explanation for the Kronecker symbol. As a consequence, we can put $h_i + h_{\bar{i}} = 2h_i$ in the exponent in equation (4.17) because the gluing map acts trivially on the Virasoro field.

Irreducibility of the zero-mode representations on the subspaces $\mathcal{H}_i^{(0)}$ and Schur's lemma imply that each matrix $A_{i\bar{i}}^\alpha$ is determined up to *one scalar factor*. Hence, if there exist several boundary conditions associated with the same bulk theory and the same gluing map Ω, they can differ only by these scalar parameters in the one-point functions. The label α (which is possibly continuous) is introduced to distinguish different choices for these parameters. In the following, we will switch freely between different notations for the set of data fixing a boundary condition and write $(\Omega, \tilde{\alpha})$, $\Omega\tilde{\alpha}$ or simply α for those data, hoping that there will be no confusion.

Once we know the $A_{i\bar{i}}^\alpha$ and the gluing automorphism, we can, in principle, apply the OPE of bulk fields and compute all correlation functions $\langle \varphi_1 \dots \varphi_N \rangle$ of bulk fields in the boundary theory; see Figure 4.2. We will see below that the partition function $Z_\alpha(q)$ of the boundary theory can also be expressed in terms of the coefficients $A_{i\bar{i}}^\alpha$.

One might suspect that, for reasons of consistency, the parameters in the one-point functions cannot be chosen at will, and in fact there are strong sewing constraints of various kinds which have been worked out by several authors [44, 100, 102, 162, 163, 204, 334, 384, 397, 398]. It turns out that not even the one-point function A_{00}^α of the identity operator can always be normalised to unity as might be expected. We will say more about these conditions in Section 4.4.

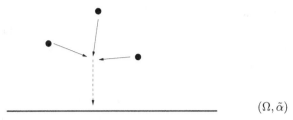

$(\Omega, \tilde{\alpha})$

Figure 4.2 Computing a three-point function $\langle \varphi_1 \varphi_2 \varphi_3 \rangle$ by first performing bulk OPEs and then evaluating the one-point functions in the presence of the boundary.

Let us once more turn to the example of free boson theories. Here, the $\widehat{U(1)}$ primary bulk fields can be written as exponentials of the free boson, and the gluing conditions for the currents (4.11, 4.12) can be integrated to obtain the boundary behaviour of $X(z, \bar{z})$ itself. Therefore, the general formulas (4.17) and (4.19) are strong enough to determine all one-point functions up to a common prefactor.

Primary fields in a $\widehat{U(1)} \times \widehat{U(1)}$ theory are labelled by their left- and right-moving charges g_L and g_R, respectively, which appear as the representation labels i and $\bar{\imath}$ in the general formula (4.17) for the one-point functions. Conjugation reverses the charges, $g^+ = -g$. Furthermore, the gluing automorphisms Ω_N and Ω_D induce the maps

$$\omega_N(g) = g, \qquad \omega_D(g) = -g$$

of the sectors. Thus, the Kronecker delta on the right-hand side of equation (4.17) implies that the one-point functions of primary fields vanish unless

$$g_L = g_R \qquad \text{for Dirichlet gluing conditions},$$
$$g_L = -g_R \qquad \text{for Neumann gluing conditions}. \qquad (4.20)$$

When dealing with an uncompactified boson, the primary vertex operators

$$\varphi_g(z, \bar{z}) = {:}\, e^{ig X(z, \bar{z})} {:} \qquad (4.21)$$

are labelled by a single real U(1) charge g, so we always have $g_L = g_R = g$. Therefore, there are no non-trivial one-point functions for Neumann boundary conditions. The Dirichlet gluing condition (4.11), on the other hand, is "solved" iff the free boson satisfies

$$X(z, \bar{z}) = x_0 \quad \text{for} \quad z = \bar{z} \qquad (4.22)$$

for some constant $x_0 \in \mathbb{R}$. We insert this into the expression (4.21) to obtain the relative residue in the one-point functions, leading to

$$\langle \varphi_g(z, \bar{z}) \rangle_{(\Omega_D, x_0)} = \kappa_D \, \frac{e^{i x_0 g}}{|z - \bar{z}|^{g^2}}, \qquad (4.23)$$

$$\langle \varphi_g(z, \bar{z}) \rangle_{\Omega_N} = \kappa_N \, \delta_{g,0}. \qquad (4.24)$$

The z-dependence follows from conformal invariance and $h = \frac{g^2}{2}$, choosing $\alpha' = 2$. The absolute normalisation of the correlators – which is simply the expectation value of the identity in the presence of the respective boundary condition – remains undetermined for the moment. We see that the general formalism reproduces the results from Chapter 1, as expected.

If the free boson takes values in a circle of radius r, the spectrum changes: the $\widehat{U(1)}$ highest weight vectors $|(k,w)\rangle$ are labelled by integer momentum and winding numbers, and their left- and right-moving charges $g_{L,R}(k,w)$ lie in a two-dimensional lattice; see equation (1.81). The primary bulk fields associated with these states are – using the convention $\alpha' = \frac{1}{2}$, which is more standard for the compact target case –

$$\varphi_{g_L, g_R}(z, \bar{z}) = \; : e^{2ig_L X_L(z) + 2ig_R X_R(\bar{z})} : \tag{4.25}$$

with left- and right-moving conformal dimensions $h_{L,R} = \frac{1}{2} g_{L,R}^2$, cf. equation (1.31).

Dirichlet gluing conditions have the same form as for the uncompactified theory, but now a continuous parameter shows up in the integrated form of Neumann conditions, too: Neumann conditions on $X(z, \bar{z}) = X_L(z) + X_R(\bar{z})$ are equivalent to Dirichlet conditions for the "dual coordinate" $X'(z, \bar{z}) := X_L(z) - X_R(\bar{z})$, so that condition (4.12) holds iff

$$X'(z, \bar{z}) = \tilde{x}_0 \quad \text{for} \quad z = \bar{z}. \tag{4.26}$$

Here, $\tilde{x}_0$ is a variable on the dual circle with radius $\frac{1}{(2r)}$; it can also be viewed as the strength of a Wilson line, an interpretation that will become clearer in Subsection 5.3.2.

One-point functions can be calculated in the same way as before, keeping in mind the selection rules (4.20): while nothing changes in the Dirichlet case, it is precisely the combination $X'(z, \bar{z})$ that enters non-vanishing one-point functions in the Neumann case. We obtain

$$\langle \varphi_{g_L, g_R}(z, \bar{z}) \rangle_{(\Omega_D, x_0)} = \kappa_D(r) \, \frac{e^{ikx_0/r}}{|z - \bar{z}|^{k^2/4r^2}} \, \delta_{w,0}, \tag{4.27}$$

$$\langle \varphi_{g_L, g_R}(z, \bar{z}) \rangle_{(\Omega_N, \tilde{x}_0)} = \kappa_N(r) \, \frac{e^{2irw\tilde{x}_0}}{|z - \bar{z}|^{r^2 w^2}} \, \delta_{k,0}. \tag{4.28}$$

It will turn out later that the normalisation factors depend on the radius of the target circle. The Kronecker conditions on the winding and momentum number, respectively, follow from relations (4.20) and the charge formula for compact bosons.

4.2 Boundary fields

In a boundary CFT on the upper half-plane with symmetry algebra $\mathcal{W}$ and boundary conditions $\alpha = (\Omega, \tilde{\alpha})$, there are additional field operators beyond the bulk fields associated to states in the parent bulk Hilbert space. In this section, we first motivate the existence of these so-called boundary fields and how they are related to the state space of a boundary CFT, then we discuss the OPE of boundary fields and their locality properties. We conclude the section with remarks on boundary-condition changing operators.

4.2.1 Bulk–boundary OPE

Let us consider the behaviour of bulk fields $\varphi_{i,\bar{\imath}}(z, \bar{z})$ when z approaches the real line. Already the one-point functions show that, generically, this limit gives singular contributions to bulk field correlators. More specifically, recall that the Ward identities for bulk fields in a boundary CFT are those for a mirror pair of chiral charges i and $\omega(\bar{\imath})$, placed on both sides of the boundary. The gluing conditions of left- and right-moving observables induce an "interaction" of these two chiral charges via chiral OPE, thus producing singularities of correlators at the "interaction point". In QFT, a common interpretation of such unexpected singularities is to ascribe them to the existence of new field operators, in our case localised at the point $x = \Re z$ of the real line.

Put differently, the observed singular behaviour of bulk fields $\varphi_{i,\bar{\imath}}(z, \bar{z})$ near the boundary may be expressed in terms of the *bulk–boundary OPE* [102]

$$\varphi_{i,\bar{\imath}}(z, \bar{z}) = \sum_k C^{\alpha}_{(i\bar{\imath}) k} \, |z - \bar{z}|^{h_k - h_i - h_{\bar{\imath}}} \, \psi_k(x) \,. \tag{4.29}$$

Here, $\psi_k(x)$ are conformal boundary fields (primaries and descendants) of dimensions h_k, and $z = x + iy$. The fields $\psi_k(x)$ are localised at the boundary and cannot, in general, be "moved" into the interior of the upper half-plane. Exceptions are provided by elements $W(z)$ of the chiral symmetry algebra itself, which can be pushed to and from the boundary without producing singularities. For these fields, the bulk–boundary OPE $W(z) = W(x) + \mathcal{O}(z - \bar{z})$ is just the Taylor expansion.

Which families $[\psi_k]$ can possibly appear on the right-hand side of (4.29) is determined by the fusion rules for the chiral halves i and $\omega(\bar{\imath})$ of the bulk fields, i.e. by $\mathcal{W}$ representation theory. Still, some of the (allowed) coefficients $C^{\alpha}_{(i\bar{\imath}) k}$ may vanish for some α, and we need to solve non-linear sewing relations (see Section 4.4) to determine the $C^{\alpha}_{(i\bar{\imath}) k}$ and see how they are connected to the one-point function coefficients $A^{\alpha}_{i\bar{\imath}}$.

Those relations are complicated in general, but the bulk–boundary OPE coefficient $C^{\alpha}_{(i\bar{\imath}) 0}$ of the identity ($k = 0$ denotes the vacuum representation) can be expressed in terms of one-point functions immediately: recall that translational and scaling invariance along the real axis are preserved; thus only the identity boundary field can have a non-vanishing expectation value. Therefore, plugging the expansion (4.29) into a vacuum expectation value gives the one-point function $\langle \varphi_{i,\bar{\imath}}(z, \bar{z}) \rangle_{\alpha}$ on the left-hand side and projects onto the identity on the right-hand side; thus we obtain

$$C^{\alpha}_{(i\bar{\imath})\,0} = \frac{A^{\alpha}_{i\bar{\imath}}}{A^{\alpha}_{00}}\,. \tag{4.30}$$

On both sides of the equation, left- and right-moving labels have to be related as $\bar{\imath} = \omega^{-1}(i^{+})$. The denominator appears because, in a boundary CFT, we are usually not allowed to fix $\langle 1 \rangle_{\alpha} = 1$, as the following sections will show.

Other bulk–boundary OPE coefficients can be computed, without resorting to non-linear sewing relations, if the model is simple enough. For an uncompacti-fied free boson theory, it is straightforward to make the calculation of one-point functions from the last subsection into a derivation of the bulk–boundary OPE, using the vertex operator representation for primary bulk fields. We merely have to "fuse" the left- and right-moving parts of $X(z,\bar{z})$ at the boundary, taking into account the gluing conditions for the free boson. In the case of Dirich-let conditions, $X(z,\bar{z})$ approaches a constant when $z \to \bar{z}$, thus only (descen-dants of) the identity boundary field occur on the right-hand side of the OPE (4.29). In the Neumann case, left- and right-moving charges of the bulk field add up:

$$C^{(\Omega_D,x_0)}_{g\,g'} = e^{ix_0 g}\,\delta_{g',0}\,, \qquad C^{\Omega_N}_{g\,g'} = e^{i\pi g^2/2}\,\delta_{g',2g}\,. \tag{4.31}$$

Here, g' is the charge of the $\widehat{U(1)}$ primary boundary field – which can of course be written as a normal-ordered exponential

$$\psi_{g'}(x) = :e^{ig'X(x,x)}:\,. \tag{4.32}$$

The phase in the second of the equations (4.31) is due to the modulus convention we use in the bulk–boundary OPE. In particular, the OPE constants (4.31) exhibit the $\widehat{U(1)}$ families which occur in the spectrum of boundary fields; compare to Chapter 1.

4.2.2 State space

We return to model independent structures and to the investigation of the prop-erties of boundary fields. First, let us note that the existence of boundary fields closes a certain gap in the exposition of boundary CFT presented thus far. We have seen that the gluing conditions guarantee that the boundary theory on the upper half-plane is covariant under a single copy of $\mathcal{W}$ – which in particu-lar means that its state space $\mathcal{H}^{(H)}_{\alpha}$ should decompose into a sum of irreducible representations $\mathcal{H}_i$ of $\mathcal{W}$,

$$\mathcal{H}^{(H)}_{\alpha} = \bigoplus_i \mathcal{H}_i^{\oplus n^i_{\alpha}}\,,$$

possibly occurring with non-trivial multiplicities n^i_{α}. The bulk fields act in $\mathcal{H}^{(H)}_{\alpha}$, but obviously cannot correspond to its states in a straightforward manner, because they transform under $\mathcal{W} \times \mathcal{W}$, not under $\mathcal{W}$. Instead, the *state–field correspondence* $\Phi^{(H)}$ assigns boundary fields to states in $\mathcal{H}_\alpha^{(H)}$, analogously to the correspondence we used in Subsection 3.2.1 for vertex operators in bulk CFTs.

Let us assume that $\mathcal{H}_\alpha^{(H)}$ contains an $SL(2, \mathbb{R})$-invariant vacuum state $|0\rangle$. Then, for any state $|v\rangle \in \mathcal{H}_\alpha^{(H)}$, there exists a boundary operator $\psi_v(x) \equiv \Phi^{(H)}(v; x)$ such that

$$\psi_v(x)|0\rangle = e^{xL_{-1}^{(H)}}|v\rangle$$

for all real x. $L_{-1}^{(H)}$ generates translations parallel to the boundary. In particular, the operator $\psi_v(0)$ creates the state $|v\rangle$ from the vacuum. If $|v\rangle$ is a Virasoro primary state of conformal weight h, it is straightforward to derive the commutators (assuming that $|0\rangle$ is separating)

$$[L_n^{(H)}, \psi_v(x)] = x^n \left(x\frac{d}{dx} + h(n+1) \right) \psi_v(x). \tag{4.33}$$

4.2.3 Correlators and boundary OPE

Once boundary fields are part of the theory, it is natural to extend the set of *correlation functions* and to consider correlators in which a number of boundary operators $\psi_k(x_k)$ are inserted along with local bulk fields:

$$\langle \psi_1(x_1)\ldots\psi_M(x_M)\,\varphi_1(z_1, \bar{z}_1)\ldots\varphi_N(z_N, \bar{z}_N)\rangle_\alpha \quad \text{for} \quad x_i > x_{i+1}\,. \tag{4.34}$$

These functions obey Ward identities which generalise those of pure bulk field correlators, and which can be derived easily from relations like (4.33) and their generalisations to other modes $W_n^{(H)}$. As far as Ward identities are concerned, boundary fields behave exactly like chiral vertex operators of a CFT on the full plane.

The correlation functions (4.34) have singularities when bulk fields approach each other or the boundary, those are controlled by the bulk OPE (3.33) and the bulk–boundary OPE (4.29). Similarly, singularities are expected whenever two boundary field insertion points coincide. These are encoded in the *boundary OPE*

$$\psi_k(x_1)\,\psi_l(x_2) = \sum_m C_{klm}^\alpha\,(x_1 - x_2)^{h_m - h_k - h_l}\,\psi_m(x_2) \tag{4.35}$$

for $x_1 > x_2$. Again, the representation theory of $\mathcal{W}$ (its fusion rules) determines which ψ_m can possibly appear on the right-hand side. The coefficients C_{klm}^α depend on the boundary condition imposed on the real line and are subject to sewing constraints (see Section 4.4), which in particular relate them to basic data of the parent bulk theory and to the one-point functions.

Away from the singularities, the correlators (4.34) are analytic in the variables z_i throughout the whole upper half-plane $\Im z_i > 0$, because the bulk fields are

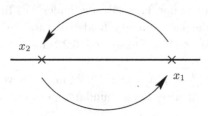

Figure 4.3 A curve along which correlation functions are analytically continued to exchange the position of two neighbouring boundary fields. In general,
the result depends on the orientation.

supposed to be local. For the variables x_k, the domain of analyticity is restricted
to the interval $x_k \in \,]\,x_{k+1}, x_{k-1}\,[$ on the boundary. In general, there is *no unique*
analytic continuation of $\psi_k(x_k)$ to other points on the real axis which lie beyond
the insertion points of the neighbouring boundary fields. In fact, if we continue
analytically along the curve shown in Figure 4.3, the result will typically depend
on whether we move the field ψ_k around ψ_{k+1} in a clockwise or anti-clockwise
direction – the usual behaviour of multi-valued chiral conformal blocks.

Boundary fields can therefore have non-trivial monodromies in correlation
functions – although they are to be regarded as physical operators. This is in fact
less disturbing than it may seem: recall that in a QFT on Minkowski space, analyticity of physical correlation functions stems from the requirement that field
operators have to commute when their arguments are separated by a space-like
vector. But the arguments of our boundary operators are always at time-like
separation, so the micro-locality criterion simply does not apply.

We have encountered simple examples of boundary OPEs in Chapter 1, and
in particular the OPE of free boson boundary vertex operators depends on their
order as soon as a B-field is present; see the B-dependent phase factors in equation (1.49). Indeed, we can now trace back the non-commutative aspects in the
target-space interpretation outlined at the end of Section 1.1 to the lack of locality of boundary fields on the worldsheet.

The above features highlight a notion of locality that will prove crucial in
our study of marginal perturbations in Chapter 5: two boundary fields $\psi_1(x_1) =
\Phi^{(H)}(v_1; x_1)$ and $\psi_2(x_2) = \Phi^{(H)}(v_2; x_2)$ are said to be *mutually local* if

$$\psi_1(x_1)\,\psi_2(x_2) = \psi_2(x_2)\,\psi_1(x_1) \quad \text{for all} \quad x_1 > x_2\,; \qquad (4.36)$$

the equation is supposed to hold after insertion into arbitrary correlation functions, and for the right-hand side to make sense it is assumed that there exists
a *unique* analytic continuation from $x_1 > x_2$ to $x_1 < x_2$. Therefore, the OPE of
two mutually local fields contains only pole singularities.

Furthermore, we will call a boundary field $\psi(x)$ *self-local* or *analytic* if it is
mutually local with respect to itself. The second expression is chosen in view of
the properties of its correlation functions and also of perturbations with self-local
marginal operators to be uncovered in Chapter 5.

Boundary fields $W(x)$ from the chiral algebra are the simplest examples of analytic fields. Since they do possess a unique analytic continuation to $x \in \mathbb{R} \setminus \{x_j\}$ and, moreover, to the interior of the upper half-plane, they are not only local with respect to themselves but to all boundary (and bulk) fields in the theory. We will come across other examples of self-local fields in Chapter 5. Usually, non-chiral self-local boundary fields ψ are only local wrt. a subset of boundary fields – which includes at least the chiral boundary fields $W(x)$ in addition to the field ψ itself.

4.2.4 Boundary-condition changing operators

For a given bulk CFT on the plane, several different (consistent) boundary conditions can usually be found; which boundary fields exist, and what their properties are, depends on the boundary condition that supports them – cf. the very different boundary spectra of Neumann and Dirichlet conditions in the free boson example. The string interpretation of boundary conditions as branes suggests that we consider open strings "stretched" between two different branes. Not surprisingly, there is an intrinsic worldsheet description of boundary fields associated with a pair of boundary conditions:

We can study CFTs on the strip instead of the upper half-plane and impose different boundary conditions α, β at both ends of the strip. Mapping the strip (with coordinate $w = \ln z$) back to the half-plane (coordinate z), we find that the boundary condition is discontinuous at $z = 0$: it jumps from β for $\Re z < 0$ to α for $\Re z > 0$; cf. Figure 4.4 in Subsection 4.3.1 below. Generic correlation functions in this theory will thus have a singularity at the origin, and it is yet again natural to think of such a jump as initiated by a boundary field inserted at $x = 0$; these fields are called *boundary-condition changing operators* (BCCOs for short) and denoted $\psi^{\alpha\beta}(x)$.

To determine their properties, correlation functions of bulk and (ordinary) boundary fields can be studied in the presence of such a jump more closely, in particular the Ward identities they satisfy. These depend on the gluing conditions which hold to the left and to the right of the discontinuity.

If the gluing conditions coincide, then still the whole symmetry algebra $\mathcal{W}$ acts covariantly on the correlation functions. The insertion of an additional field $W(x) \in \mathcal{W}$ into the correlator can then be converted into an action of $W(x)$ on the BCCO $\psi^{\alpha\beta}(0)$ via boundary OPE, or equivalently of modes $W_n^{(H)}$ analogously to relation (4.33). We conclude that there is a whole infinite space $\mathcal{H}_{\alpha\beta}$ of BCCOs, containing primaries and descendants. The space $\mathcal{H}_{\alpha\beta}$ can be decomposed into irreducible $\mathcal{W}$-representations in the same way as for $\alpha = \beta$, but does not contain an $\mathrm{SL}(2,\mathbb{R})$-invariant state for $\alpha \neq \beta$. Thus, there is no state–field correspondence in the usual sense any more. While this is slightly unusual in QFT, we note that in the context of statistical models it is very natural to consider Hamiltonians on a strip with different boundary conditions on both ends; there, $\mathcal{H}_{\alpha\beta}$ would simply enter as the space spanned by eigenstates of the Hamiltonian of the system; see, e.g., [39, 100].

If the jump in the boundary condition also changes the gluing automorphism, then a priori we cannot expect Ward identities to hold for more than the subalgebra $\mathcal{W}'$ of $\mathcal{W}$ on which the two gluing automorphisms coincide ($\mathcal{W}'$ always contains the Virasoro algebra). Accordingly, the space $\mathcal{H}_{\alpha\beta}$ will in general only decompose into $\mathcal{W}'$-irreducibles, and the computation of correlation functions is much more subtle when the gluing condition jumps.

In the relatively simple example of a Dirichlet–Neumann (DN) transition in the free boson theory, concrete expressions can still be achieved. The BCCO associated with such a DN transition can be viewed as spin field, and bulk and boundary vertex operators can be built from oscillators with half-integer modes. Correlators in the presence of DN transitions can be obtained as solutions to a suitable Ward identity of a Knizhnik–Zamolodchikov type. We refer to [195] for details.

A jump between two Dirichlet conditions (Ω_D, x_0^1) and (Ω_D, x_0^2) for a free boson poses no fundamental problems, in fact it can already be incorporated into the mode expansion: the field $X(z, \bar{z}) = X(z) + \overline{X}(\bar{z})$ with

$$X(z) = \frac{x_0^2 - x_0^1}{2\pi} \ln z + i \sum_{n \in \mathbb{Z} \setminus \{0\}} \frac{a_n}{n} z^{-n},$$

$$\overline{X}(\bar{z}) = -x_0^1 + \frac{x_0^2 - x_0^1}{2\pi} \ln \bar{z} + i \sum_{n \in \mathbb{Z} \setminus \{0\}} \frac{a_n}{n} z^{-n} \qquad (4.37)$$

satisfies the boundary conditions $X(z, \bar{z}) = x_0^1$ for $z = \bar{z} < 0$ as well as $X(z, \bar{z}) = x_0^2$ for $z = \bar{z} > 0$. Only integer oscillator modes appear, and bulk and boundary operators can be written as exponentials of this bosonic field. The BCCOs associated to a DD-jump can also be written as exponentials; see [195] and references given there.

Once we have added BCCOs to our framework, we can also consider several jumps in a row – i.e. a bulk CFT together with a whole *system of boundary conditions*. In particular, we can look at the expansion of a product of two BCCOs into other BCCOs. Fixing an order $x_1 > x_2$, this can be written as

$$\psi_k^{\alpha\beta}(x_1) \, \psi_l^{\gamma\delta}(x_2) = \sum_l \delta_{\beta,\gamma} \, C_{klm}^{\alpha\beta\delta} \, (x_1 - x_2)^{h_m - h_k - h_l} \, \psi_m^{\alpha\delta}(x_2) \,. \qquad (4.38)$$

The OPE (4.35) of ordinary boundary fields $\psi_k \equiv \psi_k^{\alpha\alpha}$ is contained as the special case where all four boundary conditions coincide, $C_{klm}^{\alpha} = C_{klm}^{\alpha\alpha\alpha}$. The Kronecker delta on the right-hand side reflects the fact that two BCCOs can be multiplied only when they "fit together", just like kinks mediating between soliton sectors [188]. It also shows that the OPE of BCCOs follows rules similar to those of matrix multiplication.

Since the result of the OPE depends drastically on the ordering of the BCCOs, they provide obvious examples for fields that are not mutually local. As we will see in Subsection 5.A.3, this simple observation will lead directly to some "non-commutative effects" in connection with branes, first pointed out in [454].

As stated before, within string theory a boundary CFT with BCCOs inserted along the worldsheet has an interpretation as a system of a number of branes $\alpha_1, \ldots, \alpha_n$. The BCCOs $\psi^{\alpha\beta}$ describe states of open strings stretched between two branes α and β. The system of branes may consist of several copies of one and the same brane (boundary condition); then the multiplication rules for BCCOs directly reproduce the *Chan–Paton matrix factors* "attached to the ends" of an open string: the boundary fields for a "stack" of N identical boundary conditions are simply given by the boundary fields for a single such boundary condition, tensored with complex $N \times N$ matrices.

Originally, the Chan–Paton degrees of freedom were introduced in a fairly ad hoc manner to allow for gauge symmetries in the low-energy effective field theories associated with open strings (cf. the brief remarks in Chapter 2), and to deal with consistency issues like tadpole cancellation. The worldsheet notion of BCCOs not only provides a conceptual explanation of the matrix factors, but also makes it clear that they are merely a rather special case of a more general structure.

4.3 The boundary state formalism

In this section we want to demonstrate that, with each boundary CFT, we may associate a so-called *boundary state* which incorporates the defining data of a boundary condition into objects built from quantities available in the "parent" CFT on the full plane.

The formulation of boundary CFT by means of boundary states is very efficient and elegant in some respects; in particular it provides deep conceptual insights into the spectra of boundary CFTs, leading to certain non-linear constraints (the Cardy conditions, see Section 4.4) along with a direct method of computing partition functions. The framework discussed before is, on the other hand, better suited to discuss correlators with boundary fields.

We will present two definitions of boundary states: one is inspired by worldsheet duality, i.e. by equating open string one-loop and closed-string tree-level diagrams; in less stringy terms, we use modular covariance of Euclidean finite-temperature CFT (or of models of statistical mechanics). The second definition only requires the zero-temperature data used up to now, such as bulk field one-point functions. We will see that both approaches lead to the same boundary states.

After giving the general (worldsheet duality) definition, we will turn towards the explicit construction of the boundary states. It involves the so-called Ishibashi states, which are generalised coherent states associated with the bulk-theory Hilbert space. While they are not normalisable with respect to the ordinary inner product, they implement the gluing conditions into the bulk CFT, and form an abstract linear space.

The full boundary states are special linear combinations of Ishibashi states. With the help of the first definition of boundary states – which goes back to Cardy [100] – it is easy to derive non-linear constraints on the coefficients in the linear combination. (Therefore, the traditional term boundary "state" is actually doubly misleading.) Cardy's constraints become especially restrictive when applied to pairs of boundary conditions. For rational CFTs, they are strong enough to determine complete sets of boundary conditions, in a sense to be made precise.

With explicit expressions for boundary states at our disposal, we can compare them to the basic data used so far: the bulk–boundary OPE combined with the study of certain limits of the finite-temperature correlators will enable us to relate boundary states to bulk field one-point functions.

Afterwards, we will present the second, zero-temperature definition of boundary states mentioned above. The tools developed until then will make it easy to establish agreement with the finite-temperature approach.

For illustration and later applications to string theory, we will briefly review boundary states for free bosons and other standard examples, and compute spectra for various configurations of boundary conditions.

4.3.1 Finite-temperature definition of boundary states

We first want to introduce boundary states in a way that is technically more demanding, but also more rewarding than the "zero-temperature" definition given later. In the present approach to boundary states, we consider some boundary CFT with the boundary condition $\alpha \equiv (\Omega, \tilde{\alpha})$ at finite temperature $1/\beta_0$; as usual in QFT, this is achieved by compactifying the (Euclidean) time, with compactification radius β_0. The finite-temperature bulk field correlators are therefore given by

$$\langle \varphi_1^{(H)}(z_1, \bar{z}_1) \cdots \varphi_N^{(H)}(z_N, \bar{z}_N) \rangle_\alpha^{\beta_0}$$
$$:= \operatorname{Tr}_{\mathcal{H}_\alpha^{(H)}} \left(e^{-\beta_0 H^{(H)}} \varphi_1^{(H)}(z_1, \bar{z}_1) \cdots \varphi_N^{(H)}(z_N, \bar{z}_N) \right). \tag{4.39}$$

The trace is over the boundary CFT Hilbert space $\mathcal{H}_\alpha^{(H)}$, and $H^{(H)} = L_0^{(H)} - \frac{c}{24}$ is the Hamiltonian. Note for later that in the special case $N = 0$, i.e. without bulk insertions, the correlator (4.39) becomes the partition function of the boundary CFT.

We may assume the arguments z_i to be radially ordered and that all the fields $\varphi^{(H)}(z, \bar{z})$ are quasi-primary, $\varphi^{(H)}(\lambda z, \bar{\lambda}\bar{z}) = \lambda^{-h}\bar{\lambda}^{-\bar{h}}\varphi^{(H)}(z, \bar{z})$; then, the above correlators are (anti-)periodic in the time variable $t = \ln|z|$ up to a scalar factor.

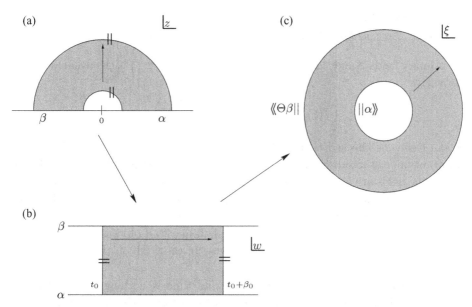

Figure 4.4 The maps from the upper half-plane to the strip and on to the ξ-plane. The arrows inside the shaded regions indicate the time flow. Inner and outer semi-circles in 4.4a are identified, and so are the intervals at t_0 and $t_0 + \beta_0$ in 4.4b. In the ξ-plane in 4.4c, we have traded boundary conditions for boundary states.

The shaded region in Figure 4.4a shows the domain the z_i are taken from; the bounding semi-circles are identified.

Now, the idea is to replace $z, \bar{z}$ by new variables $\xi, \bar{\xi}$ in terms of which the correlators (4.39) may be re-interpreted as certain correlators for a theory on the full plane. This theory will be the "parent" bulk CFT, enriched by a new object, a boundary state, which encodes the boundary condition α.

Let us first introduce the variable $w = \ln z = t + i\sigma$. For $z \in \mathbb{H}$, w sweeps out a strip, but because of the periodicity properties of the functions (4.39) the theory essentially lives on a cylinder parametrised by $\sigma \in [0, \pi]$ and $t \in [t_0, t_0 + \beta_0]$ with the two segments at $t = t_0$ and $t_0 + \beta_0$ identified; see Figure 4.4b. We recognise the underlying worldsheet geometry as that of an open-string one-loop diagram.

However, the same principle that leads to modular invariance of the partition functions of two-dimensional CFTs tells us that we are free to *interchange* Euclidean *time and space*.

After a subsequent rescaling by $2\pi/\beta_0$, we obtain a cylinder which is now periodic in space with period 2π and for which the time variable runs from 0 to $2\pi^2/\beta_0$. With the help of the exponential mapping

$$\xi = e^{\frac{2\pi i}{\beta_0} \ln z} \quad \text{and} \quad \bar{\xi} = e^{-\frac{2\pi i}{\beta_0} \ln \bar{z}}, \tag{4.40}$$

the cylinder is finally mapped onto an annulus in the full ξ-plane; see Figure 4.4c. To rewrite the original correlators in terms of these new variables, we make use of the transformation behaviour

$$\varphi(\xi, \bar{\xi}) = \left(\tfrac{dz}{d\xi}\right)^h \left(\tfrac{d\bar{z}}{d\bar{\xi}}\right)^{\bar{h}} \varphi^{(H)}(z, \bar{z}), \quad T(\xi) = \left(\tfrac{dz}{d\xi}\right)^2 T^{(H)}(z) + \tfrac{c}{12}\{z, \xi\} \qquad (4.41)$$

for primary fields φ and the energy-momentum tensor; $\{z, \xi\}$ is the usual Schwarz derivative,

$$\{f(\xi), \xi\} = \frac{f'''(\xi)}{f'(\xi)} - \frac{3}{2}\left(\frac{f''(\xi)}{f'(\xi)}\right)^2. \qquad (4.42)$$

It can easily be checked that the resulting correlators for the fields $\varphi(\xi, \bar{\xi})$ are invariant under the substitution $\xi \mapsto e^{2\pi i}\xi$ if $h - \bar{h}$ is an integer (for half-integer $h - \bar{h}$, the fields are anti-periodic, i.e. they live on a double cover of the annulus). This shows, at the level of correlation functions, that the fields on the left-hand side of equation (4.41) can be consistently defined on the full plane, $\varphi(\xi, \bar{\xi}) \equiv \varphi^{(P)}(\xi, \bar{\xi})$. Throughout this subsection, we shall mark all objects of the CFT on the full plane by an upper index $^{(P)}$ and those of the CFT on the half-plane by an index $^{(H)}$.

In string theory, the exchange of space and time is nothing but *worldsheet duality*, and identifies an open-string one-loop diagram with a tree-level closed-string diagram. In the second interpretation, the cylinder describes a closed string propagating from an in- to an out-state. In our case, these states will have to encode the boundary conditions originally imposed on the open-string endpoints. Such boundary states $\|\alpha\rangle\rangle$ cannot be ordinary states from the closed-string state space $\mathcal{H}^{(P)}$, but as an ansatz we assume that $\|\alpha\rangle\rangle$ can be written as a formal linear combination of elements of $\mathcal{H}^{(P)}$. The fact that $\|\alpha\rangle\rangle$ implements the boundary conditions is formulated by means of the defining equation [389]

$$\langle\langle \Theta\alpha \| e^{-\frac{2\pi^2}{\beta_0} H^{(P)}} \varphi_1^{(P)}(\xi_1, \bar{\xi}_1) \cdots \varphi_N^{(P)}(\xi_N, \bar{\xi}_N) \| \alpha \rangle\rangle$$

$$:= \mathcal{J}(\xi, z)\, \mathrm{Tr}_{\mathcal{H}_\alpha^{(H)}}\left(e^{-\beta_0 H^{(H)}} \varphi_1^{(H)}(z_1, \bar{z}_1) \cdots \varphi_N^{(H)}(z_N, \bar{z}_N)\right), \qquad (4.43)$$

in which the right-hand side contains objects from the boundary CFT with boundary condition α. On the left-hand side, $H^{(P)} = L_0 + \overline{L}_0 - \frac{c}{12}$ is the Hamiltonian of the bulk theory, and $\mathcal{J}$ denotes the product of Jacobians arising in equations (4.41) from the coordinate transformation $\xi = \xi(z)$. Relation (4.43) is an extension to arbitrary φ-correlators of Cardy's definition in [100] where only partition functions ($N = 0$) were considered. Moreover, we find it natural to introduce the bulk CPT operator Θ into the definition (4.43) because the two boundary components of the annulus have opposite (induced) orientation. (Note, however, that in many concrete examples of interest, including free bosons and SU(2) WZW models, Θ leaves the boundary states invariant.)

The new object $\|\alpha\rangle\rangle$ cannot be viewed as a state in the Hilbert space of the CFT on the plane. This is no surprise since, after all, imposing boundary conditions means adding new structure to a CFT. We will investigate the nature of boundary states in a moment, in particular we have to prove their existence. But let us first make some brief remarks on possible generalisations of equation (4.43).

We have motivated the existence of BCCOs by looking at CFTs on the strip with different boundary conditions imposed at both ends. If we imagine some BCCO $\psi_k^{\alpha\beta}$ inserted at $z = 0$ in Figure 4.4a, the associated annulus diagram in the ξ-plane has two different boundary states $\|\alpha\rangle\rangle$, $\|\beta\rangle\rangle$ attached to the inner and outer circle. As was mentioned in Section 4.2, there is a space $\mathcal{H}_{\alpha\beta}^{(H)}$ of such fields inducing jumps in the boundary condition from α to β. Correlation functions of BCCOs with arbitrary bulk operators $\varphi^{(H)}(z, \bar{z})$ are still well defined, and we can extend definition (4.43) to

$$\langle\langle \Theta\beta \| e^{-\frac{2\pi^2}{\beta_0} H^{(P)}} \varphi_1^{(P)}(\xi_1, \bar{\xi}_1) \cdots \varphi_N^{(P)}(\xi_N, \bar{\xi}_N) \| \alpha\rangle\rangle$$
$$:= \mathcal{J}(\xi, z) \, \mathrm{Tr}_{\mathcal{H}_{\alpha\beta}^{(H)}} \left(e^{-\beta_0 H^{(H)}} \varphi_1^{(H)}(z_1, \bar{z}_1) \cdots \varphi_N^{(H)}(z_N, \bar{z}_N) \right). \quad (4.44)$$

Anticipating applications of boundary CFT to string theory to be discussed later, we note that the correlators (4.44) describe a system of two different branes exchanging (tree-level) closed strings. We will soon see that the simplest case of definition (4.44) – without any field insertions – leads to strong constraints on the boundary states [100].

The as-yet formal definition (4.43) deals with bulk fields only; no boundary fields appear in the finite-temperature correlator on the upper half-plane. In a sense, boundary fields can be regarded as "derived objects" in a boundary CFT, so this is not a severe restriction. If we nevertheless want to "transport" boundary fields into the CFT on the ξ-plane, we immediately meet the obstacle that they have no direct operator counterpart there, in contrast to the bulk fields $\varphi^{(H)}$. But our approach to boundary states just relies on correlation functions, so it is in principle possible to "model" boundary fields by suitably modified boundary states – which, in particular, carry information about the location of the boundary fields. In a rather symbolic notation,

$$\mathrm{Tr}_{\mathcal{H}_\alpha^{(H)}} \left(e^{-\beta_0 H^{(H)}} \varphi_1^{(H)}(z_1, \bar{z}_1) \cdots \varphi_N^{(H)}(z_N, \bar{z}_N) \, \psi(x_1) \cdots \psi(x_M) \right)$$
$$= \mathcal{J}(z, \xi) \langle\langle \Theta\alpha_{\{\psi\}} \| e^{-\frac{2\pi^2}{\beta_0} H^{(P)}} \varphi_1^{(P)}(\xi_1, \bar{\xi}_1) \cdots \varphi_N^{(P)}(\xi_N, \bar{\xi}_N) \| \alpha_{\{\psi\}}\rangle\rangle \quad (4.45)$$

where $\|\alpha_{\{\psi\}}\rangle\rangle$ now depends on the points $\xi(x_j)$, $j = 1, \ldots, M$, and on the $\mathcal{W}$-representations carried by the boundary fields $\psi_j(x_j)$. Their conformal properties are to be reflected in the $\xi(x_j)$-dependence of $\|\alpha_{\{\psi\}}\rangle\rangle$. It looks, however, almost hopeless to obtain explicit formulas for such "boundary field boundary states", because already ordinary, "constant" boundary states $\|\alpha\rangle\rangle$ involve rather unwieldy expressions of bulk-theory states, as we will now see.

4.3.2 Ishibashi states

The boundary state $\|\alpha\rangle\rangle_\Omega$ – we make the gluing conditions explicit here – must implement the boundary conditions $(\Omega, \tilde{\alpha})$ into the CFT on the ξ-plane. In order to determine $\|\alpha\rangle\!\rangle_\Omega$ explicitly, let us start with the gluing conditions

$$T^{(H)}(z) = \overline{T}^{(H)}(\bar z) \quad \text{and} \quad W^{(H)}(z) = \Omega\overline{W}^{(H)}(\bar z) \quad \text{at} \ z = \bar z$$

of the upper half-plane theory. They enforce the so-called *Ishibashi conditions* on the boundary state. To derive them, we make the special choices

$$\varphi_N^{(H)}(z_N, \bar z_N) = T^{(H)}(z) - \overline{T}^{(H)}(\bar z) \quad \text{and} \quad \varphi_N^{(H)}(z_N, \bar z_N) = W^{(H)}(z) - \Omega\overline{W}^{(H)}(\bar z),$$

respectively, on the right-hand side of relation (4.43); the resulting correlators vanish at $z = \bar z$, for arbitrary bulk fields $\varphi_i^{(H)}(z_i, \bar z_i)$ inserted at the remaining $N-1$ points.

When mapping to the ξ-plane, the Jacobians from (4.41) – including the Schwarz derivative (4.42) – are crucial. We need the following simple properties of the functions $\xi(z)$ and $\bar\xi(\bar z)$, which hold for real positive $z = \bar z$:

$$\bar\xi = \xi^{-1}, \quad \left(\frac{d\bar\xi}{d\bar z}\right)\left(\frac{d\xi}{dz}\right)^{-1} = -\bar\xi^2, \quad \{\xi, z\} = \{\bar\xi, \bar z\}. \tag{4.46}$$

The restriction to positive z means that, for the moment, we are deriving Ishibashi conditions for the boundary state $\|\alpha\rangle\!\rangle_\Omega$ placed at the unit circle $|\xi| = 1$.

Using the last of relations (4.46), the vanishing of $T^{(H)}(z) - \overline{T}^{(H)}(\bar z)$ at $z = \bar z$ inside arbitrary half-plane correlators translates into

$$\left[\left(\frac{d\xi}{dz}\right)^2 T^{(P)}(\xi) - \left(\frac{d\bar\xi}{d\bar z}\right)^2 \overline{T}^{(P)}(\bar\xi)\right] \|\alpha\rangle\!\rangle_\Omega = 0$$

for $\xi = \bar\xi^{-1}$ on the unit circle. Using mode expansions $T^{(P)}(\xi) = \sum_n L_n^{(P)}\xi^{-n-2}$, analogously for $\overline{T}^{(P)}(\bar\xi)$, and the remaining relations in (4.46), we can order by powers of ξ and obtain

$$\left(L_n^{(P)} - \overline{L}_{-n}^{(P)}\right)\|\alpha\rangle\!\rangle_\Omega = 0 \tag{4.47}$$

for all $n \in \mathbb{Z}$. In the same way, we can implement $W^{(H)}(z) - \Omega\overline{W}^{(H)}(\bar z) = 0$ on the boundary state. The sign in the second equation in our list (4.46) together with the Jacobian in equation (4.41) leads to the following result:

$$\left(W_n^{(P)} - (-1)^{hw} \Omega\overline{W}_{-n}^{(P)}\right)\|\alpha\rangle\!\rangle_\Omega = 0. \tag{4.48}$$

Recall that we always require Ω to act locally, i.e. to commute with the mode expansion. The linear equations (4.47, 4.48) are called *Ishibashi conditions*, and are clearly equivalent to the gluing conditions from the upper half-plane formulation. Note that relations (4.47) for the Virasoro generators are required to hold for every conformal boundary state, while conditions (4.48) only apply to $W(z) \in \mathcal{W}^{\mathrm{op}}$, the sub-W-algebra of $\mathcal{W}$ preserved by the boundary state.

Analogous calculations can be performed to derive the Ishibashi conditions at $|\xi| = R$ with $\ln R := 2\pi^2/\beta_0$. Here, we have to move z towards the negative real axis, $z = \bar z < 0$, which modifies the formulas (4.46). We find that the boundary state at $|\xi| = R$ is annihilated by $L_n - R^{2n}\overline{L}_{-n}$, which is precisely the property of $\langle\langle\Theta\alpha\|R^{-H^{(P)}}$ as long as $\|\Theta\alpha\rangle\rangle$ satisfies the $|\xi| = 1$ Ishibashi conditions (4.47); the same is true for the other generators. The "propagator" $\exp\{(\ln R)\,H^{(P)}\}$ similarly induces a conformal transformation of Ishibashi states, and we do not obtain new conditions on the boundary states.

Let us now try to build a solution to the linear conditions (4.47, 4.48) from objects available in the plane CFT, starting with the case $\Omega = \mathrm{id}$ and assuming that $\mathcal{W}^{\mathrm{op}} = \mathcal{W}$. Ishibashi has shown [294, 295] that to each representation π_i on $\mathcal{H}_i$ of the chiral symmetry algebra $\mathcal{W}$ a solution $|i\rangle\rangle$ can be constructed of

$$\left(W_n - (-1)^{h_W} \, \overline{W}_{-n} \right) |i\rangle\rangle = 0 \, ; \qquad (4.49)$$

for simplicity, we drop the superscript $^{(P)}$ for a while. Using $|i, N\rangle$, $N \in \mathbb{Z}_+$, to denote an orthonormal basis of $\mathcal{H}_i$, we can give the (formal) expression

$$|i\rangle\rangle = \sum_{N=0}^{\infty} |i, N\rangle \otimes U |i, N\rangle \qquad (4.50)$$

for the *Ishibashi state* associated with i. Here, U denotes an anti-unitary operator on the total chiral Hilbert space $\mathcal{H}_R^{\mathrm{ch}} = \bigoplus_i \mathcal{H}_i$ which satisfies the commutation relations

$$U \, \overline{W}_n = (-1)^{h_W} \, \overline{W}_n \, U \qquad (4.51)$$

with the right-moving generators: U acts like the chiral CPT operator. (We have implicitly selected a set of real W-generators.) Note that $U|i, N\rangle \in \mathcal{H}_{i^+}$, the Hilbert space carrying the representation i^+ conjugate to i, so that the terms in the sum (4.50) are elements of $\mathcal{H}_i \otimes \mathcal{H}_{i^+}$. As always, we take left- and right-moving symmetry algebra to be the same. In this subsection, we will for simplicity assume that our boundary conditions preserve the whole symmetry algebra, i.e. that $\mathcal{W}^{\mathrm{op}} = \mathcal{W}$. For symmetry-breaking boundary conditions, where $\mathcal{W}^{\mathrm{op}} \neq \mathcal{W}$, Ishibashi states are to be built for each irreducible representation of $\mathcal{W}^{\mathrm{op}}$ contained in the $\mathcal{W}$-irreps $\mathcal{H}_i$.

Before we prove that equation (4.50) solves the conditions (4.49) and does not depend on the basis $|i, N\rangle$, let us generalise Ishibashi's expression (4.50) to non-trivial gluing automorphisms Ω. Recall the remarks around equations (4.7) and (4.8), which mean that Ω induces a representation π_i^{Ω} on $\mathcal{H}_i$, such that the $\mathcal{W}$-module is isomorphic to $\mathcal{H}_{\omega(i)}$ carrying the representation $\pi_{\omega(i)}$. We implement the isomorphism by a unitary operator

$$V_\Omega \; : \; \mathcal{H}_{\omega(i)} \longrightarrow \mathcal{H}_i \qquad (4.52)$$

such that $\pi_i^{\Omega}(A) = V_\Omega \pi_{\omega(i)}(A) V_\Omega^{-1}$. The map V_Ω is defined on the whole right-moving chiral space $\mathcal{H}_R^{\mathrm{ch}} = \bigoplus_i \mathcal{H}_i$, and we assume that $V_\Omega U = U V_\Omega$. By

$$|i\rangle\rangle_\Omega := (\mathrm{id} \otimes V_\Omega) |i\rangle\rangle \qquad (4.53)$$

we define the Ω-*twisted Ishibashi state* associated to the (left-moving) representation i of $\mathcal{W}$. It satisfies the Ishibashi conditions (4.48). The components of $|i\rangle\!\rangle_\Omega$ are elements of $\mathcal{H}_i \otimes \mathcal{H}_{\omega^{-1}(i+)}$.

We still have to show that $|i\rangle\!\rangle_\Omega$ defined in equations (4.50) and (4.53) satisfies the Ishibashi conditions (4.47) and (4.48). It is easy to generalise Ishibashi's original proof to incorporate non-standard gluing conditions.

Clearly, the sum (4.50) of normalised states has no chance to converge in $\mathcal{H}^{(P)}$. But even though $|i\rangle\!\rangle_\Omega$ is not itself an element of the plane CFT Hilbert space, scalar products with states of definite energy in $\mathcal{H}^{(P)}$ are well defined – and the collection of all such scalar products determines $|i\rangle\!\rangle_\Omega$. Therefore, it is sufficient to show that

$$\langle k, M| \otimes \langle V_\Omega U(j, L)| \left(W_n - (-1)^{h_W} \Omega \overline{W}_{-n} \right) |i\rangle\!\rangle_\Omega = 0 \qquad (4.54)$$

for all orthonormal basis vectors $|k, M\rangle \otimes |j, L\rangle \in \mathcal{H}^{(P)}$, with notations as before. We have included the chiral CPT operator and the unitary V_Ω in equation (4.54) for later convenience only. Also, the Dirac sandwich notation is to mean $\langle v|A|w\rangle := \langle v, A\, w\rangle$.

To evaluate the left-hand side of equation (4.54), we use the intertwining property $\Omega W_n V_\Omega |i, N\rangle = V_\Omega W_n |i, N\rangle$, anti-unitarity of U and its fundamental commutation relation (4.51). The calculation to perform is

$$\sum_N \langle k, M| \otimes \langle V_\Omega U(j, L)| \left(W_n - (-1)^{h_W} \Omega \overline{W}_{-n} \right) |i, N\rangle \otimes V_\Omega U|i, N\rangle$$

$$= \sum_N \Big[\langle k, M| W_n |i, N\rangle \langle V_\Omega U(j, L)|V_\Omega U|i, N\rangle$$

$$- (-1)^{h_W} \langle k, M|i, N\rangle \langle V_\Omega U(j, L)| \Omega \overline{W}_{-n} V_\Omega U|i, N\rangle \Big]$$

$$= \sum_N \Big[\delta_{j,i}\, \delta_{L,N} \langle k, M| W_n |i, N\rangle$$

$$- (-1)^{h_W} (-1)^{-h_W} \delta_{k,i}\, \delta_{M,N} \langle V_\Omega U(j, L)| V_\Omega U\, \overline{W}_{-n} |i, N\rangle \Big]$$

$$= \delta_{k,j}\, \delta_{j,i} \big[\langle i, M| W_n |i, L\rangle - \langle U(i, L)| U\, \overline{W}_{-n} |i, M\rangle \big] = 0\,.$$

The last equality holds because $\langle (Uv)| UA |w\rangle = \langle w| A^* |v\rangle$ and because $\overline{W}_n$ acts on the right-moving $\mathcal{H}_i$ just like W_n on the left-moving $\mathcal{H}_i$.

This shows that the Ishibashi states given in equations (4.50) and (4.53) implement the gluing conditions, but the formulas suggest that they depend on the choice of an orthonormal basis $|i, N\rangle$ of $\mathcal{H}_i$. However, apart from a possible overall phase, they are unique: let $|i\rangle\!\rangle_\Omega$ be any non-zero Ishibashi state as in equation (4.53), then its product $p := \langle i|i\rangle\!\rangle_\Omega$ with the highest weight state $|i\rangle \equiv |i, 0\rangle \otimes V_\Omega U|i, 0\rangle$ of $\mathcal{H}_i \otimes \mathcal{H}_{\omega^{-1}(i+)}$ must be non-zero. Here, the appearance of the anti-unitary operator U and the fact that it maps highest weight states to highest weight states are crucial. If $|i\rangle\!\rangle'_\Omega$ is another Ishibashi state for this representation with $p' \neq 0$, then $|i\rangle\!\rangle_\Omega - \frac{p}{p'}|i\rangle\!\rangle'_\Omega$ is a third such state, but with vanishing projection onto the highest weight vector. Therefore, the last Ishibashi state is zero and $|i\rangle\!\rangle_\Omega$ and $|i\rangle\!\rangle'_\Omega$ are linearly dependent.

There is a more elegant way [37, 44, 372] to see that Ishibashi states are in fact canonical objects associated with $\mathcal{H}_i \otimes \mathcal{H}_{\omega^{-1}(i^+)}$: every element of a tensor product of two Hilbert spaces $\mathcal{H}_1 \otimes \mathcal{H}_2$ can be mapped canonically to a linear operator $\mathcal{H}_2 \longrightarrow \mathcal{H}_1$, namely $|v_1\rangle \otimes |v_2\rangle \longmapsto |v_1\rangle \otimes \langle v_2|$, acting on states in $\mathcal{H}_2$ with the help of the inner product. With the caveats from above concerning questions of convergence, the same is true for the infinite sum of states in equations (4.50) and (4.53). Moreover, it follows from the Ishibashi conditions (4.48) that the resulting linear map $T_i^\Omega : \mathcal{H}_{\omega^{-1}(i^+)} \longrightarrow \mathcal{H}_i$ intertwines two actions of the chiral algebra $\mathcal{W}$, i.e. $W_n T_i^\Omega = T_i^\Omega Ad_{U_\Omega}(\overline{W}_n)$, where $U_\Omega := UV_\Omega$, and $\overline{W}_n$ means the $\mathcal{W}$-action on $\mathcal{H}_i$. As our representations are irreducible, Schur's lemma not only confirms that the right-moving "partner" of $\mathcal{H}_i$ must be $\mathcal{H}_{\omega^{-1}(i^+)}$, but also that the linear operator T_i^Ω, and therefore the Ishibashi state, is unique up to normalisation.

In the computations above we implicitly used the standard inner product for tensor products of left- and right-moving states, namely

$$\langle\, v_1 \otimes w_1 \,|\, v_2 \otimes w_2 \,\rangle = \langle\, v_1 \,|\, v_2 \,\rangle \cdot \langle\, w_1 \,|\, w_2 \,\rangle$$

for $v_i \otimes w_i \in \mathcal{H}_L \otimes \mathcal{H}_R$. When dealing with fermions, it is sometimes more natural, although not strictly necessary, to employ a graded tensor product with corresponding scalar product

$$\langle\, v_1 \otimes w_1 \,|\, v_2 \otimes w_2 \,\rangle = (-1)^{F(w_1)F(v_2)} \langle\, v_1 \,|\, v_2 \,\rangle \cdot \langle\, w_1 \,|\, w_2 \,\rangle$$

where $F(w)$ is the chiral fermion number of w, etc. In this case, it is necessary to replace the chiral CPT operator U in the definition of the Ishibashi states by the anti-unitary operator U_F satisfying $U_F W_n = (-1)^{h_W} W_n U_F(-1)^F$.

We have obtained a complete overview of the possible Ishibashi states for a bulk CFT with symmetry algebra $\mathcal{W} \times \mathcal{W}$: to each gluing automorphism Ω of $\mathcal{W}$ (including the identity) there is a set of Ishibashi conditions (4.48) between the left- and right-moving generators. And to each irreducible highest weight representation i of $\mathcal{W}$ on a Hilbert space $\mathcal{H}_i$, a unique (up to rescaling) Ishibashi state $|i\rangle\!\rangle_\Omega$ which "implements" these gluing conditions can be formed. This means that if we form a *boundary state*

$$\|\alpha\rangle\!\rangle_\Omega = \sum B_\alpha^i \, |i\rangle\!\rangle_\Omega \tag{4.55}$$

as a linear combination of (twisted) Ishibashi states, the system of bulk CFT and boundary state will have $\mathcal{W}$ as its symmetry algebra.

However, a bulk CFT is not specified by its symmetry algebra $\mathcal{W} \times \mathcal{W}$ alone, but it comes with a given modular invariant partition function on the torus – i.e., with a consistent selection of irreducible $\mathcal{W} \times \mathcal{W}$-modules making up the total bulk Hilbert space

$$\mathcal{H}_{\text{tot}} = \bigoplus_{(i,\bar{\imath})\in I} \mathcal{H}_i \otimes \mathcal{H}_{\bar{\imath}}. \tag{4.56}$$

Therefore, usually only a subset of all possible Ishibashi states $|i\rangle\rangle_\Omega$ with $i \in I_\mathcal{W}$ – where $I_\mathcal{W}$ is the set of all $\mathcal{W}$-representations – really contributes to the boundary states for a given bulk theory: the condition is that the term $\mathcal{H}_i \otimes \mathcal{H}_{\omega^{-1}(i+)}$ occurs in the bulk Hilbert space (4.56). In this sense, the coefficients B_α^i in equation (4.55) implicitly depend on the right-moving representation label and on the gluing automorphism, and therefore $B_\alpha^{i\bar{\imath}} \cdot \delta_{\bar{\imath},\omega^{-1}(i+)}$ would be a more complete notation. We will see that such simple selection rules explain various observations made in D-brane physics.

To conclude this subsection, let us list some Ishibashi conditions that appear frequently in string applications, in part repeating formulas already introduced in Chapter 1.

In the free boson case, the Ishibashi conditions for the $\widehat{U(1)} \times \widehat{U(1)}$ boundary states follow from the gluing conditions (4.11) and (4.12) and the mode expansion of the currents. For a boson compactified on a circle of radius r, we arrive at

$$\left(a_n \pm \bar{a}_{-n} \right) |(k,w)\rangle\rangle = 0. \tag{4.57}$$

The plus (minus) sign is for Neumann (Dirichlet) boundary conditions, the integers k and w are wave and winding number, respectively, which label the $\widehat{U(1)} \times \widehat{U(1)}$ highest weight representations; for an uncompactified boson, w is absent. Setting $n = 0$, we obtain $\hat{k} |(k,w)\rangle\rangle_N = 0$ (Neumann) and $\hat{w} |(k,w)\rangle\rangle_D = 0$ (Dirichlet), respectively.

Using the canonical commutation relations, it is straightforward to show that conditions (4.57) are solved by the coherent states

$$|(0,w)\rangle\rangle_N = \exp\left(-\sum_{n=1}^{\infty} \frac{1}{n} a_{-n}\bar{a}_{-n} \right) |(0,w)\rangle, \tag{4.58}$$

$$|(k,0)\rangle\rangle_D = \exp\left(\sum_{n=1}^{\infty} \frac{1}{n} a_{-n}\bar{a}_{-n} \right) |(k,0)\rangle, \tag{4.59}$$

where the kets on the right-hand side denote oscillator ground states; cf. equation (1.35) in Chapter 1 and [89, 378]. With a little combinatorics, these coherent states can be rewritten as sums over normalised basis states as in the general formulas (4.50) and (4.53); see [294] for the Neumann case.

Chapter 1 also contained boundary states for free fermions, see equations (1.62) and (1.64). The Ishibashi conditions preserving the free fermion symmetry, see equations (1.63) and (1.65), parallel those for free bosons, and the Ishibashi states can again be expressed in terms of exponentials of creation operators (which is no longer possible for more general symmetry algebras that do not have a free field realisation).

For applications to worldsheet superstring compactifications, we will need to deal with boundary states for the $N = 2$ *super Virasoro algebra*, so let us write down the form that A-type and B-type gluing conditions (4.14) and (4.15) take after passing to the annulus: the simpler B-type conditions are

$$(L_n - \overline{L}_{-n}) |j\rangle\!\rangle_B = (J_n + \overline{J}_{-n}) |j\rangle\!\rangle_B = 0 \ ,$$

$$(G_r^+ + i\eta \, \overline{G}_{-r}^+) |j\rangle\!\rangle_B = (G_r^- + i\eta \, \overline{G}_{-r}^-) |j\rangle\!\rangle_B = 0, \tag{4.60}$$

while the A-type conditions involve a twisting by the mirror automorphism,

$$(L_n - \overline{L}_{-n}) |j\rangle\!\rangle_A = (J_n - \overline{J}_{-n}) |j\rangle\!\rangle_A = 0 \ ,$$

$$(G_r^+ + i\eta \, \overline{G}_{-r}^-) |j\rangle\!\rangle_A = (G_r^- + i\eta \, \overline{G}_{-r}^+) |j\rangle\!\rangle_A = 0. \tag{4.61}$$

In general, η can be any phase, but is restricted to $\eta = \pm 1$ if we insist on an unbroken $N = 1$ subalgebra; see Subsection 4.1.3.

4.3.3 Zero-temperature definition and relation to one-point functions

We have introduced boundary states by a modular transformation of finite-temperature correlators of bulk fields in the upper half-plane; see equation (4.43). This has the virtue that "one-loop effects" and worldsheet duality are incorporated from the start, which will in turn provide a recipe to compute the spectrum and lead to new non-linear constraints, to be discussed in the next section.

There is, however, an alternative definition of the same boundary states, which is more convenient in some respects: it uses zero-temperature correlators and a simpler transformation from the half-plane with coordinate z into the full plane with coordinate ζ, namely

$$\zeta = \frac{1 - iz}{1 + iz} \quad \text{and} \quad \bar{\zeta} = \frac{1 + i\bar{z}}{1 - i\bar{z}} \ . \tag{4.62}$$

Here, the half-plane is mapped to the complement of the unit disk in the ζ-plane. Note that periodicity in the "space variable" is ensured through $\lim_{x \to +\infty} \zeta(x) = \lim_{x \to -\infty} \zeta(x)$, in contrast to the finite-temperature case of Subsection 4.3.1 where we started from a periodic "time variable" in the upper half-plane and performed a worldsheet duality transformation.

The boundary state $\|\alpha\rangle\!\rangle'$ modeling the half-plane boundary condition α within the CFT on the ζ-plane is then defined by

$$\langle 0 | \, \varphi_1^{(P)}(\zeta_1, \bar{\zeta}_1) \cdots \varphi_N^{(P)}(\zeta_N, \bar{\zeta}_N) \, \|\alpha\rangle\!\rangle'_\Omega := \mathcal{J}(\zeta, z) \, \langle \varphi_1^{(H)}(z_1, \bar{z}_1) \cdots \varphi_N^{(H)}(z_N, \bar{z}_N) \rangle_\alpha \ , \tag{4.63}$$

where $|0\rangle$ is the vacuum of the CFT on the full plane; again, $\mathcal{J}$ denotes Jacobians from the z–ζ-transformation, and formula (4.63) is to hold for arbitrary bulk field insertions.

It is not obvious that, starting from a given boundary CFT, equations (4.43) and (4.63) define the same boundary states on the plane. In the following, we will first observe that they lead to the same Ishibashi conditions for the symmetry generators in the plane CFT, i.e. to the same vector space of Ishibashi states $|i\rangle\!\rangle_\Omega$. Then, we will turn to the coefficients $B_\alpha'^i$ in the expansion

$$\|\alpha\rangle\!\rangle_\Omega' = \sum_{i\in I} B_\alpha'^i |i\rangle\!\rangle_\Omega$$

and identify them (up to a possible overall scaling) with the B_α^i occurring in finite-temperature boundary states (4.55) – by relating both of them to the one-point function coefficients $A_{i\bar{i}}^\alpha$. Taken together, this will prove that the boundary states obtained from the two definitions coincide up to an overall normalisation which, as we will see in Section 4.4, is fixed by Cardy's conditions.

The derivation of the Ishibashi conditions for $\|\alpha\rangle\!\rangle_\Omega'$ can simply be copied from Subsection 4.3.2, because the essential transformation rules (4.46) hold for the map $\zeta(z)$ just as well as for the exponential mapping $\xi(z)$ used in equation (4.40). Moreover, the Schwarz derivative vanishes for the rational map $z \mapsto \zeta$. This proves that the linear spaces of Ishibashi states coincide.

In order to relate the boundary state coefficients $B_\alpha'^i$ and B_α^i to each other, we specialise the correlator (4.63) to a single primary bulk field insertion,

$$\langle 0| \varphi_{i,\bar{i}}^{(P)}(\zeta,\bar{\zeta}) \|\alpha\rangle\!\rangle' = \left(\frac{dz}{d\zeta}\right)^{h_i} \left(\frac{d\bar{z}}{d\bar{\zeta}}\right)^{h_{\bar{i}}} \frac{A_{i\bar{i}}^\alpha}{|z-\bar{z}|^{h_i+h_{\bar{i}}}}, \tag{4.64}$$

where on the right-hand side we inserted the definition of the one-point functions on the upper half-plane, as well as the Jacobian from the conformal transformation $z = z(\zeta)$ for a primary field. We also assumed that $\bar{i} = \omega^{-1}(i^+)$, otherwise both sides of the equation vanish – compare to the "image charge" picture arising from the half-plane Ward identities in Subsection 4.1.1.

To compute the left-hand side explicitly, we use the conformal properties of the bulk vacuum $|0\rangle$ and of the boundary state: both are annihilated by the operators $l_n := L_n^{(P)} - \bar{L}_{-n}^{(P)}$ for $n = 0, \pm 1$. Since $\varphi_{i,\bar{i}}^{(P)}(\zeta,\bar{\zeta})$ is primary, the relations

$$\left[l_n, \varphi_{i,\bar{i}}^{(P)}(\zeta,\bar{\zeta})\right] = \left(\zeta^n \left(h_i(n+1) + \zeta\partial_\zeta\right) - \bar{\zeta}^{-n}\left(h_{\bar{i}}(-n+1) + \bar{\zeta}\partial_{\bar{\zeta}}\right)\right) \varphi_{i,\bar{i}}^{(P)}(\zeta,\bar{\zeta})$$

provide three linear differential equations for the functional dependence, with solution

$$\langle 0| \varphi_{i,\bar{i}}^{(P)}(\zeta,\bar{\zeta}) \|\alpha\rangle\!\rangle' = \frac{\delta_{\bar{i},\omega^{-1}(i^+)}}{(\zeta\bar{\zeta}-1)^{2h_i}}. \tag{4.65}$$

We have fixed the normalisation of $\varphi_{i,\bar{i}}$ by demanding the bulk OPE $\varphi_{i,\bar{i}}(\zeta,\bar{\zeta})\varphi_{i^+,\bar{i}^+}(0,0) \sim |\zeta|^{-4h_i}$ as usual, such that only the boundary state contributes a non-trivial prefactor to equation (4.65). Using the rational transformations (4.62), it is easy to check that the z-dependence is the same on both sides of equation (4.64), thus we find

$$B'^{i+}_{\alpha} = A^{\alpha}_{i\bar{i}}. \tag{4.66}$$

The boundary state coefficients and one-point function coefficients agree.

We would now like to establish that the same relation holds for the coefficients B^{i}_{α}, starting from the finite-temperature definition (4.43) of boundary states, in the ξ-plane. Here, the procedure is more involved, as we need to deal with the trace in formula (4.43).

For simplicity, we assume that our CFT is unitary, has the vacuum $|0\rangle$ as the unique state of lowest energy, and of course that all energy levels are only finitely degenerate. In this case the $\beta_0 \to 0$ behaviour of the correlators can be read off easily on the ξ-plane, where $\beta_0 \to 0$ means $\tilde{q} \to 0$:

$$_\Omega\langle\!\langle\Theta\alpha\|\,\tilde{q}^{L_0-\frac{c}{24}}\,\varphi^{(P)}_{i,\bar{i}}(\xi,\bar{\xi})\,\|\alpha\rangle\!\rangle_\Omega = \tilde{q}^{-\frac{c}{24}}\left(B^0_\alpha\,\langle 0|\,\varphi^{(P)}_{i,\bar{i}}(\xi,\bar{\xi})\,\|\alpha\rangle\!\rangle_\Omega + O(\tilde{q}^{\Delta h})\right)$$

where $\Delta h > 0$ is the lowest non-zero eigenvalue of L_0 in $\mathcal{H}^{(P)}$. Thus, the $\beta_0 \to 0$ limit of equation (4.43) with $N = 1$ becomes finite if we divide by

$$Z := {}_\Omega\langle\!\langle\Theta\alpha\|\,\tilde{q}^{L^{(P)}_0-\frac{c}{24}}\,\|\alpha\rangle\!\rangle_\Omega = \mathrm{Tr}_{\mathcal{H}_\alpha}\,q^{L^{(H)}_0-\frac{c}{24}} = \left(B^0_\alpha\right)^2\tilde{q}^{-\frac{c}{24}} + O(\tilde{q}^{\Delta h})$$

on both sides of equation (4.43). The remaining term in the ξ-plane reads

$$f^{\alpha}_{i\bar{i}}(\xi,\bar{\xi}) := \left(B^0_\alpha\right)^{-1}\langle 0|\,\varphi_{i,\bar{i}}(\xi,\bar{\xi})\,\|\alpha\rangle\!\rangle_\Omega\;; \tag{4.67}$$

as before, it vanishes unless $\bar{i} = \omega^{-1}(i^+)$. The concrete form of $f^{\alpha}_{i\bar{i}}(\xi,\bar{\xi})$ can be determined from the same Ward identities used in the zero-temperature setting, so we find

$$f^{\alpha}_{i\bar{i}}(\xi,\bar{\xi}) = \frac{B^{i+}_\alpha}{B^0_\alpha}\,\frac{\delta_{\bar{i},\omega^{-1}(i^+)}}{(\xi\bar{\xi}-1)^{2h_i}}. \tag{4.68}$$

We turn to the half-plane side of the equation, where we insert the bulk–boundary OPE for $\varphi^{(H)}_{i,\bar{i}}(z,\bar{z})$; this yields a sum of traces over boundary fields, namely

$$\left(\frac{dz}{d\xi}\right)^{h_i}\left(\frac{d\bar{z}}{d\bar{\xi}}\right)^{h_{\bar{i}}}Z^{-1}\,\mathrm{Tr}_{\mathcal{H}^{(H)}}\left(q^{L^{(H)}_0-\frac{c}{24}}\varphi^{(H)}_{i,\bar{i}}(z,\bar{z})\right)$$

$$= \left(\frac{dz}{d\xi}\right)^{h_i}\left(\frac{d\bar{z}}{d\bar{\xi}}\right)^{h_{\bar{i}}}Z^{-1}\sum_k C^{\alpha}_{(i\bar{i})\,k}\,|z-\bar{z}|^{h_k-h_i-h_{\bar{i}}}\,\mathrm{Tr}_{\mathcal{H}^{(H)}_\alpha}\left(q^{L^{(H)}_0-\frac{c}{24}}\psi_k(x)\right).$$

Recall that we have normalised both sides by the partition function Z.

Those boundary fields ψ_k that do not map any superselection sector of $\mathcal{H}^{(H)}_\alpha$ to itself drop out from the trace, but others (infinitely many since the k-summation runs over descendants, too) contribute to the $\beta_0 \to 0$ limit – which corresponds to $q \to 1$ in the z-half-plane. Each term would have to be treated by a modular transformation (acting on the argument x as well) as discussed in [355].

However, since we are only interested in a relation between numerical coefficients, we can restrict ourselves to the leading term in the limit when z approaches the boundary, $z \to \bar{z}$, with $\Re z > 0$, say. In this limit, because of the

$|z - \bar{z}|$-exponent, the term containing the trace of the identity field $k = 0$ dominates, coming with a factor $C^{\alpha}_{(i\bar{i})\,0}$. Again, we assume $\bar{i} = \omega^{-1}(i^+)$, otherwise this term would be zero.

Next we Taylor-expand $z(\xi)$ around $\bar{\xi}^{-1}$, which shows that

$$\frac{1}{(\xi\bar{\xi} - 1)^{2h_i}} \approx \left(\frac{dz}{d\xi}\right)^{h_i} \left(\frac{d\bar{z}}{d\bar{\xi}}\right)^{h_i} \frac{1}{|z - \bar{z}|^{2h_i}}$$

for $\xi\bar{\xi} \approx 1$ or $z \approx \bar{z}$, and we finally obtain an equation between (finite-temperature) boundary state coefficients and structure constants from the bulk–boundary OPE, namely

$$C^{\alpha}_{(i\,\omega^{-1}(i^+))\,0} = \frac{B^{i^+}_{\alpha}}{B^0_{\alpha}} . \tag{4.69}$$

In Section 4.2, equation (4.30), we have expressed the bulk–boundary coefficients $C^{\alpha}_{(i\bar{i})\,0}$ through the constants in the one-point functions. Relation (4.69) shows that the latter coincide with the boundary state coefficients up to an overall normalisation,

$$B^{i^+}_{\alpha} = \kappa \, A^{\alpha}_{i\omega^{-1}(i^+)} . \tag{4.70}$$

The methods employed here do not allow us to to determine κ, but we will see presently that the overall normalisation of boundary states is fixed by the non-linear Cardy constraints.

As an immediate application of relation (4.70), let us write down the Neumann and Dirichlet boundary states for a compactified free boson, using the results (4.27) and (4.28) for the one-point functions:

$$\| N(\tilde{x}_0) \rangle\!\rangle = \kappa_N(r) \sum_{w\in\mathbb{Z}} e^{2irw\tilde{x}_0} \, |(0, w)\rangle\!\rangle_N , \tag{4.71}$$

$$\| D(x_0) \rangle\!\rangle = \kappa_D(r) \sum_{k\in\mathbb{Z}} e^{ikx_0/r} \, |(k, 0)\rangle\!\rangle_D . \tag{4.72}$$

As in equations (4.27) and (4.28), the parameters x_0, $\tilde{x}_0$ take values in the target circle and dual target circle of radii r and $1/(2r)$, respectively. It can be shown that $\exp\{i\hat{x}/r\} \| D(x_0)\rangle\!\rangle = \exp\{ix_0/r\} \| D(x_0)\rangle\!\rangle$ where $\hat{x}$ is the centre-of-mass position.

The normalisations are undetermined at present, but with the help of Cardy's conditions they can be shown to be given by $\kappa_N(r) = \sqrt{r}$ and $\kappa_D(r) = \frac{1}{\sqrt{2r}}$.

4.4 Non-linear constraints

We have stressed now and then that the data specifying a boundary condition are subject to various consistency conditions, and we will make them explicit in this section.

To start with, in Subsection 4.4.1, we work with the boundary state formalism and discuss a class of constraints on boundary states, commonly known as the *Cardy condition*, which restrict the possible linear combinations (4.55) of Ishibashi states in a boundary state. While the Ishibashi states (for a given gluing condition) still form an abstract vector space, the new constraints are non-linear: the terminology boundary "states" is, therefore, even more misleading than for the generalised coherent states $|i\rangle\!\rangle_\Omega$. Rather, the $\|\alpha\rangle\!\rangle_\Omega$ should be viewed as labels for different ("non-perturbative") boundary sectors associated with a bulk CFT – much in the spirit of string theory, where it does not make sense, either, to form arbitrary linear combinations of D-branes as geometrical objects.

The derivation of Cardy's condition (based on modular transformations) is not only very elegant but, as a by-product, gives a very efficient method to compute the partition function of the boundary theory, i.e. the spectrum of boundary fields; various basic examples from free boson and WZW theories will be collected.

We will also review a special solution to Cardy's condition, the so-called *Cardy (boundary) states*. For these, the boundary state coefficients are expressed in terms of the modular S-matrix; the formula is applicable whenever rational CFT and maximally symmetric boundary conditions are considered, provided the gluing condition and bulk partition "match" in a certain way. In the special case of (diagonal) Virasoro minimal models, all possibly boundary states are Cardy states or superpositions thereof.

In view of the connection between boundary states and one-point functions established in the previous subsection, the Cardy constraints also provide a powerful tool to determine the coefficients A_{ii}^α in many cases of interest. Further non-linear constraints on correlation function coefficients, or equivalently on the various OPEs occurring in a boundary CFT, will be reviewed briefly in Subsection 4.4.2. In contrast to the "genus-one" Cardy condition (which involves an open string one-loop diagram), the remaining constraints are formulated in genus-zero – and more conveniently discussed using the half-plane formulation. Solutions to these sewing relations are available whenever the boundary condition is described by a Cardy state.

4.4.1 Cardy condition and Cardy boundary states

Cardy's derivation [100] of the conditions starts from a specialisation of our setting in Subsection 4.3.1: consider a boundary CFT with symmetry algebra $\mathcal{W}$ for which the boundary condition jumps at $z = 0$ from α to β (of course, $\alpha = \beta$ is permitted). Study the system at finite temperature $\beta_0 = -2\pi i\tau$, or make the "time" direction periodic with period β_0, then compute the partition function

$$Z_{\alpha\beta}(q) = \mathrm{Tr}_{\mathcal{H}_{\alpha\beta}^{(H)}} q^{H^{(H)}} \,, \tag{4.73}$$

where $H^{(H)} = L_0^{(H)} - \frac{c}{24}$ is the Hamiltonian in the $w = \ln z$ coordinate, and where $q = e^{2\pi i \tau}$. We assume that the boundary CFT on the half-plane has $\mathcal{W}$ as its symmetry algebra; therefore, the space $\mathcal{H}_{\alpha\beta}^{(H)}$ decomposes into irreducible representations $\mathcal{H}_i$ of $\mathcal{W}$,

$$\mathcal{H}_{\alpha\beta}^{(H)} = \bigoplus_i \mathcal{H}_i^{\oplus n_{\alpha\beta}^i}. \tag{4.74}$$

It is the space of boundary fields (BCCOs if $\alpha \neq \beta$) and can be identified with the state space of the boundary CFT if the two boundary conditions coincide.

The partition function $Z_{\alpha\beta}(q)$ encodes the energy spectrum of the boundary theory, and it is a correlator of the type (4.44), without insertions of bulk fields $\varphi(z, \bar{z})$. As in the general case, we can re-interpret this situation and compute the same partition function within the bulk theory by inserting boundary states $\|\alpha\rangle\rangle_\Omega$ and $\|\beta\rangle\rangle_{\Omega'}$ at the boundaries of the annulus. The re-interpretation involves an interchange of space and time which, in terms or the parameter τ, amounts to a modular transformation $\tau \longmapsto -1/\tau$. Cardy's constraints will arise from the modular covariance properties of conformal characters.

For simplicity, let us restrict ourselves to the case that the jump in the boundary conditions does not change the gluing automorphism, i.e. $\Omega' = \Omega$, and that Ω preserves the full symmetry algebra $\mathcal{W}$. Again, we denote by $I_\mathcal{W}$ the set of all possible representations of $\mathcal{W}$, and introduce

$$I_\mathcal{W}^\Omega := \{\, i \in I_\mathcal{W} \,|\, (i, \omega^{-1}(i^+)) \in I \,\}$$

where I labels the bulk modular invariant specifying the bulk state space (4.56). For the special quantity under consideration, the defining relation (4.44) for boundary states yields the expression

$$Z_{\alpha\beta}(q) = {}_\Omega\langle\langle \Theta\beta \| \tilde{q}^{\frac{1}{2}\left(L_0^{(P)} + \overline{L}_0^{(P)} - \frac{c}{12}\right)} \|\alpha\rangle\rangle_\Omega \tag{4.75}$$

with $\tilde{q} = e^{-2\pi i/\tau}$. The right-hand side describes free closed string propagation from the boundary state $\|\alpha\rangle\rangle$ at "time" 0 to the boundary state $\|\Theta\beta\rangle\rangle$ at "time" $2\pi^2/\beta_0 = \pi i/\tau$, driven by the Hamiltonian $H^{(P)} = L_0^{(P)} + \overline{L}_0^{(P)} - \frac{c}{12}$.

The innocent-looking identity (4.75) contains severe constraints on the boundary states $\|\alpha\rangle\rangle$ and $\|\beta\rangle\rangle$: using the decomposition (4.74) of the field space into irreducibles, the partition function $Z_{\alpha\beta}(q)$ becomes a sum of characters

$$Z_{\alpha\beta}(q) = \sum_{i \in I_\mathcal{W}} n_{\alpha\beta}^i \, \chi_i(q) \tag{4.76}$$

with $\chi_i(q) = \mathrm{Tr}_{\mathcal{H}_i} q^{L_0 - \frac{c}{24}}$ and *non-negative integer coefficients* $n_{\alpha\beta}^i$, which give the multiplicity of $\mathcal{H}_i$ in $\mathcal{H}_{\alpha\beta}^{(H)}$.

On the other hand, we can compute the "bulk amplitude" on the right-hand side of equation (4.75) explicitly, with the help of the expansion (4.55) of boundary states into Ishibashi states: first note that the gluing conditions (4.47) for $n = 0$ imply that on boundary states, the Hamiltonian $H^{(P)}$ acts as $L_0^{(P)} - \frac{c}{24}$.

Next, we observe that the damping factor $q^{L_0 - \frac{c}{24}}$ allows us to introduce an "inner product" for (the otherwise highly non-normalisable) Ishibashi states $|i\rangle\!\rangle_\Omega$: from their explicit form (4.50) and unitarity of V_Ω, we find that

$$_\Omega\langle\!\langle j|\, q^{L_0 - \frac{c}{24}}\, |i\rangle\!\rangle_\Omega = \delta_{i,j}\, \chi_i(q) \tag{4.77}$$

with $\chi_i(q) = \mathrm{Tr}_{\mathcal{H}_i}\, q^{L_0 - \frac{c}{24}}$ being the conformal character of the irreducible representation i. Equation (4.77) holds independently of the gluing automorphism, but if $|i\rangle\!\rangle_\Omega$ and $|j\rangle\!\rangle_{\Omega'}$ belong to different twistings $\Omega \neq \Omega'$, weighted traces of $V_\Omega V_{\Omega'}^{-1}$ will appear in the inner product and in the boundary CFT partition function calculated from it.

Combining this with the expansion (4.55) results in

$$_\Omega\langle\!\langle \Theta\beta \|\, \tilde{q}^{\frac{1}{2}(L_0^{(P)} + \overline{L}_0^{(P)} - \frac{c}{12})}\, \|\alpha\rangle\!\rangle_\Omega = \sum_{i \in I_{\mathcal{W}}^\Omega} B_\beta^{i^+} B_\alpha^i\, \chi_i(\tilde{q})\,. \tag{4.78}$$

Implicitly, we have used $\Theta\, B_\beta^j |j\rangle\!\rangle = \overline{B_\beta^j}\, |j^+\rangle\!\rangle$ – i.e. we have picked a special normalisation of Θ – as well as the fact that the conjugate representation i^+ must occur among the left-movers in $\mathcal{H}_{\mathrm{tot}}$ as soon as i does.

The modular transformation $q \mapsto \tilde{q}$ acts linearly on the characters,

$$\chi_i(\tilde{q}) = \sum_{j \in I_{\mathcal{W}}} S_{ij}\, \chi_j(q)$$

with the modular S-matrix introduced in Chapter 3, equations (3.72) and (3.73). Thus, we obtain the final form of Cardy's constraints on the coefficients B_α^i of the Ishibashi states making up a full boundary state:

$$\sum_{i \in I_{\mathcal{W}}} \sum_{j \in I_{\mathcal{W}}^\Omega} B_\beta^{j^+} B_\alpha^j\, S_{ji}\, \chi_i(q) = \sum_{i \in I_{\mathcal{W}}} n_{\alpha\beta}^i\, \chi_i(q) \tag{4.79}$$

for some set of (non-negative) *integers* $n_{\alpha\beta}^i$.

An immediate consequence of these non-linear conditions is that in general it is not allowed to multiply an acceptable boundary state by an overall factor other than a (positive) integer. Instead of a linear space, the solutions to equation (4.79) form a *lattice* over the integers, at least if we are prepared to drop the extra conditions $n_{\alpha\alpha}^0 = 1$ and $n_{\alpha\beta}^i \geq 0$.

In superstring theory, positivity of the $n_{\alpha\beta}^i$ must indeed be relaxed: some characters in the worldsheet partition function (4.76) correspond to spacetime fermions, whose loop diagrams contribute with a relative sign. We have seen this in Subsection 2.A.1.

A boundary condition (Ω, α) is called *elementary* if there is a unique vacuum state, i.e. if $n_{\alpha\alpha}^0 = 1$.

Per pair of boundary conditions, Cardy's constraints (4.79) give a single equation for a q-series. Nevertheless, they are often read as equations on the coefficients of the characters in equation (4.79), tacitly assuming that the latter are linearly independent. This is true, for example, in Virasoro minimal models or in

$\widehat{SU(2)}_k$ WZW theories, but false as soon as non-selfconjugate sectors occur: the CPT theorem shows that $\chi_j(q) = \chi_{j^+}(q)$. In such cases, however, the chiral algebra usually is an extension of the Virasoro algebra, and the maximal abelian subalgebra often contains zero modes W_0 of other generators, e.g. currents J_0. If this enlarged algebra is preserved by the boundary conditions – which was one of our initial requirements to make the construction more tractable – then we can use these zero modes to extend Cardy's worldsheet duality argument to "charged" characters such as the unspecialised affine Lie algebra characters; see the end of this subsection, and also [44] for an example.

In all cases known, the additional degrees of freedom are sufficient to lift the degeneracy among the characters. And even should charged characters not suffice, all possible degeneracies could still be lifted up by working directly with the finite-temperature definition (4.43) of boundary states, with arbitrary bulk fields inserted. In [355], it is shown that those correlation functions still transform linearly under the modular group.

Cardy actually concentrated on the special case of standard gluing conditions $\Omega = \mathrm{id}$ and of a bulk state space

$$\mathcal{H}_{\mathrm{tot}} = \bigoplus_{j \in I_\mathcal{W}} \mathcal{H}_j \otimes \mathcal{H}_{j^+}$$

corresponding to a charge-conjugate modular invariant partition function. Furthermore, he assumed that the model is rational, i.e. that $|I_\mathcal{W}| < \infty$. In that situation, Cardy gave a solution to the constraints (4.79) with the help of the Verlinde formula (3.76), relating S-matrix elements and fusion rules.

In Cardy's solution, the boundary states $\|a\rangle\rangle$ carry the same labels as the irreducible representations of $\mathcal{W}$, i.e. $a \in I_\mathcal{W}$, and the expansion of these *Cardy states* into Ishibashi states reads

$$\|a\rangle\rangle = \sum_{i \in I_\mathcal{W}} \frac{S_{ai}}{\sqrt{S_{0i}}} |i\rangle\rangle. \qquad (4.80)$$

Exploiting the Verlinde formula and the properties (3.73) of the modular S-matrix, it is easy to see that the partition function of the boundary CFT on the half-plane with boundary conditions described by states $\|a\rangle\rangle, \|b\rangle\rangle$ as in equation (4.80) is given by

$$Z_{ab}(q) = \sum_{i \in I_\mathcal{W}} N_{a^+b}^i \chi_i(q). \qquad (4.81)$$

Note that, because of the CPT operator we introduced into the finite-temperature correlators (4.43), we obtain an additional conjugation on the right-hand side compared to Cardy's original result [100]. As a consequence, there is always a unique vacuum sector if the (Cardy-type) boundary condition is constant along the real line, since $N_{a^+a}^0 = 1$.

Cardy's result (4.81) for the partition function is completely general and gives a concise expression for open string partition functions whenever the theory is rational and Cardy states (4.80) are available; therefore, not much can be learnt from spelling out explicit examples here. We defer a discussion of boundary SU(2) WZW models to Chapter 6, where both ordinary Cardy states and boundary states with non-trivial gluing conditions (involving inner automorphisms Ω) will occur naturally, along with partition functions between boundary states obeying different gluing conditions.

Other examples for open string partition functions were already given in Chapter 1 on free bosons and free fermions. Although the free boson is not rational (for generic or infinite compactification radius), its Dirichlet and Neumann boundary states still follow the general pattern in Cardy's solution (4.80) rather closely, compare the coefficients in the free boson boundary states (4.71) and (4.72) to the coefficients in the modular transformation rule

$$\chi_{\tilde{g}}(\tilde{q}) = \int_{-\infty}^{\infty} dg \, e^{2\pi i g \tilde{g}} \chi_g(q) \, . \tag{4.82}$$

Likewise, the spectrum between two Neumann or two Dirichlet boundary states follows Cardy's formula (4.81). For the Dirichlet–Neumann overlap $Z_{DN}(q)$ from equation (1.90), on the other hand, the gluing conditions of the two boundary states are different, thus the U(1)-symmetry is broken. This can be made explicit by rewriting the open string partition function with the help of the Jacobi triple product identity (and with some re-orderings of the absolutely convergent products):

$$Z_{ND}(q) = \frac{q^{\frac{1}{16}}}{\eta(q)} \frac{1}{2} \prod_{n=1}^{\infty} (1 - q^{\frac{n}{2}})(1 + q^{\frac{n}{2}})(1 + q^{\frac{n-1}{2}}) = \frac{1}{\eta(q)} \sum_{n=1}^{\infty} q^{\frac{1}{4}(n-\frac{1}{2})^2} \, . \tag{4.83}$$

This is a sum of irreducible Virasoro characters at $c = 1$ with highest weights $(n - \frac{1}{2})^2/4$. For these values, there are no null-states in the associated Verma modules. (As was mentioned in Subsection 1.A.1, the summands can still be viewed as characters of twisted $\widehat{U(1)}$-representations.) The lowest conformal dimension counted by the partition function $Z_{ND}(q)$ is $\frac{1}{16}$, in accordance with the interpretation of this BCCO as a spin field.

As an aside, let us note that there is a further free boson boundary state which satisfies Cardy's conditions, namely $\|\alpha\rangle\rangle = |0\rangle\rangle_D$, which leads to the same $Z_{\alpha\alpha}(q)$ as $\|\alpha\rangle\rangle = \|N\rangle\rangle$ from the uncompactified boson. However, as it is a superposition of ordinary Dirichlet boundary states $|0\rangle\rangle_D \sim \int dx_0 \, |D(x_0)\rangle\rangle$ a "delocalised D-brane" [123], this boundary condition violates the cluster property to be discussed in the next subsection.

Inspired by the special form of the coefficient matrix B_α^i in Cardy's solution (4.80), Pradisi, Sagnotti and Stanev [384] gave a fruitful definition of *completeness* of a set of boundary states. Following [44], it is convenient to change our conventions slightly from above and to define the coefficients in the boundary states by

$$\|\alpha\rangle\rangle_\Omega = \sum_{j \in I^\Omega_W} \frac{K^j_\alpha}{\sqrt{S_{0j}}} |j\rangle\rangle_\Omega \,, \tag{4.84}$$

where S denotes the modular S-matrix of the (rational) CFT, and where we assume that the boundary condition preserves the maximal symmetry – but not necessarily that $\|\alpha\rangle\rangle_\Omega$ is a Cardy state. The multiplicities of the characters in the boundary partition function $Z_{\alpha\beta}(q) = \sum_{i \in I_W} n^i_{\alpha\beta} \chi_i(q)$ are then given by

$$n^i_{\alpha\beta} = \sum_{j \in I^\Omega_W} \frac{S_{ij}}{S_{0j}} K^{j^+}_\beta K^j_\alpha \,.$$

Now let us define a *complete* set $\mathcal{B}$ of boundary states as one where the relation

$$\sum_{\alpha \in \mathcal{B}} K^i_\alpha K^{i^+}_\alpha = \delta^{i,j} \tag{4.85}$$

is satisfied [384]. This means in particular that we can write Ishibashi states as (complex) linear combinations of full boundary states from $\mathcal{B}$, although it does not in general exclude the existence of further boundary states for the same gluing conditions.

Note that if our set of boundary states is also *orthogonal* in the sense that

$$\sum_{j \in I^\Omega_W} K^j_\alpha K^{j^+}_\beta = \delta_{\alpha,\beta} \,, \tag{4.86}$$

then it follows that $n^0_{\alpha\beta} = \delta_{\alpha,\beta}$.

Using the properties of the S-matrix collected in equation (3.73) it is easy to show that the numbers $n^i_{\alpha\beta}$ enjoy the symmetry $n^{i^+}_{\alpha\beta} = n^i_{\beta\alpha}$.

More importantly, exploiting the fact that the ratios S_{ij}/S_{0j} provide the (one-dimensional) representations of the fusion rule algebra, the completeness relation suffices to show that also the $|\mathcal{B}| \times |\mathcal{B}|$ matrices n^i with entries $n^i_{\alpha\beta}$ obey [44]

$$n^i n^j = \sum_k N^k_{ij} n^k \,. \tag{4.87}$$

It is said that the matrices n^i furnish a non-negative integer matrix representation of the fusion algebra, a *NIM-rep* for short [220]. The n^i are normal, mutually commuting matrices and, in many cases, such representations of fusion rule algebras are under good control. Therefore, the completeness condition (4.85) and the constraint (4.87) on the spectrum of boundary theories are rather useful.

Relation (4.87) was first observed to hold for SU(2) WZW models on the half-plane in [383, 384], where the notion of a complete set of boundary states was also introduced. On the other hand, the fact that the NIM-rep condition (4.87) can be derived from Cardy's constraints for any rational CFT and without using Cardy's special solution for the boundary states, was realised only later [43, 44, 431, 466].

4.4.2 Genus-zero sewing relations

In the following, we want to discuss other non-linear constraints on the structure constants of a boundary CFT which arise already on the half-plane.

The basic idea is the same as the one behind the sewing relations for a CFT on the full complex plane: associativity of the operator product expansion. More precisely, correlation functions are evaluated in different limiting regimes (using different fundamental systems for the Ward identities), and comparison leads to equations involving OPE structure constants together with Moore–Seiberg duality matrices. For a theory on the half-plane, additional structure constants come into play because of the boundary fields, and there are also more sewing relations that have to be obeyed.

We should emphasise an important difference between sewing relations on the plane and on the half-plane: in the former case, the requirement of locality can be invoked (along with modular invariance of the bulk partition function) to constrain the possible left–right couplings of chiral vertex operators. In a boundary theory, we still demand that bulk fields are mutually local, but we cannot do so for boundary operators. Mutual locality of boundary fields is an exceptional phenomenon, cf. the discussions in Sections 4.2 and 5.2; in general, correlation functions involving boundary operators will have cuts, just like the chiral conformal blocks of a CFT on the plane.

The study of sewing constraints in boundary CFT was initiated by Lewellen in [334]. His results were revisited, extended to more general situations and enriched by useful further relations by Sagnotti and coworkers in [383, 384, 431]; see also [204]. Quite some time later, Runkel gave a solution of the sewing relations for the case of Virasoro minimal models [397, 398], while Zuber and collaborators [44] presented a very extensive discussion of the SU(2) WZW case. As was pointed out in the works [162, 163, 171, 201], many of these results hold for arbitrary rational CFTs. Another nice discussion of sewing relations for the simpler, but somewhat similar, case of topological branes was given in [325], using a category-theoretic framework.

Here, we will merely state the actual sewing relations and give hints on how to derive them. We will highlight a particular constraint, the cluster condition, which is simple and gives some rather useful insights. For more details, the reader is referred to the literature mentioned above. We should point out that in fully-fledged open string theory, non-orientable worldsheets have to be admitted too [399]. As a consequence, there are further consistency conditions for one-loop diagrams; see, e.g., [21, 79, 80, 401] for more details and references, and Subsection 4.A.1 for some brief remarks on crosscap states.

The sewing analysis starts from the sets of (primary) bulk fields $\varphi_i(z, \bar{z}) := \varphi_{i\bar{i}}(z, \bar{z})$ and of boundary fields $\psi_k^{\alpha\beta}(x)$, with associated structure constants C_{ijk} of the bulk OPE, with $C_{i,k}^{\alpha}$ of the bulk–boundary OPE, and with $C_{ijk}^{\alpha\beta\gamma}$ of the boundary OPE.

To keep notations lighter, we make only the left-moving $\mathcal{W}$-representation index of bulk fields explicit, but one should be aware of the possibility of non-trivial multiplicities or of different right-moving representations being coupled to a given left-moving one.

We assume that the constants C_{ijk} from the bulk OPE are already known from the bulk sewing relations reviewed in Section 3.3. In addition, the set of all $\varphi_{i\bar{\imath}}(z,\bar{z})$ should correspond to a modular invariant partition function, so that the bulk sector by itself defines a consistent CFT on the plane.

It is convenient to consider a parent CFT on the plane with a whole system of boundary conditions α, β, γ, ...; therefore we admit BCCOs.

Due to Cardy's conditions, we have to take non-trivial normalisations of the vev of the identity operator into account, and introduce the (non-vanishing) constants $A_0^\alpha \equiv A_{00}^\alpha = \langle \mathbf{1} \rangle_\alpha$. We have seen before that these, together with the bulk–boundary coefficients $C_{i,0}^\alpha$, determine all the one-point functions A_i^α, cf. equation (4.30) – and in fact all correlators containing bulk fields only, upon using the bulk OPE.

The normalisations of primary boundary fields are contained in the OPE coefficients $C_{ijk}^{\alpha\beta\gamma}$ and are not set to unity from the start.

Building on previous studies of sewing relations for CFTs on closed Riemann surfaces, Lewellen argued in [334] that there are just four additional sewing relations which must be satisfied in order to make a theory consistent on any Riemann surface with a boundary. In the string theory context, this implies that all open and closed string diagrams have the required factorisation properties [334]. One of the additional boundary sewing relations is Cardy's condition for the cylinder partition function. The other three relations can be formulated directly on the upper half-plane.

First of all, the OPE of boundary fields must be associative by itself – meaning that four-point functions $\langle \psi_i^{\alpha\beta}(x_1)\psi_j^{\beta\gamma}(x_2)\psi_k^{\gamma\delta}(x_3)\psi_l^{\delta\alpha}(x_4)\rangle$ with fixed ordering $x_1 > x_2 > x_3 > x_4$ of arguments are crossing symmetric. A "chiral" analogue of the sewing relations (3.67) is obtained for bulk fields, since boundary correlators obey Ward identities for a single copy of $\mathcal{W}$ – assuming that the whole system of boundary conditions preserves the full symmetry (otherwise, we have to work with Ward identities for the sub-symmetry algebra left unbroken). Accordingly, a single fusing matrix F appears in the boundary sewing relation

$$\sum_p C_{ijp}^{\alpha\beta\gamma}\, C_{klp+}^{\gamma\delta\alpha}\, C_{p+p0}^{\alpha\gamma\alpha}\, \mathsf{F}_{pq}\begin{bmatrix} j & k \\ i & l+ \end{bmatrix} = C_{jkq}^{\beta\gamma\delta}\, C_{iql+}^{\alpha\beta\delta}\, C_{l+l0}^{\alpha\delta\alpha}. \qquad (4.88)$$

On the left-hand side, we let x_1 approach x_2 and x_3 approach x_4 before evaluating the resulting two-point functions, which produces the combination $C_{p+p0}^{\alpha\gamma\alpha} A_0^\alpha$. On the right-hand side, x_2 approaches x_3 first before the results are "fused" with $\psi_i^{\alpha\beta}(x_1)$. The constant A_0^α has been divided out.

The fusing matrix (3.58) maps the chiral conformal blocks associated with these regimes into each other, leaving us with equation (4.88) on the numerical coefficients.

The remaining sewing relations connect bulk and boundary quantities. A correlation function of one bulk and two boundary fields $\langle \varphi_i(z, \bar{z})\psi_k^{\alpha\beta}(x_1)\psi_l^{\beta\alpha}(x_2)\rangle$ with $x_1 > x_2$ obeys the Ward identities of a chiral four-point function; thus, it is a linear combination of such conformal blocks. We can evaluate it in the regime $x_1 > x_2 > \Re z$, using the bulk–boundary OPE of $\varphi_i(z, \bar{z})$ with the α-boundary and then "computing" the three-point functions of boundary operators (which are of course fixed by conformal invariance and the boundary OPE coefficients). In the regime $x_1 > \Re z > x_2$, on the other hand, the bulk–boundary OPE of $\varphi_i(z, \bar{z})$ in the presence of the boundary condition β comes into play. Keeping track of the conformal blocks associated with both regimes, and of the duality matrices connecting them, we arrive at the sewing relation [334]

$$C_{i,q}^{\alpha} \, C_{kql+}^{\alpha\beta\beta} \, C_{l+l0}^{\alpha\beta\alpha} = \sum_{p,m} e^{i\pi\Delta_{p,m}} \, \mathsf{F}_{p+m}\!\left[\begin{smallmatrix} \bar{\imath} & l \\ i+ & k \end{smallmatrix}\right] \mathsf{F}_{mq}\!\left[\begin{smallmatrix} i & \bar{\imath} \\ k+ & l \end{smallmatrix}\right] C_{i,p}^{\beta} \, C_{klp+}^{\alpha\beta\alpha} \, C_{pp+0}^{\alpha\alpha\alpha}$$

(4.89)

with the abbreviation $\Delta_{p,m} := 2h_m - 2h_i - h_k - h_l + \frac{1}{2}(h_p + h_q)$. The phase comes from trading braiding matrices for fusing matrices, see equation (3.65), and also from our convention to use the modulus $|z - \bar{z}|$ in the bulk–boundary OPE (4.29). In writing equation (4.89) we have assumed $\Omega = \mathrm{id}$ for ease of notation. In the general case, the right-moving label $\bar{\imath}$ in the fusing matrix has to be replaced by $\omega(\bar{\imath})$, as follows from the Ward identities.

Finally, we need to consider correlators $\langle \varphi_i(z, \bar{z})\varphi_j(w, \bar{w})\psi_k^{\alpha\alpha}(x)\rangle$ of two bulk and one boundary field, which satisfy Ward identities of chiral five-point functions. The two limiting regimes to be compared arise, respectively, from sending both bulk fields to the boundary and then computing boundary three-point functions and from performing a bulk OPE first and using a single bulk-boundary OPE afterwards. The basis transformation connecting the different five-point conformal blocks can be split into a sequence of "duality moves" for four-point blocks, thus the same fusing and braiding matrices as above and in Section 1.1 enter the sewing relation. The outcome is

$$C_{ijm} \, C_{m,k}^{\alpha} = \sum_{p,q,r} e^{i\frac{\pi}{2}\Delta_{p,q,r}} \, \mathsf{F}_{qr}\!\left[\begin{smallmatrix} k & \bar{\jmath} \\ p+ & j \end{smallmatrix}\right] \mathsf{F}_{p+m}\!\left[\begin{smallmatrix} \bar{\imath} & r \\ i & j \end{smallmatrix}\right] \mathsf{F}_{r\bar{m}}\!\left[\begin{smallmatrix} \bar{\imath} & \bar{\jmath} \\ m & k \end{smallmatrix}\right] C_{i,p}^{\alpha} \, C_{j,q}^{\alpha} \, C_{pqk+}^{\alpha\alpha\alpha}$$

(4.90)

where $\Delta_{p,q,r} := h_k + h_p - h_q - 2h_r + h_m - h_{\bar{m}} - h_i + h_{\bar{\imath}} + h_j + h_{\bar{\jmath}}$. A very detailed derivation of this relation can be found in [398] for the minimal model case where all sectors are self-conjugate, but see also [44]. The same comment about $\omega \neq \mathrm{id}$ as above applies. While this equation looks rather involved, it yields a simple and rather useful relation, the cluster condition upon some specialisations; we will discuss this below.

The sewing relations listed here and the Cardy conditions from the previous subsection provide an overdetermined set of non-linear equations on the structure constants of a boundary CFT. In practice, we are often satisfied with determining all structure constants of interest, in particular the one-point functions from

Cardy's conditions, and then trust that they can in principle be completed to a consistent solution of the remaining sewing constraints. In many cases, circumstantial evidence that this is possible comes from other sources, e.g. when the boundary theory in question arises in a critical limit of integrable field theories or lattice models, or via some geometric construction of D-branes.

Nevertheless, substantial new insight into the structure constants of the boundary OPE has been obtained since the late 1990s. Virasoro minimal models and SU(2) WZW models (with ADE modular invariants) were treated in [397, 398] and [44], respectively, in both cases using standard gluing conditions. For the Cardy case with a diagonal partition function, the authors of [44] relate the problem of solving the boundary sewing relations to that of constructing the Moore–Seiberg duality data for the bulk theory.

We refrain from quoting the complete formulae from [44, 397, 398] here. Their main common feature is that the fusing matrices feature prominently in expressions for boundary and bulk–boundary structure constants (together with the modular S-matrix). This is perhaps not entirely surprising in view of the structure of the associativity constraint (4.88) on the boundary OPE: ignoring the coefficients with a vacuum index as normalisations, it has the rough form $\sum C \cdot C \cdot \mathsf{F} = C \cdot C$, so setting $C = \mathsf{F}$ makes it look like the pentagon relation (3.63) for the fusing matrix.

This vague reasoning can be made precise, see in particular [44, 162, 163, 201, 397]. For Cardy boundary states in a CFT with a charge-conjugate modular invariant bulk partition function, we find the relation

$$C_{ijk}^{\alpha\beta\gamma} = \mathsf{F}_{\beta k}\begin{bmatrix} i & j \\ \alpha & \gamma \end{bmatrix}; \qquad (4.91)$$

here we have suppressed any multiplicity labels distinguishing different fusion channels as in equation (3.55); see, e.g., [44] for the general case with $N_{ij}^k > 1$.

Since OPE coefficients are normally viewed as "dynamical data" of a CFT, relation (4.91) is somewhat surprising: it implies that the structure constants of the boundary OPE are representation-theoretic data of the chiral symmetry algebra, at least if Cardy-type boundary conditions hold. The analysis of the cluster condition given below will suggest that bulk–boundary and even bulk OPE coefficients can be related to data from the "duality category" as well.

In [163] and in subsequent work [171, 201], principles behind the connection between boundary and bulk data are explored (for the case of Cardy-type boundary conditions) by placing them into a unifying context of three-dimensional topological field theory. It would be very interesting to pursue these relations in greater generality, beyond the Cardy case – not only to understand the intrinsic structure of boundary CFT better, but also in view of the "geometric" role of the coefficients C_{ijk}^{α} which we will uncover in Chapter 6: in a certain limit, the C_{ijk}^{α} determine the algebra of functions on the worldvolume algebra of a D-brane, viewed as a non-commutative space.

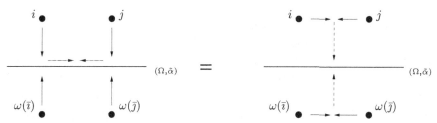

Figure 4.5 Two ways to compute a two-point function of bulk fields, leading to the cluster condition below. The fusions indicated by solid arrows are performed before those corresponding to the dashed arrows.

Together with Cardy's conditions, the genus-zero sewing relations listed above form a complete set of constraints on the structure constants of a boundary CFT, but they are rather unwieldy and can, up to now, be solved completely only for a special class of boundary conditions. We can, however, derive a much more concise constraint from the bulk–boundary sewing relation (4.90), the so-called *cluster condition*, which is almost as practical as Cardy's conditions.

To do this, we specialise to $\psi_k^{\alpha\alpha}(x) = 1$ in equation (4.90), and require $\bar{\imath} = \omega^{-1}(i^+)$; analogously for $\bar{\jmath}$. Then, the sewing relation concerns a bulk two-point function, and equates the result of the following two methods of computation: either perform a bulk OPE first and then a bulk–boundary OPE for the resulting bulk field, or perform two bulk–boundary OPEs first and then evaluate a boundary two-point function – see also Figure 4.5.

It is not too difficult to see that, since we set $k = 0$, only the terms $q = p^+$ and $r = \bar{\jmath}$ contribute to the sum on the right-hand side of relation (4.90), and that the first and the third fusing matrix become trivial (equal to 1). Moreover, the phase factor disappears in this special case.

Next one multiplies with the inverse of the remaining F, and then projects onto the vacuum fusion channel $p = 0$. At least in a unitary CFT, this can be interpreted as isolating the long-distance behaviour of the bulk field two-point function in the presence of a boundary: for $|\Re z - \Re w| \gg 1$, the vacuum channel dominates (as all other fields have higher scaling dimension); hence the name cluster condition.

After having performed these steps, we arrive at the following relation

$$C_{i,0}^{\alpha} \, C_{j,0}^{\alpha} = \sum_m \mathsf{F}_{m0}\!\left[\begin{smallmatrix} j & \bar{\jmath} \\ i & \bar{\imath} \end{smallmatrix}\right] C_{ijm} \, C_{m,0}^{\alpha} \, . \tag{4.92}$$

Using formula (4.30) to pass from bulk OPE coefficients to coefficients of the one-point functions yields the very simple equation

$$A_i^{\alpha} \, A_j^{\alpha} = \sum_m \Xi_{ijm} \, A_0^{\alpha} \, A_m^{\alpha} \tag{4.93}$$

where $A_i^{\alpha} := A_{i,\omega^{-1}(i^+)}^{\alpha}$ and where Ξ abbreviates $\mathsf{F}\,C$ from above. Note that the Ξ_{ijm} depend only on bulk data.

We will come back to the Ξ_{ijm} below, but even irrespective of their concrete values, the cluster condition (4.93) immediately provides some information which, in a sense, is complementary to Cardy's constraints.

The latter do not permit scalings of one-point functions by an arbitrary common factor, because the multiplicities $n^k_{\alpha'\alpha'}$ in the partition function must be integers; the cluster condition (4.93) is invariant under such scalings. On the other hand, Cardy's conditions allow for superpositions $\|\alpha'\rangle\rangle = \|\alpha\rangle\rangle + \|\beta\rangle\rangle$ of two consistent boundary conditions, at least if we do not insist on $n^0_{\alpha'\alpha'} = 1$. In contrast, condition (4.93) is usually violated by adding up one-point functions, $A^{\alpha'}_i = A^\alpha_i + A^\beta_i$, since it is not an additive relation. Thus, while *systems* of "coexisting" boundary conditions are perfectly alright as far as genus-zero sewing relations are concerned, *superpositions* of boundary conditions usually spoil the sewing relations.

We will make use of this feature in our discussion of deformations of boundary conditions in Chapter 5: violation of the cluster condition at certain points of the moduli space will indicate that, for these parameter values, a single brane has been converted into a superposition of several "elementary" boundary conditions. We can then try to "resolve" the superposed boundary condition into a system of elementary boundary conditions, which contains BCCOs and satisfies all consistency relations.

In some simple cases, the constraint (4.93) is even sufficient to determine the one-point functions – up to a common scaling factor. For the free boson, only a single fusion channel of charge $m = i + j$ appears on the right-hand side, and both the fusing matrix and the bulk OPE coefficients are equal to 1. Thus, equation (4.93) becomes the functional equation of the exponential,

$$\frac{A^\alpha_i}{A^\alpha_0}\frac{A^\alpha_j}{A^\alpha_0} = \frac{A^\alpha_{i+j}}{A^\alpha_0}$$

and we recover the one-point functions from Subsection 4.1.2 directly.

Let us briefly return to the coefficients Ξ_{ijk} that appear in the cluster condition. They can in principle be computed from fusing matrices and bulk OPE, but both are typically very hard to compute. The specific combination Ξ_{ijk}, on the other hand, seems to be associated with a simple algebraic structure. To see this, we change our point of view and assume that we have found an orthogonal set $\mathcal{B}$ of boundary states, with boundary state coefficients $K^i_\alpha/\sqrt{S_{0i}}$ as in equation (4.84). Then we can write the relevant bulk–boundary coefficients as $C^\alpha_{i,0} = D^{-1}_i K^i_\alpha/K^0_\alpha$, where $D_i = (S_{0i}/S_{00})^{\frac{1}{2}}$ is the quantum dimension of the sector i. It is then straightforward to derive from the orthogonality relation (4.86) that [44]

$$C^\alpha_{i,0}\, C^\alpha_{j,0} = \sum_k \frac{D_k}{D_i D_j}\, M^k_{ij}\, C^\alpha_{k,0} \quad \text{with} \quad M^k_{ij} = \sum_{\beta\in\mathcal{B}} \frac{K^i_\beta K^j_\beta K^{k^+}_\beta}{K^0_\beta}. \tag{4.94}$$

Comparison with the sewing relation (4.92) suggests a direct relation between fusing matrix and bulk OPE coefficients on the one hand with the one-point functions K_α^j on the other.

If $\mathcal{B}$ consists of Cardy boundary states for a rational CFT, i.e. $K_\alpha^i = S_{\alpha i}$, we conclude that

$$\Xi_{\mathrm{ijm}} = \left(\frac{S_{00} S_{0k}}{S_{0i} S_{0j}} \right)^{\frac{1}{2}} N_{ij}^k .\tag{4.95}$$

Equation (4.94) was first used in [383, 384] to construct boundary conditions for SU(2) WZW models with non-diagonal bulk modular invariant partition functions. This case was investigated in more detail in [44], where it was pointed out that the M_{ij}^k form the structure constants of the so-called "Pasquier algebras", generalisations of fusion rule algebras which also occur in the context of integrable lattice models. In [373], connections to the wider context of Ocneanu cell calculus were made. Furthermore, the ideas in [43, 44, 171, 201, 373] show that boundary CFT constraints can be used to draw conclusions on the possible modular invariant partition functions of the parent bulk CFT.

This concludes our exposition of the general structure of boundary CFT. The first main feature is the reduced amount of symmetry in comparison to CFT on surfaces without boundaries, with the gluing conditions (4.6) determining what symmetries are preserved by the boundary condition and how they are implemented on correlation functions. Further consequences of introducing a boundary are the possibility of non-trivial bulk fields having non-trivial one-point functions (4.17), as well as the existence of new degrees of freedom localised on the boundary, the boundary fields. Coefficients of one-point functions and of OPEs of boundary and bulk fields are subject to non-linear constraints, namely genus-zero sewing relations (among them the cluster condition (4.93)), as well as Cardy's conditions, which involve worldsheet duality in a way similar to modular invariance of bulk CFT partition functions. Much of the structure of a boundary CFT can be encoded in boundary states, which is particularly useful for the study of Cardy's conditions. There is a generic class of maximally symmetric boundary conditions, described by the Cardy boundary states (4.80), where solutions to the non-linear constraints are known. Most importantly, they come with the very simple formula (4.81) for boundary partition functions, counting open strings stretched between two branes of the Cardy type.

4.A Additional material

The common theme in this section could be summarised as "beyond the Cardy case": Cardy's general formula for boundary states allows us to produce a wealth of examples but, in applications, some of the underlying assumptions are not satisfied. We may, for example, be interested in (maximally symmetric) boundary conditions for gluing conditions which are not "properly aligned" with the given bulk partition function (using, say, $\Omega = $ id and a non-diagonal modular invariant). Or there may be reasons to look for boundary conditions which break part of the maximal symmetry, potentially rendering the boundary CFT non-rational. In those cases, some of the general prescriptions from the Cardy case may fail and may need to be revisited.

We start, however, with a very brief description of objects that are not boundary states at all, but crosscap states. These provide the worldsheet description of orientifold branes (rather than ordinary D-branes), important degrees of freedom in unoriented superstring theories. Subsection 4.A.2 reviews some essential features of the orbifold construction, which is one of the most prolific methods for producing a new CFT from an old one, requiring only that the latter carries the action of a finite group. The new orbifold CFT has the same central charge but, in general, its modular invariant bulk partition function is not of the diagonal or charge-conjugate type – so in particular we cannot rely on Cardy boundary states to describe orbifold branes. The closely related case of simple current extensions is reviewed in Subsection 4.A.3.

Permutation branes for tensor products of (diagonal) rational CFTs are discussed in Subsection 4.A.4; here, a complete list of rational boundary states for gluing conditions which permute the tensor factors can be given, as a generalisation of Cardy states. We add a few remarks on defects, which can also be viewed as boundary conditions for tensor product theories.

Subsection 4.A.5 addresses the problem of finding maximally symmetry-breaking conformal boundary states for $c = 1$ theories – so far these are the only family of CFTs, apart from Virasoro minimal models of course, where a complete list of conformal boundary states can be given. The construction in particular shows that the cluster condition can be a powerful computational tool.

4.A.1 Crosscap states and orientifolds

Having introduced boundary states, let us briefly remark on crosscap states, which are very similar at the level of concrete formulas, although they arise in a rather different context, namely from unoriented string theory (like type I), which involves worldsheets such as the Möbius strip or the Klein bottle. Starting from a closed string theory (on ordinary oriented worldsheets), an associated unoriented string theory can be obtained via orbifolding by worldsheet parity $(t, x) \mapsto (t, -x)$. This in particular maps left- to right-movers. The resulting theory loses orientation (hence the term "orientifold"), and it loses (at least) half of the symmetry. Therefore, the effect of orientifolding can again be implemented by objects which are similar to boundary states, the so-called crosscap states. Note however, that orientifolding does not introduce boundaries to the worldsheet: the process can be visualised as cutting out a disk from a closed string worldsheet followed by identifying opposite points of the disk boundary (this gives the crosscap, or $\mathbb{R}P^2$). To construct the crosscap "gluing conditions", we proceed as for boundary states except that the left-moving field $W(\xi)$ is identified with the right-mover $\overline{W}(-\bar{\xi})$, sitting at the opposite point of the disk boundary; ξ is defined as in equation (4.40). Keeping track of signs in the transformations, we arrive at the conditions

$$\left(W_n - (-1)^{h_W} (-1)^n \overline{W}_{-n} \right) |j\rangle\rangle^{\mathrm{cc}} = 0 \qquad (4.96)$$

for the crosscap Ishibashi states. These replace equations (4.49), but there is a simple relation of crosscap to ordinary Ishibashi states, namely

$$|j\rangle\rangle^{\mathrm{cc}} = e^{i\pi(L_0 - h_j)} |j\rangle\rangle .$$

Full crosscap states are linear combinations of the $|j\rangle\rangle^{\mathrm{cc}}$, with coefficients restricted by the requirement that overlaps between two crosscap states and between a boundary and a crosscap state have worldsheet dual interpretations as one-loop diagrams (the Klein bottle and the Möbius strip, respectively) – i.e. the overlaps have to decompose into integer linear combinations of characters. There is again a universal solution for bulk theories with charge-conjugate modular invariant partition function, generalising Cardy's boundary states; it involves matrix elements of the modular S-matrix and of another matrix that appears naturally in the modular transformations of the Möbius strip amplitude, namely of

$$P := T^{\frac{1}{2}} S T^2 S T^{\frac{1}{2}} ;$$

see e.g. [383] and [21] for further references. With this, a (symmetry-preserving) crosscap state is given by

$$\|0\rangle\rangle^{\mathrm{cc}} = \sum_j \frac{P_{0j}}{\sqrt{S_{0j}}} |j\rangle\rangle^{\mathrm{cc}} . \qquad (4.97)$$

This choice ensures that all worldsheet duality conditions are satisfied, see e.g. [383, 384]; to show this, we use Verlinde's formula and the fact that the matrix elements

$$Y_{ij}^k = \sum_l \frac{S_{il} P_{jl} \overline{P}_{kl}}{S_{0l}}$$

are integers and form a representation of the fusion rule algebra, too. Klein bottle and Möbius strip amplitudes are given as linear combinations of characters with certain Y_{ij}^k as coefficients. Details can be found in, e.g., [21, 36, 79, 219, 244].

If the rational CFT in question has a group of simple currents, we can construct one crosscap state per simple current, see, e.g., [79, 291, 383, 400]. In formula (4.97) P_{0j} is simpley replaced by P_{Jj} where J is the label of a simple current primary field.

Worldsheet parity can be dressed with a non-trivial W-algebra automorphism; then the crosscap states are related to twisted Ishibashi states (4.53), their existence depending on the form of the bulk modular invariant partition function as usual. If the additional automorphism comes from a geometric group action on a target manifold (here adopting a sigma-model point of view), then modding out by parity and automorphism leads to what is known as orientifold planes: extended objects in the target given as fixed point sets of the geometric group action. For example, in the CFT of a compactified free boson worldsheet parity can be dressed by an additional $X \mapsto -X$ action, and orbifolding leads from a circle to an interval target, whose endpoints correspond to two orientifold planes. Orbifolding by worldsheet parity alone leads, in a theory of ten free bosons, to so-called O9 orientifold planes (see, e.g., [118, 376]). Orientifolds have been studied in a number of non-trivial CFT backgrounds, including Gepner models [79, 80] and WZW models [71] – where orientifold planes can take the form of conjugacy classes in the Lie group target.

Similar to D-branes, orientifold planes furnish sources for the graviton, the dilaton and some Ramond–Ramond (R–R) fields, so they can be assigned tensions and R–R-charges. Accordingly, they lead to tadpoles for massless space-time fields (non-vanishing closed string one-point functions for the corresponding vertex operators, which may induce a renormalisation group (RG) flow of the background), and in order to ensure stability of the configuration those tadpoles should be cancelled by introducing D-branes: in sigma-models with compact target, uncancelled tadpoles lead to violations of Gauss' law. We will not pursue questions of tadpole cancellation, nor work with orientifolds and crosscaps in this text.

4.A.2 Orbifolds

Here we review ideas and collect formulas associated with the orbifold construction, which is a method to construct new conformal models from an old one which carries an action of some finite symmetry group G on the state and field space. As long as the energy-momentum tensor is G-invariant, a new CFT can be obtained with the same central charge by projecting onto G-invariant fields (and states), and including so-called twisted sectors if necessary. Orbifolds are a rather vast subject, and we will not attempt to give a full account here; our emphasis will be on conceptual ideas and on giving motivations for the formulas appearing in the literature, focusing on the worldsheet approach but invoking natural geometric ideas from string theory where useful.

We start by considering orbifolds in the bulk, mainly focusing on partition functions, then pass to the question of how boundary states can be constructed for orbifold models; due to the interplay of gluing conditions and details of the orbifold action, it is surprisingly difficult to make model-independent statements, let alone give a complete list even of maximally symmetric boundary conditions. But we will present a few generic constructions, and we will also try to explain terminology used in the string literature.

Bulk theory

To gain some intuition into bulk orbifold theories, let us first look at the sigma-model picture, where the CFT is viewed as a field theory of maps X from a worldsheet Σ (without boundary, for the time being) into a target space M, on which the finite group G acts geometrically. Dividing the image of a closed string in M by the G-action will certainly induce a closed string in the orbifolded sigma-model with target M/G, but in addition to this projection onto G-invariant strings, new closed strings in M/G can be built from "non-closed" strings in M that start at a point $x \in M$ and end at its image $g \cdot x$ for some $g \in G$. These strings form the g-twisted sector in the M/G theory [136]. In the free boson examples to be reviewed very briefly below, it can be seen that twisted sector states are in fact localised at fixed points under the G-action on the target. Note that M/G will have singularities if the G-action is not free, but string propagation (as in particular the worldsheet description shows) is not troubled by those.

In the worldsheet approach, we start from a bulk CFT $\mathcal{C}$ with chiral W-algebra $\mathcal{W}$ and modular invariant bulk partition function $Z_\mathcal{C}(q, \bar{q})$ on the torus. We assume that some finite group G acts on states and fields, $\varphi \mapsto g \cdot \varphi$ for all $g \in G$, as a symmetry. This action of G on states induces one on the chiral W-algebra $\mathcal{W}$. The chiral W-algebra of the orbifolded theory $\mathcal{C}/G$ is the G-invariant subalgebra, denoted $\mathcal{W}^G$, and it is required to contain the energy-momentum tensor $T, \bar{T}$ of the original theory $\mathcal{C}$.

Any irreducible representation of $\mathcal{W}$ that was present in $\mathcal{C}$ certainly induces a (not necessarily irreducible) representation of $\mathcal{W}^G$, but there may be additional irreducibles for this subalgebra which are not inherited from those in $\mathcal{C}$ – in particular so-called twisted representations. The most common way to construct such twisted sectors is to place the model on a finite cover of the cylinder and impose the boundary condition $\varphi(e^{2\pi i} z, e^{-2\pi i} \bar{z}) = g \cdot \varphi(z, \bar{z})$ for any fixed element $g \in G$. The resulting fields are still local with respect to $\mathcal{W}^G$, and the associated state space will be denoted by $\mathcal{H}^g$. For $g = e$, we recover the usual state space $\mathcal{H}^e = \mathcal{H}$ of the bulk theory. Under the action of the two commuting chiral W-algebras $\mathcal{W}^G$ and $\overline{\mathcal{W}}^G$, the spaces $\mathcal{H}^g$ decompose into a sum of irreducible sectors. We shall comment on the characters of these representations below.

The fact that these additional twist fields actually have to appear, and that (together with those from $\mathcal{C}$) they are sufficient to build a new CFT, can be seen by looking at the theory on a torus $z \equiv z + 1 \equiv z + \tau$. Here, the period τ is associated to the (Euclidean) "time-like" b-cycle and the period 1 to the "space-like" a-cycle of the torus.

The partition function $Z_\mathcal{C}(q, \bar{q})$ can be projected onto G-invariant states by inserting the projector $P = |G|^{-1} \sum_{g \in G} g$ into the partition function. The resulting function, however, is in general no longer modular invariant. In fact, the quantity

$$Z(e, g; q, \bar{q}) = \mathrm{tr}_\mathcal{H}(g\, q^{L_0 - \frac{c}{24}} \bar{q}^{\bar{L}_0 - \frac{c}{24}})$$

for $g \in G$ is naturally associated with boundary conditions $\varphi(z + \tau, \bar{z} + \bar{\tau}) = g \cdot \varphi(z, \bar{z})$ along the b-cycle of the torus; a modular S-transformation interchanges the time with the space cycle (a), so in order to ensure modular invariance we need to add representations of the orbifold W-algebra $\mathcal{W}^G$ which satisfy boundary conditions $\varphi(z + 1, \bar{z} + 1) = g \cdot \varphi(z, \bar{z})$ along the a-cycle. Fields with this property form the g-twisted sector $\mathcal{H}^g$ of the orbifold theory. Hence, from the worldsheet point of view, it is the requirement of modular invariance rather than geometric target-space considerations that enforces the presence of twisted sectors [130].

As we pointed out before, the twisted sectors $\mathcal{H}^g$ may contain several distinct irreducibles of $\mathcal{W}^G$, and it is worthwhile taking a closer look at their characters, following [130]. The G-action on $\mathcal{W}$ induces one on labels of W-irreducibles, which we shall denote by $i \mapsto g(i)$. Whenever $g(i) \neq i$, the contribution $\mathrm{tr}_{\mathcal{H}_i}(g\, q^{L_0 - \frac{c}{24}})$ to the G-symmetrised partition function vanishes and thus does not generate any twisted sectors under the S-transformation. Instead, twisted sectors arise whenever $\mathcal{H}_i$ is mapped to itself by some non-trivial G-element g. Group elements g that leave the label i fixed form a subgroup $S_i \subset G$. Generically, the set of traces $\mathrm{tr}_{\mathcal{H}_i}(g\, q^{L_0 - \frac{c}{24}})$ with $g \in S_i$, will be linearly independent. From these functions and their modular transforms we can build all the characters of the chiral W-algebra $\mathcal{W}^G$ by forming suitable linear combinations.

In the worldsheet description, the G-action on M is replaced by that on fields and representation labels, but nevertheless there is a similar link between fixed points of that action and twisted sectors as in the sigma-model picture.

The total partition function of the orbifold theory with an *abelian* orbifold group G can be written as

$$Z_{C/G}(q, \bar{q}) = \frac{1}{|G|} \sum_{g,h \in G} Z(g, h; q, \bar{q}) \tag{4.98}$$

where

$$Z(g, h; q, \bar{q}) = \mathrm{tr}_{\mathcal{H}^g}\left(h\, q^{L_0 - \frac{c}{24}} \bar{q}^{\bar{L}_0 - \frac{c}{24}}\right).$$

For *non-abelian* G, the double sum over g, h has to be restricted to those satisfying $ghg^{-1}g^{-1} = e$, where e is the neutral element of G. This can be seen by noting that $ghg^{-1}g^{-1}$ is the overall G-action a field φ picks up when transported along the cycle $a \circ b \circ a^{-1} \circ b^{-1}$ of the torus – which is null-homologous, so $ghg^{-1}h^{-1}$ must be trivial. Introducing the centraliser $N_g := \{h \in G \mid ghg^{-1}h^{-1} = e\}$ of $g \in G$, the orbifold partition function for non-abelian G takes the form

$$Z_{C/G}(q, \bar{q}) = \frac{1}{|G|} \sum_{g \in G} \sum_{h \in N_g} Z(g, h; q, \bar{q}) = \sum_{\alpha} \frac{1}{|N_\alpha|} \sum_{h \in N_{g(\alpha)}} Z(g^{(\alpha)}, h; q, \bar{q}).$$

$$\tag{4.99}$$

The second equality includes a summation over the conjugacy classes C_α of G, with $g^{(\alpha)} \in C_\alpha$ an arbitrary representative and $|N_\alpha| := |N_{g^{(\alpha)}}|$, which is independent of the choice of $g^{(\alpha)}$; we have that $|G| = |C_\alpha| |N_\alpha|$. The equality then follows using $Z(g, h) = Z(fgf^{-1}f, ghf^{-1})$. This reformulation makes it clear that twisted sectors should be labelled by the conjugacy classes of the orbifold group; see, e.g., [130, 235]. Also note that, due to the restriction $gh = hg$ in the summation over $G \times G$, projecting onto G-invariant states in the $g^{(\alpha)}$-twisted sector only involves averaging over $N_{g^{(\alpha)}}$. In the following, however, we will focus on the case of abelian orbifolds.

For some groups, non-trivial *discrete torsion* phases $\epsilon(g, h)$ can be chosen in front of the summands $Z(g, h; q, \bar{q})$; these phases have to form a representation of G,

$$\epsilon(g_1 g_2, h) = \epsilon(g_1, h)\epsilon(g_2, h),$$

and need to satisfy

$$\epsilon(g^a h^b, g^c h^d) = \epsilon(g, h)$$

for all integers a, b, c, d with $ad - bc = 1$ so as to guarantee modular invariance of

$$Z_{C/G}(q, \bar{q}) = \frac{1}{|G|} \sum_{g \in G} \sum_{h \in N_g} \epsilon(g, h)\, Z(g, h; q, \bar{q}).$$

The relations in particular imply

$$\epsilon(e, g) = \epsilon(g, g) = 1 \quad \text{and} \quad \epsilon(g, h) = \epsilon(h, g)^{-1},$$

and it can be shown [440] that the $\epsilon(g, h)$ are in one-to-one correspondence with two-cocyles $c(g, h) \in H^2(G, U(1))$ via $\epsilon(g, h) = c(g, h)\, c(h, g)^{-1}$.

We will not discuss discrete torsion in any more detail here, but refer to the literature, in particular to [117, 213, 242] for a study of branes in discrete torsion orbifolds.

Boundary theory

To discuss boundary conditions for orbifold theories, it is again beneficial to look at the sigma-model picture first: the worldvolumes of branes in an orbifold theory should be invariant under the G-action, and the simplest way to achieve this is to take a brane with worldvolume W in the covering theory, then form a configuration made up of all images of W under G, and finally project this to the orbifold target M/G. There, the G-images form a stack of coincident branes, generically made up from $|G|$ constituents. Therefore, the open strings supported by such so-called *regular branes* carry Chan–Paton factors in a natural way; moreover, the orbifold group acts on the Chan–Paton indices by permuting the stack constituents, i.e. in the regular representation. New effects (leading to so-called *fractional branes*, see below) arise in connection with fixed points of the G-action, when the worldvolume $W \subset M$ is (partially) invariant under the G-action [147, 150, 234].

To a certain extent, these ideas carry over to the worldsheet description. In particular, it is straightforward to write down boundary states for regular branes in the orbifold theory $\mathcal{C}/G$: we start with a boundary state $\| B \rangle\!\rangle^{\mathcal{C}}$ of the "covering" CFT $\mathcal{C}$, such that $g \| B \rangle\!\rangle^{\mathcal{C}} \neq \| B \rangle\!\rangle^{\mathcal{C}}$ for all $g \in G$. Then,

$$\| B \rangle\!\rangle^{\mathcal{C}/G}_{\mathrm{reg}} := \frac{1}{\sqrt{|G|}} \sum_{g \in G} g \| B \rangle\!\rangle^{\mathcal{C}} \qquad (4.100)$$

is a boundary state for $\mathcal{C}/G$. The normalisation factor ensures that the vacuum representation occurs once in the self-overlap of the state (4.100), provided that $\| B \rangle\!\rangle^{\mathcal{C}}$ is an elementary boundary state for the original CFT $\mathcal{C}$. Note that the formula (4.100) for the boundary state of the orbifold theory is very general, it is not restricted to, say, the Cardy case.

The partition function counting open strings supported by the orbifold brane (4.100) is computed as usual using a modular transformation of the self-overlap; due to the exchange of space and time direction, this leads to twisted representations. Since the expansion of the boundary partition functions in terms of $\mathcal{W}^G$ characters depends on the bulk theory and gluing conditions, we shall defer concrete expressions to a more specific setup; see our discussion of simple current extensions in Subsection 4.A.3.

The fact that the boundary state (4.100) involves a superposition of $|G|$ elementary branes in the parent theory $\mathcal{C}$ suggests the introduction of Chan–Paton labels, i.e. the organisation of the open strings of the brane configuration in matrices ψ^{rs} where $r, s = 1, \ldots, |G|$. The group G acts on these as

$$\hat{g} \cdot \psi^{rs} = \gamma(g)^{rr'} \, (U(g)\psi)^{r's'} \, \gamma(g^{-1})^{s's} \qquad (4.101)$$

where $U(g)$ denotes an "internal" operation much as in the bulk case, while the γ furnish a representation of G on the Chan–Paton labels [234, 382], as follows from consistency of the boundary OPE. For regular branes as defined in equation (4.100), γ is the regular representation of the orbifold group G, which is given by G acting on itself by left translation.

Let us now look at the opposite case, i.e. we suppose $\| B \rangle\!\rangle^{\mathcal{C}}_{\mathrm{fix}}$ is a boundary state for $\mathcal{C}$ that is invariant under the orbifold group, $g \| B \rangle\!\rangle^{\mathcal{C}}_{\mathrm{fix}} = \| B \rangle\!\rangle^{\mathcal{C}}_{\mathrm{fix}}$ for all $g \in G$. We might be tempted to use such a $\| B \rangle\!\rangle^{\mathcal{C}}_{\mathrm{fix}}$ directly as a boundary state for the orbifold theory $\mathcal{C}/G$, but this would require that normalisations can be adjusted correctly:

Computing the overlap between $\| B \rangle\!\rangle^{\mathcal{C}}_{\mathrm{fix}}$ and a regular boundary state $\| \tilde{B} \rangle\!\rangle^{\mathcal{C}/G}_{\mathrm{reg}}$ of the form (4.100) leads to the partition function $\sqrt{|G|} \, Z_{B\tilde{B}}(q)$ with $Z_{B\tilde{B}}(q)$ from the covering theory $\mathcal{C}$. Generically, this violates Cardy's conditions. To compensate, we could try to multiply the fixed-point boundary state by $\sqrt{|G|}$, but the resulting brane would no longer be elementary since it supports $|G|$ vacuum sectors instead of a single one. Dividing instead by $\sqrt{|G|}$ seems even worse, as the self-overlap would yield a partition function with fractional coefficients in front of the characters.

A way out is offered by adding extra contributions involving Ishibashi states from the twisted sectors $\mathcal{H}^g$ of the bulk theory. Schematically, orbifold boundary states associated with fixed points of the G-action have the form

$$\| B; \rho \rangle\!\rangle_{\text{fix}}^{\mathcal{C}/G} := \frac{1}{\sqrt{|G|}} \| B \rangle\!\rangle_{\text{fix}}^{\mathcal{C}} + \sum_{g \in G \backslash \{e\}} \rho(g) \, \| B \rangle\!\rangle_{\mathcal{H}^g}^{\mathcal{C}/G} . \tag{4.102}$$

Here, the subscript $\mathcal{H}^g$ indicates that the summand originates from the g-twisted sector of the orbifold theory; $\rho(g)$ denotes a g-dependent prefactor, in examples given by a suitably normalised group character. Similar expressions apply in mixed cases when the boundary state of the parent theory $\mathcal{C}$ is left invariant by a subgroup of the full orbifold group G.

To our knowledge, no completely model-independent formulas for $\rho(g)$ or for the relative scalings of Ishibashi states contributing to the twisted sector terms in equation (4.102) have been found. Building on [130], the work [58] gives expressions that apply to arbitrary orbifolds $\mathcal{C}/G$ of holomorphic CFTs $\mathcal{C}$, i.e. where $Z^{\mathcal{C}}(q, \bar{q}) = \chi_0(q) \bar{\chi}_0(q)$. Many other examples are discussed extensively in the literature, including of course free boson orbifolds – see below for the case $S^1/\mathbb{Z}^2$ – and simple current orbifolds.

Irrespective of the detailed form of the boundary states (4.102) we can make a few observations of a general nature:

First of all, adding contributions from the twisted sector to $\| B \rangle\!\rangle_{\text{fix}}^{\mathcal{C}}$ does not alter overlaps with regular branes (4.100) (which contain untwisted Ishibashi states only), thus the partition functions $Z_{B\tilde{B}}(q)$ discussed above are properly normalised.

Likewise, those additions do not affect the one-point functions of (G-invariant) bulk fields from the untwisted sector. Relative to regular branes (4.100), such one-point functions for fixed point branes will be scaled down by a factor $1/|G|$, as there is no summation over G-images. In the superstring context, this implies that their (untwisted) R–R-charges are a fraction of those of regular branes, hence the name "fractional branes" [124, 125, 147, 150]. Adding a piece from the twisted sector does, however, affect the self-overlap of the fixed-point boundary state. In fact, since twisted sectors can be viewed as arising upon modular transformation of fixed sectors in the first place, it is entirely plausible that adding a twisted sector contribution may cure any normalisation issues the partition function of $|G|^{-\frac{1}{2}} \| B \rangle\!\rangle_{\text{fix}}^{\mathcal{C}}$ had.

In typical examples (see below), the functions $\rho(g)$ in (4.102) are such that after averaging the states (4.102) over all ρ, we are left with the untwisted sector contribution $|G|^{\frac{1}{2}} \|B\rangle\!\rangle_{\text{fix}}^{\mathcal{C}}$ only. In this sense, the fractional branes (4.102) resolve the orbifold fixed point; see in particular [147]. In addition, the state $|G|^{\frac{1}{2}} \|B\rangle\!\rangle_{\text{fix}}^{\mathcal{C}}$ is a consistent – though not elementary – boundary state, i.e. it describes a consistent brane configuration comprising several elementary objects.

At the same time, since the fractional branes do not contain a stack of $|G|$ branes as the regular branes do, they allow for different G-representations to act on the Chan–Paton factors: if $\|B\rangle\!\rangle_{\text{fix}}^{\mathcal{C}}$ is invariant under all $g \in G$, the γ occurring in the action (4.101) can be any irreducible representation of G; if $\|B\rangle\!\rangle_{\text{fix}}^{\mathcal{C}}$ is replaced by a $\mathcal{C}$ boundary state that is invariant under only a subgroup of G – called the stabiliser – other representations γ can show up; see [213] for an example.

Example: free boson

Let us now look at the simple example of the orbifold $S^1/\mathbb{Z}_2$, which will help make some of the schematic remarks from before more concrete. The bulk theory arises from a compactified free boson by dividing out the left–right symmetric $\mathbb{Z}_2$-action $X \longmapsto -X$ on the compactified free boson theories, discussed in great detail, e.g., in [235]. The chiral fields are the invariant elements of the $\widehat{U(1)} \times \widehat{U(1)}$ current algebra; the bulk Hilbert space consists of an untwisted sector containing all $\mathbb{Z}_2$-invariant states of the free boson Hilbert space and of two twisted sectors $\mathcal{H}_0^{\text{tw}}$ and $\mathcal{H}_{\pi r}^{\text{tw}}$ built up over twist fields of left and right conformal dimension $h_{0,\pi r}^{\text{tw}} = \frac{1}{16}$. The subscripts refer to the endpoints of the interval $[0, \pi r]$, which can be regarded as the target space of the orbifold model at radius r. Indeed, in this simple case one can see explicitly, via a mode expansion of the free boson, that the twisted boundary conditions $X(e^{2\pi i}z) = -X(z)$ imply that the constant mode is restricted to the $\mathbb{Z}_2$ fixed points, see, e.g., [367].

We list boundary states for the $S^1/\mathbb{Z}_2$ orbifold, following Oshikawa's and Affleck's work [367]. Consider the untwisted sector first. The free boson Ishibashi states are already given as $\mathbb{Z}_2$-invariant exponentials of $\sum a_{-n}\bar{a}_{-n}$ acting on $\widehat{U(1)}$ ground states; therefore, we merely have to symmetrise in the latter to obtain *untwisted* orbifold boundary states from ordinary free boson Dirichlet or Neumann boundary states,

$$\|D(x_0)\rangle\!\rangle^{\text{orb}} := \frac{1}{\sqrt{2}} \left(\|D(x_0)\rangle\!\rangle^{\text{circ}} + \|D(-x_0)\rangle\!\rangle^{\text{circ}} \right), \qquad (4.103)$$

$$\|N(\tilde{x}_0)\rangle\!\rangle^{\text{orb}} := \frac{1}{\sqrt{2}} \left(\|N(\tilde{x}_0)\rangle\!\rangle^{\text{circ}} + \|N(-\tilde{x}_0)\rangle\!\rangle^{\text{circ}} \right). \qquad (4.104)$$

The parameters range over the intervals $0 < x_0 < \pi r$ and $0 < \tilde{x}_0 < \frac{\pi}{2r}$. In terms of one-point functions, formula (4.103) means that

$$\left\langle \cos\left(\frac{k}{r}X(z,\bar{z})\right) \right\rangle_{D(x_0)}^{\text{orb}} = \frac{1}{\sqrt{2r}} \frac{\cos\frac{kx_0}{r}}{|z-\bar{z}|^{k^2/4r^2}} \qquad (4.105)$$

and that no bulk twist fields couple to the identity on the boundary. A similar formula holds for Neumann boundary conditions of the orbifold theory. Obviously, the boundary states (4.103,4.104) are the regular branes (4.100) of this theory.

To each fixed point of the $\mathbb{Z}_2$-action on S^1, two *twisted Dirichlet* and two *twisted Neumann boundary states* are assigned, made up from the corresponding circle boundary states and the (appropriately symmetrised) Dirichlet or Neumann Ishibashi states of $\mathcal{H}_{0,\pi r}^{\rm tw}$; see [367] for the details. With $\xi = 0, \pi r$ and $\tilde{\xi} = 0, \frac{\pi}{2r}$, we write

$$\|D(\xi), \pm\rangle\!\rangle^{\rm orb} := 2^{-\frac{1}{2}} \|D(\xi)\rangle\!\rangle^{\rm circ} \pm 2^{-\frac{1}{4}} \|D(\xi)\rangle\!\rangle^{\rm tw}, \qquad (4.106)$$

$$\|N(\tilde{\xi}), \pm\rangle\!\rangle^{\rm orb} := 2^{-\frac{1}{2}} \|N(\tilde{\xi})\rangle\!\rangle^{\rm circ} \pm 2^{-\frac{1}{4}} \|N(\tilde{\xi})\rangle\!\rangle^{\rm tw}. \qquad (4.107)$$

These boundary states represent the fractional branes of the $S^1/\mathbb{Z}_2$ orbifold model. In particular, we recognise the prefactor $|G|^{-\frac{1}{2}}$ in front of the fixed boundary state from the untwisted sector, as in the general formula (4.102). Moreover, the signs in front of the twisted sector contribution come from group characters of the two irreducible representations of $\mathbb{Z}_2$, namely the trivial representation $\gamma(g) = 1$ and the only non-trivial irrep $\gamma(g) - g$ for $g = \pm 1$. The two boundary states $\|D(\xi), \pm\rangle\!\rangle^{\rm orb}$ resolve the fixed point in the sense that adding up $\|D(\xi), +\rangle\!\rangle^{\rm orb}$ and $\|D(\xi), -\rangle\!\rangle^{\rm orb}$ returns the untwisted sector fixed-point contribution $2^{\frac{1}{2}} \|D(\xi)\rangle\!\rangle^{\rm circ}$. This shows that the latter is not an elementary brane, even though the normalisations are such that all partition functions have integer coefficients.

The prefactors in equations (4.106) and (4.107) ensure proper, and minimal, normalisation of all partition functions $Z_{\alpha\beta}(q)$ with α, β taken from the two lists (4.103–4.107). In Subsection 5.3.3, we will give explicit expressions at least for the case $\alpha = \beta$, i.e. for partition functions describing the spectrum of open strings supported by one such orbifold brane. If α is a regular brane as in boundary states (4.103) and (4.104), the computation involves the overlap between two different branes in the circle theory, located at x_0 and $-x_0$, respectively; thus it should be no surprise that the spectrum of open strings in general depends on the location x_0 of the orbifold brane: after dividing out the $\mathbb{Z}_2$ action, the target is no longer translation-invariant. Likewise, the concrete expressions for the partition functions will also show that they are no longer given as sums of $\widehat{U(1)}$ characters.

4.A.3 Simple current extensions

There is another closely related method to obtain new modular invariant bulk partition functions from old ones. It can be performed whenever the original rational CFT contains a simple current; the construction is a variation on the orbifold procedure used to obtain $Z^{\mathcal{C}}/G$ from $Z^{\mathcal{C}}$ described above, but possesses some new features: most importantly, the simple current construction can lead to CFTs with an extended chiral algebra $\mathcal{W}_{\text{ext}}$, while $\mathcal{W}$ is reduced to its invariant subalgebra $\mathcal{W}^G$ in a true orbifold. The simple current construction fits into the present context because it leads to bulk partition functions which are not of diagonal or charge-conjugate type, at first sight. If, however, these can be re-interpreted as diagonal modular invariants of an extension $\mathcal{W}_{\text{ext}}$ (as is often the case), and if the modular S-matrix S_{ext} for $\mathcal{W}_{\text{ext}}$ is known, $\mathcal{W}_{\text{ext}}$ symmetric boundary states for the seemingly non-diagonal model can be written down simply as Cardy states for $\mathcal{W}_{\text{ext}}$.

Bulk theory

We first list some of the ingredients of the simple current construction in the bulk; see [405, 406] and also [293]. Suppose we have a rational CFT with (bosonic) chiral symmetry algebra $\mathcal{W}$, which admits a simple current J, i.e. an irreducible representation which is invertible at the level of the fusion algebra,

$$[J] \times [J^c] = [0] \,,$$

where J^c is the conjugate of J. This relation implies that the fusion product of J with any sector i only contains a single $\mathcal{W}$ irrep $[Ji]$, and that the OPE with the associated primary field ϕ_i reads

$$J(z)\phi_i(w) = (z - w)^{h_{Ji} - h_i - h_J} \, \phi_{(Ji)}(w) + \cdots \,;$$

here we focus on the holomorphic coordinate dependence, and the dots stand for descendants. In this way, the (sub)group of simple currents (under consideration) acts on the set of $\mathcal{W}$ representation labels, and we denote the orbit of $[i]$ by $\{i\}$ in the following.

The OPE above suggests the introduction of the *monodromy charge* of ϕ_i with respect to the simple current J,

$$Q_J(i) := h_i + h_J - h_{Ji} \,(\text{mod }1)\,. \tag{4.108}$$

This fractional number governs the effect of moving J around ϕ_i, and accordingly it can be expressed through the braiding matrix; see, e.g., [84, 348]. Due to its interpretation in terms of a monodromy, Q_J behaves additively under OPEs, i.e. $Q_J(i) + Q_J(j) = Q_J(k)$ if $N_{ij}^k \neq 0$. It can thus be viewed as associated with some "hidden" symmetry of the CFT, and it can be subjected to an orbifold procedure where G is a group of simple currents. The action of G on states is defined through the monodromy charge so that $Z^{\mathcal{C}}(q, \bar{q})$ gets projected onto states with vanishing monodromy charge (the untwisted sector) before twisted sectors are added. In simple current extensions, the latter consist of all representations that arise from the untwisted sector by fusion with the simple current.

There are different possibilities for implementing this general recipe; these choices and further model-dependent properties determine whether interesting new modular invariants, and which ones, are obtained.

For example, we can act with the product $J(z)J^c(\bar{z})$ to generate all twisted sector states from untwisted ones. This procedure works irrespectively of the conformal dimension of the simple current. If J is not self-conjugate, i.e. $J \neq J^c$, then the simple current construction with this twist field is guaranteed to yield a non-diagonal partition function from a diagonal one (see [405] for more details).

On the other hand, if J has integer conformal dimension and is self-local (i.e. $Q_J(J) = 0$), we have the option to perform a simple current construction using $J(z)$ as a chiral twist field; the right-moving partner $J(\bar{z})$ then enters as a separate field. This gives rise to partition functions that are non-diagonal with respect to the original symmetry algebra $\mathcal{W} \times \mathcal{W}$, but diagonal with respect to the extension $\mathcal{W}_{\text{ext}} \times \mathcal{W}_{\text{ext}}$ by $J(z)$ and $J(\bar{z})$. We quote the resulting partition function (in the form given in [208]) for the more general case that a whole group G of simple currents with the above properties is "orbifolded":

$$Z^{\mathcal{C}\,\text{ext}}(q, \bar{q}) = \sum_{\substack{\{i\} \\ \text{s.th. } Q_J(i) = 0\ \forall J \in G}} |S_i| \, \Big| \sum_{k \in \{i\}} \chi_k(q) \Big|^2 . \qquad (4.109)$$

The normalisation in front of the second sum is the order $|S_i| = |G|/|\{i\}|$ of the stabiliser S_i of the representation $[i]$, the subgroup of all simple currents in G that leaves $[i]$ fixed.

Famous examples of extended partition functions of this form are provided by the D-type modular invariants [97] for the $\widehat{su}(2)_k$ affine Lie algebra at level $k = 4n$; see Subsection 6.4.1 for an extensive discussion.

Non-trivial stabilisers lead to multiplicities of characters in the extended partition function, which need to be resolved, e.g. when one attempts to find the modular S-matrix of the extended theory (see, e.g., [202, 208] for useful expositions). Typically, the primary fields of the extended theory are labelled $[i, \rho^{(i)}]$ where $\rho^{(i)}$ is a character of the stabiliser group S_i. The $\rho^{(i)}$, along with the elements of the original S-matrix, then also enter S_{ext}. For some simple current orbifolds, modular properties are known from other sources, in particular for the SU(2) D-models; see, e.g., [44] and references therein.

Boundary theory

Once the extended modular S-matrix is known, it is of course straightforward to write down boundary states for the bulk CFT with the partition function (4.109), preserving the extended symmetry algebra $\mathcal{W}_{\text{ext}}$ with respect to which (4.109) is diagonal: we can simply use Cardy states built from extended Ishibashi states with coefficients expressed in terms of S_{ext}.

Alternatively, we can use the orbifold constructions outlined in the previous subsection to construct maximally symmetric boundary theories. Since the action of the orbifold group G on states is given by the monodromy charge, regular boundary states (4.100) are built by projecting on representations i of the original theory with $Q_J(i) = 0$. We label these regular branes by $\{I\}$. Using standard properties of the modular S-matrix under the action of simple currents we can then compute the partition function for regular branes (4.100) and obtain

$$Z^{\mathrm{orb}}_{\{I\}\{J\}}(q) = \sum_{g,k} N^{(gJ)}_{I\,k}\, \chi_k(q)\,. \tag{4.110}$$

This can easily be matched with expectations from the geometric picture of branes on orbifolds. The I, J can be considered as geometric labels specifying brane positions on the covering space (i.e. as labels of Cardy branes in the original CFT). To compute the spectrum of two branes $\{I\}$ and $\{J\}$ of the orbifold theory, we lift $\{I\}$ to one of its pre-images I on the covering space and include all the open strings that stretch between this fixed brane I on the cover and an arbitrary pre-image (gJ) of the second brane $\{J\}$.

As in the general case, the regular branes $\{I\}$ need not be elementary. This happens whenever the stabiliser subgroup S_I is non-trivial. In this case, it may be possible to resolve $\{I\}$, i.e. write it as a superposition of elementary boundary states with positive integer coefficients.

The resolved elementary branes carry an extra label a associated with a choice of character $\rho^{(I)}_a$ of S_I. Geometrically, this corresponds to the fact that the Chan–Paton factors of branes at orbifold fixed points can carry different representations of the stabiliser subgroup, as explained before.

Boundary states for such resolutions take the general form sketched in equation (4.102). The most interesting piece of information is given by the spectrum of open strings stretching between two such resolved branes $\{I\}_a$ and $\{J\}_b$ in a simple current extension. To spell out the result, we need some more notation: any characters $\rho^{(I)}_a : S_I \to \mathrm{U}(1)$ and $\rho^{(J)}_b : S_J \to \mathrm{U}(1)$ of the stabilisers induce characters of the subgroup $H_{IJ} := S_I \cap S_J$. Furthermore, for every fixed representation $[k]$, the monodromy charge provides another character $h \mapsto (-1)^{Q_h(k)}$ of H_{IJ}. Exploiting orthonormality of group characters, we can conclude that the numbers

$$d^k_{ab} := \frac{1}{|H_{IJ}|} \sum_{h \in H_{IJ}} \rho^{(I)}_a(h)\, (-1)^{Q_h(k)}\, \rho^{(J)}(h^{-1}) \tag{4.111}$$

are equal to 0 or 1. The partition function for two resolved branes can then be written as (see [348])

$$Z^{\mathrm{orb}}_{\{I\}_a\{J\}_b}(q) = \frac{1}{|S_I \cdot S_J|} \sum_{g,k} N^{(gJ)}_{I\,k}\, d^k_{ab}\, \chi_k(q)\,; \tag{4.112}$$

for corresponding formulas for resolved branes in general orbifold theories, see [43, 44, 207].

The coefficients appearing in front of the conformal characters $\chi_k(q)$ are in fact integers. To see this, note that by definition of the stabilisers S_I and S_J and by associativity of the fusion rules,

$$N_{Ik}^{(gJ)} = N_{Ik}^{(g_1 g g_2^{-1} J)} \quad \text{for all} \quad g_1 \in S_I \,,\ g_2 \in S_J \,.$$

Since $g = g_1 g g_2^{-1}$ for $g_1 = g_2 \in H_{IJ}$, this means that every term in equation (4.112) appears with multiplicity

$$|S_I \cdot S_J| = |S_I|\,|S_J|\,|H_{IJ}|^{-1} \,.$$

Note that the partition functions (4.112) of the resolved branes sum up to the partition function (4.110) of the unresolved projected boundary states. This follows easily from the property $\sum_{a,b} d_{ab}^k = |S_I \cdot S_J|$ of the constants d, which is in turn a consequence of the completeness of group characters.

This concludes our brief discussion of simple current orbifolds; further general results, including formulas for the boundary OPE [348], can be found in the literature. We will come back to the partition functions (4.110) and (4.112) in Chapter 6 when we discuss the special case of orbifolds of the SU(2) WZW model.

4.A.4 Permutation branes

In applications, CFTs are often encountered which are tensor products of simpler, rational models, e.g. of super Virasoro minimal models in the case of the Gepner construction. If the symmetry algebras of N of the component models are identical we have additional outer automorphisms at our disposal to formulate gluing conditions: the tensor product theory admits outer automorphisms which act by permutations of the components of $\mathcal{W}^N$, namely

$$\Omega_\pi : W^{[k]}(z) \longmapsto W^{[\pi(k)]}(z) \,,$$

where $\pi \in S_N$ is a permutation and where $W^{[k]}(z) := \mathbf{1} \otimes \cdots \otimes W(z) \otimes \cdots \otimes \mathbf{1}$ denotes the action of a $\mathcal{W}$-generator in the kth component theory, for $k = 1, \ldots, N$.

We start from a unitary rational CFT on the plane with chiral (left- and right-moving) symmetry algebras $\mathcal{W}_L = \mathcal{W}_R = \mathcal{W}$, and with a charge-conjugate partition function $Z(q, \bar{q}) = \sum_{i \in \mathcal{I}} \chi_i(q) \chi_{i^+}(q)^*$ associated with a decomposition $\mathcal{H} = \bigoplus_{i \in \mathcal{I}} \mathcal{H}_i \otimes \mathcal{H}_{i^+}$ of the state space into irreducibles. For notational convenience, let us assume that all sectors are self-conjugate, i.e. that the fusion rules satisfy $N_{ii}^0 = 1$ where 0 denotes the vacuum sector.

We can then form an N-fold tensor product of this rational CFT, with chiral algebra $\mathcal{W}^N := \mathcal{W} \times \cdots \times \mathcal{W}$ and partition function

$$Z^{(N)}(q, \bar{q}) = \big(Z(q, \bar{q}) \big)^N = \sum_{I \in \mathcal{I}^N} \chi_I(q) \chi_I(q)^* \,, \qquad (4.113)$$

where we have used the multi-index $I := (i_1, \ldots, i_N)$ to label characters of the tensor product theory, $\chi_I(q) := \chi_{i_1}(q) \cdots \chi_{i_N}(q)$. Again, the partition function $Z^{(N)}(q, \bar{q})$ is a diagonal modular invariant and, at the same time, because of our simplifying assumption $i = i^+$, also a charge-conjugate one.

We want to construct boundary conditions for the tensor product theory with Ω_π appearing as a gluing automorphism, i.e. we link left- and right-moving generators by the condition

$$W^{[k]}(z) = \overline{W}^{[\pi(k)]}(\bar{z}) \tag{4.114}$$

along the boundary $z = \bar{z}$ of the upper half-plane. These boundary conditions are conformal since Ω_π leaves the diagonal energy-momentum tensor $T = T^{[1]} + \cdots + T^{[N]}$ fixed. The gluing conditions (4.114) in fact guarantee that the full symmetry algebra $\mathcal{W}^N$ is represented on the Hilbert space of the boundary CFT even though the analytic continuation from upper to lower half-plane prescribed by equation (4.114) is not the standard one.

Ishibashi states that implement the permutation gluing conditions (4.114) can be expanded as

$$|I\rangle\!\rangle_\pi = \sum_M |i_1, M_1\rangle \otimes \cdots \otimes |i_N, M_N\rangle$$

$$\otimes\, U|i_{\pi^{-1}(1)}, M_{\pi^{-1}(1)}\rangle \otimes \cdots \otimes U|i_{\pi^{-1}(N)}, M_{\pi^{-1}(N)}\rangle \tag{4.115}$$

where $M = (M_1, \ldots, M_N)$ is used to label orthonormal bases $|i_k, M_k\rangle$ of energy eigenstates in the representations $\mathcal{H}_{i_k}$ of $\mathcal{W}$, and where the operator U in front of the right-movers is the chiral CPT operator as usual. It is important to realise that the objects $|I\rangle\!\rangle_\pi$ are available only for certain multi-indices $I = (i_1, \ldots, i_N)$: since the partition function of the bulk theory is diagonal,

$$|I\rangle\!\rangle_\pi \quad \text{exists iff} \quad i_k = i_{\pi^{-1}(k)} \quad \text{for all } k = 1, \ldots, N. \tag{4.116}$$

This implies that two $\mathcal{W}$-representations i_k, i_l showing up in the label I of a permutation Ishibashi state $|I\rangle\!\rangle_\pi$ have to coincide whenever k can be reached from l by repeatedly applying π. To formalise this, we use the fact that any permutation π can be decomposed into a number P^π of cycles

$$C_\nu^\pi = \left(n_\nu^\pi \ \pi(n_\nu^\pi) \ \ldots \ \pi^{\Lambda_\nu^\pi - 1}(n_\nu^\pi) \right)$$

of length Λ_ν^π, for $\nu = 1, \ldots, P^\pi$. The permutation then becomes

$$\pi = \left(n_1^\pi \ \pi(n_1^\pi) \ \ldots \ \pi^{\Lambda_1^\pi - 1}(n_1^\pi) \right) \left(n_2^\pi \ \pi(n_2^\pi) \ \ldots \ \pi^{\Lambda_2^\pi - 1}(n_2^\pi) \right)$$

$$\ldots \left(n_{P^\pi}^\pi \ \pi(n_{P^\pi}^\pi) \ \ldots \ \pi^{\Lambda_{P^\pi}^\pi - 1}(n_{P^\pi}^\pi) \right). \tag{4.117}$$

We have chosen some arbitrary element $n_\nu^\pi \in \{1, \ldots, N\}$ as representative of the νth cycle here, and those n_ν^π will be kept fixed in the following.

An Ishibashi state $|I\rangle\!\rangle_\pi$ exists iff $i_k = i_l$ whenever k and l belong to the same cycle of π; we will abbreviate this condition by inserting a Kronecker symbol $\delta_I^{C^\pi}$ into the following formulae. Note that, even though $i_k = i_l$, the summation indices M_k, M_l in formula (4.115) are independent of each other – which in particular makes it clear that the permutation Ishibashi states are not just superpositions of standard ones.

Full-fledged boundary states $\|a\rangle\!\rangle_\pi$ for the gluing conditions (4.114) can be written as linear combinations of the Ishibashi states (4.115),

$$\|a\rangle\!\rangle_\pi = \sum_I \delta_I^{C^\pi} B_\alpha^I \, |I\rangle\!\rangle_\pi \,. \tag{4.118}$$

In the paper [386], an ansatz for the coefficients B_α^I was presented along with two important consistency checks, concerning the cluster condition and Cardy's conditions, respectively, on the strip partition functions.

The coefficients in equation (4.118) contain a multi-index $a = (\alpha_1, \dots, \alpha_{n_P})$ with as many components as there are independent labels $i_{n_\nu^\pi}$ in the Ishibashi states $|I\rangle\!\rangle_\pi$; each α_ν is taken from the label set $\mathcal{I}$ of all $\mathcal{W}$-representations. The formula for B_α^I is multiplicative in the $i_{n_\nu^\pi}$ and reduces to Cardy's solution for the component theory in the special case $N = 1$. The coefficients are

$$B_\alpha^I = B_{\alpha_1}^{i_{n_1}} \cdots B_{\alpha_P}^{i_{n_P}} \quad \text{with} \quad B_{\alpha_\nu}^{i_{n_\nu}} = \frac{S_{\alpha_\nu \, i_{n_\nu}}}{\left(S_{0 \, i_{n_\nu}}\right)^{\frac{\Lambda_\nu}{2}}} \,; \tag{4.119}$$

we have dropped the superscript $^\pi$ from the cycle representatives n_ν^π, the cycle lengths Λ_ν^π and the number of cycles P^π. As usual, the matrix S implements modular transformation of the $\mathcal{W}$-characters; therefore, the matrix (B_α^I) is invertible, and the ansatz (4.119) will provide a complete set, in the sense of [383], of boundary states for the fixed gluing condition Ω_π – provided all sewing constraints are satisfied.

We refer to [386] for detailed proofs that the boundary states defined by equations (4.118) and (4.119) satisfy cluster condition and Cardy constraints. The cluster condition was discussed under the additional assumption $N_{ij}^k < 2$ on the fusion rules of the component CFT in [386], for notational convenience; due to the multiplicative nature of the coefficients (4.119), it can essentially be traced back to the cluster condition for Cardy states of the component theory, in the form (4.95).

The proof that the Cardy constraints are satisfied is much more technical, but it is still worthwhile to spell out some of the partition functions describing open strings stretched between two permutation branes $\|a\rangle\!\rangle_\pi$ and $\|\beta\rangle\!\rangle_\sigma$. Their form depends crucially on the "relative position" of the cycles of π and σ. If both permutations coincide, we obtain an expression involving characters of the

N-fold tensor product algebra $\mathcal{W}^N$,

$$Z_{\alpha_\pi \beta_\pi}(q) := {}_\pi\langle\!\langle \beta \| \tilde{q}^{L_0 - \frac{c}{24}} \| a \rangle\!\rangle_\pi$$

$$= \sum_{j_1,\dots,j_N} \prod_{\nu=1}^{P^\pi} \left(\prod_{k \in C_\nu^\pi} N_{j_k} \right)_{\alpha_\nu \beta_\nu} \chi_{j_1}(q) \cdots \chi_{j_N}(q). \quad (4.120)$$

On the other hand, if one of the permutations is trivial, we arrive at characters of the P-fold tensor product algebra $\mathcal{W}^P$, where $P = P^\pi$ is the number of cycles of the other permutation π, and the arguments of the characters involve the cycle lengths:

$$Z_{\alpha_\pi \beta_{\mathrm{id}}}(q) = \sum_{j_1,\dots,j_P} \left[\prod_{\nu=1}^{P} \left(\prod_{k \in C_\nu^\pi} N_{\beta_k} \right)_{\alpha_\nu j_\nu} \right] \chi_{j_1}(q^{\frac{1}{\Lambda_1}}) \cdots \chi_{j_P}(q^{\frac{1}{\Lambda_P}}). \quad (4.121)$$

In both cases, the multiplicities involve the fusion rules; but in contrast to simple tensor products of Cardy boundary states (i.e. $\pi = \mathrm{id}$), they enter by way of cycle-wise products of fusion matrices, leading to novel open string spectra. The appearance of arguments $q^{\frac{1}{\Lambda_\mu}}$ in the characters, on the other hand, can be related to cyclic orbifolds (see the remarks in [386]).

Other consistency conditions for permutation branes such as associativity of OPEs have not been discussed in the literature as yet, except for the simplest case $N = 2$, where we can make an amusing observation concerning the OPE of boundary fields supported by the special permutation brane $\| 0 \rangle\!\rangle_{(1\,2)}$: the spectrum of this brane

$$Z_{0_\pi 0_\pi}(q) = \sum_{j \in \mathcal{I}} \chi_j(q) \chi_j(q) \quad (4.122)$$

"coincides" with that of the original bulk theory if we "identify" the right-moving charges of bulk fields with the second tensor factors of (chiral) boundary fields. Furthermore, it is easy to see that the associativity condition for the boundary OPE for $\| 0 \rangle\!\rangle_{(1\,2)}$ indeed coincides with that for the bulk OPE of the original bulk theory – suggesting that the OPE coefficients of the latter also provide a solution to the OPE of boundary fields that live on the brane $\| 0 \rangle\!\rangle_{(1\,2)}$. Even though other sewing relations remain to be checked, in particular those involving BCCOs between this brane and two-fold tensor products of Cardy branes, there is some evidence from the geometric sigma-model point of view that the special boundary condition $\| 0 \rangle\!\rangle_{(1\,2)}$ is indeed related to the target space of the (one-component) bulk theory; see [386] for some further remarks.

The construction of permutation branes outlined above can be extended and adapted to more general cases than the tensor-product theories (4.113), in particular to Gepner models, which can be viewed as orbifolds of tensor products. In this way, a large set of additional boundary states is obtained, including some of prime importance from the geometrical point of view (namely D0-branes); see Chapter 7 for more details.

Let us conclude this subsection with a few brief remarks on *defects*, objects whose technical description displays some common features with that of permutation branes. A defect is an interface between two conformal field theories, a ("disorder") line separating two regions of the plane which support two (potentially different) bulk theories. For example, the defect line could be the real axis, with CFT_1 given on the upper and CFT_2 on the lower half-plane. In this situation, we can fold the plane back into a two-sheeted upper half-plane with the defect residing at its boundary, and we realise that (conformal) defects are precisely given by (conformal) boundary conditions for the tensor product theory $CFT_1 \otimes \overline{CFT}_2$. Here, $\overline{CFT}_2$ denotes CFT_2 with holomorphic and anti-holomorphic degrees of freedom interchanged: note that the boundary has opposite relative orientation with respect to the upper and lower half-plane, and to account for this, we should pass from (t, x) to $(t, -x)$ in the lower half-plane before folding; see, e.g., [30, 458].

Special boundary states for a tensor product theory can of course be obtained by taking tensor products of boundary states for the two factors. The corresponding defects are called "totally reflective", they separate the two CFTs completely without any coupling. If the two CFTs are identical, we can consider the opposite extreme of "totally transmissive" defects, which simply correspond to transposition branes associated with the permutation $(1\,2)$ as introduced above.

Alternatively, a defect can be described as an operator X_D between the bulk state spaces of the two CFTs separated by the defect. (One way to see this is to consider a defect along a non-contractible loop on a cylindrical worldsheet; when viewed as a closed string diagram, the cylinder is a matrix element between closed string states of an operator associated to the disorder line.) This description was first used in [372], and it becomes particularly useful in the special case of two identical CFTs and defect lines which are invariant under infinitesimal diffeomorphisms, i.e. which satisfy $[L_n, X_D] = [\overline{L}_n, X_D] = 0$. Such defects are called "topological defects", and it can be shown that at least in rational CFTs X_D can be written as a linear combination of projection operators onto $\mathcal{H}_i \otimes \mathcal{H}_{\bar{\imath}}$, with coefficients satisfying consistency relations very similar to Cardy's conditions.

Defects come equipped with additional degrees of freedom, defect fields, in complete analogy to boundary fields supported by branes. A new feature is that defects can be "multiplied", and can act on boundary conditions. These *defect fusion* operations are obvious at an intuitive level:

Take two parallel defect lines D_{12} and D_{23} and move one on top of the other; this leads to a defect D_{13} between CFT_1 and CFT_3. Or take a boundary condition B_2 for CFT_2 and move a defect D_{12} on top of it; this results in some boundary condition B_1 for CFT_1:

$$D_{12} \star D_{23} = D_{13}, \quad D_{12} \star B_2 = B_1.$$

Normally, however, correlators diverge in the limit of coinciding points, so making these intuitive "fusion" operations $\star$ precise is a subtle problem (see [32] for a discussion in a special case). Topological defects, on the other hand, can be deformed and moved freely in the worldsheet, and fusion with other defects or boundary conditions [81] can be made well defined (as the correlators do not depend on the location of the topological defect). It can, for example, be used to study boundary RG flow patterns [246] or the effect of bulk RG flows on boundary conditions [82]. Sometimes, so-called group-like defects exist (those which have an inverse under defect fusion), and their structure gives insights into the symmetries of a CFT; we refer to [29, 191] for further discussion of these and related issues.

The pictorial realisation of defects as lines in the plane suggests another structure that has no analogue for boundary conditions: defects can end on other defects, so that *junctions* can be formed with three or more lines emanating from the junction point. Discussions can be found in, e.g., [30, 92, 107].

The study of interfaces between different materials, thus of defects, is obviously of great interest to condensed matter physics – see [367] for some early remarks and [32] for a detailed discussion treating both structural issues and applications to the Kondo problem. The full potential of defects in the study of abstract CFT and of string theory is still being explored, and we will leave it at these introductory remarks. We will, however, very briefly return to defects in the context of topologically twisted $N = 2$ superconformal field theories in Section 7.4. In such models, all branes and defects are topological and can be described in rather simple terms using matrix factorisations.

4.A.5 The conformal boundary states for $c = 1$ models

In this subsection, we briefly review some results of the papers [215, 216, 298] – see also the unpublished notes [184] for some preliminary ideas – that address the classification of branes for $c = 1$ CFTs corresponding to a single free boson compactified on a circle. In this case, new boundary conditions can be constructed "from scratch", resulting in a complete list of all *conformal* boundary states.

This construction of boundary conditions relies on the detailed knowledge about decomposition of irreducible $\widehat{U(1)}$ representations into Virasoro modules available at central charge $c = 1$, and on a careful analysis of the cluster condition, which in this case provides a stronger computational tool than Cardy's conditions. To keep the presentation short, we will quote and exploit one result on $\widehat{SU(2)}_1$ boundary states with gluing conditions twisted by an inner automorphism, which is a by-product of the general investigations of chiral marginal boundary deformations presented in Section 5.3.

The starting point of the analysis in [215, 216] is the decomposition of $\widehat{U(1)}$ modules into irreducible Virasoro representations (for $m \in \frac{1}{2}\mathbb{Z}$)

$$\mathcal{H}^{\widehat{U(1)}}_{\sqrt{2}m} = \bigoplus_{l=0}^{\infty} \mathcal{H}^{\text{Vir}}_{(|m|+l)^2} \,. \tag{4.123}$$

The subscript of $\widehat{U(1)}$ modules is the charge, that of Virasoro modules the conformal dimension. We first concentrate at the self-dual radius $r_{\text{s.d.}}$ with an enhanced
$\widehat{SU(2)}_1$ symmetry. Any conformal boundary state can be written in the form

$$\|B\rangle\rangle = \frac{1}{2^{\frac{1}{4}}} \sum_{j \in \frac{1}{2}\mathbb{Z}_+} \sum_{m,n} B^j_{m,n} \, |j, m, n\rangle\rangle \tag{4.124}$$

with m, n running from $-j$ to j in integer steps. The coefficients $B^j_{m,n}$ have to
satisfy the cluster condition

$$B^{j_1}_{m_1,n_1} \, B^{j_2}_{m_2,n_2} = \sum_{j;m,n} \Xi_{(j_1;m_1,n_1),(j_2;m_2,n_2)}^{(j;m,n)} \, B^j_{m,n} \tag{4.125}$$

with Ξ being a combination of fusing matrix elements and bulk OPE coefficients for Virasoro primary fields. A priori, the structure constants Ξ may not
be known explicitly, but we can determine them indirectly if we exploit the fact
that Cardy's boundary states for $\widehat{SU(2)}_1$ are consistent, along with their chiral
deformations; see Chapter 5 and also [215, 216]. These chiral deformations are
simply $\widehat{SU(2)}_1$-symmetric boundary states whose gluing conditions have been
twisted by an inner automorphism Ad_g for $g \in SU(2)$:

$$\left(\text{Ad}_{(g \cdot \iota)}(J^u_m) + J^a_{-m} \right) \|B\rangle\rangle_g = 0 \qquad \text{where} \qquad \iota = \begin{pmatrix} 0 & 1 \\ -1 & 0 \end{pmatrix}.$$

It turns out that their expansion into Virasoro Ishibashi states according to
equation (4.124) involves $SU(2)$ representation matrices:

$$\|B\rangle\rangle_g = \frac{1}{2^{\frac{1}{4}}} \sum_{j,m,n} D^j_{m,n}(g) \, |j, m, n\rangle\rangle$$

where $D^j_{m,n}(g)$ is the matrix element of g in the representation j. If we
parametrise $g \in SU(2)$ as $g = \begin{pmatrix} a & b \\ -b^* & a^* \end{pmatrix}$, the explicit formula is (see, e.g., [266])

$$D^j_{m,n}(\Gamma) = \sum_{l=\max(0,n-m)}^{\min(j-m,j+n)} \frac{[(j+m)!(j-m)![(j+n)!(j-n)!]^{\frac{1}{2}}}{(j-m-l)!(j+n-l)!l!(m-n+l)!} \tag{4.126}$$

$$\times \, a^{j+n-l}(a^*)^{j-m-l}b^{m-n+l}(-b^*)^l.$$

Returning to the cluster condition, consistency of these boundary states means
that setting $B^j_{m,n} = D^j_{m,n}(g)$ for any $g \in SU(2)$ provides a solution to the non-
linear equation (4.125). But since the $D^j(g)$ are matrix representations of $SU(2)$
of arbitrary spin, we may conclude that the structure constants Ξ are necessarily
given by products of Clebsch–Gordan coefficients (for left- and right-movers),

$$\Xi_{(j_1;m_1,n_1),(j_2;m_2,n_2)}^{(j;m,n)} = (j_1 m_1, j_2 m_2 | j m) \, (j n | j_1 n_1, j_2 n_2).$$

Now we can ask whether there are further solutions $B^j_{m,n}$ to the cluster con-
ditions (4.125). Because of the relation $(j' n' | j n, 0, 0) = \delta_{j',j} \, \delta_{n',n}$, any non-
trivial solutions satisfy $B^0_{0,0} = 1$. More importantly, the four elements $B^{\frac{1}{2}}_{m,n}$

fix all $B^j_{m,n}$ for $j \geq 1$, simply because the spin $\frac{1}{2}$ representation generates all higher-dimensional SU(2) representations. Finally, inserting explicit values for the Clebsch–Gordan coefficients shows that

$$B^{\frac{1}{2}}_{-\frac{1}{2},-\frac{1}{2}} B^{\frac{1}{2}}_{\frac{1}{2},\frac{1}{2}} - B^{\frac{1}{2}}_{-\frac{1}{2},\frac{1}{2}} B^{\frac{1}{2}}_{\frac{1}{2},-\frac{1}{2}} = B^0_{0,0} = 1$$

i.e. the matrix $\left(B^{\frac{1}{2}}_{m,n}\right)_{m,n=\pm\frac{1}{2}}$ must be in $SL(2,\mathbb{C})$. It can be argued that this is the only remaining consistency condition [216], and therefore

$$\| g \rangle\rangle = \frac{1}{2^{\frac{1}{4}}} \sum_{j \in \frac{1}{2}\mathbb{Z}_+} \sum_{m,n} D^j_{m,n}(g) \, |j,m,n\rangle\rangle \qquad \text{with} \quad g \in SL(2,\mathbb{C}) \qquad (4.127)$$

provide *all* the (clustering) conformal boundary conditions for $c = 1$ at the self-dual radius.

Let us now check Cardy's condition by computing overlaps between two boundary states in the family (4.127). We find

$$Z_{g_1 g_2}(q) = \langle\langle \Theta g_1 \| \tilde{q}^{L_0 - \frac{c}{24}} \| g_2 \rangle\rangle = \frac{1}{\sqrt{2}} \sum_{j,m,n} (-1)^{m-n} \, D^j_{-m,n}(g_1) \, D^j_{m,n}(g_2) \, \chi_{j^2}(\tilde{q})$$

$$= \frac{1}{\sqrt{2}} \sum_{j,n} D^j_{n,n}(g_1^{-1} g_2) \, \chi_{j^2}(\tilde{q}) \,.$$

Here we have inserted an extra $\Theta \equiv \Theta_{\text{CPT}}$ into the out-boundary state, and used the convention that $\Theta |j,m,n\rangle\rangle = (-1)^{m-n} |j,m,n\rangle\rangle$. The $D^j_{m,n}$ have certain properties (see [216]) which allow the trading of these signs for the inverse, and moreover they furnish representations not only of SU(2), but also of $SL(2,\mathbb{C})$, which leads to the last equality. Now recall that by conjugation the argument of $D^j_{n,n}$ can always be brought into Jordan normal form,

$$g_1^{-1} g_2 \sim \begin{pmatrix} \hat{a} & * \\ 0 & \hat{a}^{-1} \end{pmatrix} \,;$$

thus we have

$$\text{tr}\, D^j \left(g_1^{-1} g_2 \right) = \frac{\sinh((2j+1)\alpha)}{\sinh \alpha} \qquad \text{with} \quad \hat{a} = e^\alpha \,.$$

To proceed, we express the Virasoro characters $\chi^{\text{Vir}}_{j^2}(\tilde{q})$ as differences of $\widehat{U(1)}$ characters

$$\chi^{\text{Vir}}_{m^2}(q) = \chi^{\widehat{U(1)}}_{\sqrt{2}m}(q) - \chi^{\widehat{U(1)}}_{\sqrt{2}(|m|+1)}(q) \,, \qquad (4.128)$$

and apply modular transformation and resummation tricks [216], which leads to the expression

$$Z_{g_1 g_2}(q) = \sum_{n \in \mathbb{Z}} \chi^{\widehat{U(1)}}_{\frac{i\alpha}{\sqrt{2}\pi} + \sqrt{2}n}(q) \,. \qquad (4.129)$$

This shows that all the boundary states (4.127) satisfy Cardy's condition. Note, however, that for $g_1^{-1} g_2 \in \mathrm{SL}(2, \mathbb{Z}) \setminus \mathrm{SU}(2)$, the partition function (4.129) contains open string states with complex conformal dimensions.

The procedure from [216] can be extended to any $r = \frac{M}{N} r_{\mathrm{s.d.}}$ with coprime integers M, N. It can be shown that the most general solutions to the cluster condition for these radii is given by [215]

$$\| g \rangle\rangle_{M,N} = \frac{1}{2^{\frac{1}{4}} \sqrt{MN}} \sum_{j,m,n} \sum_{l=0}^{M-1} \sum_{k=0}^{N-1} D^j_{m,n} \left(\Gamma^k_N \Gamma^l_M \, g \, \Gamma^{-l}_M \Gamma^k_N \right) |j, m, n\rangle\rangle \quad (4.130)$$

with g as before and the SU(2) element

$$\Gamma_L := \begin{pmatrix} e^{\frac{i\pi}{L}} & 0 \\ 0 & e^{-\frac{i\pi}{L}} \end{pmatrix}$$

for any $L \in \mathbb{Z}_+$. Sandwiching g between the Γ-powers and summing over l, k simply projects onto those U(1) charges m, n actually present in the spectrum of the $r = \frac{M}{N} r_{\mathrm{s.d.}}$ theory. It can be seen that, due to the projection, multiplying g by Γ^2_N or Γ^2_M leaves the boundary state $\| g \rangle\rangle_{M,N}$ invariant, thus the family has the topology of $\mathrm{SL}(2, \mathbb{C})/\mathbb{Z}_M \times \mathbb{Z}_N$.

Proof of cluster condition (in which the structure constants Ξ only depend on conformal weights, not on the compactification radius) for generic g and computation of partition functions proceed as before, with a slightly more complicated (M- and N-dependent) distribution of the charges $i\alpha$ occurring in equation (4.129); see [215]. In particular, notice that the spectra of the $\mathrm{SL}(2, \mathbb{C})/\mathbb{Z}_M \times \mathbb{Z}_N$ boundary states are in general not invariant under addition of U(1) charges, so the boundary states $\| g \rangle\rangle_{M,N}$ do indeed break U(1) symmetry.

For generic compactification radii r, only the vacuum module has a non-trivial decomposition into irreducible representations of the Virasoro algebra,

$$\chi_0^{\widehat{U(1)}}(q) = \sum_{l=0,1,\ldots} \chi_{l^2}^{\mathrm{Vir}}(q),$$

and we can use the associated Ishibashi states $|l\rangle\rangle$ to construct new conformal boundary states $\| x \rangle\rangle$, which are parametrised by a real number $-1 < x < 1$, namely [298]

$$\| x \rangle\rangle = \sum_{l=0}^{\infty} P_l(x) \, |l\rangle\rangle, \quad (4.131)$$

where $P_l(x)$ are the Legendre polynomials. If we take x to -1 or 1, we obtain (continuous) superpositions of, respectively, the usual Dirichlet and Neumann branes (integrated over all possible locations and Wilson lines, respectively). The coefficients in the expansion (4.131) were again obtained by a careful study of the cluster conditions, exploiting some fusing matrix elements that are known explicitly and imply the recursion relation for the $P_l(x)$; see also [215] for a derivation starting from the explicit expression for the matrix elements $D^j_{m,n}(g)$.

The spectra of open strings between two such branes were also computed in [298]: the partition functions can be written as integrals over Virasoro characters, and the conformal dimensions of boundary fields come in bands of finite width.

The conformal boundary states for an uncompactified free boson theory or for circle theories at generic radii $r \neq \frac{M}{N} r_{\text{s.d.}}$ can be regarded as a limit of the rational radii $r = \frac{M}{N} r_{\text{s.d.}}$; see [215]. The projection onto the sublattice of available charges is implemented by an integration

$$\frac{1}{\pi} \int_0^\pi d\theta \; D^j_{m,n} \left(\begin{pmatrix} e^{i\theta} & 0 \\ 0 & e^{-i\theta} \end{pmatrix} g \begin{pmatrix} e^{-i\theta} & 0 \\ 0 & e^{i\theta} \end{pmatrix} \right) \tag{4.132}$$

rather than the discrete sums occurring in the boundary states (4.130). Note that, because of the integration over θ, the argument of the D-matrix in equation (4.132) only depends on a and $|b|^2$ rather than the entire $\mathrm{SL}(2,\mathbb{Z})$-element g.

The boundary spectra of these branes can be computed with the same methods as above; for the simple case of $g_1 = g_2$ being a 2×2 rotation matrix, we arrive, in a more straightforward manner, at the band spectrum that will show up as a result of non-chiral boundary deformation in Subsection 5.3.1.

Let us add some brief remarks on the interpretation of this complete list of $c = 1$ conformal boundary states. Those labelled by $g \in \mathrm{SL}(2,\mathbb{Z}) \setminus \mathrm{SU}(2)$ are most likely unphysical, due to the complex conformal dimensions in the boundary spectrum. Some of the boundary states labelled by $g \in \mathrm{SU}(2)$ will make an appearance in Section 5.3, where they are obtained via boundary deformations of ordinary Dirichlet or Neumann boundary conditions for the free boson. But the set (4.130) also contains boundary states not found by deforming a single Dirichlet or Neumann brane (as the procedure will require $M = 1$ or $N = 1$ to begin with). For example, we have [215]

$$\| \begin{pmatrix} 1 & 0 \\ 0 & 1 \end{pmatrix} \rangle\!\rangle_{M,N} = \sum_{p=0}^{M-1} \| \, D, \, \tfrac{2\pi R\, p}{M} \, \rangle\!\rangle, \qquad \| \begin{pmatrix} 0 & 1 \\ -1 & 0 \end{pmatrix} \rangle\!\rangle_{M,N} = \sum_{s=0}^{N-1} \| \, N, \, \tfrac{\pi s}{R\, N} \, \rangle\!\rangle. \tag{4.133}$$

This means that equation (4.130) describes a continuous family of boundary states containing a superposition of M equidistant Dirichlet branes as well as a superposition of N Neumann branes. At these points (and their translates), the cluster condition is of course violated. The whole $\mathrm{SU}(2)/\mathbb{Z}_M \times \mathbb{Z}_N$ family can probably be thought of as deformations of those systems of branes, more precisely as deformations with BCCOs (states of stretched open strings). It is worth noting that, apart from the special boundary conditions (4.133), all members of the family (4.130) are elementary in the sense that the vacuum sector occurs precisely once in $Z_{g\,g}(q)$. Turning on the BCCO deformations in question seems to "fuse" N Dirichlet branes into a single brane (without a ready geometric interpretation), which is "resolved" into M Neumann constituents at a special value of the deformation parameter: a first example of "brane topology change" via deformations which we will encounter in Chapter 5.

For completeness, let us mention that virtually the same pattern of boundary states and flows arises for $c = \frac{3}{2}$ when all $N = 1$ superconformal boundary conditions are classified [215]. This was also discussed in the context of tachyon condensation by Sen in [422], both from the CFT and from the effective field theory point of view.

5
Perturbations of boundary conformal field theories

We have gathered quite a large amount of technology to construct conformal boundary conditions and study their properties. So far, most of the tools relied on detailed knowledge of the underlying algebraic structure of conformal field theories (CFTs), such as representations of symmetry algebras and modular transformations. In almost all cases, this route will lead to a restricted class of boundary conditions only, favouring those that preserve a lot of symmetry.

In "real life applications" of boundary CFT, arising in string theory, in condensed matter physics, or in the study of integrable models, we are often faced with situations where we start from a simple given boundary condition and then "deform" by "turning on perturbations". In this chapter, we will discuss methods aimed at making this procedure well defined and at determining the perturbed boundary theory. Most of the time, we will restrict ourselves to pure boundary perturbations, which leave the bulk CFT untouched.

We will sometimes be led to new boundary conditions that would have been difficult to find directly. The total space of boundary CFTs takes the form of a fibration over the moduli space of bulk theories, with the fibre over each point made up from all boundary conditions (branes) that exist for that bulk theory. Marginal boundary deformations, if they exist, will generate, starting from a single boundary condition, whole continuous families within this boundary moduli space. Even for simple CFTs, like the free boson, which have a description in terms of classical geometry, the shape of these moduli spaces can be surprisingly complex and far removed from geometric intuition.

Another reason to be interested in marginal deformations is that the deforming boundary operators correspond to massless fields in the string theory context, and our findings will also be relevant for considerations of low-energy effective field theories. Renormalisation group (RG) flows triggered by relevant fields, on the other hand, are believed to be related to questions like tachyon condensation, brane decay and bound state formation.

We start from some boundary CFT with state space $\mathcal{H} = \mathcal{H}^{(H)}_{(\Omega, \tilde{\alpha})}$, where $\alpha = (\Omega, \tilde{\alpha})$ denotes an (elementary) boundary condition along the real line. Boundary operators $\psi(x)$ associated with states in $\mathcal{H}$ may be used to define a new perturbed theory whose correlation functions are constructed from the unperturbed ones by the, as yet informal, prescription

$$\langle \varphi_1(z_1, \bar z_1) \cdots \varphi_N(z_N, \bar z_N) \rangle_{\alpha;\, \tilde\lambda\psi} = Z^{-1} \cdot \langle I_{\tilde\lambda\psi}\, \varphi_1(z_1, \bar z_1) \cdots \varphi_N(z_N, \bar z_N) \rangle_\alpha \quad (5.1)$$

$$:= Z^{-1} \sum_n (-\tilde\lambda)^n \int \cdots \int_{x_i > x_{i+1}} dx_1 \dots dx_n \langle \psi(x_1) \dots \psi(x_n)\, \varphi_1 \cdots \varphi_N \rangle_\alpha$$

$$= Z^{-1} \sum_n \frac{(-\tilde\lambda)^n}{n!} \sum_{\sigma \in S_n} \int \cdots \int_{x_{\sigma(i)} > x_{\sigma(i+1)}} dx_1 \dots dx_n \langle \psi(x_{\sigma(1)}) \dots \psi(x_{\sigma(n)}) \varphi_1 \cdots \varphi_N \rangle_\alpha \,,$$

where $\tilde\lambda$ is a real parameter. In the last line, we have introduced a sum over the permutation group S_n, which shows that $I_{\tilde\lambda\psi}$ should be understood as a path-ordered exponential of the perturbing operator,

$$I_{\tilde\lambda\psi} = P \exp\left\{ -S_{\tilde\lambda\psi} \right\} := P \exp\left\{ -\tilde\lambda \int_{-\infty}^{\infty} \psi(x)\, dx \right\}. \quad (5.2)$$

The minus sign is natural from the Euclidean path integral point of view. The normalisation Z is defined as the (normalised) expectation value $Z = (A_0^\alpha)^{-1} \langle I_{\tilde\lambda\psi} \rangle_\alpha$. The expressions above address boundary deformations of bulk correlators only; if there are extra boundary fields present in the correlation function, the formulas need to be modified in an obvious manner: the boundary operators have to be included into the path-ordering since we cannot, in general, move boundary fields around each other (recall the discussion in Section 4.2).

To keep the notation simpler, we have introduced only a single perturbing field with coupling $\tilde\lambda$; generalisation to perturbations $\sum_k \tilde\lambda_k \psi_k$ involving several boundary fields from $\mathcal{H}$ is straightforward. Likewise, it is not difficult to see what happens when we pass from an elementary boundary condition α to a superposition $N \cdot \alpha$, i.e. a stack of N identical branes of type α: in this situation, boundary fields and couplings come with an $N \times N$ Chan–Paton factor, and the perturbation prescription involves a trace over the Chan–Paton indices (just like the path ordering, this includes any boundary fields already present in the unperturbed correlator).

For the above expressions to make sense beyond the formal level, it is necessary to regularise the integrals (introducing ultraviolet (UV) and infrared (IR) cutoffs) and, in general, to renormalise couplings and fields. The methods we will use for concrete computations depend crucially on whether the perturbing fields are marginal ($h_\psi = 1$) or relevant ($h_\psi < 1$). It is worth emphasising that boundary perturbations can only induce changes of the boundary condition: the local properties in the bulk are not affected by the perturbing "condensate" along the boundary.

In Section 5.1, we will discuss relevant boundary perturbations, using perturbative renormalisation adapting (in Subsection 5.1.1) the procedure given by Cardy in [101] to the boundary case; this will include a definition of beta functions and the associated renormalisation group equations. These conformal perturbation theory techniques are closely related to computations of low-energy effective actions in open string theory and will be used in Chapter 6.

In 5.1.2, the general formulas will be applied to special perturbations of minimal models. Some non-perturbative results on the c- and the g-theorems are collected in Subsection 5.1.3.

Section 5.2 is devoted to marginal boundary perturbations (also referred to as marginal deformations), for which far stronger results can be obtained than for relevant perturbations (or for bulk perturbations, for that matter); the main reason is that we need not resort to renormalisation group equations here but can instead exploit contour integration techniques to evaluate deformed correlators. We will start with some general observations on truly marginal boundary fields (those that stay marginal to all orders in the coupling constant) in Subsection 5.2.1; in particular, the important criterion of "self-locality" (also called "analyticity" in this text) will be presented. Then we will briefly review the simplest class of self-local fields, namely chiral marginal operators, in Subsection 5.2.2; a complete picture of the effects of chiral marginal deformations can be given, even in terms of boundary states. For non-chiral marginal deformations, to be addressed in 5.2.3, it is more difficult to obtain general statements, so that we will sometimes be restricted to leading-order perturbative considerations, but the effects on symmetries and spectra will be more interesting than for the chiral case.

Examples for marginal boundary deformations will be discussed extensively in Section 5.3. We study various kinds of deformations of $c = 1$ models, free bosons as well as $\mathbb{Z}_2$-orbifolds thereof. We construct the complete moduli space of boundary conditions that can be reached from ordinary Dirichlet or Neumann boundary conditions via marginal deformations. This moduli space is surprisingly complicated and shows several features which cannot be predicted from the geometry of the classical target, e.g. the fact that the moduli space of a single brane is connected to that of a system of several branes.

The final section is a collection of various additional topics: we first review some general facts on deformations of $N = 2$ supersymmetric theories, then return to the $c = 1$ brane moduli space in Subsection 5.A.2 and point out some string theory applications in the context of brane creation and decay processes; in Subsection 5.A.3 we outline that RG equations can often be recovered as equations of motion from effective actions which typically have a (classical or non-commutative) geometric interpretation.

5.1 Relevant boundary perturbations

We first discuss the case where the perturbing boundary fields ψ_k have conformal dimensions $h_k < 1$, even though it will be easier to obtain exact results for marginal deformations. The methods employed in the study of relevant perturbations are close to those familiar from standard textbooks on quantum field theory

(QFT). In particular, we often have to contend with a leading order perturbative analysis.

To start with, we imitate the bulk methods explained in [101], presenting the setup of UV-regularised perturbed boundary CFTs and recalling how to implement renormalisation group (RG) transformations. These lead to a definition of the (perturbative) beta function; the zeroes of which allow (perturbative) RG fixed points to be determined (here: IR fixed points).

Evaluating the one-point function of the identity operator at the fixed point then provides a first (perturbative) glimpse of the g-theorem [5]. This result will be quoted in the next subsection, where we also discuss an application to a special class of boundary deformations of boundary Virasoro minimal models. It turns out that in these examples the value of $\langle \mathbf{1} \rangle_{\alpha;\tilde{\lambda}\psi}$ at the fixed point is already sufficient to determine the new boundary condition at the IR fixed point exactly.

The last subsection addresses the only (to the authors' knowledge) general result on relevant boundary perturbations that has been proven without resort to perturbative techniques, namely the g-theorem. Our discussion follows the formulation given in [185]. In analogy to Zamolodchikov's c-theorem [462], the g-theorem states that in unitary CFTs, the boundary entropy decreases along the RG-flow to the infrared.

5.1.1 Perturbative treatment: renormalisation group, beta functions

We return to equations (5.1,5.2) which served as an "informal" definition of the correlators in the perturbed theory. The IR divergences will play no role in the following, and indeed they are absent altogether when working on a disk of circumference $2\pi R$. But the UV divergences need to be regulated, e.g. by a *short-distance cutoff* $\varepsilon > 0$. Thus, we demand that any integrated perturbing boundary fields keep a distance ε from each other and any other boundary operators. This can be implemented by Heaviside step-functions, see below.

For perturbing fields with $h \neq 1$, the "interaction term" $I_{\tilde{\lambda}\psi}$ will automatically contain dimensionful quantities and thus break scale invariance: $S_{\tilde{\lambda}\psi}$ is dimensionless if the length-dimension of the coupling $\tilde{\lambda}_k$ for a perturbing field ψ_k with conformal weight h_k is $h_k - 1$. We can use the cutoff ε to pass to the more convenient dimensionless couplings λ_k via

$$\tilde{\lambda}_k := \varepsilon^{h_k - 1} \lambda_k .$$

Thus, the perturbing action we consider is

$$S_{\tilde{\lambda}\psi} = \sum_k \varepsilon^{1-h_k} \lambda_k \int dx \, \psi_k(x) . \tag{5.3}$$

We focus on $h_k < 1$ in the following, as perturbations with $h_k > 1$ are irrelevant: their RG flows lead back to the original boundary theory.

To arrive at the perturbed correlation functions, in principle short-distance singularities have to be subtracted, e.g. using a minimal subtraction scheme, before the limit $\varepsilon \to 0$ can be taken. However, even with a finite cutoff ε, important information can be drawn from analysing the behaviour under renormalisation group transformations, i.e. under scalings

$$\varepsilon \mapsto (1 + dl)\,\varepsilon\,. \tag{5.4}$$

Here, l is the dimensionless *renormalisation group parameter*, which can be thought of as $l = \ln(\varepsilon/R)$, where R is an IR scale (e.g. the disk radius). We can ask (in a perturbative expansion) how the couplings λ_k have to be changed under the transformation (5.4) for the regularised perturbation (5.1,5.2) to keep its form.

An infinitesimal rescaling (5.4) first of all affects $S_{\tilde{\lambda}\psi}$ because of the explicit ε-dependence in the action (5.3): $\lambda_k \mapsto (1 + (1 - h_k)\,dl)\lambda_k$. In addition, ε enters perturbed correlators via the short-distance regularisation: consider a term

$$\sim \int dx_1 dx_2\, \langle \dots \psi_i(x_1)\psi_j(x_2)\dots\rangle\, H(|x_1 - x_2| - \varepsilon) \tag{5.5}$$

in the perturbation expansion, where ψ_i and ψ_j are perturbing boundary fields, where the dots stand for any other bulk or boundary fields in the correlator, and where H is the Heaviside function implementing the short-distance cutoff. To proceed, we insert the boundary operator product expansion (OPE)

$$\psi_i(x_1)\psi_j(x_2) = \sum_k C^{\alpha}_{ijk}\,(x_1 - x_2)^{h_k - h_i - h_j}\,\psi_k(x_2)$$

into the correlator (5.5), which shows that a second-order term $\sim \lambda_i\lambda_j$ may affect the first-order term $\sim \lambda_k$ if $C^{\alpha}_{ijk} \neq 0$. Taking the prefactors (dimensionless couplings and ε-powers) along, it is then straightforward to compute the infinitesimal change of the correlator (5.5) under the transformation (5.4), which gives the second-order contribution to the change of λ_k. To this order, we find the response of λ_k to infinitesimal changes of ε as in equation (5.4) to be

$$\frac{d\lambda_k}{dl} \equiv -\beta_k = (1 - h_k)\,\lambda_k + C^{\alpha}_{ijk}\,\lambda_i\,\lambda_j + \mathcal{O}(\lambda^3)\,; \tag{5.6}$$

these are the *renormalisation group equations*.

In writing this equation, we have implicitly defined the *beta functions* β_k of the couplings λ_k for an RG flow to the IR,

$$\beta_k(\lambda) := \frac{d\lambda_k}{d\ln R} = -\frac{d\lambda_k}{dl}\,; \tag{5.7}$$

R denotes the IR scale of the theory (e.g. the radius of the disk on which the boundary CFT is defined): scaling up that scale corresponds to scaling down the UV cutoff ε, hence the minus sign in the beta function (5.7). The beta functions in general depend on all the couplings $\lambda := (\lambda_1, \lambda_2, \dots)$.

The RG equations (5.6) describe the running of the coupling constants with the scale. The λ_k stop running when

$$\beta_k(\lambda_*) = 0\,, \tag{5.8}$$

which yields equations for the (perturbative) IR RG fixed point λ_*. There, we have a new boundary CFT which captures the low-energy effective physics of the perturbed theory.

The equations (5.6) and (5.8) can be exploited as follows: upon integration of the ordinary differential equations (5.6), we obtain a relation allowing us to express (in perturbation theory) the UV-cutoff-dependent couplings $\lambda(\varepsilon)$ through the renormalised couplings $\lambda(R)$ at the IR scale; this allows us to eliminate the UV cutoff from the perturbed correlators. Then, equation (5.8) is solved for the couplings λ_* at the IR fixed point and takes the RG flow limit $\lambda(R) \to \lambda_*$ everywhere.

An instructive and detailed exposition of this procedure can be found in [6], where Affleck and Ludwig used it to compute the change, to second-order in the coupling, of the one-point function $g_\alpha := B^0_\alpha$ of the identity field under a (nearly marginal) relevant perturbation ψ. We will quote their result, which amounts to a perturbative version of the g-theorem, in the next subsection, where it will also be applied to Virasoro minimal models.

Note that there are variants of the method sketched above (which is sometimes called "Wilsonian scheme") to compute renormalised correlation functions. In particular, we can explicitly subtract contributions which diverge as $\varepsilon \to 0$, by adding suitable counterterms to the perturbing action ("minimal subtraction scheme"). A detailed comparison of these schemes and observations concerning the scheme dependence of the beta functions in perturbed boundary CFTs can, e.g., be found in [217].

We briefly note that the RG equations (5.6) can be generalised to the case when *bulk perturbations*

$$S_{\lambda^B \phi} = \sum_{i,\bar{\imath}} \lambda^B_i \int d^2 z\, \varphi_{i,\bar{\imath}}(z, \bar{z}) \tag{5.9}$$

are switched on in a boundary CFT (alone or accompanied by boundary perturbations as above); we assume that the $\varphi_{i,\bar{\imath}}$ satisfy $h_{\varphi_{i,\bar{\imath}}} = \bar{h}_{\varphi_{i,\bar{\imath}}}$. The bulk field integrations need to be UV-regulated near the boundary, too, e.g. by demanding that integrations run over $\Im z > \frac{\varepsilon}{2}$ only. Using the methods of [101] as outlined above, it is not difficult to work out the perturbative RG equations, see [179]. The key difference is that a perturbing bulk field $\varphi_{i,\bar{\imath}}$ can trigger a boundary perturbation ψ_k if the bulk–boundary OPE coefficient $C^\alpha_{(i\bar{\imath})\,k}$ is non-zero – in contrast to pure boundary deformations, which will not change the bulk theory. We arrive at a coupled system of bulk and boundary RG equations [179]:

$$\frac{d\lambda^B_k}{dl} = (2 - 2h_{\varphi_{k,\bar{k}}})\,\lambda^B_k + \pi\, C_{ijk}\,\lambda^B_i\,\lambda^B_j + \cdots$$

$$\frac{d\lambda_k}{dl} = (1 - h_{\psi_k})\,\lambda_k + C^\alpha_{ijk}\,\lambda_i\,\lambda_j + \tfrac{1}{2}\, C^\alpha_{(i\bar{\imath})\,k}\,\lambda^B_i + \cdots \tag{5.10}$$

The bulk flow will in general lead to a new IR fixed point CFT in the bulk, and therefore the boundary condition necessarily has to change.

This is true even for marginal bulk deformations. Recall the exceptional conformal boundary conditions for $c = 1$ circle theories which existed only for a discrete set of circle radii $r = \frac{M}{N} r_{\text{s.d.}}$ from Subsection 4.A.5. The compactification radius r can be changed continuously by the marginal bulk field $\partial X \bar{\partial} X$, so any exceptional boundary condition must flow to Dirichlet or Neumann branes (or superpositions thereof) under this deformation – see [179] for a more detailed discussion.

5.1.2 Perturbative g-theorem and boundary RG flows in Virasoro minimal models

The simplest correlation function the general procedure from the preceding subsection can be applied to is the one-point function of the identity operator, often called the *g-factor* and denoted $g_\alpha(\lambda) = \langle \mathbf{1} \rangle_{\alpha; \lambda\psi}$ in the perturbed boundary CFT. This quantity is interesting for various reasons: it has an interpretation as a (non-integer) ground-state degeneracy, and its perturbative computation by Affleck and Ludwig [5, 6] gave a first hint at the g-theorem. Also, knowledge of $g_\alpha(\lambda)$ is in some cases sufficient to determine the boundary condition at the new (perturbative) IR fixed point completely; examples will be given shortly. Finally, looking back at the definition (5.1,5.2), we realise that the expectation value of the identity field in a perturbed boundary CFT is closely related to the path integral for an effective action whose "degrees of freedom" are the couplings λ; accordingly, $g_\alpha(\lambda)$ shows up when computing D-brane effective actions in string theory (see Section 5.A below, and also Chapter 6 for an example).

To start with, let us quote the result of the perturbative computation of the g-factor $g_\alpha(\lambda)$ presented in [5, 6], which used the Wilsonian scheme described in Subsection 5.1.1. (See also [318] for a useful review and additional perturbative results, employing slightly different ideas.)

Affleck and Ludwig considered perturbations by a single relevant boundary field ψ with $y := 1 - h_\psi \ll 1$. For simplicity, they assumed that ψ is the only relevant field occurring in the OPE of ψ with itself. To second order in the coupling λ, the beta function is given by

$$\beta(\lambda) = y\,\lambda + C^\alpha_{\psi\psi\psi}\,\lambda^2 + \mathcal{O}(\lambda^3)$$

so that the fixed point is $\lambda^* = -y/C^\alpha_{\psi\psi\psi}$. It is seen, in particular, that λ^* is small for small y.

This fixed point can be inserted in the perturbative expansion of the one-point function of the identity,

$$\ln g_\alpha(\lambda) = \ln g_\alpha(0) - \pi^2\, y\, C^\alpha_{\psi\psi 1}\,\lambda^2 - \frac{2\pi^2}{3}\, C^\alpha_{\psi\psi\psi}\, C^\alpha_{\psi\psi 1}\,\lambda^3 + \mathcal{O}(\lambda^4)\,, \qquad (5.11)$$

expressed in terms of the renormalised coupling as explained in [6] and above. The factors π arise from a coordinate transformation between the upper half-plane and disk, the $C^\alpha_{\psi\psi\,1}$ from evaluating correlators after performing OPEs. At the fixed point, we obtain

$$\Delta \ln g := \ln g_\alpha(\lambda^*) - \ln g_\alpha(0) = -\frac{\pi^2}{3} \frac{C_{\psi\psi\,1}}{\left(C_{\psi\psi\psi}\right)^2} y^3 + \mathcal{O}(\lambda^4). \tag{5.12}$$

We note that, in unitary theories where $C_{\psi\psi\,1}$ is positive, and at least for small λ^*, we always have $\ln g_\alpha(\lambda^*) < \ln g_\alpha(0)$, suggesting that "the ground state degeneracy decreases along the boundary RG flow to the infrared". This is the claim of the perturbative g-theorem.

For unitary minimal models of the Virasoro algebra, we have a complete overview of all possible boundary conditions: they are given by Cardy states labelled $\alpha = (r, s)$, or by integer superpositions thereof. This simplicity makes it possible to determine the IR fixed points of boundary perturbations simply by computing the quantity $\ln g_\alpha(\lambda^*)$ – at least for flows driven by the least relevant boundary field $\psi_{1,3}(x)$, which has conformal dimension $h_{1,3} = \frac{m-1}{m+1} =: 1 - y$ – cf. the Kac table (3.25) for the notation.

The strategy applied in [388] to find the fixed-point boundary condition is rather elementary: start with a minimal model boundary condition $\alpha = (r, s)$; compute the first few terms in the perturbative expansion (5.11) of the g-factor explicitly, using the boundary OPE structure constants from the unperturbed theory, see Section 4.4.2; then expand everything in $\frac{1}{m}$ – which yields [388]

$$\ln\left(\frac{g_{(r,s)}(\lambda^*)}{g_{(r,s)}(0)}\right) = -\frac{\pi^2}{3} (s^2 - 1) (m+1)^{-3} + \mathcal{O}\big((m+1)^{-4}\big).$$

Now compare this to the $\frac{1}{m}$-expansion of $\ln(g_\beta/g_\alpha)$, where $\beta = (\tilde{r}, \tilde{s})$ ranges over all possible conformal boundary conditions for the mth unitary minimal model:

$$\ln\left(\frac{g_{(\tilde{r},\tilde{s})}}{g_{(r,s)}}\right) = \ln\left(\frac{\tilde{r}\tilde{s}}{rs}\right) + \frac{\pi^2}{6} (r^2 + s^2 - \tilde{r}^2 - \tilde{s}^2) (m+1)^{-2}$$

$$+ \frac{\pi^2}{3} (r^2 - \tilde{r}^2) (m+1)^{-3} + \mathcal{O}\big((m+1)^{-4}\big).$$

Each order in $(m+1)$ gives an algebraic equation on the natural numbers $\tilde{r}$ and $\tilde{s}$, and (apart from some spurious solutions existing for special values or r, s only) the following generic RG flow pattern is found for boundary perturbations by $\psi_{1,3}$:

$$\alpha = (r, s) \longrightarrow \alpha_{\lambda^* \psi} = \sum_{l=1}^{N} (r + s + 1 - 2l, 1), \tag{5.13}$$

with $N = \min(r, s)$. The somewhat unexpected result of the computations in [388] is that already the least relevant boundary field triggers a flow to stable fixed points (boundary Virasoro models which no longer contain any nontrivial relevant boundary operators), and that furthermore the new boundary conditions are generally described by superpositions of "elementary" Cardy boundary states.

These perturbative results have been checked by means of the "truncated conformal space approach" (TCSA), a computational method which uses a cutoff in energy levels and explicitly diagonalises the off-critical Hamiltonian in this finite-dimensional state space, without resorting to a perturbative expansion in λ. These TCSA methods in particular showed that the perturbative flow pattern (5.13) holds even for small values of m (where the $\frac{1}{m}$ expansion is not too trustworthy).

Moreover, Watts' computation [444] revealed another pattern of flows, which applies for the opposite sign of λ:

$$\alpha = (r, s) \longrightarrow \alpha_{\lambda_-^* \psi} = \sum_{l=1}^{N} (r + s - 2l, 1), \qquad (5.14)$$

with $N = \min(r, s - 1)$. The latter result is not accessible with conformal perturbation theory. In subsequent work, Graham, Runkel and Watts combined CFT and TCSA techniques to predict RG flows for a much wider class of relevant perturbations of unitary minimal models [245]. The flows from above also feature in the phase diagram for the tricritical Ising model worked out in [8]. Let us add that several non-trivial examples of relevant boundary flows have been studied in the literature before (see [112, 140, 141, 168, 332, 333, 443] and references therein), partially with the help of the thermodynamic Bethe ansatz or TCSA, typically with rather model-dependent techniques.

The method from [388] can also be employed to study marginally relevant boundary perturbations of SU(2) Wess–Zumino–Witten (WZW) models, as long as it can be argued that the boundary condition at the new IR fixed point is maximally symmetric (i.e. a translate of a Cardy state). Due to the close relation between SU(2) WZW and Virasoro minimal models, the actual calculations are in fact very similar. Moreover, the computations necessary to determine the RG flow of the g-factor also govern the perturbative computation of the *low-energy effective action* of open strings. Therefore, the findings of [388] more or less encode the low-energy effective action for (systems of) branes in SU(2), see [13]. We will review this briefly in the next chapter. The paper [178] further extends these methods and provides a vast number of boundary RG flow patterns in more general coset theories. More recently, the ideas from [388] were adapted to study the behaviour of minimal model boundary conditions under perturbations by the least relevant bulk field [180].

5.1.3 Non-perturbative results: the c- and the g-theorem

For relevant perturbations of CFTs, only few exact results have been established. Here, we recapitulate (very briefly) the most famous ones, namely Zamolodchikov's c-theorem [462] for bulk perturbations and its "boundary analogue", the g-theorem. In both cases, it can be shown without resorting to perturbative expansions that, at least in unitary theories, certain quantities, which can be viewed as entropies, decrease along the RG flow to the IR.

In describing the c-theorem, we follow [101], but see also [96] for a detailed discussion (closer to the proof of the g-theorem given in [185]). Regarding the g-theorem, we will not attempt to present a full proof here, but refer to the literature for more details.

To obtain Zamolodchikov's c-theorem, we consider deformations of a bulk CFT by relevant bulk fields, $S_{\text{pert}} = \sum_i \mu_i \int d^2 w\, \varphi_i(w, \bar{w})$ with $h_{\varphi_i} = \bar{h}_{\varphi_i} < 1$, and analyse the behaviour of two-point functions of the energy-momentum tensor. In the perturbed theory, its trace Θ does not vanish but is given by

$$\Theta(z, \bar{z}) = \sum_i \beta_i(\mu)\, \varphi_i(z, \bar{z}) = -4\pi \sum_i \mu_i (1 - h_{\varphi_i})\, \varphi_i(z, \bar{z}) + \mathcal{O}(\mu^2)\,, \quad (5.15)$$

where $\beta_i(\mu) = -\frac{d\mu_i}{dl}$ are the bulk beta functions. However, even away from conformal points, the energy-momentum tensor is still conserved, therefore

$$\bar{\partial} T(z, \bar{z}) + \tfrac{1}{4} \partial \Theta(z, \bar{z}) = 0\,. \quad (5.16)$$

This allows us to find relations between the various two-point functions of $T(z, \bar{z})$ and $\Theta(z, \bar{z})$, which are a-priori constrained by rotational covariance:

$$\langle\, T(z, \bar{z}) T(0, 0)\,\rangle = \frac{F(z\bar{z})}{z^4}\,,$$

$$\langle\, \Theta(z, \bar{z}) T(0, 0)\,\rangle = \frac{G(z\bar{z})}{z^3 \bar{z}}\,, \quad (5.17)$$

$$\langle\, \Theta(z, \bar{z}) \Theta(0, 0)\,\rangle = \frac{H(z\bar{z})}{z^2 \bar{z}^2}\,.$$

Defining the c-function, depending on $\mu := (\mu_1, \mu_2, \ldots)$ and $z\bar{z}$, by

$$C \equiv C(z\bar{z}; \mu) := 2F - G - \tfrac{3}{8} H \quad (5.18)$$

and using the conservation law (5.16), it is found that

$$\dot{C} := z\bar{z}\, C'(z\bar{z}) = -\tfrac{3}{4} H\,. \quad (5.19)$$

Both H and C depend on the coupling constants μ, in fact C satisfies a Callan–Symanzik equation (with $\rho := (z\bar{z})^{\frac{1}{2}}$)

$$\left(\rho \frac{\partial}{\partial \rho} + \sum_i \beta_i(\mu) \frac{\partial}{\partial \mu_i} \right) C(z\bar{z}; \mu) = 0\,.$$

Together with equation (5.19), this implies that the derivative of the c-function along the RG flow to the IR is non-positive as long as the perturbed CFT is unitary,

$$\frac{dC}{dl} = -\tfrac{3}{2} H \leq 0\,.$$

The function C is stationary precisely when H and, thus, Θ vanish, i.e. when scale invariance is restored, namely at an RG fixed point. There, F is constant, and C coincides with the central charge c of the CFT at the new fixed point. Hence the folklore formulation of Zamolodchikov's c-theorem: the central charge decreases along the RG flow.

Using the first-order contribution to equation (5.15), a gradient formula is obtained:

$$\frac{\partial}{\partial \mu_j} C(1; \mu) = \beta_i \, G_{ij}$$

for the quantity C in the vicinity of the UV fixed point $\mu = 0$, where

$$G_{ij} = \langle \varphi_i(1,1)\varphi_j(0,0) \rangle$$

is often called the *Zamolodchikov metric* and can be used to introduce a metric structure on the moduli space of worldsheet theories.

Before turning to relevant boundary perturbations of (unitary) boundary CFTs and the g-theorem, let us take a different look at the c-theorem by recalling that the central charge of a CFT can be seen as a measure of the asymptotic density of states at high energy. Namely, if we consider the partition function

$$Z(q) = \sum_n d_n \, q^{E_n}$$

of a unitary (chiral) CFT – for simplicity, we ignore the right movers here – then the degeneracies d_n of the energy levels E_n can be expressed, for very large E_n, through the central charge c as

$$d_n \sim E_n^{-\frac{3}{4}} e^{2\pi \sqrt{\frac{c}{6} E_n}} .$$

This relation, also known as *Cardy's formula*, follows using the modular transformation properties of the characters along with a saddle-point approximation to determine the d_n-asymptotics; we refer to [103] for a discussion of the saddle-point derivation, but see also [129] for an elaborate exact expression (due to Rademacher) for the coefficients d_n in modular forms as they occur in CFT partition functions. In string theory, the leading exponential term in Cardy's formula was famously applied [434] to obtain a microscopic derivation of the Bekenstein entropy of extremal black holes.

Having thus related the central charge to a measure for the degeneracy of states, in fact to an entropy of the system, the c-theorem seems hardly surprising: renormalisation in the Wilsonian picture involves a coarse-graining procedure, a thinning-out of degrees of freedom, so the central charge should decrease along the RG flow.

The g-theorem for boundary perturbations states, loosely speaking, that (in unitary models) the one-point function of the identity operator decreases along the boundary RG flow to the IR. In view of the preceding paragraph, the g-theorem becomes plausible once the one-point function of the identity operator has been connected to an entropy; the proof remains subtle nevertheless as one has to deal with subleading contributions. Indeed, while the conjecture was formulated already in 1991 by Affleck and Ludwig, who also provided a perturbative proof in [6] – see also the formula for $\ln g_\alpha(\lambda)$ given in Subsection 5.1.2 – a full non-perturbative proof formulated entirely in terms of conformal field theory language was worked out by Friedan and Konechny [185]. These authors were at least partially inspired by ideas from string field theory [324, 424, 451] suggesting which off-critical extension of g_α to study and how to rewrite the string gradient formula.

The analysis given in [185] starts from a system on a cylinder of length L and radius R, which can be viewed as an IR scale, or as an inverse temperature $R \sim \frac{1}{T}$. It is assumed that the system is conformal in the bulk and subjected to a perturbation on the boundary (for simplicity, we take both cylinder ends to carry the same boundary condition).

Extending the finite-size scaling results of [2, 60] to systems with a boundary, we find that the logarithm of the partition function has the following universal behaviour in the large L limit [5]

$$\ln Z \approx \frac{\pi c}{6} \frac{L}{R} + \ln Z_{\text{bdy}}^2 \quad \text{for} \ \ L \to \infty . \tag{5.20}$$

The first term is extensive in the cylinder size, the second term is the one of interest here: Z_{bdy} is the contribution from one cylinder boundary and can be identified with $\langle I_{\lambda\psi} \rangle \equiv \langle 1 \rangle_{\alpha;\lambda\psi} \equiv g_\alpha(\lambda)$. From a path-integral point of view, it is simply the QFT's partition function on a disk of radius R. If the boundary condition α is conformal, we can make the further identification of $\ln Z_{\text{bdy}}$ with the g-factor $\langle 0 \| \alpha \rangle\rangle$, using the same ideas that led to Cardy's conditions in Chapter 4: compute the partition function on the cylinder as the overlap of the boundary state $\| \alpha \rangle\rangle$ with itself,

$$Z_{\alpha\alpha}(q) = \langle\langle \alpha \| \ \tilde{q}^{L_0 - \frac{c}{24}} \ \| \alpha \rangle\rangle \ ,$$

where $\tilde{q} = \exp\{-4\pi L/R\}$. In the limit of a long cylinder, $L \to \infty$, and assuming unitarity, the leading contribution comes from the (closed string) ground state $|0\rangle$, so that

$$Z_{\alpha\alpha}(q) \approx e^{\frac{\pi c}{6} \frac{L}{R}} |\langle\langle \alpha \| \ 0 \rangle|^2 \ ,$$

to be compared to equation (5.20).

From the non-extensive contribution to $\ln Z$ and to the free energy $F = -T \ln Z$, we pass to the *boundary entropy*

$$S_{\text{bdy}} \equiv S_{\text{bdy}}(R, \lambda) := -\frac{\partial (T F_{\text{bdy}})}{\partial T} = \left(1 - R\frac{\partial}{\partial R}\right) \ln Z_{\text{bdy}} \ ,$$

where we have used $R = \frac{1}{T}$.

This boundary entropy S_{bdy} is defined at and away from the critical point, it coincides with the g-factor at conformal fixed points – and it decreases along boundary RG flows. The full details of the proof given in [185] are rather subtle, so we mention only some of the key points here: as before, the generator of scalings is the trace of the energy-momentum tensor, now concentrated on the boundary (as the bulk is conformal throughout); it is given by $\beta_a(\lambda)\psi_a$, up to contributions proportional to the identity. This makes it possible to express $\frac{d}{dR}\ln Z_{\text{bdy}}$ as an integral over expectation values of $\beta_a\psi_a$, which can be related to $\partial_a \ln Z_{\text{bdy}}$, as Z_{bdy} is also the generating functional of the perturbed QFT in question. We still need a careful treatment of contact terms and integrations by parts before arriving at the gradient formula

$$\frac{\partial S_{\text{bdy}}}{\partial \lambda^a} = -\beta_b(\lambda)\, g_{ab}(\lambda)\,, \tag{5.21}$$

where g_{ab} can be viewed as a boundary Zamolodchikov metric on a disk of radius R,

$$g_{ab} = \int_0^{2\pi} \frac{R\, d\omega}{2\pi} \sin^2(\tfrac{\omega}{2})\, \langle\, \psi_a(Re^{i\omega})\psi_b(R)\,\rangle_c$$

with $\langle\cdot\rangle_c$ denoting normalised connected two-point functions.

The gradient formula (5.21) in particular implies that, in unitary theories where the Zamolodchikov metric is positive, the entropy decreases along boundary RG flows to the IR:

$$R\frac{d}{dR}\, S_{\text{bdy}} = \beta_a \frac{\partial}{\partial\lambda^a} S_{\text{bdy}} = -\beta_a\beta_b\, g_{ab} \leq 0\,.$$

5.2 Marginal boundary deformations

Let us now turn to deformations which preserve conformal symmetry for all values of the perturbation parameter λ. Here, the deforming boundary field ψ must necessarily be marginal, $h_\psi = 1$, otherwise a length scale would be introduced.

The methods we will apply to study marginal deformations will be rather different from those for relevant perturbations. The renormalisation group equations will play no role, instead contour integration techniques will feature prominently in removing the UV cutoff ε and obtaining explicit results about the perturbed boundary condition.

5.2.1 Truly marginal operators

To begin with, let us investigate the change of the two-point function $\langle\psi(x_1)\psi(x_2)\rangle_\alpha$ of the perturbing marginal boundary field ψ itself. In our short-distance cutoff regularisation, the first-order contribution is given by

$$\int_{-\infty}^{x_2-\varepsilon} dx\, \langle\psi(x_1)\psi(x_2)\psi(x)\rangle_\alpha + \int_{x_2+\varepsilon}^{x_1-\varepsilon} dx\, \langle\psi(x_1)\psi(x)\psi(x_2)\rangle_\alpha$$

$$+ \int_{x_1+\varepsilon}^{\infty} dx\, \langle\psi(x)\psi(x_1)\psi(x_2)\rangle_\alpha\,.$$

Using the general form of the three-point function (with $x_1 > x_2 > x_3$)

$$\langle \psi(x_1)\psi(x_2)\psi(x_3)\rangle_\alpha = \frac{C_{\psi\psi\psi}^\alpha}{(x_1 - x_2)(x_1 - x_3)(x_2 - x_3)}$$

it is clear that the first-order term in the perturbation expansion is logarithmically divergent unless the structure constant $C_{\psi\psi\psi}^\alpha$ from the boundary OPE vanishes. Since there is no $O(\lambda)$-counterterm available from the factor Z^{-1} in equation (5.1), such a divergence would force the conformal weight of the field ψ away from the initial value $h_\psi = 1$ as we turn on the perturbation. In other words, the marginal field ψ is not truly marginal unless $C_{\psi\psi\psi}^\alpha = 0$. If the theory contains other marginal boundary fields, degenerate perturbation theory gives a stronger condition: a marginal field ψ is *truly marginal only if* $C_{\psi\psi\psi'}^\alpha = 0$ for all marginal boundary fields ψ' in the theory – cf. [132] for the analogous result in the bulk case.

Note that this is merely a necessary condition and by no means sufficient to guarantee true marginality in higher orders of the perturbation series.

For marginal *bulk* perturbations, there are hardly any model-independent results beyond this first-order condition. In the boundary case, we can obtain a formula for the deformation of bulk correlators to all orders in λ – which even admits a generalisation to certain boundary fields – as long as we restrict ourselves to *self-local* marginal boundary operators. We will show that these are indeed truly marginal to all orders and therefore generate families of boundary CFTs.

Recall from Section 4.2 that a boundary field is self-local if it commutes with itself inside correlation functions. The OPE of a self-local boundary field ψ with conformal dimension $h_\psi = 1$ is determined up to a constant K to be

$$\psi(x_1)\psi(x_2) = \frac{K}{(x_1 - x_2)^2} + \text{reg}; \tag{5.22}$$

any other type of singularity would lead to phases upon interchanging x_1 and x_2. In particular, this shows that self-local marginal boundary operators satisfy the first-order condition.

To study higher orders, let us return to the expression (5.1) for the deformed correlators of bulk fields, which can be rewritten (after regularisation) as

$$\langle \varphi_1(z_1, \bar{z}_1) \cdots \varphi_N(z_N, \bar{z}_N) \rangle_{\alpha; \lambda\psi}^\varepsilon$$
$$= Z^{-1} \sum_n \frac{\lambda^n}{n!} \int_{-\infty}^\infty \cdots \int_{-\infty}^\infty \frac{dx_1}{2\pi} \cdots \frac{dx_n}{2\pi} \langle \psi(x_1) \ldots \psi(x_n)\, \varphi_1 \cdots \varphi_N \rangle_\alpha, \tag{5.23}$$

where all integrals run over the real line but with the regions $|x_i - x_j| < \varepsilon$ left out. Note that we here use a different normalisation of the coupling constant (namely $-\lambda/2\pi$ in place of the $\tilde{\lambda}$ from above), i.e. we work with the definition

$$I_{\lambda\psi} = P \exp\{ S_{\lambda\psi} \} := P \exp\left\{ \lambda \int_{-\infty}^\infty \psi(x) \frac{dx}{2\pi} \right\} \tag{5.24}$$

from now on; this will prove convenient later, when contour integrals arise.

Assume now that the perturbing marginal field $\psi(x)$ is self-local. With the help of the OPE (5.22) of ψ, it is not difficult to see that the divergences in ε from the numerator cancel those from the denominator (order by order in λ) so that the limit $\varepsilon \to 0$ of the deformed bulk field correlator can be taken. Moreover, recall that correlators of local boundary fields have a unique analytic continuation which interchanges boundary operators. Therefore, we can deform the integration contour into the upper half-plane, and the limit in question can be written as

$$
\langle \varphi_1(z_1, \bar{z}_1) \cdots \varphi_N(z_N, \bar{z}_N) \rangle_{\alpha;\, \lambda\psi} = \lim_{\varepsilon \to 0} \langle \varphi_1(z_1, \bar{z}_1) \cdots \varphi_N(z_N, \bar{z}_N) \rangle^{\varepsilon}_{\alpha;\, \lambda\psi}
$$

$$
= \sum_n \frac{\lambda^n}{n!} \int_{\gamma_1} \cdots \int_{\gamma_n} \frac{dx_1}{2\pi} \cdots \frac{dx_n}{2\pi} \langle \psi(x_1) \ldots \psi(x_n)\, \varphi_1 \cdots \varphi_N \rangle_\alpha , \qquad (5.25)
$$

where γ_p is the straight line parallel to the real axis with $\Im \gamma_p = i\varepsilon/p$. The integrals can be computed with contour techniques. The expression on the right-hand side is finite, and independent of ε as long as $\varepsilon < \min(\Im z_i)$ where z_i denote the insertion points of bulk fields. Thus, we are in position to construct perturbed bulk correlators to all orders in perturbation theory – and in particular we can determine the deformation of bulk one-point functions and hence the deformation of the structure constants A_φ^α which parametrise the possible boundary theories along with the gluing map.

Extension of these ideas to the deformation of correlation functions which contain boundary fields $\psi_1, \ldots, \psi_M$ along with the bulk fields is a more difficult problem. Formula (5.23) has an obvious generalisation to that case as long as all the $\psi_1, \ldots, \psi_M$ are *local* with respect to the perturbing field ψ. As we have seen before, this is usually a strong constraint on boundary fields, so in general we again need to resort to perturbative renormalisation to compute deformed boundary correlators.

However, what we can study along the lines of equation (5.23) are deformations of n-point functions of the (local) perturbing boundary field ψ itself (without bulk insertions): since the OPE (5.22) contains no first-order poles, all perturbative corrections to the n-point function of ψ vanish. This means that *self-locality* of a marginal boundary field is *sufficient* to guarantee that it is *truly marginal* to all orders in perturbation theory and thus defines a one-parameter family of boundary CFTs [390]. It is likely that the locality criterion is also a necessary one, although we do not have a general proof for this assertion.

5.2.2 Chiral marginal deformations

By far the most explicit results can be obtained for marginal self-local fields which are currents from the *chiral algebra* $\mathcal{W}$. By definition, these *chiral marginal fields* are local with respect to all bulk and boundary fields. Owing to this, the contour integration method set up in the previous subsection becomes especially simple, and it is straightforward to analyse changes of gluing conditions and of one-point functions under chiral marginal deformations. This was carried out in great detail in [390], so we merely highlight here that pure boundary correlators are not affected by chiral marginal deformations (there are no singularities in the upper half-plane and integration contours can be "pulled off" to infinity), and that the cluster property (4.93) of bulk two-point functions is preserved (the proof involves deforming the integration contours to circles around the bulk insertions; see [390]).

In contrast to arbitrary boundary fields, chiral fields are defined in the bulk, as well, and thus chiral marginal deformations can be studied in the *boundary state formalism* – indeed in a more elegant and concise manner than using the general contour integration techniques from above.

Let us work with the "zero-temperature definition" of boundary states $\|\alpha\rangle\rangle$ from Subsection 4.3.3, based on the coordinate transformation

$$\zeta = \frac{1 - iz}{1 + iz}, \tag{5.26}$$

mapping the upper half-plane to the complement of the unit disk in the ζ-plane. The perturbation operator is to be inserted at the unit circle, and using $h_J = 1$, we obtain

$$\int_{\mathbb{R}} \frac{dx}{2\pi} J^{(H)}(x) = \int_{|\zeta|=1} \frac{d\zeta}{2\pi} J^{(P)}(\zeta) = iJ_0^{(P)}. \tag{5.27}$$

Since chiral currents are analytic, we need not worry about possible divergences, as they can be avoided by deforming the integration contour. It is the last equality in (5.27) that makes it easy to treat perturbations by chiral currents in the boundary state formalism: we could *not* conclude that $\int_{\mathbb{R}} dx \, J^{(H)}(x) = J_0^{(H)}$ on the half-plane because of the different integration contour in the definition of half-plane modes, see equation (4.9) in Subsection 4.1.1. Using boundary states, however, the effect of chiral marginal boundary perturbations on a half-plane theory reduces to the action of current zero-modes on the bulk Hilbert space.

The boundary states which describe the boundary conditions before and after the chiral perturbation are related by a simple "rotation". Correlators of the deformed boundary CFT can be obtained upon replacing $\|\alpha\rangle\rangle \equiv \|(\Omega, \tilde{\alpha})\rangle\rangle$ by

$$\|(\Omega, \tilde{\alpha})\rangle\rangle_{\lambda J} \equiv \|(\Omega, \tilde{\alpha}); \lambda J\rangle\rangle = e^{i\lambda J_0} \|(\Omega, \tilde{\alpha})\rangle\rangle, \tag{5.28}$$

where $J_0 \equiv J_0^{(P)}$ is the zero-mode of the left-moving current on the plane.

We have made the unperturbed gluing map Ω explicit in equation (5.28). Indeed, from this formula, we can immediately derive the change of the gluing conditions under the marginal deformation by $J(x)$: using the fact that left- and right-movers commute, along with the simple relation $J_n \|(\Omega, \alpha)\rangle\rangle = -\Omega \bar{J}_{-n} \|(\Omega, \alpha)\rangle\rangle$, the *new gluing conditions* are

$$\left[W_n - (-1)^{h_W} \Omega \circ \gamma_{\bar{J}} (\bar{W}_{-n}) \right] \|(\Omega, \alpha)\rangle\rangle_{\lambda J} = 0 \tag{5.29}$$

with $\gamma_{\bar{J}}(\bar{W}) := \exp(-i\lambda \bar{J}_0) \bar{W} \exp(i\lambda \bar{J}_0)$.

The change of *one-point functions* under chiral marginal deformations can also be read off immediately from equation (5.28): they are simply multiplied by $\exp(i\lambda q_\varphi)$, where q_φ is the charge of φ under J. Alternatively, using the terminology introduced in Subsection 4.1.2, equation (4.18), we can write

$$\langle\varphi_{i,\bar{\imath}}(z,\bar{z})\rangle_{\alpha;\,\lambda J} = \frac{e^{i\lambda X^i_j}A^\alpha_{i\bar{\imath}}}{|z-\bar{z}|^{2h_i}}. \qquad (5.30)$$

The term X^i_j denotes the action of the zero-mode J_0 in the lowest-energy subspace of the $\mathcal{W}$-representation $\mathcal{H}_i$.

The boundary state formalism also allows us to verify – without resorting to perturbation theory – that the *partition function* $Z_\alpha(q) \equiv Z_{\alpha\alpha}(q)$ of a boundary CFT with boundary condition α along the real line *stays invariant under marginal perturbations by a chiral boundary field* $J(x)$.

To see this, we need to compute the transition amplitude between $\|\alpha;\,\lambda J\rangle\rangle$ and $\langle\langle\Theta\,(\alpha;\,\lambda J)\|$. But $\langle\langle\Theta\,\exp(i\lambda J_0)\,\alpha\| = \langle\langle\Theta\alpha\|\,\exp(-i\lambda J_0)$, and $\exp(i\lambda J_0)$ commutes with $L_0^{(P)}$. So the spectrum of the boundary theory does not change, $Z_{(\alpha;\lambda J),(\alpha;\lambda J)}(q) = Z_{\alpha,\alpha}(q)$.

If we have two boundary conditions α and β and deform each of them separately by chiral marginal operators J_1 and J_2, respectively, then the spectrum of boundary-condition changing operators will, in general, change: $Z_{(\alpha;\lambda_1 J^1),(\beta;\lambda_2 J^2)}(q) \neq Z_{\alpha,\beta}(q)$. After the deformations, the partition function will involve "twisted characters" of the symmetry algebra – more precisely, characters of representations twisted by inner automorphisms Ad_U with $U = \exp\{i(\lambda_1 J_0^1 - \lambda_2 J_0^2)\}$.

5.2.3 Non-chiral marginal deformations

Let us now discuss deformations induced by marginal boundary fields $\psi(x)$ that are self-local (or analytic) in the sense that equation (5.22) holds, but are *not* in the chiral algebra. We have shown in Subsection 5.2.1 that such fields are truly marginal to all orders; here we will determine the change of gluing conditions and collect various general statements on the one-point functions and the spectrum of the deformed theory.

It is much more difficult to obtain general results on non-chiral deformations, but their effects will turn out to be more interesting and drastic than those arising in the chiral case.

A. Change of the gluing map

Again we assume that the unperturbed theory enjoys symmetry under some observable algebra $\mathcal{W}$, i.e. that gluing conditions $W = \Omega(\overline{W})$ hold for the generators W of $\mathcal{W}$.

First note that analytic deformations do not change the conditions $T = \overline{T}$ for the energy-momentum tensor. This follows from contour integration and from the OPE of $T(z)$ and $\psi(x)$: for a field ψ of conformal dimension $h = 1$, the singular part of the OPE is a total derivative,

$$T(z)\,\psi(x) = \frac{1}{(z-x)^2}\,\psi(x) + \frac{1}{z-x}\,\partial_x\psi(x) + \text{reg} = \partial_x\left(\frac{1}{z-x}\,\psi(x)\right) + \text{reg}\,.$$

The gluing condition for the Virasoro field is probed by inserting $T(z)$ into the correlation function (5.25) with $\Im z > \varepsilon$, and then moving $T(z)$ towards the real axis where it can be compared to $\overline{T}(\bar{z})$. Whenever we pass through one of the contours γ_p, we pick up a term

$$\int_C \frac{dx}{2\pi}\,T(z)\,\psi(x) = \int_C \frac{dx}{2\pi}\,\partial_x\left(\frac{1}{z-x}\,\psi(x)\right) = 0\,,$$

where C is a small circle surrounding the insertion point of the Virasoro field. The contour integral along C vanishes, which means that the Virasoro field T cannot feel the presence of the perturbation and hence the gluing condition stays intact.

This argument can of course be extended to other symmetry generators: *the gluing condition for a chiral field $W(z)$ stays invariant to all orders in λ under deformation by a self-local marginal field $\psi(x)$ if the singular part of the OPE $W(z)\,\psi(x)$ is a total derivative with respect to x.*

We have seen that deformations by chiral currents J typically lead to a non-trivial twisting (5.29) of the gluing condition of a symmetry generator $W(z)$, so that after deformation the W-symmetry is realised by twisted Ward identities. For non-chiral analytic deformations, this is impossible.

As sketched above for $T(z)$ and described in detail in [390], the change in the $W(z)$ gluing condition is obtained by moving the chiral field $W(z)$ through a stack of integration contours (which do not affect $\Omega\overline{W}$). Applying this same procedure to non-chiral analytic deformations leads, after some combinatorics, to

$$W(z)\,e^{\lambda\int\frac{dx}{2\pi}\,\psi(x)} = e^{\lambda\int\frac{dx}{2\pi}\,\psi(x)}\left[e^{\lambda\psi}\,W\right](z)\,,$$

(which is understood in the limit $\Im z \to 0$) with the abbreviation

$$\left[e^{\lambda\psi}\,W\right](z) := \sum_{n=0}^{\infty}\frac{\lambda^n}{n!}\oint_{C_1}\frac{dx_1}{2\pi}\cdots\oint_{C_n}\frac{dx_n}{2\pi}\,W(z)\,\psi(x_1)\cdots\psi(x_n)\,. \qquad (5.31)$$

The curves C_i encircle the insertion point z in the upper half-plane; see Figure 5.1. To each order n, the integrals will pick some term $\psi^{(n)}$ from the OPE of $W(z)$ with the product of perturbing fields. But at least some of the $\psi^{(n)}$ cannot be in the chiral algebra (otherwise ψ would be), thus the new "gluing condition" would relate $\Omega\overline{W}$ to a mixture of fields from the chiral algebra and true boundary fields which cannot be defined away from the boundary. From such a condition, Ward identities cannot be derived.

This shows that a non-chiral self-local perturbation either breaks the Ward identity associated with a given generator of $\mathcal{W}$ or leaves it invariant altogether.

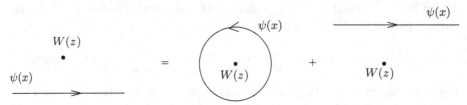

Figure 5.1 The "basic move" when pushing a chiral field towards the boundary through the integration contours.

In general, this leads to a new conformally invariant boundary theory whose Ward identities are governed by a subalgebra $\mathcal{U}$ of the original chiral algebra $\mathcal{W}$.

Examples of non-chiral analytic deformations will be given in Section 5.3. These will reveal that, even though the "gluing" (5.31) of a chiral field $W(z)$ to boundary operators destroys the Ward identity for $W(z)$, some remnant of the symmetry may survive in that the partition function of the deformed theory decomposes into characters of (twisted) representations of the original $\mathcal{W}$.

B. One-point functions, spectrum, and the cluster property

General statements on the behaviour of *one-point functions* under non-chiral analytic deformations are harder to obtain than those on gluing conditions. One reason is that, after the deformation, one-point functions need to be computed for all primary fields of the smaller ("unbroken") subalgebra $\mathcal{U} \subset \mathcal{W}$ associated to the reduced set of Ward identities that may survive after turning on the perturbation – put differently, the new boundary state is built up from $\mathcal{U}$-Ishibashi states. Therefore one has to exploit additional model-dependent features to construct deformed one-point functions to all orders – see below for some cases of non-trivial analytic deformations where this is possible.

What is easy to find is a general first-order criterion for the *invariance* of a one-point function: let $\varphi(z, \bar{z})$ be an arbitrary (spinless) quasi-primary bulk field, e.g. a primary field for the reduced chiral algebra $\mathcal{U} \subset \mathcal{W}$ with conformal weights $h = \bar{h}$. Conformal transformation properties fix the correlator of $\varphi(z, \bar{z})$ with the perturbing field $\psi(x)$ up to a constant,

$$\langle \varphi(z, \bar{z}) \psi(x) \rangle_\alpha = \frac{C^\alpha_{\varphi\psi}}{|z - \bar{z}|^{2h-1}(z - x)(\bar{z} - x)} .$$

The bulk–boundary OPE coefficient $C^\alpha_{\varphi\psi}$ here is the one for the original boundary condition α. Using the residue theorem, the first-order correction of the perturbed one-point function is computed as

$$\langle \varphi(z, \bar{z}) \rangle_{\alpha;\, \lambda\psi} = \langle \varphi(z, \bar{z}) \rangle_\alpha + \lambda \frac{C^\alpha_{\varphi\psi}}{|z - \bar{z}|^{2h}} + O(\lambda^2) .$$

To leading order, a one-point function $\langle \varphi(z, \bar{z}) \rangle_\alpha$ is invariant under a perturbation with ψ iff $C^\alpha_{\varphi\psi} = 0$. Obviously, this is a necessary (and not sufficient) condition for the invariance of a given one-point function under any truly marginal perturbation (analytic or otherwise).

Computations of deformed *partition functions* would require complete knowledge of all one-point functions, thus seem accessible only in special cases. However, we can make some interesting general statements about the behaviour of $Z_\alpha(q)$ under analytic deformations.

The key feature is locality of boundary fields. At the end of Subsection 5.2.1, we pointed out that a straightforward analogue of formula (5.25) can be used to define deformed correlators of boundary fields ψ_i which are *local* with respect to the perturbing field ψ. Contour integration then shows that conformal weights of such fields ψ_i are *invariant* under the deformation.

In general, many fields will not satisfy this criterion (in contrast to the case of chiral deformations where all boundary fields are local with respect to J), and the conformal weights of these fields may well change under non-chiral analytic deformations, but *part of the spectrum is protected.* In particular, all *chiral* (bosonic) fields W are local with respect to ψ, therefore the partition function will always contain the (bosonic) vacuum character of the original chiral algebra. This is true even if gluing conditions and Ward identities are broken down to a subalgebra $\mathcal{U} \subset \mathcal{W}$.

The *cluster property* is even more difficult to analyse in any generality. A few statements have been collected in [390], but let us only mention the main difference to chiral deformations here: for non-chiral self-local deformations, it can happen that the cluster property is preserved for an open neighbourhood of $\lambda = 0$ but breaks down at certain finite values of the perturbation parameter λ. This is indeed confirmed by the examples we will present below, where the failure of the cluster property in fact indicates interesting physical or geometric phenomena.

5.3 An example: boundary deformations for c = 1 theories

Theories built from a single free boson $X(z, \bar{z})$ are good examples to illustrate the general methods from above. In every variant, uncompactified, compactified and orbifolded, there are chiral and non-chiral marginal boundary deformations; their effect can be computed explicitly and leads to interesting phenomena.

We study Neumann and Dirichlet boundary conditions in the uncompactified theory first, then perform the analogous analysis for compactified models where the bosonic field takes values on a circle with radius r. In the third subsection we investigate boundary deformations of $c = 1$ orbifold theories.

5.3.1 The uncompactified theory

The basic facts about the free boson were given in Chapters 1 to 3; we only repeat a few formulas here for convenience. The $\widehat{U(1)} \times \widehat{U(1)}$ symmetry algebra of the bulk theory is generated by the chiral currents

$$J(z) = i\,\partial X(z,\bar z) = \sum_n a_n\, z^{-n-1} \quad \text{and} \quad \bar J(\bar z) = i\,\bar\partial X(z,\bar z) = \sum_n \bar a_n\, \bar z^{-n-1} \,.$$

The bulk Hilbert space realising an uncompactified free boson is $\mathcal{H}^{(P)} = \int^{\oplus}_{g \in \mathbb{R}} \mathcal{H}_g \otimes \mathcal{H}_g$ with corresponding $\widehat{U(1)}$ primary bulk fields

$$\varphi_g(z,\bar z) \equiv \varphi_{g,g}(z,\bar z) = \,:\exp(ig X(z,\bar z)):$$

of conformal weights $h = \bar h = \tfrac{1}{2}g^2$. We use $\alpha' = 2$ in the present subsection, because many of the subsequent formulas look simpler with this choice.

There are two possible boundary conditions that preserve the chiral symmetry algebra generated by the U(1) current J:

Neumann boundary condition: $J(z) = \Omega_N \bar J(\bar z) \equiv \bar J(\bar z)$,   (5.32)

Dirichlet boundary condition: $J(z) = \Omega_D \bar J(\bar z) \equiv -\bar J(\bar z)$.   (5.33)

In the Neumann case, $X(z,\bar z)$ is realised on a state space $\mathcal{H} = \int^{\oplus}_g \mathcal{H}_g$, and the one-point functions of $\varphi_g(z,\bar z)$ are given by, see equation (1.30),

$$\langle \varphi_g(z,\bar z) \rangle_N = \delta_{g,0}\,.$$

For Dirichlet conditions, the value of the bosonic field along the boundary, $X(z,\bar z) \equiv X(z) + X(\bar z) = x_0$ for $z = \bar z$, provides a free real parameter x_0. The one-point functions are

$$\langle \varphi_g(z,\bar z) \rangle_{D\,x_0} = \frac{e^{ig x_0}}{|z - \bar z|^{2h_g}}\,.$$

Later on, we will in particular employ the cluster property (4.93) to test whether a boundary condition generated by non-chiral marginal perturbations is admissible. Therefore, let us recall that the unperturbed Dirichlet boundary condition does satisfy this constraint: as can be inferred from the brief discussion of free boson braiding and fusing in Subsection 3.2.3, the bulk structure constants Ξ that occur in the sewing constraints (4.92) and (4.93) are given by $\Xi_{g',g_1,g_2} = \delta_{g',g_1+g_2}$ for the free boson; therefore, the numbers $A_g^{x_0} = \exp(ig x_0)$ solve equation (4.93). On the other hand, superpositions ("mixtures") of "pure" Dirichlet boundary conditions do not.

A. Chiral deformations

Let us first look at marginal deformations by the chiral current J. Its zero-mode a_0 commutes with everything in the symmetry algebra $\mathcal{W}$, so J generates the trivial inner automorphism $\gamma_J = id$, see equation (5.29), and the gluing conditions are invariant under the deformation. The chiral boundary perturbation can therefore at most affect the one-point functions.

Neumann boundary conditions imply $A_g^N = \delta_{g,0}$, so they must stay invariant under deformations with $J(x)$. For Dirichlet boundary conditions, we plug

$A_g^{x_o} = e^{igx_0}$ into formula (5.30) and see that the one-point functions change according to

$$A_g^{x_0} \longrightarrow e^{i\lambda g} e^{igx_0} = A_g^{x_0+\lambda}$$

when we turn on the chiral perturbation, i.e. the deformation results in a shift of the parameter x_0 by λ – i.e. the D-brane is displaced.

B. Non-chiral deformations

Specialising the computations of boundary spectra from Chapter 1 to one target dimension, the brane partition functions read

$$Z_{NN}(q) = \eta(q)^{-1} \int dk\, q^{\frac{1}{2}k^2}, \quad Z_{Dx_0,Dx_0}(q) = \eta(q)^{-1}.$$

They tell us that apart from the chiral current, there are two more boundary fields of conformal dimension $h = 1$ for the Neumann boundary condition, and none for Dirichlet boundary conditions. We will consider perturbations of the Neumann boundary condition by the combinations

$$\psi_1(x) := \sqrt{2}\,\cos\{\sqrt{2}X(x)\} \quad \text{and} \quad \psi_2(x) := \sqrt{2}\,\sin\{\sqrt{2}X(x)\}, \quad (5.34)$$

i.e. by switching on periodic boundary potentials, using the language of Callan *et al.* and of Polchinski and Thorlacius, who studied this problem in some detail in [88, 380], using different and more model-dependent methods. We will see that the ψ_a break the chiral symmetry down to the Virasoro algebra.

From the usual free boson expressions for the correlators of such vertex operators, we see that the boundary fields $\psi_a(x)$, $a = 1, 2$, are self-local and mutually local and therefore truly marginal to all orders in the coupling λ. However, the boundary Hilbert space $\mathcal{H} = \int_g^\oplus \mathcal{H}_g$ of the Neumann theory contains fields which are non-local with respect to $\psi_a(x)$, therefore the spectrum of boundary fields should change when the boundary potential is turned on. Only the scaling dimensions of operators with charges in the lattice $\sqrt{2}\mathbb{Z}$ are protected, according to the general observations from Section 5.2.

Let us focus on perturbations by $\psi_1(x)$ and study the behaviour of the $\widehat{U(1)}$ *gluing conditions*. In Subsection 5.2.3 we found that Ω stays invariant as long as the singular part of the OPE between a chiral symmetry generator $W(z)$ and $\psi_1(x)$ is a total ∂_x-derivative. In the present case, this is true for the Virasoro field $W(z) = T(z)$, but not for the current $W(z) = J(z)$:

$$J(z)\,\psi_1(x) = \frac{i\sqrt{2}}{z - x}\,\psi_2(z) + \text{reg}. \quad (5.35)$$

We can use this OPE to determine the effect of pushing $J(z)$ through the x_i-integration contours – which we need to make formula (5.31) for the change of the original gluing condition $J(z) = \bar{J}(\bar{z})$ explicit. Whenever $J(z)$ passes one of the contours, we pick up a term $i\sqrt{2}\psi_2(z)$, and the effect of moving this field

$\psi_2(z)$ through the remaining contours is determined by the OPE

$$\psi_2(z)\,\psi_1(x) = \frac{-i\sqrt{2}}{z-x}\,J(z) + \text{ reg}. \tag{5.36}$$

We can now evaluate the expression (5.31) and derive the following closed expression for the ψ_1-deformed gluing conditions:

$$J(z) = \sin(\sqrt{2}\lambda)\,\psi_2(x) + \cos(\sqrt{2}\lambda)\,\overline{J}(\bar{z}) \tag{5.37}$$

for $z = \bar{z} = x$. This equation was given before in [88]; an analogous formula with ψ_2 replaced by $-\psi_1$ describes the effect of perturbations with ψ_2 on the Neumann gluing condition.

After the perturbation, the boundary reflects left-moving into right-moving currents only at the expense of marginal boundary fields, and the correlation functions no longer obey Ward identities for the U(1) current J.

This is true for generic coupling λ, but the values $\lambda = n\pi/\sqrt{2}$ with integer n are exceptional: for these values, the $\psi_a(x)$ disappear from equation (5.37), so that U(1) symmetry is restored; moreover, if n is odd, the original Neumann conditions for $J(z)$ are turned into Dirichlet conditions. We will refer to the latter special values of the perturbation parameter as *Dirichlet-like points*. Note that these effects are only visible because the full perturbation expansion is under control.

Now let us try to determine all *one-point functions*, i.e. the full boundary state $\|N; \lambda\psi_a\rangle\!\rangle$ after perturbing the Neumann boundary condition by ψ_a. Since the $\widehat{U(1)}$ symmetry is broken, we need to consider Virasoro primary fields, which is possible for central charge $c = 1$ thanks to the decomposition of irreducible $\widehat{U(1)}$ modules into $c = 1$ Virasoro modules:

$$\mathcal{H}_{g=\sqrt{2}m}^{\widehat{U(1)}} = \bigoplus_{l=0}^{\infty} \mathcal{H}_{h=(|m|+l)^2}^{\text{Vir}} \quad \text{if } m \in \tfrac{1}{2}\mathbb{Z},$$

$$\mathcal{H}_g^{\widehat{U(1)}} = \mathcal{H}_{h=g^2}^{\text{Vir}} \quad \text{otherwise}. \tag{5.38}$$

This allows the state space $\mathcal{H}^{(P)}$ of the bulk theory to be rewritten as

$$\mathcal{H}^{(P)} = \int_{g \notin \frac{1}{2}\mathbb{Z}}^{\oplus} \mathcal{H}_g^{\widehat{U(1)}} \otimes \mathcal{H}_g^{\widehat{U(1)}} \oplus \bigoplus_{m \in \frac{1}{2}\mathbb{Z}} \mathcal{H}_{\sqrt{2}m}^{\widehat{U(1)}} \otimes \mathcal{H}_{\sqrt{2}m}^{\widehat{U(1)}}$$

$$= \int_{g \notin \frac{1}{2}\mathbb{Z}}^{\oplus} \mathcal{H}_{\frac{g^2}{2}}^{\text{Vir}} \otimes \mathcal{H}_{\frac{g^2}{2}}^{\text{Vir}} \oplus \bigoplus_{j \in \frac{1}{2}\mathbb{Z}_+} \left(\mathcal{H}_{j^2}^{\text{Vir}} \otimes \mathcal{H}_{j^2}^{\text{Vir}}\right)^{\oplus 2j+1} \oplus \cdots. \tag{5.39}$$

The dots stand for terms with $h \neq \bar{h}$, which we can ignore because they cannot couple to a conformal boundary state.

Clearly, there are two classes of spinless (i.e. $h = \bar{h}$) Virasoro primary bulk fields, namely

(1) $\quad \varphi_{g,g}(z,\bar{z}) \quad$ with $\quad g \notin \frac{1}{\sqrt{2}}\mathbb{Z}$,

(2) $\quad \varphi_{m,m}^{j}(z,\bar{z}) \quad$ with $\quad j \in \frac{1}{2}\mathbb{Z}_+, \quad m = -j, -j+1, \ldots, j-1, j.$

The one-point functions of the fields $\varphi_{g,g}$ in the first family are not perturbed, i.e. for any λ we still have

$$\langle \varphi_{g,g}(z,\bar z) \rangle_{N;\lambda\psi_a} = 0 \quad \text{for} \quad g \notin \tfrac{1}{\sqrt{2}}\mathbb{Z}. \tag{5.40}$$

This is due to the $U(1)$ charge conservation in the unperturbed theory and the fact that the perturbing fields span the charge lattice $\sqrt{2}\mathbb{Z}$, so any correction in the perturbation series vanishes.

The fields of the second family, however, themselves carry $U(1)$ charges $g = \bar g = \sqrt{2}m \in \tfrac{1}{\sqrt{2}}\mathbb{Z}$ under J_0.

To compute their one-point functions, we first observe the theory enjoys a useful "hidden" $SU(2)$ symmetry, as is suggested by the OPEs among J, ψ_1, ψ_2: these fields do not form an algebra of true local currents for the full boundary CFT, but a finite-dimensional $su(2)$ Lie algebra can still be extracted from them, spanned by J_0 and $(\psi_a)_0 := \int \frac{dx}{2\pi}\,\psi_a(x)$.

This $su(2)$ Lie algebra symmetry is in fact already present in the decomposition of the state spaces above: for any $j \in \tfrac{1}{2}\mathbb{Z}$, the Virasoro highest weight vectors at energy $h = j^2$ span an $su(2)$ multiplet of length $2j+1$.

Furthermore, this $su(2)$ symmetry governs the perturbed one-point functions of the fields $\varphi^j_{m,m}$ from the second family above: correlation functions of $\varphi^j_{m,m}(z,\bar z)$ with several perturbing fields have only pole singularities, so we can analytically continue in the ψ_1-insertion points and evaluate the deformed correlators by the usual contour integrals. This results in an action of the $SU(2)$ generators J_0 and $(\psi_a)_0$ on the left-moving index of the fields, and we can write

$$\langle \varphi^j_{m,m}(z,\bar z) \rangle_{N;\,\lambda\psi_a} = \sum_{m'=-j}^{j} D^j_{m,m'}(\Gamma^a_\lambda)\,\langle \varphi^j_{m',m}(z,\bar z) \rangle_N. \tag{5.41}$$

Here, $\Gamma^a_\lambda := \exp(i\lambda(\psi_a)_0)$ is regarded as an $SU(2)$ element, and $D^j_{m,m'}(\Gamma^a_\lambda)$ are the entries of its spin j representation matrix expressed in a J_0-eigenbasis – the rather cumbersome formula (4.127) for the matrix elements was quoted in Subsection 4.A.5.

The "correlator" on the right-hand side of equation (5.41) is to be understood as the function

$$\langle \varphi^j_{m',m}(z,\bar z) \rangle_N = \delta_{m',-m}\,\frac{1}{|z-\bar z|^{m^2}},$$

irrespective of whether $\varphi^j_{-m,m}(z,\bar z)$ actually occurs in the uncompactified free boson theory. To summarise, we have derived the following formula for the ψ_a-perturbed flat Neumann boundary state:

$$\| N;\lambda\psi_a \rangle\!\rangle = \sum_{j\in\frac{1}{2}\mathbb{Z}_+} \sum_{m=-j}^{j} D^j_{m,-m}(\Gamma^a_\lambda)\,|j,m,m\rangle\!\rangle \tag{5.42}$$

where $|j, m, m\rangle\rangle$ are Virasoro Ishibashi states associated with the primaries $\varphi^j_{m,m}(z, \bar{z})$. Note that the deformed boundary states fit into the classification of $c = 1$ conformal boundary conditions reviewed in Subsection 4.A.5.

Equation (5.42) gives complete information on the perturbed boundary theory, and in particular allows the computation of the perturbed partition function $Z_\alpha(q)$ directly via a modular transformation of the overlap of $\|N; \lambda\psi_a\rangle\rangle$ with itself. Since we reviewed very similar computations in Subsection 4.A.5, following [215, 216], we will merely make some observations on the results here.

As a special case, consider the Dirichlet-like points $\lambda \in \frac{2k+1}{\sqrt{2}}\pi$ with $k \in \mathbb{Z}$. For these λ, the coefficients $D^j_{m,-m}$ simplify considerably, and we obtain the partition function

$$Z_{\alpha_D}(q) \sim \sum_{n \in \mathbb{Z}} \frac{q^{n^2}}{\eta(q)} \tag{5.43}$$

for Dirichlet-like boundary conditions $\|\alpha_D\rangle\rangle := \|N; \frac{\pi}{\sqrt{2}}\psi_a\rangle\rangle$. We see that the continuous Neumann spectrum from before the deformation is reduced to a discrete one – which furthermore happens to coincide with the boundary spectrum of a free boson compactified at the self-dual radius $r_{\text{s.d.}} = 1/\sqrt{2}$; see equations (1.88) and (1.89). This suggests that the Dirichlet-like boundary condition should be viewed as a superposition of infinitely many flat D-branes located at the minima of the boundary potential; see [88]. According to these authors, the boundary fields with non-zero U(1) charge should be attributed to "solitons" interpolating between different minima.

The above is the extreme case of the *band structure* the spectrum develops for perturbations $\lambda\psi_a$ with arbitrary λ. This was first computed in [380] using a free fermion representation, but can be reproduced using modular transformations as outlined in Subsection 4.A.5 and detailed in [215]. The spectrum of a perturbed Neumann boundary state $\|\alpha\rangle\rangle_\lambda := \|N; \lambda\psi_1\rangle\rangle$ is given by

$$Z_{\alpha_\lambda}(q) = \eta(q)^{-1} \sum_{m \in \mathbb{Z}} \int_0^2 d\zeta \, q^{(2m+f_\lambda(\zeta))^2} \tag{5.44}$$

with

$$f_\lambda(\zeta) = \frac{1}{\pi} \arcsin\left[\cos\frac{\lambda}{\sqrt{2}} \sin \pi\zeta\right] ; \tag{5.45}$$

the arcsin-branch is to be chosen such that $\lim_{\lambda \to 0} f_\lambda(\zeta) = \zeta$; the integral is over the half-open interval, which becomes important only for the compactified variants to be discussed below.

Band spectra as in equation (5.44) are typical in theories of electrons moving in a crystal. As soon as an infinitesimal periodic potential is turned on, the continuous spectrum rips apart at the values $h = (2n + 1)^2/4$, and the gaps open up when the strength λ of the potential grows. The bands are reduced to points when the Dirichlet-like value $\lambda = \frac{\pi}{\sqrt{2}}$ is reached, where only primaries of dimensions $h = n^2, n \in \mathbb{Z}$, remain. At first sight, we might expect to encounter this tight-binding limit for $\lambda = \infty$, but in fact the period of the potential introduces a scale $r_{\text{s.d.}} = \frac{1}{\sqrt{2}}$ into the problem so that special effects occur when λ is in resonance with $r_{\text{s.d.}}$.

The structure of the spectrum is in line with our general arguments from Subsection 5.2.2: the states of $U(1)$ charge in the lattice $\sqrt{2}\mathbb{Z}$ – which correspond to boundary fields that are local with respect to the perturbing fields – stay in the boundary theory for all values of λ. Moreover, $Z_{\alpha\lambda}(q)$ decomposes into (twisted) $\widehat{U(1)}$ characters even though the Ward identities for the currents are destroyed.

Formulas (5.43) and (5.44) show that the perturbed boundary states (5.42) satisfy Cardy's conditions, and we can infer from the constructions in Subsection 4.A.5 and [215] that the cluster condition is obeyed as long as $|\lambda| < \frac{\pi}{\sqrt{2}}$. Our study of orbifold models in Subsection 5.3.3 will provide a different point of view confirming this.

However, for the Dirichlet-like values $\lambda = \frac{2k+1}{\sqrt{2}}\pi$ of the coupling, the cluster relation together with the Dirichlet Ward identities for the $U(1)$ current would imply the sewing constraint

$$A^{\alpha_D}_{g_1} A^{\alpha_D}_{g_2} \overset{?}{=} A^{\alpha_D}_{g_1+g_2} .\tag{5.46}$$

But if we choose g_1, g_2 such that $g_i \notin \frac{1}{\sqrt{2}}\mathbb{Z}$ and $g_1 + g_2 = \frac{n}{\sqrt{2}}$, the structure constants $A^{\alpha_D}_{g_i}$ vanish as in the original Neumann boundary theory, cf. equation (5.40), while $A^{\alpha_D}_{g_1+g_2} \neq 0$. Thus, the cluster condition is violated at the Dirichlet-like points – which is completely in line with the physical interpretation that for these special values the periodic boundary potential yields superpositions of elementary Dirichlet conditions. We therefore have reason to view failure of the cluster condition as a signal of interesting effects rather than a short-coming.

5.3.2 Compactified circle theories

If we take a circle of radius r as the target space for the free boson $X(z, \bar{z}) = X_L(z) + X_R(\bar{z})$, the bulk spectrum is discrete: the $\widehat{U(1)}$ primary fields $\varphi_{g,\bar{g}}(z, \bar{z}) := \; e^{2igX_L(z)} e^{2i\bar{g}X_R(\bar{z})} :$ can carry different left- and right-moving charges with respect to the current zero-modes a_0 and $\bar{a}_0$, namely $g = k/2r + rw$ and $\bar{g} = k/2r - rw$, where $k := rp$ and the winding number w take integer values. (As in previous chapters, we use the convention $\alpha' = \frac{1}{2}$ when discussing the compactified boson.)

No new methods are required to study boundary deformations in this situation – in fact, discreteness of the spectrum simplifies the computations. We will therefore just give a few formulas and point out some special features that do not occur in the uncompactified case. Details can be found in [390].

Again Dirichlet and Neumann boundary conditions can be imposed, and we arrive at the one-point functions

$$\langle \varphi_{g,\bar{g}}(z,\bar{z})\rangle_{D\,x_0} = \delta_{g,\bar{g}}\,\frac{e^{ikx_0/r}}{|z-\bar{z}|^{k^2/(4r^2)}}\,,\,\langle \varphi_{g,\bar{g}}(z,\bar{z})\rangle_{N\,\tilde{x}_0} = \delta_{g,-\bar{g}}\,\frac{e^{2irw\tilde{x}_0}}{|z-\bar{z}|^{r^2w^2}}\,.$$

In the Dirichlet case, the parameter x_0 (the location of the brane) takes values $x_0 \in \mathbb{R} \pmod{2\pi r}$; the Neumann case is obtained via T-duality with $\tilde{x}_0 \in \mathbb{R} \pmod{\frac{\pi}{r}}$.

As explained before, equations (1.89) and (1.88), the partition functions of the theories with boundary conditions $(D\,x_0)$ and $(N\,\tilde{x}_0)$, respectively, along the real line are

$$Z_{D\,x_0}(q) = \frac{1}{\eta(q)}\sum_{k\in\mathbb{Z}} q^{2r^2k^2}\,, \qquad Z_{N\,\tilde{x}_0}(q) = \frac{1}{\eta(q)}\sum_{w\in\mathbb{Z}} q^{\frac{w^2}{2r^2}}\,. \tag{5.47}$$

They depend on the compactification radius (i.e. on the bulk modulus), but not on the boundary parameter x_0 or $\tilde{x}_0$.

A. Chiral deformations

Marginal deformations with the chiral current $J(x)$ can be treated in complete analogy to the uncompactified case. With the conventions used above, they result in translations of x_0 and $\tilde{x}_0$, respectively, by $\lambda/2$, in particular they are periodic with period $4\pi r$ and $2\pi/r$, respectively.

Let us, however, point out that in the Neumann case the perturbation of the action can be rewritten as an integral of a (constant) gauge field over the D-brane worldvolume,

$$\lambda \int_{\partial\Sigma} dx\,\partial_x X(x) = \int_{\partial\Sigma} A(\lambda)\cdot dX$$

with $A(\lambda) = \lambda$. As before, x is the coordinate along worldsheet boundary $\partial\Sigma$, while X is viewed as a sigma-model field taking values in the target $\mathbb{R}$. This reasoning, which also applies in the case of higher-dimensional targets, explains why the parameter $\tilde{x}_0$ is often referred to as the "strength of a Wilson line".

For the "rational radii" $r = \sqrt{M/N}$ with positive coprime integers M, N, the symmetry algebra of the bulk theory is enlarged. With respect to this W-algebra, generated by the additional chiral local fields (g_{loc} is $2\sqrt{MN}$ if N is odd and $\sqrt{MN}$ if N is even)

$$W^{\pm}_{g_{\text{loc}}}(z) = \,:e^{\pm 2ig_{\text{loc}}X_L(z)}:\quad \text{and}\quad \overline{W}^{\pm}_{g_{\text{loc}}}(\bar{z}) = \,:e^{\pm 2ig_{\text{loc}}X_R(\bar{z})}:; \tag{5.48}$$

the theory becomes rational and is referred to as "rational Gaussian model"; see [132] and references therein.

It is interesting to note that Dirichlet and Neumann boundary conditions automatically respect this extended symmetry, with parameter-dependent gluing conditions: in the Dirichlet case, we have

$$W^{\pm}_{g_{\text{loc}}}(z) = \Omega_D \left[\overline{W}^{\pm}_{g_{\text{loc}}} \right](\bar{z}) := e^{\pm 2 i g_{\text{loc}} x_0} \overline{W}^{\mp}_{g_{\text{loc}}}(\bar{z}) \tag{5.49}$$

for $z = \bar{z}$. In the Neumann case $X_L(x) = X_R(x) + \tilde{x}_0$ leads to

$$W^{\pm}_{g_{\text{loc}}}(z) = \Omega_N \left[\overline{W}^{\pm}_{g_{\text{loc}}} \right](\bar{z}) := e^{\pm 2 i g_{\text{loc}} \tilde{x}_0} \overline{W}^{\pm}_{g_{\text{loc}}}(\bar{z}) \tag{5.50}$$

along the boundary. Chiral marginal boundary deformations by J change the phases but respect the extended symmetry.

Specifically, at the *self-dual radius* $r_{\text{s.d.}} = 1/\sqrt{2}$, the additional local fields $J^{\pm}(z) = W^{\pm}_{\sqrt{2}}(z)$ have conformal dimension $h_{\pm} = 1$, and together with $J^3(z) :=$ $J(z)$, $\overline{J}(\bar{z})$ they generate the non-abelian current algebra $\widehat{SU(2)}_1$ (analogously for their right-moving counterparts). It is not difficult to show that marginal boundary deformations by all available chiral fields lead to the three-dimensional family (4.127) from Subsection 4.A.5, with $g \in SU(2)$, which we can more simply write as

$$\| \Gamma \rangle\rangle_{\text{s.d.}} = \Gamma \, \| N(0) \rangle\rangle_{\text{s.d.}} \qquad \text{with } \Gamma = e^{i \lambda_a J_0^a} . \tag{5.51}$$

Observe that a perturbation with λJ^1, say, changes the gluing condition $J^3(z) = \pm \overline{J}^3(\bar{z})$ to

$$J^3(z) = \pm (\cos \sqrt{2} \lambda) \, \overline{J}^3(\bar{z}) \pm (\sin \sqrt{2} \lambda) \, \overline{J}^2(\bar{z}) ;$$

cf. the general formula (5.29) and also equation (5.37) for the non-chiral deformation ψ_1. In particular, at the self-dual radius, we can *rotate Neumann conditions for* $J = J^3$ *into Dirichlet conditions*,

$$\| D(0) \rangle\rangle_{\text{s.d.}} = e^{i \frac{\pi}{\sqrt{2}} J_0^1} \, \| N(0) \rangle\rangle_{\text{s.d.}} . \tag{5.52}$$

We obtain one connected $SU(2)$ family of boundary conditions for $r_{\text{s.d.}}$, and in contrast to the Dirichlet-like point of the non-chiral deformation $\lambda \psi_1$, the cluster property here stays intact for all values of λ.

Let us also remark on the relation to rational Cardy boundary states for $\widehat{SU(2)}_1$: this theory has two Ishibashi states $|0\rangle\rangle$ and $|1\rangle\rangle$, and the Cardy boundary states

$$\| \alpha \rangle\rangle^{\text{SU(2)}} = 2^{\frac{1}{4}} |0\rangle\rangle + (-1)^{\alpha} \, 2^{\frac{1}{4}} |1\rangle\rangle$$

(see equation (4.80)) with $\alpha = 0, 1$ are related to those in the family (5.51) as

$$\| 0 \rangle\rangle^{\text{SU(2)}} = \| N(0) \rangle\rangle_{\text{s.d.}}, \quad \| 1 \rangle\rangle^{\text{SU(2)}} = \| N(\pi \sqrt{2}) \rangle\rangle_{\text{s.d.}} ;$$

see [390] for more details on this identification.

B. Non-chiral deformations

We have seen that the uncompactified free boson theory with Neumann boundary conditions admits additional non-chiral marginal deformations $\psi_{1,2}$. In the compactified case, analogous fields are encountered for both Dirichlet and Neumann boundary conditions, at special values of the compactification radius. Inspection of the partition functions (5.47) shows that these radii are $r = \frac{N}{\sqrt{2}}$ in the Neumann and $r = \frac{1}{\sqrt{2N}}$ in the Dirichlet case, with integer $N \geq 2$.

Deformations by these self-local and mutually local boundary fields can be treated by the same methods we used in the uncompactified Neumann theory, so we merely collect a few formulas; more details can be found in [390].

Choosing a Neumann boundary condition as a starting point, the deformed boundary state can be written as

$$\|N(\tilde{x}_0); \lambda\psi_a\rangle\rangle = \sum_{j\in\frac{1}{2}\mathbb{Z}_+} \sum_{w,k'\in\mathbb{Z}} D^j_{\frac{k'+Nw}{2}, \frac{-k'+Nw}{2}}(\Gamma^a_{\tilde{x}_0,\lambda}) \, |j, \tfrac{k'+Nw}{2}, \tfrac{k'-Nw}{2}\rangle\rangle. \quad (5.53)$$

The Ishibashi states here are those corresponding to the decomposition of the bulk Hilbert space at $r = \frac{N}{\sqrt{2}}$ into Virasoro irreps. In contrast to the uncompactified situation, the SU(2) element $\Gamma^a_{\tilde{x}_0,\lambda} = e^{i\lambda(\psi_a)_0} e^{2i\tilde{x}_0 J_0}$ now depends on the perturbation parameter λ as well as the parameter $\tilde{x}_0$ which specifies the original Neumann condition. The expression (5.53) is, however, in line with the formulas given in Subsection 4.A.5 and in [215].

Again, there are Dirichlet-like values $\lambda = \frac{2k+1}{\sqrt{2}}\pi$ of the coupling constant where perturbations of a single Neumann brane by ψ_a results in a superposition of N Dirichlet boundary states. Moreover, due to T-duality for compactified targets, single Dirichlet branes can be deformed into superpositions of Neumann branes.

The open string spectra of the deformed boundary states can be computed as before. Let us merely give the result in one special case that will play a role in connection with orbifold models below: at the special radius $r = \sqrt{2}$, the boundary partition function of $\|\alpha\rangle\rangle_\lambda := \|N(\tilde{x}_0); \lambda\psi_a\rangle\rangle$ reads

$$Z_{\alpha_\lambda}(q) = \eta(q)^{-1} \sum_{m\in\mathbb{Z}} \left(q^{m^2} + q^{\left(m+\frac{1}{2}+\frac{\lambda}{\sqrt{2}\pi}\right)^2} \right). \quad (5.54)$$

This is a discrete analogue of equation (5.44), and the charges $g \neq n\sqrt{2}$ indeed follow the flow prescribed by the function (5.45), as the corresponding fields are precisely those that are not local with respect to the perturbing field. At $\lambda = \frac{\pi}{\sqrt{2}}$, only the charges $g = n\sqrt{2}$ are left.

5.3.3 c = 1 orbifold theories

Let us also collect a few facts about marginal boundary deformations of $S^1/\mathbb{Z}_2$ orbifold models. No entirely new techniques are needed, so we focus on the differences from and noteworthy points of contacts with circle theories here. More details can be found in [390].

Boundary states for the $S^1/\mathbb{Z}_2$ orbifold theories were discussed in Subsection 4.A.2: equations (4.103,4.104) give the untwisted (or regular) Dirichlet and Neumann boundary states, while equations (4.106,4.107) contain explicit formulas for twisted (or fractional) branes.

In order to study marginal boundary deformations of the orbifold boundary theories, we first need to compute the partition functions $Z_{\alpha\alpha}(q)$ for boundary states α from the above two sets of equations. In the case of Dirichlet gluing conditions, we have

$$Z_{\alpha\alpha}(q) = \sum_{k\in\mathbb{Z}} \frac{q^{2r^2k^2}}{\eta(q)} + \sum_{k\in\mathbb{Z}} \frac{q^{2(rk+\frac{x_0}{\pi})^2}}{\eta(q)} \qquad \text{for } \|\alpha\rangle\rangle = \|D(x_0)\rangle\rangle^{\text{orb}}, \qquad (5.55)$$

$$Z_{\beta\beta}(q) = \sum_{k=1}^{\infty} \frac{q^{2r^2k^2}}{\eta(q)} + \sum_{n=0}^{\infty} \chi^{\text{Vir}}_{4n^2}(q) \qquad \text{for } \|\beta\rangle\rangle = \|D(\xi), \pm\rangle\rangle^{\text{orb}}; \qquad (5.56)$$

the first line refers to regular, the second to fractional branes. The Neumann partition functions are obtained when r is replaced with $\frac{1}{2r}$. The Virasoro characters are given by $\chi^{\text{Vir}}_h(q) = \eta^{-1}q^h$ if $h \neq m^2$ for any $m \in \frac{1}{2}\mathbb{Z}$ and by the difference (4.128) of $\widehat{U(1)}$ characters otherwise. Note that the second sum in equation (5.56) is the vacuum character of the symmetry algebra that survives the orbifold procedure, namely the $\mathbb{Z}_2$-invariant subalgebra of $\widehat{U(1)}$.

This subalgebra contains no dimension-one field, and twisted boundary conditions admit a marginal boundary operator only for the special radii $r = \frac{1}{\sqrt{2N}}$ with some $N \in \mathbb{Z}_+$. This boundary field $\cos(2\sqrt{2}X(x))$, leads to similar effects as the $\psi_a(x)$ from the circle case.

An untwisted Dirichlet brane $\|D(x_0)\rangle\rangle^{\text{orb}}$ can be viewed as a superposition of two branes in the circle theory located at x_0 and $-x_0$. It nevertheless obeys the cluster property (4.93) with respect to the reduced set of bulk fields present in the orbifold theory. In accordance with this picture, the partition function (5.55) consists of an x_0-dependent part and a more generic term depending only on the radius r. The former terms contribute $h = 1$ states if the distance of the two pre-image branes on the circle – i.e. the length of the stretched open string – is appropriately adjusted, but it can be shown that these marginal fields arise from the bulk–boundary OPE of a twist field [367]. Therefore, they have non-trivial monodromy with respect to that bulk field and with respect to themselves and are not truly marginal.

Among the fields counted by the first sum in equation (5.55), on the other hand, we do find analytic marginal operators. For radii $r = \frac{1}{\sqrt{2N}}$, we have the familiar fields $\psi_a(x)$, $a = 1, 2$ corresponding to periodic boundary potentials. Perhaps more interestingly, one marginal operator $J(x)$ is present for arbitrary r, which can be viewed as the boundary value of the original current of the circle theory. While the bulk current is removed by the orbifolding, the boundary field

$J(x)$ re-appears via the bulk–boundary OPE of the bulk fields $\cos(\frac{k}{r}X)$ with $k \neq 0$,

$$\cos\left(\frac{k}{r}X(z,\bar{z})\right) = \frac{\cos\frac{k x_0}{r}}{|z-\bar{z}|^{\frac{k^2}{4r^2}}} \; 1 - \frac{\frac{k}{2r}\sin\frac{k x_0}{r}}{|z-\bar{z}|^{\frac{k^2}{4r^2}-1}} \; J(x) + \cdots; \tag{5.57}$$

compare to the one-point function (4.105) given in Subsection 4.A.2.

The field $J(x)$ is local with respect to all other boundary fields from the x_0-independent part of the spectrum, but non-local with respect to the others, which arise from the bulk–boundary OPE of twist fields. Consequently, the second part of the boundary spectrum (5.55) can be expected to change under a perturbation with J. We find that

$$\|D(x_0)\rangle\rangle^{\mathrm{orb}}_{\lambda J} = \tfrac{1}{\sqrt{2}} e^{i\lambda J_0} \|D(x_0)\rangle\rangle^{\mathrm{circ}} + \tfrac{1}{\sqrt{2}} e^{-i\lambda J_0} \|D(-x_0)\rangle\rangle^{\mathrm{circ}} = \|D(x_0 + \tfrac{\lambda}{2})\rangle\rangle^{\mathrm{orb}},$$
$$\tag{5.58}$$

i.e. the marginal operator $J(x)$ moves the untwisted orbifold brane along the interval $]0, \pi r[$. When the endpoints $\xi = 0, \pi r$ are reached, we arrive at super-positions $\|D(\xi), +\rangle\rangle^{\mathrm{orb}} + \|D(\xi), -\rangle\rangle^{\mathrm{orb}}$, and the cluster condition is violated (see [390] for a detailed discussion).

Let us add a few remarks on the special radius $r_{\mathrm{orb}} = \frac{1}{\sqrt{2}}$. It is well known that this orbifold theory is isomorphic to a free boson compactified on a circle with radius $r_{\mathrm{circ}} = \sqrt{2}$. The isomorphism is explicit (see, e.g., Ginsparg's review [235]); in particular, the current $J^3_{\mathrm{circ}}(z)$ from the circle theory corresponds to the field $J^1_{\mathrm{orb}}(z) := \sqrt{2}\cos 2\sqrt{2}X(z)$ that appears as an additional chiral current in the orbifold model at $r_{\mathrm{orb}} = \frac{1}{\sqrt{2}}$, and the twist fields σ_0 and $\sigma_{\pi r}$ of the latter are identified with $\sin(\frac{1}{\sqrt{2}}X)$ and $\cos(\frac{1}{\sqrt{2}}X)$ in the $r_{\mathrm{circ}} = \sqrt{2}$ theory.

As the bulk theories are equivalent, it must be possible to match boundary conditions, as well. It is straightforward to find identities for boundary partition functions, e.g.

$$Z^{\sqrt{2}}_{\alpha\alpha}(q) = Z^{1/\sqrt{2}}_{\beta\beta}(q) \quad \text{for } \|\alpha\rangle\rangle = \|N(\tilde{x}_0)\rangle\rangle^{\mathrm{circ}}, \; \|\beta\rangle\rangle = \|G(\tfrac{\pi}{2\sqrt{2}})\rangle\rangle^{\mathrm{orb}}, \tag{5.59}$$

$$Z^{\sqrt{2}}_{\alpha\alpha}(q) = Z^{1/\sqrt{2}}_{\beta\beta}(q) \quad \text{for } \|\alpha\rangle\rangle = \|D(x_0)\rangle\rangle^{\mathrm{circ}}, \; \|\beta\rangle\rangle = \|G(\xi), \pm\rangle\rangle^{\mathrm{orb}}. \tag{5.60}$$

The gluing conditions "G" in the orbifold theory can be either Neumann or Dirichlet, and the circle parameters take values $x_0 \in [0, 2\sqrt{2}\pi]$ and $\tilde{x}_0 \in [0, \frac{\pi}{\sqrt{2}}]$ as usual.

Families of deformed boundary states, too, can be easily matched at the level of partition functions: we first note that the isomorphism also maps marginal boundary fields, e.g. $J^3_{\mathrm{orb}}(x) := J_{\mathrm{orb}}(x)$ to $\psi_{2,\,\mathrm{circ}}(x)$ and $J^2_{\mathrm{orb}}(x) := \sqrt{2}\sin(2\sqrt{2}X(x))$ to $\psi_{1,\,\mathrm{circ}}(x)$. Equations (5.54) and (5.55) then show that the following partition functions coincide,

$$Z^{\sqrt{2}}_{\alpha\alpha}(q) = Z^{1/\sqrt{2}}_{\beta\beta}(q) \quad \text{for } |\alpha\rangle = |N(\tilde{x}_0); \lambda\psi_1\rangle^{\mathrm{circ}} \text{ and } \quad |\beta\rangle = |G(\tfrac{\pi}{2\sqrt{2}} + \tfrac{\lambda}{2})\rangle^{\mathrm{orb}}.$$
$$\tag{5.61}$$

By computing bulk field one-point functions in the circle theory and comparing them to those of the appropriate fields in the orbifold model, the boundary conditions can be matched themselves. All in all, we find that the family of orbifold boundary states generated from $|N(\frac{\pi}{2\sqrt{2}})\rangle^{\mathrm{orb}}$ by J_{orb}^3, J_{orb}^1 and J_{orb}^2 correspond to the family of circle boundary states generated from $|N(0)\rangle^{\mathrm{circ}}$ by J_{circ}^3 and $\psi_{a,\mathrm{circ}}$.

This correspondence in particular provides an elegant argument as to why perturbations by $\lambda\psi_{a,\mathrm{circ}}$ in the $r_{\mathrm{circ}} = \sqrt{2}$ preserve the cluster property for $|\lambda| < \frac{\pi}{\sqrt{2}}$: the corresponding orbifold boundary deformations obviously have this property. The subfamilies of boundary conditions generated by $\psi_{1,\mathrm{circ}}$ or $\psi_{2,\mathrm{circ}}$ form open intervals in the moduli space of the circle model: from the point of view of the orbifold model with target space $S^1/\mathbb{Z}_2$, the emergence of such a geometry is hardly surprising, while it would be hard to predict, on geometric grounds, within the circle model.

In Subsection 5.A.2, we will summarise our findings on the moduli space of $c = 1$ boundary conditions and discuss some direct applications in string theory. We emphasise that we obtain a global description of this brane moduli space because we were able to analyse marginal deformations to all orders in the perturbation parameter; first-order results would have allowed for a local picture only, giving the dimension of the moduli space but not its topology.

5.A Additional material

Here we discuss some additional aspects of perturbation theory which are of interest mainly in applications to string theory. Models with $N = 2$ superconformal symmetry play an important role for superstring compactifications, so it is natural to collect some statements on bulk and boundary deformations of such theories. In Subsection 5.A.2, we return to the $c = 1$ moduli space and point out how the underlying computations can be used to analyse brane condensation processes in string theory. We conclude with a brief discussion of how worldsheet RG flows can often be given a natural physical and geometric interpretation via spacetime effective actions.

5.A.1 Remarks on deformations of $N = 2$ SCFTs

The chiral symmetry algebra of $N = 2$ superconformal field theories (SCFTs) contains two dimension-$\frac{3}{2}$ operators $G^\pm(z)$ as well as an abelian current $J(z)$. Their (anti-)commutation relations, see equation (3.16), admit the non-trivial "mirror" automorphism, and accordingly there are two types of gluing conditions in $N = 2$ boundary SCFTs, called A-type and B-type. They were given in equations (4.14,4.15), but we repeat them here for convenience:

$$\text{A-type:} \quad J(z) = -\overline{J}(\bar{z}), \quad G^\pm(z) = \eta\,\overline{G}^\mp(\bar{z})\,, \tag{5.62}$$

$$\text{B-type:} \quad J(z) = \overline{J}(\bar{z}), \quad G^\pm(z) = \eta\,\overline{G}^\pm(\bar{z})\,. \tag{5.63}$$

The parameter η may only take the values ± 1 if it is required – as is indispensable in the superstring theory context – that the boundary condition preserves an $N = 1$ subalgebra, with generating supercurrent $G(z) := G^+(z) + G^-(z)$ or $G'(z) := i(G^+(z) - G^-(z))$.

As $N = 2$ boundary CFTs always contain the chiral marginal field $J(z)$, it is natural to look at deformations by this field. According to our general formulas from Subsection 5.2.2, such deformations change the gluing conditions to

$$\text{A-type:} \quad G^\pm(z) = e^{-i\lambda}\eta\,\overline{G}^\mp(\bar{z}), \quad \text{B-type:} \quad G^\pm(z) = e^{i\lambda}\eta\,\overline{G}^\pm(\bar{z}). \qquad (5.64)$$

We see that the $N = 1$ symmetry is broken unless the deformation parameter λ is a multiple of π; thus, we cannot generate a family of supersymmetric string vacua by perturbing an $N = 2$ model by its U(1) boundary current J.

In many models of interest, there is another type of marginal fields which may preserve supersymmetry and may lead to interesting families. These fields are built from so-called *chiral primaries* or *anti-chiral* primaries; both types of fields are primary with respect to the $N = 2$ super Virasoro algebra and moreover the former satisfy $G^+_{-\frac{1}{2}}\phi_c = 0$, the latter $G^-_{-\frac{1}{2}}\phi_a = 0$. It should be emphasised that the term "chiral" here originates from older supersymmetry terminology and does not refer to left- or right-moving coordinates; in fact, we can consider bulk fields $\phi_{(c,c)}(z,\bar{z})$ which are annihilated by both $G^+_{-\frac{1}{2}}$ and $\overline{G}^+_{-\frac{1}{2}}$, or bulk fields $\phi_{(a,c)}(z,\bar{z})$ which are annihilated by $G^-_{-\frac{1}{2}}$ and $\overline{G}^+_{-\frac{1}{2}}$, and so on.

Chiral and anti-chiral primaries display various interesting properties, some of which will be discussed further in Chapter 7; for our present purposes, just note that if ϕ_c (or ϕ_a) has conformal dimension $h = \frac{1}{2}$, then the descendant $G^-_{-\frac{1}{2}}\phi_c$ (or $G^+_{-\frac{1}{2}}\phi_a$, respectively) is a marginal field of dimension $h = 1$. Moreover, these descendants are uncharged under J_0, because charge and conformal dimension are related by $|q| = 2h$ for (anti-)chiral primaries.

Let us study how deformations by such marginal descendants $\psi(x)$ of (anti-) chiral primaries affect the $N = 2$ boundary conditions, to first order in the perturbation parameter. The Virasoro gluing conditions are preserved because of $h_\psi = 1$. Since $\psi(x)$ carries no charge, the singular contribution to the operator product of the current $J(w)$ with $\psi(x)$ vanishes; thus, the gluing condition for the current J is unbroken as well. As for the supercurrents $\mathsf{G}^\pm(z)$, note that the $N = 2$ relations (3.16) imply

$$\mathsf{G}^-(z)\left(G^-_{-\frac{1}{2}}\psi_c\right)(x) = \text{reg}, \quad \mathsf{G}^+(z)\left(G^-_{-\frac{1}{2}}\psi_c\right)(x) = \partial_x\left(\frac{2\,\psi_c(x)}{z-x}\right) + \text{reg},$$

along with analogous relations for perturbations $G^+_{-1/2}\psi_a(x)$ built on anti-chiral fields. The first equation already holds when $\psi_c(x)$ is any $N = 2$ primary, whereas in the second it is crucial that $\psi_c(x)$ is chiral. The total-derivative criterion found in Subsection 5.2.3 allows us to conclude that deformations with marginal fields $G^-_{-1/2}\psi_c(x)$ or $G^+_{-1/2}\psi_a(x)$ preserve the $N = 2$ gluing conditions, at least to first order. If $\psi(x)$ is a self-local marginal field, they are invariant to all orders. It seems, however, that there are no model-independent statements on self- (or mutual) locality of (families of) marginal boundary fields and on boundary correlators that do not receive perturbative corrections; see [72], for a discussion of a number of simple Gepner models and for further details.

It is interesting to compare this to the analogous question in the bulk case, where we can make a few important model-independent statements on deformations. In an $N = 2$ SCFT on a worldsheet without a boundary, we can have marginal bulk fields $G^+_{-\frac{1}{2}} \overline{G}^+_{-\frac{1}{2}} \varphi_{(c,c)}(z, \bar{z})$ built over fields that are chiral with respect to left- and right-moving symmetry algebras. Likewise, there may be $h = \bar{h} = 1$ fields of the form $G^+_{-\frac{1}{2}} \overline{G}^-_{-\frac{1}{2}} \varphi_{(c,a)}(z, \bar{z})$, etc. In an $N = (2,2)$ sigma model, these fields are related to complex structure and Kähler parameters of the Calabi–Yau target manifold, respectively – see Chapter 7 for some further remarks.

To investigate deformations by such marginal fields, we use ideas from Dixon's discussion of effective superstring potentials in [135], but see also [134]. The key ingredient are the $N = 2$ superconformal Ward identities.

For the time being, we may concentrate on the left-moving coordinates; so let us suppress $\bar{z}$-dependencies at first and introduce the shorthand notation $\varphi^1_{c,i}(z) := G^+_{-\frac{1}{2}} \varphi_{c,i}(z)$ and $\varphi^1_{a,j}(z) := G^-_{-\frac{1}{2}} \varphi_{a,j}(z)$ for the marginal descendants of (anti)-chiral primaries with dimension $h = \frac{1}{2}$. We want to study correlation functions

$$F_{I;J} := \langle \varphi^1_{c,i_1}(z_1) \cdots \varphi^1_{c,i_m}(z_m) \cdot \varphi^1_{a,j_1}(w_1) \cdots \varphi^1_{a,j_n}(w_n) \rangle \tag{5.65}$$

for $m + n \geq 3$. Upon integrating over suitable insertion points, $F_{I;J}$ provides information about perturbative corrections of various bulk correlators that are of interest for string compactifications.

We will exploit the fusion product (3.52), suitably generalised to more than two insertion points, in order to rewrite the correlators $F_{I;J}$. We have

$$\Delta_{u_1,\dots,u_N}\left(G^\pm_{-\frac{1}{2}}\right) = \sum_{k=1}^N G^{\pm\,[k]}_{-\frac{1}{2}},$$

$$\Delta_{u_1,\dots,u_N}\left(G^\pm_{\frac{1}{2}}\right) = \sum_{k=1}^N \left(u_k\, G^{\pm\,[k]}_{-\frac{1}{2}} + G^{\pm\,[k]}_{\frac{1}{2}}\right),$$

where $G^{\pm\,[k]}_r := \mathbf{1} \otimes \cdots \otimes \mathbf{1} \otimes G^\pm_r \otimes \mathbf{1} \otimes \cdots \otimes \mathbf{1}$, with non-trivial action in the kth tensor factor only. We arrive at these formulas by splitting up integration contours, as described before equation (3.52).

Inserting the fusion product actions of $G^\pm_{\pm\frac{1}{2}}$ into the correlators $F_{I;J}$ gives zero as the in- and out-vacua are annihilated by these modes, and therefore we have two conformal Ward identities to work with. We use them to shuffle a $G^\pm_{-\frac{1}{2}}$ from one of the marginal fields onto the other operators in $F_{I;J}$, where already a $G^+_{-\frac{1}{2}}$ or $G^-_{-\frac{1}{2}}$ is present.

The super Virasoro algebra relations (3.16) imply that $G^\pm_{-\frac{1}{2}} G^\pm_{-\frac{1}{2}} = 0$ and that on any primary field $G^\pm_{-\frac{1}{2}} G^\mp_{-\frac{1}{2}} \phi = 2L_{-1}\phi$. Moreover, all chiral (and, respectively, anti-chiral) primaries are annihilated by $2L_0 - J_0$ (and $2L_0 + J_0$, respectively). If, in addition, the conformal dimension of such an (anti-)chiral primary is $h = \frac{1}{2}$, we have $G^+_{-\frac{1}{2}} G^-_{-\frac{1}{2}} \phi_c = 2\phi_c$ and $G^-_{-\frac{1}{2}} G^+_{-\frac{1}{2}} \phi_a = 2\phi_a$, respectively.

Altogether, we arrive at the following two relations [135] for the correlator (5.65):

$$F_{I;J} = -2 \sum_{l=1}^{n} \partial_{w_l} F^{1;l}_{I;J},$$

$$F_{I;J} = -2 \sum_{l=1}^{n} \partial_{w_l} \left(\frac{w_l}{z_1} F^{1;l}_{I;J} \right),$$

where $F^{1;l}_{I;J}$ denotes the function $F_{I;J}$ from equation (5.65) with the $G^\pm_{-\frac{1}{2}}$ stripped off the (anti-)chiral primaries at z_1 and at w_l; we have assumed $n \geq 2$ without loss of generality, i.e. that at least two of the marginal fields are based on anti-chiral primaries.

The fusion product relations can be applied once again to the functions $F^{1;l}_{I;J}$, to show that $F_{I;J} = 0$ for $m = 1$, $n \geq 2$, and moreover that for $m \geq 2, n \geq 2$ the correlator can be expressed as a sum of derivatives:

$$F_{I;J} = \sum_{l=2}^{n} \partial_{w_l} \left(\frac{2(w_l - w_1)}{(w_1 - z_1)} F^{1;l}_{I;J} + \frac{4(w_l - z_1)}{(z_1 - z_2)(w_1 - z_1)} F^{1,2;1,l}_{I;J} \right). \quad (5.66)$$

Using a straightforward extension of the notation above, the very last correlator contains two chiral and two anti-chiral primaries of $h = \frac{1}{2}$. The main point is that equation (5.66) expresses $F_{I;J}$ as a total derivative. It is also useful to note that an analogous formula holds for the right-moving coordinates.

Because $F_{I;J}$ is a total derivative with respect to (left- and right-moving) integration variables, the integrals over the worldsheet will vanish provided the worldsheet has no boundary (the effect of closed string deformations on branes is more difficult to analyse) *and* provided there are no short-distance singularities severe enough to spoil the argument.

The further analysis, and also the interpretation of $F_{I;J}$ in the context of string theory on Calabi–Yau manifolds, depends on the choice of m and n. Let us first focus on the case $m > 1$ and $n = 3$ or $m = 3$ and $n > 1$; then the $F_{I;J}$ describe corrections to Yukawa couplings.

The short-distance singularities occurring in $F_{I;J}$ are controlled by the OPE of the perturbing fields. In particular, it is useful to know that the OPE between two anti-chiral primaries (and that between two chiral primaries) is regular; see [331] and also Chapter 7. For the case at hand, this is sufficient to conclude that for certain perturbations there is no obstruction against exploiting the fact that $F_{I;J}$ is a total derivative: Specifically, the work [134] showed that three-point functions of marginal fields built over $\varphi_{(c,c)}$ receive no corrections from marginal fields built over $\varphi_{(c,a)}$, and vice versa. The importance of this result stems from the fact that it implies [134] that certain Yukawa couplings of massless spacetime fields computed in the classical sigma model receive neither α' nor worldsheet instanton corrections; combining this with the mirror automorphism led to a fast algorithm to compute the number of rational curves in Calabi–Yau manifolds [93].

If we choose $m \geq 1$ and $n \geq 1$ in $F_{I;J}$, we can compute the order $m + n - 2$ correction to the two-point function $\langle \varphi^1_{c,i_1}(z_1)\varphi^1_{a,j_1}(w_1)\rangle$ by integrating $F_{I;J}$ over $\prod_{k,l\geq 2} \int_\Sigma d^2 z_k \int_\Sigma d^2 w_l$. The field inserted at w_1 is supposed to be the conjugate of the one at z_1.

Now we have to deal with more serious singularities than for Yukawa couplings, of order up to $|z_1 - w_2|^{-4}$. The latter comes from the OPE channel of the identity field and can be absorbed into a redefinition of the vacuum. More generally, any singularity arising from the OPE channel of any marginal field can be taken care of by a suitable renormalisation of the corresponding coupling constant. However, to our knowledge, no general model-independent CFT argument exists to ensure that these are the only troublesome terms, and thus that we have a whole family of perturbing fields that are exactly marginal to all orders.

In [135], Dixon argues that a related statement holds if the model describes a superstring background with $D = 4$ external flat dimensions. Then, each of the marginal fields from the internal CFT has to be multiplied by a massless tachyon vertex operator $\exp(i\, k_r X)$ with external momentum k_r such that $k_r^2 = 0$ for $r = 1,\dots,m+n$. These external vertex operators contribute terms like $|z_1 - w_2|^{k_1 k_2}$ to the OPE and can thus be used to improve the singularities of the integrand. Moreover, after integrating by parts, the (holomorphic and anti-holomorphic) derivatives appearing in front of the internal correlation function (5.66) produce factors like $k_1 k_2$ from acting on the external vertex operators. Following Dixon, this shows that, after integration, no scattering amplitudes without external momenta remain – in other words, the low-energy effective action (LEEA) of such a string compactification contains no non-trivial potential which could obstruct some of the moduli associated with $h = \bar{h} = \frac{1}{2}$ chiral and anti-chiral primaries of the internal bulk CFT.

5.A.2 The $c = 1$ brane moduli space, and string theory applications

In this subsection, we summarise the previous results on boundary conditions for $c = 1$ theories from Sections 5.3 and 4.A.5, and we point out how they bear on string theory processes such as brane annihilation and bound state formation.

Let us first give a "picture" of the moduli space $\mathcal{M}_B$ of $c = 1$ branes. Since our findings on deformations were not confined to finite-order perturbation theory, they capture the global topology of this moduli space (viewed as given by the standard topology on the space of deformation parameters). We will, however, not attempt to discuss metric properties. From the worldsheet point of view, this would involve the Zamolodchikov metric

$$g_{ij}(\lambda) \sim \langle\, \psi_i(1)\psi_j(0)\,\rangle_{\alpha;\,\lambda\cdot\psi} \tag{5.67}$$

where $\psi_1(x),\ldots,\psi_n(x)$ are truly marginal boundary deformations of given boundary condition α, spanning the tangent space to $\mathcal{M}_B$, and where we have used conformal invariance to place the perturbing fields at two preferred points on the boundary (taken to be the real line here).

The space of branes $\mathcal{M}_B$ can be viewed as a fibration $\mathcal{M}_B = \bigcup_{m\in\mathcal{M}_S}(\mathcal{M}_B)_m$ over the (closed string) moduli space $\mathcal{M}_S$ of $c = 1$ bulk CFTs, which is, e.g., described in [235]. We focus on the connected part $\mathcal{M}_S = \mathcal{M}_S^{\mathrm{circ}} \cup \mathcal{M}_S^{\mathrm{orb}}$ and ignore the three exceptional orbifold points. Both branches of $\mathcal{M}_S$ are parametrised as half-lines $\mathbb{R}_{\geq 1/\sqrt{2}}$, since radii below the self-dual one lead to equivalent theories upon T-duality $r \leftrightarrow \frac{1}{2r}$.

Different points α, β in $\mathcal{M}_B$ will indeed correspond to *inequivalent boundary theories*, because we have $Z_{\alpha\alpha}(q) \neq Z_{\alpha\beta}(q)$ for the spectra of open strings stretched between the two branes. It should be noted that this criterion provides a higher resolution of the space of boundary conditions than looking at bulk field one-point functions, i.e. boundary states, only: an automorphism can often be found of a bulk CFT that intertwines the gluing automorphisms and the sets of one-point functions for two boundary conditions that are inequivalent by the spectral criterion. For example, $\mathrm{SU}(2) \times \mathrm{SU}(2)$ symmetry of the $c = 1$ circle model at the self-dual radius allows us to map every symmetry-preserving boundary state onto one of the two $\widehat{\mathrm{SU}(2)}_1$ Cardy states, at the level of one-point functions. Likewise, when restricting to interactions with bulk fields, we could use translational invariance of the target to "identify" all Dirichlet boundary states, irrespective of the location x_0 of the brane.

The topological type of the fibre $(\mathcal{M}_B)_m$ depends on the base point $m \in \mathcal{M}_S$: for $m = r_{\mathrm{circ}} \in \mathcal{M}_S^{\mathrm{circ}}$, we find the following elementary branes (i.e. boundary conditions with a single vacuum character in the spectrum):

$$(\mathcal{M}_B)_{r_{\mathrm{circ}}} = \begin{cases} S_r^1 \cup S_{\frac{1}{2r}}^1 & r_{\mathrm{circ}} \neq \frac{N}{\sqrt{2}} \\ S_r^1 \cup T_{\frac{1}{2r}}/\mathbb{Z}_N & r_{\mathrm{circ}} = \frac{N}{\sqrt{2}},\ N \geq 2 \\ S^3 & r_{\mathrm{circ}} = \frac{1}{\sqrt{2}} \\ \mathbb{R} \cup D_{\frac{\pi}{\sqrt{2}}} & r_{\mathrm{circ}} = \infty \ . \end{cases} \tag{5.68}$$

The fibres over generic radii are disjoint unions of two circles. Points x_0 in S_r^1 are positions of Dirichlet branes, while the Neumann parameters $\tilde{x}_0 \in S_{\frac{1}{2r}}^1$ label Wilson lines.

The spaces $T_{\frac{1}{2r}} \simeq \overset{\circ}{D}{}^2_{\frac{\pi}{\sqrt{2}}} \times S^1_{\frac{1}{2r}}$ have the topology of a solid (filled) two-torus with the boundary taken away. The boundary of the two-disk $D^2_{\frac{\pi}{\sqrt{2}}}$, or indeed of the full two-torus, corresponds to Dirichlet-like mixtures of elementary boundary conditions that violate the cluster property. The $\mathbb{Z}_N$-identification was already pointed out in Subsection 4.A.5.

In taking the limit $r_{\text{circ}} \to \infty$, we lose one dimension of the fibre: the matrix elements $D^j_{m,-m}$ contributing to infinite radius boundary states depend only on one of the entries of the SU(2) matrices, cf. the remarks after equation (4.132).

The S^3 over the self-dual radius can also be viewed as follows: the chiral marginal deformations induce twists of the standard $\widehat{\text{SU}(2)}_1$ gluing conditions by inner automorphisms, generating an SO(3) family of gluing conditions Ω since central elements of SU(2) yield trivial γ_J in equation (5.29). But there are two different twisted Cardy boundary states $\| \alpha \rangle\!\rangle_\Omega$, $\alpha = 0, 1$, for each point of this SO(3) – making up the full SU(2) moduli space we found before.

The fibres over the bulk moduli space of orbifold models have the following form:

$$(\mathcal{M}_B)_{r_{\text{orb}}} = \begin{cases} \widehat{I}_r \cup \widehat{I}_{\frac{1}{2r}} & r_{\text{orb}} \neq \frac{N}{\sqrt{2}} \\ \widehat{I}_r \cup \widehat{C}_{\frac{1}{2r}}/\mathbb{Z}_N & r_{\text{orb}} = \frac{N}{\sqrt{2}}, \ N \geq 2 \\ S^1_{\frac{1}{\sqrt{2}}} \cup T_{\frac{1}{\sqrt{8}}} & r_{\text{orb}} = \frac{1}{\sqrt{2}} . \end{cases} \qquad (5.69)$$

where $\widehat{I}_r$ denotes the disjoint union of the open interval $\overset{\circ}{I} =]0, \pi r[$ with four extra points for the twisted boundary states. The spaces $\widehat{C}_r$ arise from the non-chiral orbifold deformations we did not discuss in detail above: $\widehat{C}_r$ consists of five disjoint parts; one has the topology of an open ball $D^3_r \simeq \overset{\circ}{D}{}^2_{\frac{\pi}{\sqrt{2}}} \times \overset{\circ}{I}$ (from the action of $\psi_{a,\text{orb}}$ and J_{orb} on the untwisted Neumann boundary states), the four remaining components are open intervals (from the action of $\sqrt{2}\cos(2\sqrt{2}X)$ on the twisted Neumann boundary states). These four intervals would form a single circle (and in fact do at $r_{\text{orb}} = \frac{1}{\sqrt{2}}$) were it not for the four Dirichlet-like points where clustering is violated. Note that the fibres over $r_{\text{circ}} = \sqrt{2}$ and $r_{\text{orb}} = \frac{1}{\sqrt{2}}$ coincide, as they should because these bulk theories are identical.

The construction of conformal boundary conditions in Subsection 4.A.5 shows that the lists in equations (5.68) and (5.69) do not cover all possible $c = 1$ boundary CFTs: for any rational multiple $r_{\text{circ}} = \frac{M}{N} r_{\text{s.d.}}$ of the self-dual radius, there are further conformal boundary states $\| g \rangle\!\rangle_{M,N}$ with $g \in \text{SU}(2)/\mathbb{Z}_M \times \mathbb{Z}_N$ – ignoring the cases with $g \in \text{SL}(2, \mathbb{C})$ as their spectrum is not real. However, these branes are not continuously connected to single Dirichlet or Neumann branes. Instead, they interpolate between superpositions of M Dirichlet and N Neumann branes. Within each such set, only the (J_3-translates of) these superpositions violate the cluster condition, all other members correspond to elementary branes.

The "Legendre branes" $\| x \rangle\rangle$ given in equation (4.131), which exist for any radius [298], show a similar feature: they are parametrised by $x \in]-1, 1[$ with the endpoints corresponding to continuous superpositions of Neumann and Dirichlet branes, respectively.

We note that, whenever we encountered a violation of the cluster property in our study of $c = 1$ models, it could simply be attributed to the fact that we have arrived at a superposition $\alpha = \alpha_1 + \cdots + \alpha_n$ of several elementary branes, and if we were to "resolve" the superposition into a system of boundary conditions, the issues with clustering would disappear – cf. Subsection 4.4.2. After having determined the constituents α_i, passing from a superposition to a system amounts to computing all correlators

$$\langle \psi_0^{\alpha_{i_0}\alpha_{i_1}}(\infty)\, \psi_1^{\alpha_{i_1}\alpha_{i_2}}(x_1) \cdots \psi_m^{\alpha_{i_m}\alpha_{i_0}}(x_m)\, \varphi_1 \cdots \varphi_N \rangle \qquad (5.70)$$

of bulk fields and arbitrary boundary fields. The BCCOs yield the (generalised) Chan–Paton matrix structure of boundary OPE and correlation functions. The correlators of the mixed boundary condition α are recovered by summing over all possible $\{\alpha_{i_l}\}$ in equation (5.70); this trace over Chan–Paton indices causes violations of factorisation constraints.

It is perhaps the most interesting conclusion arising from the $c = 1$ exercise treated in Section 5.3 that *the moduli space of a single brane can be continuously connected to that of several branes*. At boundaries of the moduli space patches listed in equations (5.68) and (5.69) – where clustering is violated – the single- and multiple-brane moduli spaces are glued together. If we reach such a point via deforming a single brane and then resolve the superposition into a system, we can turn on all available boundary operators (BCCOs and others) one by one. (This allows us, e.g., to vary the relative distance of the equidistant M Dirichlet branes that arise at certain points in the $SU(2)/\mathbb{Z}_M \times \mathbb{Z}_N$.)

Obviously, the moduli space of boundary conditions or of D-branes is much richer than that of bulk theories, and it is more complicated than what could have been predicted from classical geometry. The classical target is what a quantum mechanical point particle would map out. Closed strings already "perceive" the target geometry in a different way due to effects like T-duality, which means that momentum and winding modes are interchangeable. Open strings "see" a more complicated picture still, because even the clear distinction between intrinsic geometric characteristics of the target and externally imposed structure is lost: moduli due to global or local isometries (such as the marginal boundary fields $J_{\text{circ}}(x)$ or $J_{\text{orb}}(x)$ from Section 5.3) and moduli corresponding to turning on periodic boundary potentials (such as the marginal boundary fields $\psi_{a,\text{circ}}(x)$) appear on the same footing, and can sometimes even be transmuted into one another (cf. the identification of J_{orb} with $\psi_{2,\text{circ}}$ at radius $r_{\text{circ}} = \sqrt{2}$). We are forced to conclude that the geometry of string and brane moduli spaces is rather non-classical by nature, and that we ought not to trust geometric intuition too readily.

On the other hand, this non-classical behaviour allows for interesting effects within string theory. The topology of a brane can change – and in a more drastic fashion than is familiar from closed strings in Calabi–Yau spaces (see [253] and references therein): neither is the worldvolume dimension of a brane an invariant (cf. the smooth deformation from Dirichlet to Neumann conditions at the self-dual radius), nor is the number of branes (cf. the family connecting M Dirichlet to N Neumann branes).

Indeed, some of the computations from the $c = 1$ "toy model" can be used, almost without change, to discuss the question of whether branes can form bound states, and whether there is brane creation or annihilation. These are rather difficult questions in general, but some fundamental cases can be treated exactly in the worldsheet approach, as shown in a series of works by Sen [416, 417, 419–421].

As a second point, we would like to emphasise that our analysis of $c = 1$ examples shows the possibility of what we might call "brane topology change" – in analogy to phenomena in the closed string moduli space of Calabi–Yau compactification: there, effects from stringy geometry allow for transitions between targets of different topology without producing singularities in physical quantities (string amplitudes); see [253] for a brief review and for further references on the Calabi–Yau case. The identification of the bulk theories with $r_{\text{circ}} = \sqrt{2}$ and $r_{\text{orb}} = \frac{1}{\sqrt{2}}$ is, of course, another example.

In [420], Sen studied a system of two D1-branes on a circle of radius $r = \frac{r_{\text{s.d.}}}{2}$, with one of the branes carrying half of a unit Wilson line – this is the T-dual of having two D0-branes at opposite points on a $r = 2r_{\text{s.d.}}$ circle. The open strings stretched between the two D-strings have a tachyonic mode which, at the radius in question, happens to become massless. This string state is described by a marginal BCCO of the two-brane system, which has the form

$$\psi_a^{NN}(x) = \psi_a(x) \otimes \begin{pmatrix} 0 & 1 \\ 1 & 0 \end{pmatrix}. \tag{5.71}$$

Since this BCCO is simply an ordinary boundary field times a Chan–Paton matrix, perturbations of the Neumann–Neumann (N–N) system by $\psi_a^{NN}(x)$ can be studied quite explicitly. In fact, it is found that the computations performed to this end in [420] coincide with those involved with the non-chiral deformation $\psi_a(x)$ of a single Neumann condition, see equation (5.34). The result found by Sen is that for a certain value of the perturbation coupling (what we called "Dirichlet-like value" in Section 5.3) the N–N system has been deformed to a single D0-brane – of course after taking the trace over the Chan–Paton indices. we can say that the deformation annihilates one brane, or that the D0-brane arises as a bound state of two D1-branes with a Wilson line profile. The process is the reverse of the perturbation leading from a single brane to a superposition of two (or more) branes encountered in Section 5.3.

In the supersymmetric context, brane moduli spaces are expected to be closer to intuitions from classical geometry, but brane creation or annihilation occur just as well. Again, we sketch ideas taken from Sen's work; see [416, 417]. The main concern there is formation of stable non-BPS bound states of branes, a question that cannot be studied in great detail using spacetime methods since they rely heavily on supersymmetry.

First, let us expand a bit on the string theory context, which is reviewed in [423]: Sen's motivation to search for non-BPS branes was the conjectured strong–weak coupling duality between SO(32) heterotic and type I string theory in ten dimensions; see [279] and references therein. The former theory enjoys only $N = 1$ supersymmetry (which does not allow for central charges in the spacetime supersymmetry algebra) and has no BPS states, unless higher supersymmetry is introduced via compactification. For these cases, BPS spectra could be compared [222, 347] to BPS branes in the type I theory – which can be regarded as type IIB string theory modulo the orientifold projection with worldsheet parity Ω (such that, in particular, type I branes arise from type IIB branes). The BPS property protects states against decaying into others – but the same is true for a non-BPS state that is the lightest one with a given conserved charge, e.g. for the lowest-energy state in the SO(32) spinor representation in (uncompactified) SO(32) heterotic string theory. If S-duality is to hold in ten dimensions, too, this spinor state must have a stable non-BPS partner in the type I theory.

Sen constructed this partner as bound state of a type I D1-$\overline{\text{D1}}$-pair. Since the Ramond–Ramond (R–R) charges of the D-string and the anti-D-string cancel and their masses are non-zero, the bound state cannot be BPS, even after compactification. Sen first [416] relied on spacetime arguments for the existence of such a bound state (which should be a tachyonic kink on the D-string worldvolume), but soon gave an exact and much more explicit CFT formulation [417]. The CFT techniques used to describe the "tachyon condensation" leading to bound state formation are essentially identical to those used above: after a bosonisation procedure (at a "critical" radius r_{crit}), the operator (5.71) again becomes a marginal perturbing field. (For an analogous scenario concerning decay of unstable branes in type II string theory at the critical radius, see [215, 422].) In contrast to the bosonic case, the additional orientifold projection removes some boundary fields from the excitation spectrum of the resulting D-particle, among them those which would become tachyonic at higher radii, and also those which could induce relative displacements of the two D1-constituents. In summary, the "Dirichlet-like point" of the ψ^{NN}-perturbation corresponds to a stable bound state of the D1-$\overline{\text{D1}}$ pair. A very similar object was constructed by Bergman and Gaberdiel in [48] for a type IIB orbifold theory: here, it is the orbifold instead of the orientifold projection which removes tachyonic and relative displacement modes so that a stable non-BPS D-particle in type IIB theory is left over. The complete boundary state for this D-particle involves the NS–NS Ishibashi contribution (2.47) to the boundary state of a BPS D-particle along with a contribution from the twisted sector.

Sen's construction involves a marginal bulk deformation from infinite radius to r_{crit}, then a marginal boundary deformation, then another marginal bulk deformation back to infinite radius. It is found that the resulting SO(32) spinor obtained as a stable bound state is lighter than the D1–$\overline{\text{D1}}$ pair one starts from. This means that the chain of marginal deformations has lowered the g-factor of the boundary condition at $r = \infty$ – i.e. the chain replaces a relevant boundary perturbation of the uncompactified theory. It would be interesting to find a general characterisation of situations where a relevant boundary perturbation can be replaced by marginal deformations, which are easier to handle to all orders.

The techniques outlined above can be applied to other backgrounds, including branes wrapped around non-supersymmetric cycles of Calabi–Yau manifolds (see [341] for some simple examples). Generalisation to "tachyon condensation" on higher Dp-branes is also straightforward, leading to a pattern of descent relations among BPS and non-BPS configurations of various dimensions. In particular, this motivated Witten's proposal to interpret branes in the framework of K-theory [457].

5.A.3 Remarks on RG flows and effective actions

When studying CFT perturbations in a string theory context, it is typically found that the equations for a fixed point $\beta(\lambda) = 0$ can be reproduced as the equations of motion $\delta S_{\text{eff}}^{\text{RG}}[\lambda] = 0$ of an effective action $S_{\text{eff}}^{\text{RG}}$. In string theory, we are normally dealing with continuous families of perturbing fields, such as $\partial X e^{ikX}$, $k \in \mathbb{R}^d$ in the open string case, and the associated coupling constants λ_k can be viewed as Fourier modes of some effective field $\lambda(x)$ on spacetime, $x \in \mathbb{R}^d$. The action $S_{\text{eff}}^{\text{RG}}[\lambda]$ underlying the beta function equations then becomes an integral over spacetime, and very often has a natural geometric interpretation. In this subsection, we will present some examples.

But first, recall that we have already introduced an effective action S_{eff} in Section 2.1, arising from a change of variables in the path integral, from fields in the sigma model (the worldsheet CFT) to the tower of string excitations (the effective fields $\lambda(x)$ on spacetime). This S_{eff} was to formally reproduce the (connected) string amplitudes (the scattering matrix) from an effective QFT on the target, instead of computing those amplitudes from correlators in the worldsheet CFT.

We should not expect that $\ln S_{\text{eff}}$ and $S_{\text{eff}}^{\text{RG}}$ coincide (the logarithm occurs because we have implicitly restricted ourselves to connected diagrams before), because there are already many actions leading to the same equations of motion; such "off-shell ambiguities" are discussed, e.g., in [438]. However, the following observations should at least make it plausible that the two actions share a common general structure:

The order $n \geq 3$ terms in S_{eff} have to reproduce string amplitudes with $n \geq 3$ legs. These are determined from n-point CFT correlators; in particular, the

Yukawa couplings ($n = 3$) in S_{eff} are given by OPE coefficients. The higher-order terms in the beta functions, and thus in $S_{\text{eff}}^{\text{RG}}$, have the same origin; in particular, the OPE coefficients directly give the cubic terms in $S_{\text{eff}}^{\text{RG}}$. At a more generic level, recall that the path integral with S_{eff} is dominated by the classical configurations $\delta S_{\text{eff}} = 0$. On the other hand, the RG flow drives a perturbed string background towards a fixed point $\beta = 0$, or $\delta S_{\text{eff}}^{\text{RG}} = 0$, and the theory will settle at such a saddle point.

A more complete understanding of the relation between different ways to introduce effective actions requires boundary string field theory. We will not pursue this issue in this text, beyond mentioning the suggestion [424] that, with a suitable choice of off-shell completion, the effective action is related to the partition function of the perturbed boundary CFT on the disk as

$$S_{\text{eff}}^{\text{RG}}[\lambda] = \left(1 - \beta_i \frac{\partial}{\partial \lambda_i}\right) Z[\lambda];$$

see also [232, 324, 451] for earlier and subsequent discussions.

We will from now on treat the effective actions arising from RG flows and scattering amplitudes as equal – except that we consider the latter S_{eff} as generating connected diagrams. Taking this into account, the above relation between effective action and boundary partition functions in particular leads to the relation $\ln g(\lambda) = S_{\text{eff}}[\lambda]$, which will be exploited in Chapter 6, when we consider effective actions for D-brane systems in WZW models.

In the remainder of this subsection, let us list some examples of effective actions associated with perturbations. The general relation above applies to the open string case, and most of our examples will be taken from this context; nevertheless, the most important and most famous example of a string effective action arises from closed string theory in a (weakly) curved D-dimensional target. Treating this sigma model as a perturbation around the CFT of D free bosons leads to the beta functions (2.14)–(2.16) given in Chapter 2. These contain Einstein's equations together with stringy corrections, and they can be obtained as equations of motion from the (super)gravity action (2.18). Finally, after passing to the Einstein frame (by setting $G_{\mu\nu}^E \sim e^{4\phi/(2-D)} G_{\mu\nu}^S$, see, e.g., [377]), the Einstein–Hilbert action of general relativity can be obtained (along with stringy α' corrections). In particular, it is remarkable that the results of an abstract worldsheet computation organise themselves (at least to leading order in α') into a structure that is natural from the point of view of (classical) target-space geometry.

The same is true for open strings in a (weakly) curved target. The analogous computation of beta functions for a sigma model on the upper half-plane [328] can be reproduced as equations of motion of the Born–Infeld action (2.22), which is again a natural geometric quantity, depending on the induced metric (2.23) on the p-brane worldvolume and also on the (abelian) field strength F on the brane; see Section 2.2.

The next example involves a stack of identical branes and matrix-valued space-time fields; it leads to (non-abelian) gauge theory and allows a first glimpse of non-commutative geometry.

Let us start with some general boundary CFT observations: given a self-local marginal boundary operator ψ in some boundary CFT with boundary condition α, what can we say about marginal deformations in the system of N copies of α (identical branes "on top of each other")? We have already seen in Section 4.2 that the OPE properties of boundary-condition "changing" operators in such a system are captured by Chan–Paton factors, and $\Lambda \cdot \psi = \psi \otimes \Lambda$ is a self-local boundary field in the brane system since it commutes with itself inside correlation functions (this is trivial for the matrix part $\Lambda \in M_N(\mathbb{C})$ and true for the entry ψ by assumption). So each $\Lambda \cdot \psi$ is a truly marginal deformation of the system, but two such deformations $\Lambda_1 \cdot \psi$ and $\Lambda_2 \cdot \psi$ associated with the same ψ are mutually local iff [391]

$$[\Lambda_1, \Lambda_2] = 0. \tag{5.72}$$

Two perturbing fields are expected to spoil each other's marginality if they are not mutually local, and therefore the single field ψ associated with the original "elementary" boundary condition α generically provides an N-parameter family $\mathrm{diag}(\lambda_1, \ldots, \lambda_N) \cdot \psi$ of deformations of the N-brane system, because the maximal commuting subalgebra of $M_N(\mathbb{C})$ is N-dimensional.

For concreteness, we discuss the situation treated in [454]. Let α be a Dp-brane in flat D-dimensional spacetime, and $\psi_a = \partial X^a$, for $a = p+1, \ldots, D$ marginal boundary fields inducing motions perpendicular to the worldvolume, as in Subsection 5.3.1. Passing to a stack of N such branes, the above considerations apply. The condition (5.72) on mutual locality, i.e. on marginality, can be encoded in the LEEA by adding a potential term

$$V \sim \sum_{a,b} \mathrm{tr}\, [\Lambda_a, \Lambda_b^\dagger]^2 \tag{5.73}$$

where the trace is taken over the Chan–Paton indices.

Contact to geometry is made when observing that this potential coincides with what was obtained from dimensional reduction of supersymmetric Yang–Mills theory from ten dimensions, the Λ_a becoming Higgs fields (in the adjoint representation of U(N)) on the worldvolume. The flat directions of V are given by commuting matrices, whose eigenvalues $x \in \mathbb{R}^{9-p}$ are interpreted as transverse positions of the stack of branes [454]. We might then speculate about extending this interpretation beyond the commutative case, and that the branes in fact move in a non-commutative target with "co-ordinates" Λ_a.

In the next chapter, we will encounter an example that is somewhat similar, although the technical details are more involved and the non-commutative aspects more compelling: the LEEA of open strings on a stack of N identical Cardy branes in an SU(2)$_k$ WZW model. The perturbing fields are

J^a_{-1}-descendants of primary boundary fields; they become marginally relevant in the limit $k \to \infty$ (related to $\alpha' \to 0$), i.e. their dimension approaches $h = 1$, but are not mutually local just as in the example of the fields $\Lambda \cdot \psi$ discussed above. It turns out that the OPE of these fields can be conveniently encoded in terms of matrix multiplication (due to special properties of the SU(2) fusing matrix; see Chapter 6 for more details), and thus the variables Λ of $S_{\text{eff}}[\Lambda]$ are matrix-valued even before tensoring by a $M_N(\mathbb{C})$ Chan–Paton factor.

The first few terms in the effective action can be computed from two-, three- and four-point functions on the worldsheet. The same calculations underlie the results of [6], where Affleck and Ludwig derive the perturbative g-factor in their study of the Kondo effect. This is in keeping with the identification $\ln g(\lambda) \sim S_{\text{eff}}[\lambda]$ argued for above. We shall also exploit this relation to identify perturbative RG fixed points using classical solutions Λ_{cl} of the LEEA: as long as the full SU(2)$_k$ symmetry is restored at the new fixed point, $\Delta \ln g \sim S_{\text{eff}}[\Lambda_{\text{cl}}]$ can be compared to the discrete list of possible g-factors for Cardy branes (in the same way as for minimal model perturbations discussed at the end of Subsection 5.1.2) and read off the Cardy label of the RG end product. In this way, further examples of "topology changing" boundary flows are encountered, e.g. stacks of D0-branes may evolve into D2-branes, via condensation of the open string tachyons associated with marginally relevant boundary fields.

As in the examples outlined before, the LEEA for SU(2) WZW branes can be given a geometric interpretation, namely as a combination of Yang–Mills and Chern–Simons actions. However, this theory lives on a non-commutative space, a fuzzy sphere which quantises the classical worldvolume associated with Cardy branes in the SU(2) group manifold. Details of this interpretation and of the computations will be given in the next chapter.

6

The Wess–Zumino–Witten model on SU(2)

In the preceding chapters, we have discussed general ideas and results of boundary conformal field theory (CFT), but we had only a few relatively simple examples available to illustrate them. Our aim now is to apply some of these techniques to a very instructive case, namely to the so-called Wess–Zumino–Witten (WZW) model on the group manifold SU(2).

In the first section, we introduce the model from its classical action and review properties of its chiral symmetry algebra. This is followed, in Section 6.2, by an algebraic solution of both the bulk and the boundary theory in the spirit of Chapters 3 and 4. In particular, we describe the bulk spectrum and three-point couplings, boundary states, boundary spectra and the three-point couplings of boundary operators.

The third section is devoted to a geometric interpretation of these abstract algebraic constructions. We review two different lines of arguments which suggest that Cardy-type boundary conditions of the SU(2) WZW model are associated with spherical branes, which are localised along special (integer) conjugacy classes of SU(2). Boundary operator products for these branes will be compared to the quantisation of functions on S^2. Finally, we compute the low-energy effective action of open strings on SU(2) Cardy branes, which takes the form of a special non-commutative gauge theory on the fuzzy sphere. We study its classical solutions, which provide an insight into boundary perturbations and the dynamics of branes on SU(2).

WZW models are ideal examples in themselves to illustrate the power of the theory we have outlined, but they are also of fundamental importance for CFT model building. In fact, a large variety of highly relevant theories can be obtained through coset and orbifold constructions. Section 6.4 gives a quick review of these methods and shows how some of the results on the geometry and dynamics of WZW branes descend to those in orbifold and coset models.

6.1 Basic material

The goal of this section is to introduce the action functional of a WZW model on a group manifold SU(2) and to study its symmetries. Once the structure of the chiral algebra is understood, we provide all the necessary representation theoretic data. These include the set of integrable highest weight representations, their characters, the modular S-matrix and the fusion and fusing matrices.

6.1.1 Action and chiral symmetries

The action functional of the SU(2) WZW model involves a single SU(2)-valued field on the worldsheet $\Sigma \cong \mathbb{R} \times S^1$

$$g : \begin{cases} \Sigma & \longrightarrow \text{SU}(2) \\ (z, \bar{z}) & \longmapsto g = \exp(it_a X^a) \end{cases}$$

where t_a, $a = 1, 2, 3$, are the three generators of the Lie algebra of SU(2) and $X^a : \Sigma \to \mathbb{R}$. The simplest action we can write down is of the form

$$S_0[g] = -\frac{\mathsf{k}}{4\pi} \int_\Sigma d^2z \; \text{tr}\big(g^{-1}\partial g g^{-1}\bar{\partial}g\big) , \tag{6.1}$$

where for the moment $\mathsf{k} := 1/\alpha'$ is just a notation. A few comments on the ingredients of this action are in order. To begin with, the elements $g^{-1}\partial g$ and $g^{-1}\bar{\partial}g$ are both in the Lie algebra of SU(2), i.e. they are elements of the tangent space at the identity. The trace of the product of two elements u, v of the Lie algebra is, up to a constant multiplicative factor, equal to the Killing product $\langle u, v \rangle$ which is the canonical scalar product for Lie algebras. In order to fix normalisations, we agree to evaluate the trace in the adjoint representation. Let us also note that the action (6.1) is invariant under global left or right translations by $h \in \text{SU}(2)$, i.e. $S_0[gh] = S_0[hg] = S_0[g]$. We can get a better intuition about equation (6.1) if we expand g in powers of fields X^a. To leading order in X we find

$$S_0[g] = -\frac{\mathsf{k}}{4\pi} \int_\Sigma d^2z \; \big[\kappa_{ab} \, \partial X^a \bar{\partial} X^b + \mathcal{O}(X^3)\big] ,$$

where we have set $\kappa_{ab} := \text{tr}\big(t_a t_b\big)$. The action (6.1) describes the propagation in the canonical SU(2) invariant metric on S^3. The equations of motion for the full non-linear theory (6.1) are

$$\bar{\partial}\big(g^{-1}\partial g\big) + \partial\big(g^{-1}\bar{\partial}g\big) = 0 .$$

In the case of free bosonic fields, the bulk equations of motion implied the existence of independent chiral currents $J(z)$ and $\bar{J}(\bar{z})$. Unfortunately, analogous holomorphic currents cannot be defined in the theory (6.1). In addition, whereas this model is classically invariant under conformal transformations, quantisation leads to a non-zero beta function and thus to scale dependence. It turns out that both problems have the same root and are cured by adding an appropriate modification to the action.

Scale or conformal invariance in the quantum theory can, at least perturbatively, be read off from the beta functions for the components of target-space metric. To lowest order in α', these are given by the general formula (see equation (2.14))

$$\beta_{ab} \sim R_{ab} - \frac{1}{4}H_{acd}H_b{}^{cd} .$$

where R_{ab} is the Ricci tensor and $H = dB$ is the three-form field strength. The target space $SU(2) \cong S^3$ of the model (6.1) has constant curvature with a Ricci tensor $R_{ab} = \frac{2}{R^2} G_{ab}$, where R is the radius and G_{ab} the metric of S^3. Hence, the only way to ensure the vanishing of the beta function is to add a Kalb–Ramond two-form field B in such a way as to obtain a cancellation. Since $H = dB$ is a three-form and S^3 has only one non-trivial such form, namely the volume form

$$\omega_3 = \mathrm{tr}\big(g^{-1}dg \wedge g^{-1}dg \wedge g^{-1}dg\big)\,,$$

we must set $H \sim \omega_3$. This leads to a modified action [448]

$$S[g] = S_0[g] + \frac{ik}{12\pi} \int_M \mathrm{tr}\big(\tilde{g}^{-1}d\tilde{g} \wedge \tilde{g}^{-1}d\tilde{g} \wedge \tilde{g}^{-1}d\tilde{g}\big)\,. \tag{6.2}$$

Here, M is a three-dimensional manifold whose boundary ∂M is equal to the worldsheet Σ, and $\tilde{g}$ is an extension of g to M such that $\tilde{g}|_{\partial M} \equiv \tilde{g}|_\Sigma = g$. This extension is not unique, but since $d\omega_3 = 0$, we can use Gauss' theorem locally to express the second part of the action as an integral over Σ, so that the equations of motion are actually independent of the extension $\tilde{g}$. In the quantum theory, however, this ambiguity of the choice of $\tilde{g}$ has observable effects since the action itself enters the path integral. It is a non-trivial fact that by choosing the *level* k to be an integer, two different extensions lead to actions which differ from one another by $2\pi n$ with $n \in \mathbb{Z}$. Consequently, the quantity $e^{iS[g]}$ that appears in the path integral is well defined. From now on, we assume $k \in \mathbb{Z}_+$.

The equations of motion resulting from the WZW model action $S[g]$ are

$$\partial\left(g^{-1}(z, \bar{z})\bar{\partial}g(z, \bar{z})\right) = 0\,. \tag{6.3}$$

They should be considered as a non-linear version of the Laplace equation, which governs the string motion in flat backgrounds. Using $\partial(g^{-1}) = -g^{-1}\,\partial g\, g^{-1}$, it follows that

$$\bar{\partial}\big(\partial g\, g^{-1}\big) = g^{-1}\partial\left(g^{-1}\,\bar{\partial}g\right)g^{-1} = 0\,.$$

Hence, if we define the two currents $J(z)$, $\bar{J}(\bar{z})$ as

$$J(z) := -k\,\partial g(z, \bar{z})\, g^{-1}(z, \bar{z})\,, \qquad \bar{J}(\bar{z}) := k\, g^{-1}(z, \bar{z})\bar{\partial}g(z, \bar{z})\,, \tag{6.4}$$

their (anti-)holomorphicity follows from the equation of motion. Since the chiral fields J and $\bar{J}$ take values in the Lie algebra $su(2)$, we can expand each of the two currents into three component fields, i.e. $J(z) =: J^a(z)t_a$ and similarly for the anti-holomorphic partner $\bar{J}$.

6.1.2 Representation theory of affine $\widehat{su}(2)$

Here we repeat some facts about affine Lie algebras from Chapter 3, and we supply some further data such as the fusing matrix for $\widehat{su}(2)_k$, which will play a role later when we study branes in the $SU(2)$ WZW model.

The (anti-)holomorphic currents $J^a(z)$ and $\overline{J}^a(\bar{z})$ generate the left- and right-moving chiral symmetry algebras $\mathcal{W}$ and $\overline{\mathcal{W}}$ of the WZW model. In the quantum theory, $\mathcal{W}$ is the Lie algebra of a centrally extended loop group, i.e. an affine Lie algebra or Kac–Moody algebra $\hat{g}$ associated with a finite-dimensional (and, say, semi-simple, simply connected, simply laced) Lie algebra g; see, e.g., [241] for a detailed description, and also [121, 198, 385]. In a suitable g-basis, the modes J_n^a, $a = 1, \ldots, \dim g$, satisfy

$$[J_n^a, J_m^b] = i\, f^{abc}\, J_{n+m}^c + \mathsf{k}\, n\, \delta^{a,b}\, \delta_{n+m,0}\,, \tag{6.5}$$

where f^{abc} are structure constants of g, and k is the value the central element of $\hat{g}$ takes in the model under consideration; as before, we assume $\mathsf{k} \in \mathbb{Z}_{>0}$, which is required in a unitary theory.

The fact that WZW models are CFTs can be seen most directly by noting that the Sugawara construction

$$L_n = \frac{1}{2}\frac{1}{\mathsf{k}+h^\vee} \sum_a \sum_{m \in \mathbb{Z}} :J_m^a J_{n-m}^a: \tag{6.6}$$

defines generators of the Virasoro algebra with central charge $c = \frac{\mathsf{k}\,\dim g}{\mathsf{k}+h^\vee}$, where $h^\vee$ is the dual Coxeter number of g and where the normal ordering prescription simply means placing the higher mode to the right. The $J^a(z)$ are primary fields of conformal dimension one with respect to L_n.

The formula for the central charge is another indication that WZW models are fundamentally different from free bosonic field theories, where the central charge equals the dimension of the classical target space. Here, for finite k, the central charge is strictly smaller than $\dim g$.

Let us now specialise to the three-dimensional target SU(2). We shall often use a particular basis of elements $J^\pm, J^0$ in which the relations of the affine Lie algebra $\widehat{su(2)}$ or $\widehat{sl}_2$ with level k take the form

$$[J_n^0, J_m^0] = \frac{\mathsf{k}}{2}\, n\, \delta_{n+m,0}\,, \quad [J_n^0, J_m^\pm] = \pm J_{n+m}^\pm\,,$$
$$[J_n^+, J_m^-] = 2 J_{n+m}^0 + \mathsf{k}\, n\, \delta_{n+m,0}\,.$$

We want to investigate the lowest-energy representations of this current algebra, i.e. representations for which the L_0-eigenvalue is bounded from below. Such representations are built up on irreducible representations H_j of SU(2), parametrised by a spin $j \in \frac{1}{2}\mathbb{Z}_+$. We denote the states in these finite-dimensional su(2) representations by $|j; l\rangle \in H_j$ with $|l| \leq j$. Let us define the Verma modules $\mathcal{V}_j$ as representations generated from the $|j; l\rangle$ by applying the creation operators J_n^a for $n < 0$, and imposing $J_n^a |j; l\rangle = 0$ for $n > 0$. According to the formula (6.6), the ground states $|j; l\rangle$ have conformal weight

$$h_j = \frac{j(j+1)}{\mathsf{k}+2}\,. \tag{6.7}$$

In general, these Verma modules are neither unitary nor irreducible representations. First of all, it turns out that the Verma modules $\mathcal{V}_j$ contain vectors with negative norm iff $j > \frac{k}{2}$.

To show this, we observe that the three elements J^+_{-1}, J^-_1 and $J^0_0 - \frac{k}{2}$ generate an sl_2 algebra, with respect to which the Verma $\mathcal{V}_j$ module decomposes into an infinite direct sum of indecomposable representations. It is not too difficult to prove that these representations are free from states with negative norm as long as the eigenvalues of $J^0_0 - \frac{k}{2}$ on ground states are negative half-integer or zero. Then the claim follows because the $|j; l\rangle$ are eigenstates of J^0_0 with eigenvalues $l = -j, -j+1, \ldots, j$.

The Verma modules $\mathcal{V}_j$ with $j \leq \frac{k}{2}$ still contain states of zero norm. Again, the embedded finite-dimensional sl_2 algebra from above is useful to determine them. The singular states with lowest energy take the form

$$\psi^j_{\text{sing}} = \left(J^+_{-1}\right)^{k+1-2j} |j; j\rangle \, .$$

It is easy to show that they are annihilated by all J^a_n with $n > 0$. Hence, the states that are generated from ψ^j_{sing} by application of $J^a_n, n < 0$, form an invariant submodule $\mathcal{N}_j$ of the Verma module $\mathcal{V}_j$, and it can be shown that the quotient $\mathcal{H}_j := \mathcal{V}_j / \mathcal{N}_j$ is an irreducible representation of the $\widehat{su}(2)_k$ current algebra.

In the $\widehat{su}(2)_k$ highest weight representations $\mathcal{H}_j$ with $j = 0, \frac{1}{2}, \ldots, \frac{k}{2}$, the operators L_0 and J^3_0 are simultaneously diagonalisable. Thus, "charged" or "unspecialised" characters can be defined:

$$\chi_j(z, \tau, u) := e^{-2\pi i k u} \, \text{Tr}_{\mathcal{H}_j} \left[q^{L_0 - \frac{c}{24}} e^{-2\pi i z J^3_0} \right] = \left(\frac{\Theta^{(k+2)}_{2j+1} - \Theta^{(k+2)}_{-2j-1}}{\Theta^{(2)}_1 - \Theta^{(2)}_{-1}} \right)(z, \tau, u) \, .$$

$$(6.8)$$

On the right, we have expressed these in terms of the theta functions defined in Appendix A. The characters we have spelt out encode the full information on the position of singular vectors. The modular transformation properties

$$\chi_i \left(\frac{z}{\tau}, -\frac{1}{\tau}, u + \frac{z^2}{2\tau} \right) = \sum_j S_{ij} \, \chi_j(z, \tau, u)$$

of the unspecialised characters are of course well-known in the mathematical literature, and the modular S-matrix of the SU(2) WZW model at level k reads

$$S_{ij} = \sqrt{\frac{2}{k+2}} \, \sin\left(\frac{(2i+1)(2j+1)}{k+2} \pi \right) \qquad \text{for} \quad i, j = 0, \frac{1}{2}, \ldots, \frac{k}{2} . \qquad (6.9)$$

This matrix does not depend on z and u, and therefore it also applies to the specialised characters, which merely count L_0 degeneracies and are defined by setting $u = 0$ and taking the limit $z \to 0$ in equation (6.8):

$$\chi_j(\tau) = \text{Tr}_{\mathcal{H}_j} \, q^{L_0 - \frac{c}{24}} = \frac{1}{\eta(q)^3} \sum_{m \in \mathbb{Z}} N_j(m) \, q^{N_j(m)^2 / (4k+8)} \, , \qquad (6.10)$$

with $N_j(m) := 2(k + 2)m + 2j + 1$.

The fusion rules of the affine Lie algebra $\widehat{su}(2)$ at level k can either be obtained from the tensor product of representations or from the modular S-matrix with the help of Verlinde's formula. The result is

$$N^{j_3}_{j_1 j_2} = \begin{cases} 1 & \text{if } |j_1 - j_2| \le j_3 \le \min(j_1 + j_2, k - j_1 - j_2) \text{ and } j_1 + j_2 + j_3 \in \mathbb{Z} , \\ 0 & \text{otherwise} . \end{cases}$$

(6.11)

These are just the decomposition rules for tensor products of SU(2) representations, except for the level-dependent truncation in the first line.

The final piece of information we shall need from the representation theory of the $\widehat{su}(2)_k$ current algebra is the fusing matrix. This was initially computed by Zamolodchikov and Fateev in [463]. It may be found through an analysis of solutions to the Knizhnik–Zamolodchikov equations

$$\partial_{z_i} \mathcal{F}_{v_1 \ldots v_N}(\mathbf{z}) = \frac{1}{2} \frac{1}{k + h^\vee} \sum_{j; j \neq i} \frac{t_a^{(j)} t_a^{(j)}}{z_i - z_j} \mathcal{F}_{v_1 \ldots v_N}(\mathbf{z}) ;$$

cf. equation (3.50). We have indicated in Chapter 3 how this system of linear first-order differential equations for conformal blocks of WZW models arises from the Sugawara construction and the basic Ward identity (3.44) for theories with Kac–Moody symmetry.

Close inspection of these equations and their solutions lead to the following formula for the fusing matrix of $\widehat{su}(2)_k$:

$$F_{PQ}\begin{bmatrix} J & K \\ I & L \end{bmatrix} = \sqrt{\lfloor 2P + 1 \rfloor \lfloor Q + 1 \rfloor} \Delta(J, K, P) \Delta(I, L, P) \Delta(K, L, Q) \Delta(I, J, Q)$$

$$\times (-1)^{K+J-L-I-2P} \sum_{s=0}^{\infty} (-1)^s \lfloor s + 1 \rfloor!$$

$$\times \left(\lfloor s - J - K - P \rfloor! \lfloor s - I - L - P \rfloor! \lfloor s - K - L - Q \rfloor! \right.$$

$$\times \lfloor s - I - J - Q \rfloor! \lfloor I + J + K + L - s \rfloor! \lfloor I + K + P + Q - s \rfloor!$$

$$\times \left. \lfloor J + L + P + Q - s \rfloor! \right)^{-1} ,$$

(6.12)

where

$$\Delta(I, J, K) := \sqrt{\frac{\lfloor I + J - K \rfloor! \lfloor J + K - I \rfloor! \lfloor K + I - J \rfloor!}{\lfloor J + I + K + 1 \rfloor!}}$$

and where we have used the symbols $\lfloor n \rfloor$ and $\lfloor n \rfloor!$ for $n \in \mathbb{Z}_+$ to denote

$$\lfloor n \rfloor = \frac{\sin\left(\frac{\pi n}{k+2}\right)}{\sin\left(\frac{\pi}{k+2}\right)} , \quad \lfloor n \rfloor! = \prod_{m=1}^{n} \lfloor m \rfloor , \quad \lfloor 0 \rfloor! = 1 .$$

The formula (6.12) is only to be used for arguments that are allowed by the fusion rules and the summation over s is restricted to values for which none of the arguments of $\lfloor . \rfloor$ are negative. Note that the fusing matrix depends on the normalisation of the chiral vertex operators which were introduced after equation

(3.55). If we renormalise chiral vertex operators by constant factors $\gamma\left({}^{\,i}_{kj}\right)$, the fusing matrix acquires an overall factor of the form

$$N_{PQ}\left[{}^{\,J\,\,K}_{\,I\,\,L}\right] = \frac{\gamma\left({}^{\,P}_{IJ}\right)\gamma\left({}^{\,L}_{PK}\right)}{\gamma\left({}^{\,Q}_{JK}\right)\gamma\left({}^{\,L}_{IQ}\right)}.$$

Special entries of the fusing matrix for the $\widehat{su}(2)_k$ current algebra were first computed in [463]. The formula displayed above was obtained through its relation with 6J symbols of the quantum deformed $U(sl_2)$ in [314] and with the help of Coulomb gas representations for conformal blocks in [287]. From the re-interpretation as 6J symbols of a quantum deformed universal enveloping algebra we can conclude

$$\lim_{k\to\infty} F_{PQ}\left[{}^{\,J\,\,K}_{\,I\,\,L}\right] = \left\{{}^{\,J\,\,K\,\,P}_{\,I\,\,L\,\,Q}\right\}. \qquad (6.13)$$

Here, the limit is expressed in terms of the 6J symbols of the ordinary (unde-formed) Lie algebra su(2). The latter can be found in many texts on representation theory; see, e.g., [203].

6.2 SU(2) WZW model as bulk and boundary CFT

With the representation theory of the chiral symmetry algebra $\widehat{su}(2)_k$ being sufficiently under control, we can now let our general theory work for us to construct the bulk and boundary SU(2) WZW model as full CFTs. In the first subsection we provide expressions for the state space and the operator product expansions (OPEs) of the bulk theory. Then we turn to the discussion of boundary conditions. In particular, we provide explicit formulas for boundary states and one-point functions of bulk fields, for boundary spectra and for boundary three-point functions.

6.2.1 Bulk structure constants and spectrum

We choose diagonal combinations of left- and right-moving sectors to form the bulk state space of our SU(2) WZW model on the full plane, i.e.

$$\mathcal{H}^{(P)} = \bigoplus_j \mathcal{H}_j \otimes \overline{\mathcal{H}}_j \,.$$

Note that for $\widehat{su}(2)_k$ diagonal and charge-conjugate combinations are identical since its representations are self-conjugate, i.e. $j^+ = j$. Using equation (6.9) for the S-matrix of the current algebra, it is not difficult to check explicitly that the associated (diagonal or charge-conjugate) partition function

$$Z(q, \bar{q}) = \sum_{j=0}^{k/2} |\chi_j(q)|^2 \qquad (6.14)$$

is invariant under modular transformations although this already follows from the general setup of CFT as discussed in Chapter 3.

As any other CFT, the WZW model contains a field for every state. In particular, we will work with the fields that are associated with the ground states $|j; m\rangle$ of the irreducible representations $\mathcal{H}_j$ of the left- and right-moving chiral algebras. These will be denoted by

$$\varphi_{j,j}^{mn}(z, \bar{z}) \equiv \Phi\big(|j; m\rangle \otimes \overline{|j; n\rangle}; z, \bar{z}\big); \tag{6.15}$$

cf. equation (3.27) above. The $\varphi_{j,j}^{mn}$ are $Vir \times Vir$ primary for all m, n, and $\widehat{su}(2)_k \times \widehat{su}(2)_k$ primary for $m = n = j$.

The operator product expansions of the fields (6.15) possess the form (3.45) with conformal dimensions (6.7), and with appropriate choice of normalisations, the OPE coefficients for current algebra primaries are given by [463]:

$$C_{j_1,j_2}^{k;j_3} = (J+1)! \, P(J+1) \, P(1)^{\frac{1}{2}} \prod_{\nu=1}^{3} \frac{P(\hat{\jmath}_\nu) \, \hat{\jmath}_\nu!}{(2j_\nu + 1)^{\frac{1}{2}} (2j_\nu)! \, P(2j_\nu)^{\frac{1}{2}} P(2j_\nu + 1)^{\frac{1}{2}}},$$

where $J = i_1 + j_2 + j_3$ and $\hat{\jmath}_\nu = J - 2j_\nu$; the symbol $P(j)$ is defined by $P(0) = 1$ and

$$P(j) = \prod_{n=1}^{j} \frac{\Gamma(\frac{n}{k+2})}{\Gamma(1 - \frac{n}{k+2})}.$$

The OPE of descendants, and the contributions of descendants to the OPE of primaries, are determined by the current algebra Ward identities, cf. the statements around equation (3.45). For later purposes, we need coefficients for the fields (6.15) associated with ground states, i.e. the fields that have $N(\boldsymbol{n}) = 0 = N(\overline{\boldsymbol{n}})$ in the notations of Subsection 3.2.2. For these fields, the coefficients β_{ijk}^{n} from equation (3.45) and their anti-holomorphic counterparts $\beta_{ijk}^{\overline{n}}$, respectively, are constrained by the Ward identities for global left and right su(2) transformations. These constraints from global symmetries guarantee that they are proportional to the Clebsch–Gordan coefficients of the Lie algebra su(2) and their complex conjugates, respectively:

$$\beta_{j_1 j_2 j_3}^{m_1 m_2 m_3} = \begin{bmatrix} j_1 & j_2 & j_3 \\ m_1 & m_2 & m_3 \end{bmatrix}.$$

The precise numerical values of these Clebsch–Gordan coefficients can be found in the standard literature. Altogether, the OPE of fields associated to ground states of the SU(2) WZW model reads

$$\varphi_{j_1,j_1}^{m_1 n_1}(z, \bar{z}) \, \varphi_{j_2,j_2}^{m_2 n_2}(w, \bar{w})$$

$$= \sum_{j_3, m_3, n_3} |z - w|^{2h_{j_3} - 2h_{j_1} - 2h_{j_2}} \, C_{j_1,j_2}^{k;j_3} \, \beta_{j_1 j_2 j_3}^{m_1 m_2 m_3} \, \beta_{j_1 j_2 j_3}^{n_1 n_2 n_3 \, *} \, \varphi_{j_3,j_3}^{m_3 n_3}(w, \bar{w}) + \cdots .$$

6.2.2 One-point functions and boundary states

Let us now turn to the boundary theory. Since our bulk CFT is rational with a charge-conjugate modular invariant partition (which here is diagonal at the same time), all assumptions needed to form Cardy boundary states as discussed in Chapter 4 are satisfied as long as we choose trivial gluing conditions $\Omega =$ id.

To begin with, we learn from Cardy's solution that there are $k + 1$ possible boundary conditions (with maximal symmetry and trivial gluing). We label them $J = 0, \frac{1}{2}, \ldots, \frac{k}{2}$, just as we enumerate the sectors of the corresponding affine Lie algebra. Combining our general formula (4.80) with the expressions (6.9) for the modular S-matrix of the $\widehat{su}(2)_k$ current algebra, we obtain

$$\| J \rangle\!\rangle = \sum_{j=0}^{\frac{k}{2}} B_J^j \, |j\rangle\!\rangle := \left(\frac{2}{k+2} \right)^{\frac{1}{4}} \sum_j \frac{\sin \frac{\pi(2j+1)(2J+1)}{k+2}}{\left(\sin \frac{\pi(2j+1)}{k+2} \right)^{\frac{1}{2}}} \, |j\rangle\!\rangle \qquad (6.16)$$

where $|j\rangle\!\rangle$ are the Ishibashi states for trivial gluing conditions

$$(J_n^a + \bar{J}_{-n}^a)|j\rangle\!\rangle = 0 \, .$$

From the boundary states, we can in particular extract one-point functions of the ground state bulk fields (6.15), namely

$$\langle \varphi_{j,j}^{mn}(z, \bar{z}) \rangle_J = \left(\frac{2}{k+2} \right)^{\frac{1}{4}} \frac{\sin \frac{\pi(2j+1)(2J+1)}{k+2}}{\left(\sin \frac{\pi(2j+1)}{k+2} \right)^{\frac{1}{2}}} \frac{\delta^{m,n}}{|z - \bar{z}|^{2h_j}} \qquad (6.17)$$

with $h_j = \frac{j(j+1)}{(k+2)}$ as before. The superscripts mn label different components within the tensor multiplet $\varphi_{j,j}$, each of them running over a basis $|j;m\rangle$ in the representation space H_j of su(2). Since Ω is trivial, the intertwiner $\mathcal{U}_{jj}$ between the left and right representation of su(2) is trivial as well, i.e. $\mathcal{U}_{jj}^{mn} = \delta^{m,n}$.

There is a simple extension of the above in which we allow non-trivial gluing conditions given conjugation with an element g of SU(2), i.e. $\Omega_g = \mathrm{Ad}_g$, where Ad stands for the adjoint action of SU(2) on its Lie algebra. Without loss of generality we can assume that g takes the special form $g = \exp(-i\lambda J_0^3)$, i.e. we consider the gluing automorphism

$$\Omega_g = \mathrm{Ad}_g \quad \text{with } g = e^{-i\lambda J_0^3}, \quad \lambda \in \mathbb{R}. \qquad (6.18)$$

Applying the results of Subsection 4.3.2, especially equation (4.53), we build the associated Ishibashi states with the help of the unitary operator

$$V_{\Omega_g} : \begin{cases} \mathcal{H}_j \longrightarrow \mathcal{H}_j \\ |\xi\rangle \longmapsto e^{i\lambda q_j} \, g \, |\xi\rangle \end{cases},$$

which implements the inner automorphism Ω_g. Here, $q_j = j$ is the J_0^3-eigenvalue of the highest weight state $|j; j\rangle$ in the $\widehat{su}(2)_k$ representation $\mathcal{H}_j$, the phase ensuring that V_{Ω_g} maps this state to itself. The twisted Ishibashi states can then be expressed through the standard Ishibashi states $|j\rangle\rangle$ as

$$|j\rangle\rangle_{\Omega_g} = e^{i\lambda q_j}\, e^{-i\lambda \bar{J}_0}\, |j\rangle\rangle . \tag{6.19}$$

The boundary states $\|\alpha = J\rangle\rangle_{\Omega_g}$ for the twisted gluing (6.18) are again subject to Cardy's conditions. It is easy to check that the linear combinations

$$\|J\rangle\rangle_{\Omega_g} = \sum_{j=0}^{\frac{k}{2}} B^{j}_{\Omega_g\,J}\, |j\rangle\rangle_{\Omega_g} \quad \text{with} \quad B^{j}_{\Omega_g\,J} = e^{-i\lambda q_j}\, B^{j}_{J}$$

provide solutions if B^{j}_{J} are the coefficients (6.16) for standard gluing, i.e. given by Cardy's formula (4.80).

6.2.3 Boundary spectra

The partition functions for two boundary conditions of the same gluing type Ω_g are particularly easy to find. Since $\Theta g \Theta^{-1} = g$ and g commutes with L_0, the partition functions

$$Z_{(\Omega_g\ I)\,(\Omega_g\ J)}(q) = \sum_j N^{j}_{IJ}\, \chi_j(q)$$

coincide with those from the untwisted case and hence with the general expression in Cardy's solution. Note that the characters on the right-hand side are the usual specialised characters (6.10) for the $\widehat{su}(2)_k$ current algebra.

We also want to consider the case in which the boundary conditions at either side of the strip involve different gluing automorphisms. To be specific, let us assume that one of them is the identity automorphism $\Omega = \mathrm{id}$ while the other is taken to be of the form Ω_g as discussed in the previous subsection. In this case, the partition functions are computed as follows:

$$Z_{(\mathrm{id}\ I)\,(\Omega_g\ J)}(q) = {}_{\Omega_g}\langle\langle\Theta J\|\, \tilde{q}^{L_0 - \frac{c}{24}}\, \|I\rangle\rangle = \sum_j B^{j}_J B^{j}_I\, \langle\langle j|\, \tilde{q}^{L_0 - \frac{c}{24}}\, e^{i\lambda \bar{J}_0^3}\, |j\rangle\rangle$$

$$= \sum_j B^{j}_J B^{j}_I\, \chi_j\Big(\frac{\lambda}{2\pi}, -\frac{1}{\tau}, 0\Big),$$

where, for the last equality, we have exploited the gluing condition $\bar{J}_0^3|j\rangle\rangle = -J_0^3|j\rangle\rangle$ on the standard Ishibashi states. We arrive at unspecialised affine characters as defined in equation (6.8). The S-matrix for the transformation of such characters under $(z,\tau,u) \longmapsto \big(\frac{z}{\tau}, -\frac{1}{\tau}, u + \frac{z^2}{2\tau}\big)$ was given in equation (6.9) and is the same as for specialised characters, therefore we find

$$Z_{(\mathrm{id}\ I)\,(\Omega_g\ J)}(q) = \sum_j N^{j}_{IJ}\, \mathrm{Tr}_{\mathcal{H}_j}\, q^{L_0 - \frac{\lambda}{2\pi}J_0^3 + \frac{\lambda^2}{8\pi^2}k - \frac{c}{24}} . \tag{6.20}$$

We see that the pair of boundary conditions does lead to a proper open string partition function, but due to the fact that two different gluing conditions hold at the two string endpoints, equation (6.20) is not the sum of ordinary affine characters. As the non-trivial gluing automorphism (6.18) is inner, there is still a relation to twisted representations, but we will not pursue this here.

6.2.4 Boundary OPE

The state space associated with two Cardy-type boundary conditions for gluing automorphism $\Omega = \mathrm{id}$ is

$$\mathcal{H}_{IJ} := \bigoplus_{j=0}^{\frac{k}{2}} N_{IJ}^j \, \mathcal{H}_j \,,$$

and by the state–field correspondence Ψ of the boundary CFT, each state $|\psi\rangle \in \mathcal{H}_{IJ}$ gives rise to a boundary field $\psi(x) = \Psi(|\psi\rangle; x)$. Ground states in $\mathcal{H}_j$ furnish a multiplet of Virasoro primary boundary fields

$$\psi_{j,m}^{IJ}(x) := \Psi(|j;m\rangle; x) \tag{6.21}$$

where $|m| \leq j$ and $j = 0, 1, \ldots, j_{\max}$. Here $j_{\max}$ is the largest spin j such that $N_{IJ}^j \neq 0$, and m is the J_0^3-eigenvalue. Since the fusion multiplicities of the $\widehat{\mathrm{su}}(2)_k$ current algebra are either 0 or 1, a given label j can at most appear once for a given pair I, J of boundary conditions. As discussed in Subsection 4.2.2, the boundary fields satisfy a boundary OPE of the form (4.38),

$$\psi_{i,m}^{IJ}(x_1)\, \psi_{j,p}^{JK}(x_2) \sim \sum_{k,r} (x_1 - x_2)^{h_k - h_i - h_j} \, C_{ijk}^{IJK} \begin{bmatrix} i & j & k \\ m & p & r \end{bmatrix} \psi_{k,r}^{IK}(x_2) + \cdots .$$

$$\tag{6.22}$$

In comparison with the general OPE (4.38), we have written the formula for all the fields that are associated with ground states. The OPE depends on m, p and r through the Clebsch–Gordan coefficients. The constants C_{ijk}^{IJK} are obtained from the general equation (4.91) for the boundary OPE coefficients in Cardy-type boundary conditions along with the formula (6.12),

$$C_{ijk}^{IJK} = \mathsf{F}_{Jk}\begin{bmatrix} i & j \\ I & K \end{bmatrix} . \tag{6.23}$$

6.3 Geometric interpretation at large volume

In the previous section we have applied our general machinery to find boundary conditions for the SU(2) WZW model. The formulas provide an explicit solution of the theory valid for arbitrary level, in particular deep in the stringy regime of small k. Now, however, we would like to obtain a geometric understanding of these branes, and we will therefore focus on the limit $\mathsf{k} \to \infty$, in which the volume of the target three-sphere (measured in units of the string length) becomes large. We will first make a connection between brane worldvolumes and SU(2) conjugacy classes, then show that these classical submanifolds become quantised, in the sense that the dynamics of open strings has a natural description in terms of non-commutative geometry.

6.3.1 Brane worldvolumes from one-point functions

In a free boson theory, it is easy to to read off the location of a brane with Dirichlet and Neumann boundary conditions from the set of one-point functions, see equation (1.30) in Chapter 1. Let us now try to do the same for Cardy branes on SU(2), following [162, 344]. We first collect a few facts on functions on the group, then relate those to one-point functions.

According to the Peter–Weyl theorem for compact groups, the space of functions on SU(2) is spanned by the matrix elements $D^j_{m,n}(g)$ of finite-dimensional unitary representations D^j; see equation (4.127). More precisely, the functions

$$\phi^{mn}_j(g) := \sqrt{2j+1}\, D^j_{m,n}(g) \tag{6.24}$$

form a complete orthonormal basis of $Fun(S^3)$, just as the exponentials $\exp(ikx)$ do for $Fun(\mathbb{R}^D)$. Writing $\phi^{mn}_j(g)$ in terms of three Euler angles on S^3, we find in particular that

$$\sum_{m=-j}^{j} \phi^{mm}_j(g) = \sqrt{2j+1}\, \frac{\sin\vartheta(2j+1)}{\sin\vartheta}$$

where $\vartheta \in [0,\pi]$ parametrises the polar angle, with the group identity having $\vartheta = 0$. From the completeness of $\sin(n\vartheta)$ on the interval $[0,\pi]$, we conclude that

$$\frac{1}{\sin\vartheta_0}\delta(\vartheta-\vartheta_0) = \frac{4}{\pi}\sum_{j}\sum_{m=-j}^{j}\frac{\sin(\vartheta_0(2j+1))}{\sqrt{2j+1}}\,\phi^{mm}_j(g). \tag{6.25}$$

Now let's turn to the one-point functions (6.17) of the bulk fields $\varphi^{mn}_{j,j}(z,\bar z)$ in the presence of a Cardy brane with label J. We want to take the limit $\mathsf{k}\to\infty$ and at the same time scale J as $J \equiv J(\mathsf{k}) = \frac{\vartheta_0}{2\pi}\mathsf{k}$ for some constant angle $\vartheta_0 \in [0,\pi]$. We then find that the result agrees with the coefficients in the expansion (6.25), except for a proportionality constant not involving j. We can phrase this observation in a form similar to equation (1.30),

$$\langle\varphi^{mn}_{j,j}(z,\bar z)\rangle_{J(\mathsf{k})} \overset{\mathsf{k}\to\infty}{\sim} \int_{SU(2)} d\mu(g)\,\rho_0\,\delta(\vartheta(g)-\vartheta_0)\,\phi^{mn}_j(g)$$

where ρ_0 is some constant and $d\mu(g)$ denotes the Haar measure on SU(2).

To give a catchier interpretation of this result, recall that the two-spheres $\{\vartheta_0 = \text{const}\} \subset S^3$ are precisely the conjugacy classes of SU(2). Thus we arrive at the statement that the worldvolumes of Cardy branes in the SU(2) WZW model are the spherical conjugacy classes of the group target. This holds in the large volume limit $k \to \infty$, while for finite level k the graphs of one-point functions presented in [162] show that the brane worldvolumes are still concentrated around the conjugacy classes, but display a finite "thickness".

6.3.2 Gluing conditions, conjugacy classes and B-field

A more direct way towards uncovering the geometry of Cardy branes in the SU(2) WZW model starts from the gluing conditions $J(z) = \bar{J}(\bar{z})$ which hold along the worldsheet boundary $z = \bar{z}$. Re-expressing them in terms of the group-valued field will not only provide classical worldvolumes, but at the same time indicates that string excitations will induce non-commutative effects. We shall largely follow the presentation in [14], but see also [225, 428] for slightly different analyses of the gluing conditions.

Let us then rewrite the standard gluing conditions for the currents using their expressions (6.4) in terms of the group-valued field $g(z, \bar{z})$ taking values in SU(2). This leads to an equation that implicitly determines the classical worldvolume $g(\partial\Sigma)$ swept out by our maximally symmetric branes, namely

$$-(\partial g)g^{-1} = g^{-1}\bar{\partial}g \quad \text{for} \quad z = \bar{z}. \tag{6.26}$$

To make this more explicit, we use $z = x + iy$ and split ∂, $\bar{\partial}$ into derivatives ∂_x, ∂_y tangential and normal to the boundary (which extends along the x-direction). Then we can rewrite equation (6.26) in the form

$$(\mathrm{Ad}(g) - 1) \, g^{-1}\partial_y g = -i \, (\mathrm{Ad}(g) + 1) \, g^{-1}\partial_x g \quad \text{for} \quad z = \bar{z}. \tag{6.27}$$

Here, we use $\mathrm{Ad}(g)$ to denote the adjoint action of SU(2) on its Lie algebra su(2). The next step is to decompose the tangent space $T_h\mathrm{SU}(2)$ at each point $h \in \mathrm{SU}(2)$ into a part $T_h^{\|}\mathrm{SU}(2)$ tangential to the conjugacy class through h, and its orthogonal complement $T_h^{\perp}\mathrm{SU}(2)$ with respect to the Killing form.

Now note that $\mathrm{Ad}(g)|_{T_g^{\perp}} = 1$. Therefore, equation (6.27) implies that the endpoints of open strings on SU(2) are forced to move along conjugacy classes, i.e.

$$(g^{-1}\partial_x g)^{\perp} = 0.$$

Except for two degenerate cases, namely the points e and $-e$ on the group manifold, all conjugacy classes, and therefore the classical worldvolumes of all Cardy branes on SU(2), are two-spheres centred at e.

Furthermore, equation (6.27) shows that the branes wrapping conjugacy classes of SU(2) carry a B-field. This follows by comparison with the boundary conditions (1.3) we used in the flat background for Neumann conditions with a B-field switched on. In the present case, it is given by

$$B = \frac{1 + \mathrm{Ad}(g)}{1 - \mathrm{Ad}(g)}. \tag{6.28}$$

The associated two-form can be written as $\mathrm{tr}(g^{-1}dg \, B \, g^{-1}dg)$ and a short computation shows that it provides a potential for the Neveu-Schwarz–Neveu-Schwarz

(NS–NS) three-form $H \sim \omega_3$ where ω_3 denotes the volume form on S^3 that appears in the WZW action (6.2).

In the case of flat branes with B-field, we have seen that the brane's non-commutative geometry is not encoded in B itself but rather in another antisymmetric tensor Θ constructed from B through equation (1.13). Carrying over this relation to the present context, and choosing appropriate conventions for α' and the target metric, we obtain the expression

$$\Theta(g) = \frac{2}{B - B^{-1}} = \frac{1}{2}\left(\text{Ad}(g^{-1}) - \text{Ad}(g) \right).$$

This object simplifies in the large-level limit, where the three-sphere grows and approaches flat three-space $\mathbb{R}^3$. Points on SU(2) can be parametrised by elements X in the Lie algebra su(2), such that near the group unit $g \approx 1 + X$. Inserting this gives the simple formula

$$\Theta = -\text{ad}(X). \tag{6.29}$$

Here, ad denotes the adjoint action of su(2) on itself. If we expand $X = y^a t_a$ into Lie algebra generators, we can evaluate the matrix elements of Θ more explicitly,

$$\Theta_{ab} = -(t_a, \text{ad}(X)t_b) = f_{abc}\, y^c,$$

where $(\cdot, \cdot)$ denotes the Killing form on su(2), the generators t_a are normalised such that $(t_a, t_b) = \delta_{ab}$, and f_{abc} are the structure constants of su(2).

The antisymmetric tensor Θ gives rise to a Poisson structure on $\mathbb{R}^3$,

$$\{y^a, y^b\} = \Theta^{ab}(y) = f^{ab}{}_c\, y^c. \tag{6.30}$$

Due to the linear dependence on the coordinates, this is more complicated than the Poisson bracket (1.50) we met in our discussion of flat branes. On the other hand, the Poisson algebra defined by equation (6.30) possesses a large centre: any function of $c(y) := \sum_a (y^a)^2$ has vanishing Poisson brackets with any other function on $\mathbb{R}^3$ so that formula (6.30) induces a Poisson structure on the two-spheres defined by

$$c(y) := \sum_a (y_a)^2 \overset{!}{=} c \tag{6.31}$$

for any constant c. Summarising, the semi-classical analysis of the gluing conditions suggests that the worldvolumes of Cardy branes in the SU(2) WZW model are given by conjugacy classes, and that those two-spheres carry non-trivial magnetic fields, which should lead to a quantisation of the worldvolume algebras.

6.3.3 Boundary fields and fuzzy harmonics

In the following, we will make the non-commutative geometry of WZW branes more concrete by studying the properties of boundary fields on branes in the SU(2) WZW model. The connection is, however, more easily established once we have gathered some further mathematical facts about the quantised spaces we expect to play a role.

In Subsection 1.1.3, we saw how the Moyal–Weyl product arises from quantising the constant Poisson bracket (1.50) on $\mathbb{R}^d$. By analogy, quantising the Poisson geometry (6.30) and (6.31) on the spherical conjugacy classes in SU(2) requires finding operators $\mathsf{Y}^a = Q(y^a)$ acting on some space V such that

$$[\,\mathsf{Y}^a\,,\,\mathsf{Y}^b\,] \;=\; i\,f^{ab}{}_c\,\mathsf{Y}^c\,, \tag{6.32}$$

$$\mathsf{C} := \sum (\mathsf{Y}^a)^2 = c\,\mathbf{1}\,, \tag{6.33}$$

where $\mathbf{1}$ denotes the identity operator on V.

Due to the commutation relation (6.32), the operators Y^a have to form a representation of su(2). Condition (6.33) states that in this representation, the quadratic Casimir element C must be proportional to the identity $\mathbf{1}$, which is true iff the representation on V is irreducible (or equivalent to a direct sum of several copies of one and the same irrep). Hence, any irreducible representation of su(2) can be used to quantise our Poisson geometry, on a finite-dimensional vector space V.

On the other hand, since the set of equivalence classes of su(2) irreps is discrete, only a discrete set of two-spheres in $\mathbb{R}^3$ is quantisable; their radii increase with the value of the quadratic Casimir in the corresponding irreducible representation.

For each quantisable two-sphere $S_J^2 \subset \mathbb{R}^3$ (which corresponds to a single irreducible representation), we obtain a state space V_J of dimension $\dim V_J = 2J + 1$ equipped with an action of the "quantised coordinate functions" Y^a. In fact, applying a theorem of the Artin–Wedderburn type, it can be seen that the associative algebra generated by $\mathbf{1}$ and the Y^a is the full matrix algebra $\mathrm{Mat}(2J + 1)$.

To proceed, it is useful to look at the classical two-sphere for a moment. Recall that the spherical harmonics Y_m^j, $j \in \mathbb{Z}_+$, $m = -j, \ldots, j$, form a linear basis of the space $Fun(S^2)$ of functions on the two-sphere. The elements of the vector multiplet Y_a^1 may be identified with the restriction of the three-coordinate functions y^a to the two-sphere in $\mathbb{R}^3$. Since the Y_m^j form a basis, products of spherical harmonics can be written as a linear combination

$$Y_m^i\,Y_p^j = \sum_{k,r} c_{ijk} \begin{bmatrix} i & j & k \\ m & p & r \end{bmatrix} Y_r^k\,. \tag{6.34}$$

The $[:::]$ denote the Clebsch–Gordan coefficients of su(2), and the structure constants c_{ijk} are simple expressions that can be found in many places in the literature.

Moreover, the algebra $Fun(S^2)$ admits an action of su(2) which is generated by infinitesimal rotations in $\mathbb{R}^3$,

$$L_a := f_{ac}{}^b\,y^c\partial_b\,.$$

Under the action of L_a, spherical harmonics Y_m^j transform in the representation j of su(2).

This classical symmetry survives quantisation, i.e. there is an analogous action of su(2) on $\mathrm{Mat}(2J+1)$ given by

$$L_a\, \mathsf{A} := [t_a^J, \mathsf{A}] \quad \text{for all} \quad \mathsf{A} \in \mathrm{Mat}(2J+1). \tag{6.35}$$

Here t_a^J denote the generators of su(2) evaluated in the $(2J+1)$-dimensional irreducible representation, in other words, $t_a^J = \mathsf{Y}_a$ from above.

This action of su(2) on $\mathrm{Mat}(2J+1)$ is reducible; its decomposition into irreducible sub-representations leads to the following equivalence of su(2) representations,

$$\mathrm{Mat}(2J+1) \cong \bigoplus_{j=0}^{2J} H_j, \tag{6.36}$$

where the sum runs over integer spins j. We have chosen the notation H_j for the irreducible components on the right-hand side to avoid confusion with the space V_J from above. Each H_j has a linear basis consisting of $(2j+1)$ matrices, which we denote by Y_m^j.

In the case $J = \frac{1}{2}$, only the scalar and the vector multiplet appear and explicit expressions for the corresponding 2×2 matrices are of course well-known: they are given by the identity and the Pauli matrices, respectively.

Since the Y_m^j span $\mathrm{Mat}(2J+1)$ linearly, the product of any two such matrices may be expressed as a linear combination of Y_c^k,

$$\mathsf{Y}_m^i\, \mathsf{Y}_p^j = \sum_{k \leq 2J, r} \left\{ {i \atop j}\, {j \atop j}\, {k \atop j} \right\} \left[{i \atop m}\, {j \atop p}\, {k \atop r} \right] \mathsf{Y}_r^k. \tag{6.37}$$

Here $\{:::\}$ denote the re-coupling coefficients (or 6J symbols) of su(2). This relation should be considered as a quantisation of the expansions (6.34), and the classical expression is indeed recovered from equation (6.37) upon taking the limit $J \to \infty$ [278].

Hence, the matrices Y_m^j in the quantised theories are a proper replacement for spherical harmonics. Note, however, that the angular momentum $j \leq 2J$ is bounded from above. This may be interpreted as "fuzziness" of the quantised two-spheres on which short distances cannot be resolved [339]. We shall sometimes refer to the Y_m^j as "fuzzy spherical harmonics".

We will now make the link to CFT and show that objects with the algebraic properties of the Y_m^j arise naturally from Cardy branes in the SU(2) WZW model [12].

The state space of the Jth boundary theory is determined by equation (4.81) and has the form

$$\mathcal{H}_{JJ} = \bigoplus_{j=0}^{j_{\max}} N_{JJ}^j\, \mathcal{H}_j, \tag{6.38}$$

where the $\mathcal{H}_j$ denote irreducible highest weight representations of the affine Lie algebra $\widehat{\mathrm{su}}(2)_k$ as before, and where N_{JJ}^j are the fusion rules (6.11). Note that only integer spins j appear on the right-hand side of (6.38) and that the sum is truncated at $j_{max} = \min(2J, \mathsf{k} - 2J)$. In the limit $\mathsf{k} \to \infty$, we obtain $j_{max} = 2J$ so that the decomposition of $\mathcal{H}_{JJ}$ corresponds to the decomposition (6.36) of $\mathrm{Mat}(2J + 1)$. More precisely, at the level of $\mathrm{su}(2)$ multiplets, there is a unique correspondence between fuzzy spherical harmonics $Y_a^j \in \mathrm{Mat}(2J + 1)$ and ground states in $\mathcal{H}_j \subset \mathcal{H}_{JJ}$. This generalises the relation between the Weyl operators $\exp(ik\hat{x})$ and the boundary fields $: \exp(ikX(x)):$ that we found for flat branes in Chapter 1.

To compare algebraic properties, we consider the OPE (6.22) of open string vertex operators associated with these ground states in $\mathcal{H}_{JJ}$, which reads

$$\psi_{i,m}(x_1)\, \psi_{j,p}(x_2) = \sum_{k,r} x_{12}^{h_k - h_i - h_j} \begin{bmatrix} i & j & k \\ m & p & r \end{bmatrix} F_{Jk} \begin{bmatrix} i & j \\ J & J \end{bmatrix} \psi_{k,r}(x_2) + \cdots, \qquad (6.39)$$

where $h_j = \frac{j(j+1)}{\mathsf{k}+2}$ is the conformal dimension of $\psi_{j,m} \equiv \psi_{j,m}^{JJ}$, and the square bracket symbol denotes the Clebsch–Gordan coefficients of $\mathrm{su}(2)$.

When we send the level k to infinity (while keeping the boundary label J fixed), the conformal dimensions h_j tend to zero so that the OPE (6.39) of boundary fields becomes regular as in a topological model. At the same time, the fusing matrix turns into a 6J symbol, cf. equation (6.13). Therefore,

$$\lim_{x_1 \to x_2} \lim_{\mathsf{k} \to \infty} \left(\psi_{i,m}(x_1)\, \psi_{j,p}(x_2) \right) = \sum_{k,r} \begin{bmatrix} i & j & k \\ m & p & r \end{bmatrix} \begin{Bmatrix} i & j & k \\ J & J & J \end{Bmatrix} \psi_{k,r}(x_2). \qquad (6.40)$$

In other words, the large k limit of the OPE of the boundary fields $\psi_{i,m}$ exactly reproduces the multiplication of matrices Y_m^i. It is therefore justified to say that fuzzy spheres arise naturally as the worldvolumes of $\mathrm{SU}(2)$ WZW branes.

The relation between boundary fields and fuzzy harmonics can be exploited to evaluate the $\mathsf{k} \to \infty$ limit of arbitrary n-point functions of the boundary fields $\psi_{j,m}$ in a very simply way. Let us introduce the notation

$$\psi[A](x) := \sum a_{j,m}\, \psi_{j,m}(x) \quad \text{for all} \quad A = \sum a_{j,m} Y_m^j \in \mathrm{Mat}(2J + 1).$$

The OPE formula (6.40) then implies that in the limit $\mathsf{k} \to \infty$ we have, for an arbitrary set of matrices $A_r \in \mathrm{Mat}(2j + 1)$,

$$\langle\, \psi[A_1](x_1)\, \psi[A_2](x_2)\, \cdots\, \psi[A_n](x_n)\,\rangle = \mathrm{tr}(A_1\, A_2\, \cdots\, A_n). \qquad (6.41)$$

The trace appears because, just as the vacuum expectation value, it projects matrices to their $\mathrm{SU}(2)$ invariant component.

The expression (6.41) for the correlators of open string vertex operators shares many features with the Moyal product formula (2.39) that we encountered for branes in flat space. In both cases, the vertex operators are in one-to-one correspondence with elements of the non-commutative algebra of "functions" on the worldvolume of the brane. Furthermore, in a limiting regime, the correlators are independent of the insertion points u_r, and they can be evaluated using the multiplication and integration (trace) over the non-commutative worldvolume algebra. There remains, however, one important difference between the two cases: for branes on SU(2), the worldvolume algebra is cut off at some angular momentum $2J$ so that there are only finitely many linearly independent functions (or boundary primary fields). It is likely that similar truncation quantisations occur whenever (an appropriate decoupling limit of) branes are studied in a compact target, or in a rational CFT.

6.3.4 Gauge theory on fuzzy spheres, and boundary flows

From the string theory perspective, it is natural to ask what the low-energy effective action (LEEA) of open strings on WZW branes is. We will give the result in the large k and small α' limit (such that $\alpha'\mathsf{k} \to \infty$), which corresponds to the decoupling limit (2.37) that was suitable for the discussion of branes in flat space. There, the LEEA for a stack of flat branes was a non-commutative Yang–Mills theory. Here, the underlying worldvolume geometry is described by finite-dimensional algebras, and the gauge theories relevant for SU(2) WZW branes are matrix models.

We only sketch their derivation [13], because all the concepts needed are familiar from computations of string scattering amplitudes in flat targets and can be found, e.g., in [377]. We will, however, study some classical solutions of the resulting LEEA in more detail, since they have an interesting interpretation in terms of brane bound state formation. We then finish this subsection with some brief remarks on brane dynamics in the stringy regime, which turns out to be a problem studied in condensed matter physics, and its relation with twisted K-theory.

Non-commutative gauge theory

We want to study the low-energy effective physics of open strings on Cardy branes in the limit $\mathsf{k} \to \infty$, $\alpha' \to 0$, $\alpha'\mathsf{k} \to \infty$. The boundary fields that become massless in this limit are given by normal-ordered products

$$:J^a\psi[\mathsf{A}_a]:(x) = \sum_{a;j,m} a_{j,m;\,a} :J^a\psi_{j,m}:(x) \tag{6.42}$$

of a WZW current with a linear combination of ground state fields (6.21); summation over $a = 1, 2, 3$ is understood, indices are raised and lowered with the open string metric $G^{ab} = (2/\mathsf{k})\,\delta^{ab}$; see [10, 13] for details about normalisations. We can check that the physical state condition imposes the relation $L^a\mathsf{A}_a = 0$ on the admissible linear combinations, where we use $L_a\mathsf{A} = [\mathsf{Y}_a^1, \mathsf{A}]$ as in equation (6.35).

Computation of the LEEA requires the computation of three- and four-point functions of the fields (6.42), to leading order in α' and $\frac{1}{\mathsf{k}}$. The main ingredients needed are the OPE

$$J^a(x_1)\,\psi[\mathsf{A}](x_2) = \frac{\alpha'}{x_1 - x_2}\,\psi[L^a\mathsf{A}](x_2) + \text{reg}$$

expressing the current algebra symmetry, and the OPE of currents, which in suitable normalisations ($\mathsf{f}_{ab}{}^c := f_{ab}{}^c/\sqrt{2\alpha'}$) reads

$$J^a(x_1)\, J^b(x_2) = \frac{\alpha'}{2}\, \frac{G^{ab}}{(x_1 - x_2)^2} + \alpha'\, \frac{i\, \mathsf{f}^{ab}{}_c}{(x_1 - x_2)}\, J^c(x_2) + \text{reg}. \tag{6.43}$$

Equipped with these relations, we can essentially guess the outcome of following the usual rules of computing string scattering amplitudes, which involves multiplication with ghost propagators, summing over all orderings of insertion points of boundary fields, etc. The overall structure is as in the flat-target case, with the following changes: Moyal–Weyl products are replaced by matrix multiplications, due to equation (6.41). Spacetime derivatives $i\partial^a$ are substituted by L^a, which plays the role of "momentum" (rotation generator) on the fuzzy sphere. The most important change is due to the term $\mathsf{f}^{ab}{}_c J^c$ in the OPE (6.43) of currents, which is absent in the flat-target case. It results in an extra contribution of the form $\mathsf{f}_{abc}\mathsf{A}^a\mathsf{A}^b\mathsf{A}^c$ in the scattering amplitude of three massless open string modes. (The non-abelian term does not affect the quartic interactions in the LEEA, to lowest order in α'.)

This rough description of the derivation should make the end result for the LEEA of open strings on SU(2) Cardy branes obtained in [13] plausible. Up to a numerical overall factor, we find

$$\mathcal{S}_{(1,J)} = \mathcal{S}_{\text{YM}} + \mathcal{S}_{\text{CS}} = \frac{1}{4}\, \text{tr}\left(\mathsf{F}_{ab}\, \mathsf{F}^{ab}\right) - \frac{i}{2}\, \text{tr}\left(\mathsf{f}^{abc}\, \mathsf{CS}_{abc}\right) \tag{6.44}$$

where we have defined the field strength F_{ab} by the expression

$$\mathsf{F}_{ab}(\mathsf{A}) = i\, \mathsf{L}_a\mathsf{A}_b - i\, \mathsf{L}_b\mathsf{A}_a + i\, [\mathsf{A}_a\, ,\, \mathsf{A}_b] + \mathsf{f}_{abc}\mathsf{A}^c \tag{6.45}$$

and a non-commutative analogue of the Chern–Simons form by

$$\mathsf{CS}_{abc}(\mathsf{A}) = \mathsf{L}_a\mathsf{A}_b\, \mathsf{A}_c + \frac{1}{3}\, \mathsf{A}_a\, [\mathsf{A}_b\, ,\, \mathsf{A}_c] - \frac{i}{2}\, \mathsf{f}_{abd}\, \mathsf{A}^d\, \mathsf{A}_c\, . \tag{6.46}$$

The subscript $(1, J)$ of the effective action indicates that it describes open strings on a single Cardy brane $\| J \rangle\rangle$ as in equation (6.16); the "gauge fields" A_a are matrices in $\text{Mat}(2J + 1)$.

The generalisation to a stack of M branes of the same type J is straightforward: the action $\mathcal{S}_{(M,J)}$ is given by the same expression, except that now $\mathsf{A}_a \in \text{Mat}(M) \otimes \text{Mat}(2J + 1)$, and the L_a act exclusively on the fuzzy spherical harmonics Y_m^j and commute with the Chan–Paton coefficients in $\text{Mat}(M)$. It is also worth noting that the same expression (6.44) results as the bosonic part of the LEEA in a supersymmetric WZW model [13].

A direct computation shows that the action (6.44) is invariant under the gauge transformations

$$\mathsf{A}_a \longmapsto \mathsf{A}_a + i\, \mathsf{L}_a\lambda + i\, [\mathsf{A}_a\, ,\, \lambda] \quad \text{with} \quad \lambda \in \text{Mat}(M) \otimes \text{Mat}(2J + 1)\, .$$

Indeed, the quadratic "mass term" in the Chern–Simons form (6.46) guarantees the gauge invariance of $\mathcal{S}_{\text{CS}}$ by itself. On the other hand, the effective action

(6.44) is the unique linear combination of $\mathcal{S}_{\mathrm{YM}}$ and $\mathcal{S}_{\mathrm{CS}}$ from which mass terms cancel. As we shall see below, it is this special feature of our action that allows for solutions describing translations of the branes on the group manifold. The Yang–Mills action $\mathcal{S}_{\mathrm{YM}}$ had already been considered in the non-commutative geometry literature [104, 258, 340], where the expression was derived from a Connes spectral triple [115] and viewed as describing Maxwell theory on the fuzzy sphere. In a similar spirit, the work [315] studied arbitrary linear combinations of non-commutative Yang–Mills and Chern–Simons terms, but no guiding principle from string theory was available to single out the special action (6.44).

Classical solutions and brane dynamics

In view of the general ideas collected in Subsection 5.A.3, stationary points of the action (6.44) are expected to describe endpoints of condensation processes of a brane configuration $Q = (M, J)$, which drive the whole system into another configuration Q'. From the worldsheet point of view, these processes are RG flows triggered by a boundary perturbation in the CFT of the brane configuration Q, and the computations in [5] are highly useful to investigate brane condensation in SU(2) WZW models.

In order to identify the configuration Q' from a classical solution, we have two different types of information at our disposal, coming from the closed and the open string channel, respectively. On the one hand, we can compare the tension of D-branes in the final configuration Q' with the value of the action $\mathcal{S}_Q(\Lambda)$ at the classical solution Λ. On the other hand, we can look at fluctuations around the chosen stationary point and compare their dynamics with the low-energy effective theory $\mathcal{S}_{Q'}$ of the brane configuration Q'. In formulas, this means that

$$\mathcal{S}_Q(\Lambda + \delta\mathbf{A}) \overset{!}{=} \mathcal{S}_Q(\Lambda) + \mathcal{S}_{Q'}(\delta\mathbf{A}) \quad \text{with} \quad \mathcal{S}_Q(\Lambda) \overset{!}{=} \ln\frac{g_{Q'}}{g_Q}. \tag{6.47}$$

In writing the second condition, we have used the fact that brane tensions are proportional to the g-factors [5] of the respective boundary CFTs. The g-factor of the starting configuration is of course known from Cardy's solution,

$$g_{(M,J)} := M\, g_J := M\, \langle \varphi_{0,0}(z, \bar{z}) \rangle_J = M\, \frac{S_{J0}}{\sqrt{S_{00}}}. \tag{6.48}$$

All equalities in (6.47) must hold to the same order in $\frac{1}{k}$ to which the effective action was computed. We say that the brane configuration Q decays into Q' if Q' has lower mass, i.e. whenever $g_{Q'} < g_Q$.

From the effective action (6.44), we obtain the following equations of motion for the gauge fields $\mathbf{A}^a \in \mathrm{Mat}(M) \otimes \mathrm{Mat}(2J + 1)$

$$\mathsf{L}_a\, \mathsf{F}^{ab} + [\mathsf{A}_a, \mathsf{F}^{ab}] = 0. \tag{6.49}$$

These equations may admit many solutions whose realisation as conformal boundary conditions is unknown. We will come back to more general solutions later, but it turns out that even when we restrict ourselves to a simple subclass of gauge configurations, we can still find two (rather different) types of solutions with interesting interpretations. That subclass is given by "constant" gauge fields, i.e. those satisfying $L_a A^b = 0$; only the Chan–Paton degrees of freedom play a role. Note that if we start from a stack of D0-branes, only this type of gauge fields appear anyway.

Let us use the letter $S^a \in \mathrm{Mat}(M) \otimes 1_{2J+1}$ for such constant gauge fields, to avoid confusion. It is easy to see that with $L_a S^b = 0$, equation (6.49) takes the simpler form

$$\left[S^a, \, [S_a, S_b] - i\, \mathsf{f}_{abc}\, S^c \right] = 0 \,, \tag{6.50}$$

and the two types of solutions we have in mind are obvious.

For the first one, we choose three pairwise commuting Chan–Paton matrices $a^a_{00} \in \mathrm{Mat}(M)$ to form $S^a = a^a_{00} \otimes Y^0_0$. These solutions describe translations of the M branes in the three-dimensional group manifold, the $3M$ parameters in the family being coordinates of the M branes in $\mathrm{SU}(2)$. The boundary CFT corresponding to these constant solutions can be constructed using chiral marginal deformations, leading to twisted gluing conditions (6.18). Such continuous families also appear for branes in flat backgrounds; in the present context, the existence of these solutions is guaranteed by the absence of the mass term in the full effective action (6.44).

Much more interesting processes are described by the second type of solutions to equations (6.50) and (6.49): any three matrices S_a that form a representation of $\mathrm{su}(2)$ satisfy (6.50). To arrive at their interpretation [13], let us specialise to a starting configuration consisting of a stack of M branes of type $J = 0$, i.e. of M point-like branes at the origin of $\mathrm{SU}(2)$. Also, let us choose the M-dimensional irreducible representation of spin $J_M = \frac{(M-1)}{2}$ for the S_a. We claim that this solution corresponds to an RG fixed point where the initial stack of M point-like branes has been driven to a final configuration containing only a single brane that wraps the sphere of type J_M, i.e.

$$(M, \, J = 0) \longrightarrow \left(1, \, J_M = \frac{(M-1)}{2} \right) .$$

Support for this statement comes from both the open string sector and the coupling to closed strings. In the open string sector small fluctuations δA_a of the fields $A_a = S_a + \delta A_a \in \mathrm{Mat}(M)$ can be studied around the stationary point $S_a \in \mathrm{Mat}(M)$. If S_a form an irreducible representation of $\mathrm{su}(2)$, we find indeed

$$\mathcal{S}_{(M,0)}(S_a + \delta A_a) = \mathcal{S}_{(1,J_M)}(\delta A_a) + \mathrm{const} \,.$$

In the closed string channel, the leading term (in the ($\frac{1}{k}$)-expansion) from the exact "mass" formula (6.48) gives [13]

$$\ln \frac{g_{(1,J_M)}}{g_{(M,0)}} = -\frac{\pi^2}{6} \frac{M^2 - 1}{k^2} = \mathcal{S}_{(M,0)}(S) \,.$$

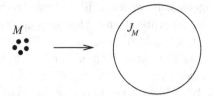

M

J_M

Figure 6.1 A configuration of M point-like branes stacked at the origin of a weakly curved $S^3 \sim \mathbb{R}^3$ is unstable against decay into a single spherical brane with label $J_M = \frac{(M-1)}{2}$.

Note that the mass of the final state is lower than the mass of the initial configuration. This means that a stack of M point-like branes on a three-sphere is unstable against condensation into a single spherical brane; see Figure 6.1. Note also that, from the worldsheet point of view, the perturbations triggering the renormalisation group (RG) flow described by this second type of solutions are by marginally relevant boundary fields, rather than the chiral marginal ones leading to translations in the target.

In [13], the final brane configurations Q' associated to reducible su(2) representations S_a were described as well, and it was moreover shown that any $M(2J+1)$-dimensional representation A_a of the Lie algebra su(2) can be used to solve the equations of motion, even if the gauge field is non-constant in the sense that $\mathsf{L}_a\mathsf{A}_b \neq 0$.

Stationary points of the action (6.44) and the formation of spherical branes on S^3 were also discussed in [272, 276, 299]. Similar effects have been described for branes in Ramond–Ramond (R–R) background fields [352]. The advantage of our scenario with NS–NS background fields is that it can be treated on the worldsheet so that stringy corrections may be taken into account (see [15, 181, 182] and below).

Kondo effect, dynamics in the stringy regime, and K-theory

The effective action underlying the preceding discussion could only be computed in a large k, thus large volume approximation. With the methods available at present a similarly good control of brane dynamics cannot be achieved in the stringy regime where k is finite.

A less ambitious question is whether some of the solutions we found in the large volume limit persist in the small volume theory and, if so, which boundary CFTs they correspond to. It turns out that all RG fixed points described by constant gauge field solutions S_a occur in the finite k theory, as well. Somewhat surprisingly, this answer to a string theory problem was given in the condensed matter literature on the Kondo effect [3, 4, 6].

The Kondo model is designed to understand the effect of magnetic impurities on the low-temperature properties of a conductor in three dimensions. The latter can have electrons in a number k of conduction bands. If the impurities are far apart, their effect may be understood within an s-wave approximation of scattering events between a conduction electron and the impurity. This allows the formulation of the whole problem on a two-dimensional worldsheet for which the coordinates (x, y) are the time and the radial distance from the impurity. One of the currents that can be built from the basic fermion fields is a *spin current* $\vec{J}(x, y)$, which satisfies the relations (6.43) of $\widehat{su}(2)_k$. It is this current that couples to the magnetic impurity of spin J_M sitting at the boundary $y = 0$,

$$S_{\text{pert}} \sim \lambda \int_{-\infty}^{\infty} dx \, \Lambda_\alpha J^a(x, 0) \,. \tag{6.51}$$

Here, Λ_a is a $(2J_M + 1)$-dimensional irreducible representation of $su(2)$, and the parameter λ controls the strength of the impurity.

The term (6.51) is identical to the coupling of open string ends to a background gauge field $S_a = \Lambda_a \in \text{Mat}(2J_M + 1)$. Hence, Λ_a may be interpreted as a constant gauge field on a Chan–Paton bundle of rank $M = 2J_M + 1$.

The study of the Kondo effect has a long history in the condensed matter community, and using Wilsonian renormalisation group techniques, two different cases were distinguished. The "under-screened" Kondo model (where $2J_M > k$) shows a low-temperature RG fixed point only at infinite values of λ. On the other hand, the fixed point is reached at a finite value $\lambda = \lambda^*$ of the renormalised coupling constant λ if $2J_M \leq k$ (exact or over screening); then the fixed points are described by non-trivial (interacting) CFTs. In particular, the results for the spectrum at the fixed point can be summarised in the so-called "absorption of the boundary spin" principle established by Affleck and Ludwig [4]

$$\text{tr}_{H_{J_M} \otimes \mathcal{H}_j} \left(q^{H_{\text{unpert}} + H_{\text{pert}}} \right)_{\lambda = \lambda^*} = \sum_l N^l_{j \, J_M} \, \chi_l(q) \,. \tag{6.52}$$

Here, $H_{\text{unpert}} = L_0 - \frac{c}{24}$ is the unperturbed Hamiltonian, and H_{J_M} denotes the finite-dimensional $su(2)$ irrep for the spin J_M impurity, while $\mathcal{H}_j$ is an affine Lie algebra representation in the state space of the boundary WZW model.

Formula (6.52) means that perturbing by a boundary spin drives the configuration at the ultraviolet fixed point $\lambda = 0$ – which consists of $M := 2J_M + 1$ copies of the original boundary WZW model so that the spin J_M irrep can be coupled as in equation (6.51) – to an infrared fixed point theory where the boundary spin has been "absorbed" (only a single copy is left) according to the $\widehat{su}_k$ fusion rules. In terms of characters, we have

$$M \chi_j(q) \longrightarrow \sum_l N^l_{j \, J_M} \, \chi_l(q) \,. \tag{6.53}$$

With this, it is straightforward to determine the decay product of a stack of M point-like branes, even for finite level $k \geq M - 1$. Since each of the two string ends can be attached to any of the M branes, the partition function of the whole stack is M^2 times the partition function for a single brane. For this system we find

$$Z_{(M,0)}(q) = M^2 Z_{(1,0)}(q) = M^2 \chi_0(q)$$
$$\longrightarrow M \chi_{J_M}(q) \longrightarrow \sum_j N^j_{J_M J_M} \chi_j(q) = Z_{(1,J_M)}(q).$$

We have applied the rule (6.53) twice because both endpoints of an open string couple to the background field. The result then is that Kondo model studies show that the brane condensation process

$$(M,\, 0) \longrightarrow (1,\, J_M) \tag{6.54}$$

we found in the large k regime persists in the stringy regime as long as the level satisfies $2J_M \leq k$.

Let us conclude this section with some brief remarks on brane charges. As has been pointed out in many places, the classical concept of measuring brane charges through their coupling to R–R closed string modes sometimes seems to fail in string theory on curved backgrounds. For example, on S^3 these couplings are not quantised [31] for finite level. On the whole, K-theoretic ideas to classify D-brane charges [457] seem to be more appropriate, although it is not always obvious which particular kind of the K-theories that can be associated with a classical target space, if any, will capture all the stringy aspects of D-branes.

A worldsheet approach to the question would start from RG flows and define charges as quantities which are invariant under such flows. More precisely, two brane configurations Q and Q' can be called *dynamically equivalent* if there is boundary perturbation that relates the two boundary theories. The space of all brane configurations modulo this dynamical equivalence should define a group of brane charges. The latter is a property of the background.

One obvious practical problem with such a definition is that the set of all admissible (super)conformal boundary conditions for a given background bulk CFT is not known except for very few cases. Still, let us follow [15] and try to achieve a partial understanding of the charge group for branes in a (supersymmetric) SU(2) WZW model, by exploiting the rule (6.54) for RG flows between maximally symmetric brane configurations. These branes can wrap $k+1$ different conjugacy classes labelled by $J = 0, \frac{1}{2}, \ldots, \frac{k}{2}$. If we stack more and more point-like branes at the origin, the radius of the two-sphere that is wrapped by the condensation product will first grow, then decrease, and finally a stack of $k+1$ point-like branes at e will merge into a single point-like object at $-e$ (see Figure 6.2). By taking orientations into account, one can see that the final point-like object is the translate of an anti-brane at e. From this, we conclude that the stack of $k+1$ point-like branes at e has decayed into a single point-like anti-brane at $-e$. Assigning charges $+1$ and -1 to a single brane at e and $-e$, respectively, this suggests that the D-brane charge group of a supersymmetric SU(2) model contains a finite group $\mathbb{Z}_{k+2}$.

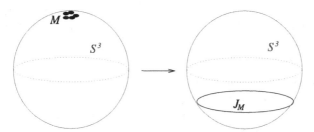

Figure 6.2 Brane dynamics on S^3: a stack of point-like branes at the north pole e can decay into a single spherical object. The distance of the latter increases with the number of branes in the stack until a single point-like object is obtained at the south pole $-e$.

This observation fits with the proposal of Bouwknegt and Mathai [69] that brane charges on a background X with a non-vanishing NS–NS three-form field H take values in some twisted K-groups $K_H^*(X)$ which feel the presence of $H \in H^3(X, \mathbb{Z})$; for $X = S^3$ this twisted K-group is a finite cyclic group.

Related proposals for the case that H is torsion have been put forward in [302, 457]. Many more details about twisted K-theory, its computation through spectral sequences and the relation with string theory can be found in [345]. Geometric construction of brane charges on S^3 and other group manifolds have also been discussed in [15, 170, 429, 435].

6.4 WZW cosets and orbifolds

The final section is devoted to two types of CFTs that can be obtained from the SU(2) WZW model. We begin by exploiting the orbifold construction to build strings and branes on the orbifold space $\mathrm{SU}(2)/\mathbb{Z}_2 = \mathrm{SO}(3)$. Aspects of the general theory were described in Subsection 4.A.2, and we mainly need to apply them to the case at hand. We then turn to the so-called coset construction, which can be interpreted as producing sigma models with target space G/H, given a WZW model on the group G together with a continuous subgroup H of G. We shall sketch the general theory in the second subsection, and then conclude the chapter with remarks on one of the most important examples of coset models, the $N = 2$ superconformal minimal series. This series will play a central role in the following chapter.

6.4.1 Orbifolds: WZW model on SO(3)

Throughout this subsection let us assume that the level of the SU(2) WZW model is of the special form $\mathsf{k} = 4n$ for some integer n. The orbifold model we are about to discuss describes strings and branes on $\mathrm{SU}(2)/\mathbb{Z}_2$. It belongs to the class of simple current orbifolds sketched in Subsection 4.A.3. The bulk spectrum of the parent theory with charge-conjugate modular invariant was spelt out in equation (6.14). Using the formula (6.11) for fusion of any two sectors j_1, j_2 and the general definition (3.54) of simple currents, we can read off that those of the $\widehat{su}(2)_k$ current algebra are given by $j = 0$ and $j = 2n =: g$. These two sectors form the group $G = \mathbb{Z}_2 = \{0, g\}$ which we will use in the orbifold construction. Note that the conformal dimension $h_{2n} = n$ of the non-trivial simple current is integer, so we can extend the chiral symmetry algebra by this field.

The general expression for the bulk partition function of a simple current orbifold model has already been stated in equation (4.109). Let us specialise this formula to the case at hand. First note that the monodromy charges Q defined in equation (4.108) are

$$ Q_0(j) = 0, \quad Q_g(j) = j \ (\text{mod } 1). $$

To determine these, we have inserted the expression (6.7) for the conformal weights in the SU(2) WZW model and used the fact that g acts on sectors as $g(j) = 2n - j$, which follows when formula (6.11) is applied to the fusion of $g \sim j_1 = 2n = \frac{k}{2}$ and $j_2 = j$. Orbits of $G = \mathbb{Z}_2$ consist of pairs $\{j, 2n - j\}$ as long as $j \neq n$, while the sector $j = n$ leads to an orbit of length 1 with stabiliser group $\mathcal{S}_{[n]} = \mathbb{Z}_2$. An important condition for the general theory to apply is that the members of a G-orbit have the same monodromy charge Q. For the orbits $[j] = \{j, 2n - j\}$ this amounts to $j \equiv (2n - j) \ (\text{mod } 1)$ which holds since $2n$ and $2j$ are both integer. Finally, the orbits $[j]$ have vanishing monodromy charge Q if j is integer and the orbit $[n]$ of length 1 belongs to this set. Now we have all the ingredients required to spell out equation (4.109) for the example of SU(2) and the simple current group $\mathbb{Z}_2 = \{e = 0, g = 2n = \frac{k}{2}\}$:

$$ Z^{\text{SO}(3)}(q, \bar{q}) = \sum_{j=0}^{n-1} |\chi_j(q) + \chi_{2n-j}(q)|^2 + 2|\chi_n(q)|^2 , $$

where the summation is over integer j only. The term $Z^{\text{SO}(3)}$ is known as the D_{even} modular invariant of the SU(2) WZW model.

According to the general facts collected in Subsection 4.A.3, our orbifold model has branes labelled by $J = [0], [\frac{1}{2}], \ldots, [n-1]$ and two additional branes $[n]_\pm$ associated with the two characters γ_+ and γ_- of the stabiliser group $\mathcal{S}_n = \mathbb{Z}_2$. Boundary states can be written down as Cardy boundary states using the S-matrix S_{ext} of the simple current extension of the $\widehat{su}(2)_k$ current algebra; see, e.g., [59] for details on the modular properties of the extension. We will not discuss boundary spectra in full detail here, but rather restrict ourselves to the most interesting case which appears when the open strings have both ends on branes sitting at the fixed point. Before resolution, the partition function (4.110) for these open strings reads

$$ Z^{\text{SO}(3)}_{[n][n]}(q) = \sum_{j=0}^{2n} 2 \chi_j(q) , \tag{6.55} $$

where the summation runs over integer j. To split this $Z^{SO(3)}$ into partition functions for resolved branes $[n]_\pm$, we use $\mathbb{Z}_2$-characters to compute the associated object d defined in equation (4.111). In a matrix notation $d^j = (d^j_{ab})$ it is given by

$$d^{4p} = \begin{pmatrix} 1 & 0 \\ 0 & 1 \end{pmatrix} \quad \text{and} \quad d^{4p+2} = \begin{pmatrix} 0 & 1 \\ 1 & 0 \end{pmatrix},$$

with $p = 0, \ldots, n$. When inserted into equation (4.112), the resolved partition functions become

$$Z^{SO(3)}_{[n]_\pm [n]_\pm}(q) = \sum_{p=0}^{n} \chi_{2p}(q), \quad Z^{SO(3)}_{[n]_\pm [n]_\mp}(q) = \sum_{p=1}^{n} \chi_{2p-1}(q). \tag{6.56}$$

These expressions are well known from studies of boundary conditions in the D_{even} theory; see, e.g., [44].

Let us briefly discuss geometric aspects of branes in the $SU(2)/\mathbb{Z}_2$ orbifold, in particular of the fixed point branes $[n]_\pm$. As we have seen above, Cardy branes on $SU(2)$ can be viewed as localised along the $\mathsf{k}+1$ "integer" conjugacy classes of $SU(2) = S^3$, which are two-spheres with centres at the origin $e \in SU(2)$. The generator of our orbifold group acts by reflection $g \mapsto -g$ about the equatorial plane of S^3. Hence the nth two-sphere, which wraps the equator of S^3, is located along the fixed surface of the group action, in agreement with the algebraic results above.

The algebra $Fun(S^2)$ of functions on the equatorial two-sphere $S^2 \subset S^3$ is spanned by spherical harmonics Y^j_m where $|m| \le j$ and $j \in \mathbb{Z}$. This algebra inherits an involution ϑ from the reflection on S^3, acting on spherical harmonics as $\vartheta(Y^j_m) = (-1)^j Y^j_m$. The involution ϑ and its square $\vartheta^2 = \text{id}$ give rise to an action of $\mathbb{Z}_2$ on $Fun(S^2)$, and we can use these data to pass to the *crossed product* $Fun(S^2) \times_\vartheta \mathbb{Z}_2$, generated by the Y^j_m and one additional element θ with relations $\theta^2 = 1$ and

$$\theta \, Y^j_m = \vartheta(Y^j_m) \, \theta = (-1)^j \, Y^j_m \, \theta.$$

Hence the crossed product contains two $SU(2)$ multiplets for each integer j.

All this is very similar to the structure of the partition function (6.55), which also contains each integer spin $\widehat{su}(2)_\mathsf{k}$ character twice. For finite k, however, labels $j > 2n$ do not appear in the partition function, a truncation of the spectrum related to the quantisation of the spherical branes we found in the $SU(2)$ WZW model. But in the limit $\mathsf{k} \to \infty$, we also have $n \to \infty$, and arbitrary high-integer spins occur. Moreover, it can be shown that the B-field vanishes in this limit, at least on the equatorial two-sphere, which from our previous experience suggests that the world-volume geometry of this brane should be more or less commutative, apart from possible signs arising from the orbifold projection. It is therefore not too surprising that, in the large k limit, the OPE of boundary fields on the brane $[n]$ reproduces the multiplication of $Fun(S^2) \times_\vartheta \mathbb{Z}_2$; see [348] for more details. The relation between orbifold constructions and crossed products is of course well-known in the theory of operator algebras; see, e.g., [157].

We conclude this section with a few remarks on the dynamics of branes in $SU(2)/\mathbb{Z}_2$. For a spherical brane with fixed label J, the computation of the effective action carries over from [13]. This means that we still find a linear combination of a Yang–Mills and a Chern–Simons term on a fuzzy S^2 to control the behaviour of these branes in the limit $k \to \infty$. In particular, a stack of D0-branes at the origin is unstable and will expand into a spherical brane. Following the reasoning of [15, 181], it can be shown that these (unresolved) branes contribute a term $\mathbb{Z}_{n+1}$ to the charge group of the background. In addition, the resolved branes at the equator carry one extra charge that can be measured through the coupling of the closed string states in the twisted sector when the level k becomes large. More details can be found in [178, 348].

6.4.2 Cosets: some general results

Let us now briefly discuss aspects of the coset construction; for the bulk theory see [240] and also textbooks like [121] and references therein. Assume that $H \subset G$ is some simply connected subgroup of G. In a sense, the coset construction can be viewed as an algebraic method to produce a model with target G/H.

The affine Lie algebras associated with G and H – which we will for simplicity take to have the same level k – both contain Virasoro algebras L_n^G, L_n^H due to the Sugawara construction, and it is not difficult to see that the operators

$$L_n^{G/H} = L_n^G - L_n^H . \tag{6.57}$$

also obey the Virasoro algebra with central charge given by $c^{G/H} = c^G - c^H$. Moreover, the $L_n^{G/H}$ commute with the currents in $\widehat{h}_k$.

Let us label the sectors $\mathcal{H}_\lambda^G$ of the current algebra $\widehat{g}_k$ by $\lambda \in I_G$, those of $\widehat{h}_k$ by $\lambda' \in I_H$. Since the denominator current algebra acts on the representations of the numerator current algebra, we have decompositions

$$\mathcal{H}_\lambda^G = \bigoplus_{\lambda'} \mathcal{H}_{(\lambda,\lambda')} \otimes \mathcal{H}_{\lambda'}^H ,$$

and the infinite-dimensional spaces $\mathcal{H}_{(\lambda,\lambda')}$ are interpreted as sectors of the coset chiral algebra, which contains the Virasoro modes $L_n^{G/H}$ from above.

Note that some of the spaces $\mathcal{H}_{(\lambda,\lambda')}$ may vanish simply because a given sector $\mathcal{H}_{\lambda'}^H$ of the denominator theory does not appear as a subsector in a given $\mathcal{H}_\lambda^G$. This motivates the introduction of the set

$$\mathcal{E} = \{ (\lambda, \lambda') \in I_G \times I_H \mid \mathcal{H}_{(\lambda,\lambda')} \neq 0 \}.$$

Furthermore, different elements of $\mathcal{E}$ may correspond to the same sector, i.e. there is an equivalence relation (the so-called field identification)

$$(\lambda, \lambda') \sim (\mu, \mu') \iff \mathcal{H}_{(\lambda,\lambda')} \cong \mathcal{H}_{(\mu,\mu')} .$$

Let us assume that all the equivalence classes we find in $\mathcal{E}$ contain the same number N_0 of elements. It can be shown that in this case the sectors of the coset theory are simply labelled by the equivalence classes, i.e. $I_{G/H} = \mathcal{E}/\sim$; see [407] for more discussion and further references. (For more general cases, if the field identification does not arise from a free group action on $\mathcal{E}$, there could be further coset model sectors that cannot be constructed within the sectors of the numerator theory.)

With these assumptions, we can also spell out explicit formulas for the fusion rules and the S-matrix of the coset model. These are given by

$$N^{(k,k')}_{(\lambda,\lambda')\,(\mu,\mu')} = \sum_{(\nu,\nu')\sim(k,k')} N^{G;\nu}_{\lambda\,\mu}\, N^{H;\nu'}_{\lambda'\,\mu'} \,, \tag{6.58}$$

$$S_{(\lambda,\lambda')(\mu,\mu')} = N_0\, S^G_{\lambda\,\mu}\, \overline{S}^H_{\lambda'\,\mu'} \,, \tag{6.59}$$

where the sum runs over all elements (ν,ν') of $\mathcal{E}$ that are equivalent to (k,k'), and where the bar over the second S-matrix denotes complex conjugation.

Let us note that (λ,λ') is an element of $\mathcal{E}$ if (but not only if) the representation λ' of the finite-dimensional Lie algebra $\mathbf{h}$ appears as a sub-representation of the representation λ for $\mathbf{g}$. The equivalence classes of such special pairs form a subset $I^r_{G/H} \subset I_{G/H}$. One distinguishing feature of sectors in the subset $I^r_{G/H}$ is that the conformal dimension of their ground state satisfies the equality

$$h^{G/H}_{(\lambda,\lambda')} = h^G_\lambda - h^H_{\lambda'} + n$$

with $n = 0$, while for other pairs $(\lambda,\lambda') \in I_{G/H}$, the offset n is a (non-vanishing) positive integer. This means in particular that fields associated with the sectors in $I_{G/H} \setminus I^r_{G/H}$ have conformal dimension $h > 1$ and hence are irrelevant operators. Therefore, the sectors labelled by elements of $I^r_{G/H}$ play a special role in the analysis of brane geometry and dynamics in coset models.

We will now study boundary theories associated with the charge-conjugate modular invariant bulk partition function of the coset theory,

$$Z(q,\bar{q}) = \sum_{(\lambda,\lambda')} \chi_{(\lambda,\lambda')}(q)\, \chi_{(\lambda^+,\,\lambda'^+)}(\bar{q}) \,.$$

We want to impose trivial gluing conditions along the boundary and are restricted to the maximally symmetric case, so that the boundary states are given by Cardy's solution (4.80) as in Subsection 4.4.1; we label them by $(\Lambda,\Lambda') \in I_{G/H}$. The couplings of closed string modes to the boundary are given by

$$\langle\, \varphi_{(\lambda,\lambda')}(z,\bar{z})\,\rangle_\Lambda = \frac{S^G_{\lambda\Lambda}\overline{S}^H_{\lambda'\Lambda'}}{\sqrt{S^G_{0\lambda}\overline{S}^H_{0\lambda'}}}\, \frac{1}{|z-\bar{z}|^{2h_{(\lambda,\lambda')}}} \,, \tag{6.60}$$

where we have used formula (6.59) for the S-matrix of the coset theory. The spectrum of open strings stretched between two branes (Λ_1, Λ_1') and (Λ_2, Λ_2') follows from equation (4.81) as

$$Z^{(\Lambda_2,\Lambda_2')}_{(\Lambda_1,\Lambda_1')}(q) = \sum_{(l,l')} N^{G;\Lambda_2}_{\Lambda_1\, l}\, N^{H;\Lambda_2'}_{\Lambda_1'\, l'}\, \chi_{(l,l')}(q) \,. \tag{6.61}$$

This formula involves the fusion rules (6.58) of the coset model. Let us point out that we can think of elementary Cardy branes as being labelled by a set of irreducible representations $[\Lambda, \Lambda'^+]$ of $G \times H$, while more complicated brane configurations involve reducible representations.

A description of the geometry of Cardy-type branes in coset models was given in [226]; see also [155]. We assume that the quotient G/H is formed with respect to the adjoint action of H on G, i.e. two points on G are identified if they are related by conjugation with an element of $H \subset G$. We denote the projection from G to the space G/H of H-orbits by $\pi^G_{G/H}$. Furthermore, we use C^G_Λ to refer to the conjugacy class of G along which the brane with label Λ is localised, and similarly for $C^H_{\Lambda'}$. The latter is a conjugacy class in H, but through the embedding of H into G, we can regard it as a subset of G. Now we construct the set $C_{(\Lambda,\Lambda')}$ of all elements in G which are of the form uv^{-1} for some $u \in C^G_\Lambda$ and $v \in C^H_{\Lambda'}$. This set is invariant under conjugation with elements of H and hence it can be projected down to G/H. The claim of [226] is that the brane (Λ, Λ') is localised along the resulting subset $C^{G/H}_{(\Lambda,\Lambda')}$ of G/H,

$$C^{G/H}_{(\Lambda,\Lambda')} = \pi^G_{G/H}\left(C^G_\Lambda\, (C^H_{\Lambda'})^{-1}\right) \subset G/H \,. \tag{6.62}$$

This assertion can be verified by analysing the one-point functions (6.60), as we explained for branes in SU(2) in Subsection 6.3.1. Details can be found, e.g., in the appendix of [183].

Aspects of the dynamics of coset theory branes can be described using an extension [178, 181] of the principle of boundary spin absorption that encodes brane dynamics in the SU(2) WZW model, as outlined at the end of Subsection 6.3.4. Exploiting this and the branching rules of affine Lie algebra representations, a rule can be established for the RG flow of coset brane configurations $(\Lambda, \sigma|_H \times \Lambda')$. Here, $\times$ denotes the fusion product for representations of $\widehat{g}_k \times \widehat{h}_k$, and $\sigma|_H$ is defined as follows: pick a representation σ of the finite-dimensional Lie group G that extends to a sector (also called σ) of the corresponding current algebra $\widehat{g}_k$; we can also restrict the G-representation σ to the subgroup $H \subset G$, and then lift the resulting representation of H to a representation $\sigma|_H$ of the current algebra $\widehat{h}_k$. Then the rule found in [181] predicts the following flow between boundary conditions:

$$(\Lambda, \sigma|_H \times \Lambda') \longrightarrow (\Lambda \times \sigma, \Lambda') \,. \tag{6.63}$$

Note that in most cases, the boundary labels on both sides of the flow (6.63) involve reducible representations. To identify the configurations as mixtures of elementary boundary conditions, we have to decompose the representations into irreducibles. A detailed analysis of this rule along with several examples can be found in [177, 178].

6.4.3 Cosets: N = 2 minimal models

Let us now turn to a particular class of coset models, namely the $N = 2$ supersymmetric minimal models

$$\left(\widehat{su}(2)_k \oplus \widehat{u}(1)_2\right)/\widehat{u}(1)_{k+2} .$$

The $\widehat{u}(1)$ factors here are to be understood as rational Gaussian models, where the abelian current algebra is extended by the additional chiral fields (5.48) of spin k.

We need two numbers to label the sectors $\lambda = (l, s)$ of the numerator and one label $\lambda' = m$ for the denominator. The usual convention in the context of $N = 2$ minimal models is to take the parameters l, m, s to be integers and have them run over $l = 0, \ldots, k$, $m = -k - 1, \ldots, k + 2$ and $s = -1, 0, 1, 2$; they are subject to the selection rule that $l + m + s$ has to be even.

If we start from a diagonal modular invariant bulk partition function and choose A-type gluing conditions, the maximally symmetric branes are of the Cardy type and, as before, parametrised by $(\Lambda, \Lambda') \sim (L, M, S)$ from the same set as (λ, λ'). We restrict our attention to the cases with $S = 0$ and drop this label from now on.

It is straightforward to adjust the formulas (6.60) for the one-point functions and (6.61) for the open string spectrum to the $N = 2$ supersymmetric minimal series. The necessary input from the representation theory of $\widehat{su}(2)_k$ can be found in Subsection 6.1.2, equations (6.9) and (6.11). The fusion rules for the rational $\widehat{u}(1)_{k+2}$ theory are simply given by addition of m-values modulo $k + 2$, and the entries of the modular S-matrix are suitably normalised roots of unity,

$$S_{m_1 m_2} = \frac{1}{\sqrt{2k + 4}} \, e^{-\frac{i\pi m_1 m_2}{k+2}} .$$

Let us briefly discuss the geometric interpretation of the branes (L, M), following [178, 344] and the ideas discussed in the previous subsection. First note that the target space of a SU(2)/U(1) coset model, where U(1) acts by conjugation, is a disk. One way to see this is to parametrise $g \in \text{SU}(2)$ and $h \in \text{U}(1)$ as

$$g = \begin{pmatrix} \alpha & \beta \\ -\beta^* & \alpha^* \end{pmatrix}, \quad h = \begin{pmatrix} \omega & 0 \\ 0 & \omega^* \end{pmatrix}$$

with $|\alpha|^2 + |\beta|^2 = |\omega|^2 = 1$, and to notice that the entry α in the upper-left corner of g is invariant under the action of h; thus the space of orbits is parametrised by the complex coordinate α with $|\alpha|^2 = 1 - |\beta|^2 \leq 1$.

In order to pass from the $SU(2)/U(1)$ coset model (the parafermion CFT; see, e.g. [230]) to an $N = 2$ minimal model, we have to take a product with the circle theory corresponding to the $\hat{u}(1)_2$ factor, but this additional target dimension will only play a spectator role for the branes we consider, and can be ignored for the following.

Conjugacy classes C_L^G in the $SU(2)$ group manifold are two-spheres centred around the group unit e. Each conjugacy class C_M^H of the abelian denominator group $H = U(1)$ consists of a single element $h_M \in U(1) \subset SU(2)$. Hence, the product $C_L^G (C_M^H)^{-1}$ describes a sphere in $SU(2)$ that is centred around the point h_M^{-1}. It remains to push this sphere down to the disk G/H with the help of the projection map $\pi_{G/H}^G$. The resulting brane is localised along a one-dimensional straight line that connects two points on the boundary of the disk [344]. The length of this line is determined by the integer L, while shifting M by one unit amounts to rotating the brane by the discrete angle $\pi/(k+2)$.

It is quite instructive to work out the consequences of the RG flow rule (6.63) in this coset theory. As an example, let us choose $\Lambda = L = 0$ and $\Lambda' = M$, and let us insert an irreducible $su(2)$ representation of dimension $P + 1 \leq k + 1$ for σ. Then formula (6.63) becomes

$$\sum_{\nu=0}^{P} \left(L = 0, \; M + \nu - \frac{P}{2} \right) \longrightarrow (L = \sigma_P, \; M).$$

This means that a chain of $P + 1$ adjacent branes with $L = 0$ (those of minimal length) condenses into a single brane with $L = P$. The process admits a very suggestive pictorial presentation in which a chain of branes, each of minimal length, condenses into a brane forming the shortest line between the ends of the chain. In [281] similar pictures occur using the Landau–Ginzburg realisation of $N = 2$ minimal models.

7

Gepner model boundary states and Calabi–Yau branes

In this chapter, we apply some of the general methods outlined in earlier chapters to a class of conformal field theories (CFTs) which are of particular interest for string theory: Gepner models [228] provide an algebraic formulation of superstring compactifications from ten to lower dimensions, formulated in terms of rational "internal" CFTs. Gepner models are considered to be the most relevant perturbative string vacua as far as phenomenology is concerned. It has turned out that they are closely connected to σ-model compactifications on Calabi–Yau manifolds and that many properties can be studied equally well in this geometric setting. The Gepner model description is more appropriate in the stringy regime, while the Calabi–Yau picture can be associated with the large-volume regime. While many (not all) correlation functions may receive corrections when varying Kähler moduli to pass from one regime to another, some aspects (such as the number and type of fields in the effective low-energy theory) do not change, due to the high amount of spacetime supersymmetry: $N = 2$ for a closed string compactification to four dimensions. A brane, on the other hand, breaks at least half of the supersymmetry, and thus it may well be that the low-energy physics of the geometric regime differs from the one obtained at stringy distances. Therefore, it is clearly of some interest to study D-branes in the exact CFT framework of Gepner models.

We first review Gepner's construction of superstring vacua from superconformal minimal models, and in Subsection 7.1.2 we briefly discuss the target-space interpretation of Gepner models. The main emphasis will be on their connection to Calabi–Yau sigma models, established via Landau–Ginzburg models; many of the algebraic notions from $N = 2$ superconformal field theories (SCFTs), notably their chiral rings, have a fairly direct translation into topological and geometric data, e.g. the Dolbeault cohomology of Kähler manifolds.

In Section 7.2, we discuss the construction of Gepner model boundary states, at first in a simplified situation, which displays aspects like orbifolding and the difference between A- and B-type boundary conditions, and which allows for a straightforward translation between algebraic and geometric results. Subsection 7.2.2 is devoted to boundary states in arbitrary Gepner models, mainly following [386, 389].

The interpretation of these abstract boundary states in terms of low-energy effective physics and, mainly, of Calabi–Yau geometry is the topic of Section 7.3. We will briefly review supersymmetric cycles, which govern the large-volume (target-space) picture of branes on Calabi–Yau manifolds, and also some aspects of the work of Douglas and collaborators, who showed how to link Gepner branes to supersymmetric cycles. Their results confirm that, for branes, the transition from the large-volume regime to stringy distances is not as smooth as for closed strings ("walls" of marginal stability are crossed), and suggest the development of a more refined category-theoretic picture of branes.

So far, such a picture has mainly been developed for branes in topologically twisted models, which can be viewed as simplified "skeletons" of full SCFTs. Section 7.4 briefly reviews topological twisting and then introduces matrix factorisations, which provide an extremely efficient formulation of (B-type) topological branes.

7.1 Gepner models in the bulk

In [228], Gepner introduced a construction of supersymmetric string compactifications based on minimal models of the $N = 2$ super Virasoro algebra. In this algebraic approach, the compactification is achieved not by "curling up" $10 - D$ dimensions of ten-dimensional flat spacetime into a compact Calabi–Yau manifold, but rather by replacing the $10 - D$ free superfields with some "internal CFT" of central charge $15 - \frac{3D}{2}$; the full CFT is a tensor product of external free fields and an internal CFT.

Certain conditions have to be met, most of which serve to maintain spacetime supersymmetry ([228]; see also Greene's review [253] for a useful discussion):

(1) The internal CFT must at least have $N = 2$ worldsheet supersymmetry.
(2) The total U(1) charges in the tensor product of internal and external theory must be odd integers for both left and right movers – here, total refers to internal charges plus charges of the $D - 2$ free external superfields associated with transverse uncompactified directions; this condition implements the (generalised) GSO projection.
(3) The left-moving states must be taken from the NS-sectors of each subtheory (external and, in Gepner's models, various internal subtheories) or from the R-sectors in each subtheory; analogously for the right-moving states.
(4) The torus partition function must be modular invariant.

We discuss the worldsheet side first, then collect some remarks on the spacetime interpretation in Subsection 7.1.2.

7.1.1 The CFT description

To start with, let us recall some general features of the worldsheet theories underlying superstring compactifications. Their symmetry algebra contains the $N = 2$ super Virasoro algebra, whose commutation relations we repeat for convenience:

$$[L_n, L_m] = (n - m) L_{n+m} + \frac{c}{12}(n^3 - n) \delta_{n+m,0},$$

$$[L_n, J_m] = -m J_{n+m},$$

$$[L_n, G^{\pm}_{r\pm a}] = \left(\frac{n}{2} - (r \pm a)\right) G^{\pm}_{n+r\pm a},$$

$$[J_n, J_m] = \frac{c}{3} n \delta_{n+m,0}, \tag{7.1}$$

$$[J_n, G^{\pm}_{r\pm a}] = \pm G^{\pm}_{n+r\pm a},$$

$$\{G^{+}_{r+a}, G^{-}_{s-a}\} = 2 L_{r+s} + (r - s + 2a) J_{r+s} + \frac{c}{3}\left((r + a)^2 - \frac{1}{4}\right) \delta_{r+s,0}.$$

All the indices n, m, r, s are integer, the parameter a takes the values $a = \frac{1}{2}$ for the Neveu-Schwarz (NS) sector and $a = 0$ for the Ramond (R) sector. The NS and R algebras are isomorphic via the spectral flow by half a unit (see equation (3.17)), which can be implemented by $U_{\frac{1}{2}} = e^{i\sqrt{\frac{c}{12}}\phi}$ where ϕ bosonises the U(1) current $J = i\sqrt{\frac{1}{3}\partial\phi}$. As mentioned in Chapter 3 after equation (3.17), in string theory the field $U_{\frac{1}{2}}$ has an interpretation as spacetime supersymmetry charge.

$N = 2$ SCFTs have many interesting properties absent for general CFTs; see in particular the work [331]. The graded Lie algebra (3.16) has a two-dimensional Cartan subalgebra spanned by L_0 and J_0 (besides the central charge of the Virasoro algebra), thus we can choose, in every irreducible representation, a basis of eigenstates $|\psi\rangle$ labelled by the conformal dimension h_ψ and the *charge* q_ψ. Within the NS sector of a unitary $N = 2$ SCFT, the two satisfy the simple inequality

$$h_\psi \geq \frac{|q_\psi|}{2}, \tag{7.2}$$

as follows from computing the norms of $G^{\pm}_{-\frac{1}{2}}|\psi\rangle$. (Note that $(G^{+}_r)^* - G^{-}_{-r}$ in a unitarisable representation.) It is easy to see that whenever equality holds in equation (7.2), then $|\psi\rangle$ is an $N = 2$ highest weight state, and moreover

$$h_{\psi_c} = \frac{1}{2} q_{\psi_c} \quad \Longleftrightarrow \quad G^{+}_{-\frac{1}{2}}|\psi_c\rangle = 0,$$

$$h_{\psi_a} = -\frac{1}{2} q_{\psi_a} \quad \Longleftrightarrow \quad G^{-}_{-\frac{1}{2}}|\psi_a\rangle = 0. \tag{7.3}$$

Such states $|\psi\rangle_{c,a}$ and the corresponding primary fields $\psi_{c,a}$ are called *chiral* and *anti-chiral*, respectively. The supermultiplets associated with (anti-)chiral fields are half as long as usual. This terminology stems from supersymmetry and is not related to left- or right-movers. We hope not to create further confusion by only treating "left-moving chiral halves" – in the CFT sense – of chiral and anti-chiral primaries in the following.

From a similar analysis with $G^{\pm}_{-\frac{3}{2}}|\psi\rangle$, we obtain the bound $h \leq \frac{c}{6}$ for the conformal dimensions of chiral and anti-chiral primaries. This means that in most models of interest, there are only finitely many such fields.

In the R sector, the distinguished states are those with conformal dimension $\frac{c}{24}$, the *Ramond ground states*. They are precisely the images of chiral states under the spectral flow $U_{\frac{1}{2}}$. Applying this operator once more, we reach the anti-chiral states from the Ramond ground states.

Chiral states can be viewed as "harmonic" representatives of cohomology classes of the nilpotent operator $G^{+}_{-\frac{1}{2}}$: an arbitrary state $|\psi\rangle$ in the (left-moving) NS–Hilbert space can be decomposed as

$$|\psi\rangle = |\psi_c\rangle + G^{+}_{-\frac{1}{2}}|\psi_1\rangle + G^{-}_{\frac{1}{2}}|\psi_2\rangle, \tag{7.4}$$

with $|\psi_c\rangle$ chiral [331]. Since $G^{-}_{\frac{1}{2}}$ is the adjoint of $G^{+}_{-\frac{1}{2}}$, this gives a Hodge decomposition of the NS sector. Accordingly, polynomials $P(t,\bar{t}) \sim \text{tr}_{(c,c)} t^{J_0} \bar{t}^{\bar{J}_0}$ can be defined as traces over left- and right-moving chiral states which share the abstract properties of Poincaré polynomials of Kähler manifolds; see [331].

In addition, cohomology rings of classical manifolds carry a multiplicative structure, which finds its CFT analogue in the operator product expansion (OPE) of two chiral primary fields,

$$\psi_i(z)\psi_j(w) = \sum_{k} C_{ijk}\,(z-w)^{h_k - h_i - h_j}\,\psi_k(w)\,.$$

Because of charge conservation, the conformal fields on the right-hand side all have U(1) charge $q_k = q_i + q_j = 2(h_i + h_j)$. The bound (7.2) implies that $z - w$ occurs with positive powers only, and that the limit $z \to w$ projects onto chiral ψ_k on the right-hand side. This provides the multiplication table of the *chiral ring*

$$\psi_i \cdot \psi_j = \sum_{k} C_{ijk}\,\psi_k\,.$$

Analogously, a commutative ring structure is defined on the set of anti-chiral fields.

The simplest *examples* of $N = 2$ SCFTs are free theories and the minimal models of the super Virasoro algebra; both are needed for Gepner's construction. Starting from a complex free boson $X = X_1 + iX_2$ and a complex free fermion $\psi = \psi_1 + i\psi_2$, an $N = 2$ superconformal algebra with central charge $c = 3$ is obtained; the generators were already given in Subsection 1.2.3:

$$
\begin{aligned}
T_{\text{free}} &= -\partial X \partial X^* - \tfrac{1}{4}\psi\partial\psi^* - \tfrac{1}{4}\psi^*\partial\psi\,, \\
J_{\text{free}} &= -\tfrac{1}{2}\psi^*\psi\,, \\
G^{+}_{\text{free}} &= i\,\psi\partial X^*\,, \qquad G^{-}_{\text{free}} = i\,\psi^*\partial X\,.
\end{aligned}
\tag{7.5}
$$

The unitary minimal models of the super Virasoro algebra have central charges

$$c_k = \frac{3k}{k+2} \ , \qquad k = 1, 2, \ldots . \tag{7.6}$$

The agreement with SU(2) Wess–Zumino–Witten (WZW) models is no coincidence, as minimal models can be obtained via a coset construction from SU(2) and U(1); see Subsection 6.4.3. The super minimal models are rational, and it is useful (e.g. for the GSO projection to be carried out later) to consider irreps of the bosonic subalgebra of the $N = 2$ super Virasoro algebra (7.1), labelled by three integers (l, m, s) with

$$l = 0, 1, \ldots, k \ , \qquad m = -k-1, -k, \ldots, k+2 \ , \qquad s = -1, 0, 1, 2$$

and

$$l + m + s \quad \text{even} . \tag{7.7}$$

Triples (l, m, s) and $(k-l, m+k+2, s+2)$ give rise to the same representation: the so-called "field identification", arising naturally in the coset construction.

The conformal dimension h and charge q of the highest weight state with labels (l, m, s) are given by

$$h^l_{m,s} = \frac{l(l+2) - m^2}{4(k+2)} + \frac{s^2}{8} \pmod 1 \ , \tag{7.8}$$

$$q^l_{m,s} = \frac{m}{k+2} - \frac{s}{2} \pmod 2 . \tag{7.9}$$

For many purposes, it is sufficient to know h (and q) up to (even) integers. The values in equations (7.8) and (7.9) are exact (without the modulo-prescription) if (l, m, s) lies in the *standard range* [200]:

$$l = 0, 1, \ldots, k \ , \qquad |m - s| \leq l \ , \qquad s = -1, 0, 1, 2 \ , \qquad l + m + s \quad \text{even} \tag{7.10}$$

or

$$l = 1, \ldots, k \ , \qquad m = -l \ , \qquad s = -2 . \tag{7.11}$$

Every (l, m, s) can be brought into this form with the help of the field identification and of the transformations $(l, m, s) \longmapsto (l, m + 2k + 4, s)$ and $(l, m, s) \longmapsto (l, m, s + 4)$.

Representations with an even value of s are part of the NS sector, while those with $s = \pm 1$ belong to the R sector. Within the two sectors, representations can be grouped into pairs $(l, m, s) \& (l, m, s + 2)$ which make up a full $N = 2$ super Virasoro module, with all states in a bosonic subrepresentation having the same fermion number modulo two.

Note that the chiral primaries, defined as in equation (7.3), are labelled by the triples $(l, l, 0)$, while the anti-chiral primaries are the representations labelled by $(l, -l, 0)$.

Also, it will be useful later on to observe that in a tensor product of $N = 2$ theories, the chiral primaries of the diagonal super Virasoro algebra are simply the tensor products of chiral primaries in each subtheory: it is easy to see, from the inequality relating h and q in unitary SCFTs, that $\bigotimes_i |h_i, q_i\rangle$ can only be a chiral or anti-chiral state in the tensor product theory if all q_i have the same sign; in that case, we have

$$h_{\text{tot}} = \sum_i h_i \geq \sum_i \frac{1}{2}|q_i| = \frac{1}{2} \left| \sum_i q_i \right| = \frac{1}{2} |q_{\text{tot}}|$$

and equality for the total dimensions and charges can only hold if $h_i = \frac{q_i}{2}$ in each subtheory.

In Gepner's string compactifications to $D =: d + 2$ external dimensions, tensor products of $N = 2$ minimal models are used as internal CFTs, with levels k_j, $j = 1, \ldots, r$, chosen in such a way that the central charges sum up to the desired value of the internal CFT,

$$c_{\text{int}} = 12 - \frac{3}{2}d \; ; \tag{7.12}$$

The value 12 appears instead of 15 because we work in the light-cone gauge, and we assume for later convenience that the number d of transverse flat directions is equal to 2 or to 6. The case $d = 4$ can be treated analogously, but some of the formulas below undergo minor modifications, mostly signs.

We would now like to write down *Gepner's* result for *partition functions* that satisfy all the requirements for string compactifications listed above and therefore describe superstring compactifications. We need to add external fermions and bosons for the d flat directions to the minimal models, and to employ an orbifold-like procedure which enforces spacetime supersymmetry and modular invariance [200, 228, 253]. But we need some further notation first. For a compactification involving r minimal models, we use

$$\boldsymbol{\lambda} := (l_1, \ldots, l_r) \quad \text{and} \quad \boldsymbol{\mu} := (s_0; m_1, \ldots, m_r; s_1, \ldots, s_r) \tag{7.13}$$

to label the tensor product of representations: l_j, m_j, s_j are taken from the range (7.7), and $s_0 = 0, 2, \pm 1$ characterises the irreducible representations of the $SO(d)_1$ current algebra that is generated by the d external fermions (the latter also contain an $N = 2$ algebra for each even d and, again, the NS sector has s_0 even).

Accordingly, we write

$$\chi^{\boldsymbol{\lambda}}_{\boldsymbol{\mu}}(q) := \chi_{s_0}(q) \, \chi^{l_1}_{m_1, s_1}(q) \cdots \chi^{l_r}_{m_r, s_r}(q) \tag{7.14}$$

for the conformal characters of these tensor products of internal minimal model and external fermion representations, with $\chi^l_{m,s}(q) = \text{Tr}_{\mathcal{H}^l_{m,s}} q^{L_0 - \frac{c}{24}}$, etc. We refer to [228] and [330] for the explicit expressions.

Let us introduce the special $(2r + 1)$-dimensional vectors $\boldsymbol{\beta}_0$ with all entries equal to 1, and $\boldsymbol{\beta}_j$, $j = 1, \ldots, r$, having zeroes everywhere except for the 1st and the $(r + 1 + j)^{\text{th}}$ entry which are equal to 2. Then we can consider the following products of $2\,\beta_0$ and β_j with a vector μ as above:

$$2\,\beta_0 \bullet \mu := -\frac{d}{2}\frac{s_0}{2} - \sum_{j=1}^{r} \frac{s_j}{2} + \sum_{j=1}^{r} \frac{m_j}{k_j + 2} \, , \qquad (7.15)$$

$$\beta_j \bullet \mu := -\frac{d}{2}\frac{s_0}{2} - \frac{s_j}{2} \, . \qquad (7.16)$$

It is easy to see that $q_{\text{tot}} := 2\,\beta_0 \bullet \mu$ is precisely the total U(1) charge of the highest weight state in $\chi_{\mu}^{\lambda}(q)$, and therefore projection onto states with odd $2\,\beta_0 \bullet \mu$ will implement the GSO projection: if the spectral flow operator $U_{\frac{1}{2}}$ of the total theory is to serve as a spacetime supersymmetry charge, it should be a semi-local worldsheet field, enforcing all U(1) charges q_{tot} to be integer – the OPE with $U_{\frac{1}{2}}$ has a $(z - w)^{\frac{q_{\text{tot}}}{2}}$ dependence; restricting further to odd total charge removes the vacuum, which would correspond to a tachyon in spacetime.

Similarly, projecting onto states with $\beta_i \bullet \mu$ integer ensures that only states in the tensor product of $r + 1$ NS sectors (or of $r + 1$ R sectors) are admitted (recall that we assumed $d = 2$ or $d = 6$).

Modular invariance of the partition function can be achieved if the above projections are accompanied by adding "twisted" sectors (in a way similar to orbifold constructions). To state Gepner's result, we put $K := \text{lcm}(4, 2k_j + 4)$ and let $b_0 \in \{0, 1, \dots, K - 1\}$, $b_j \in \{0, 1\}$ for $j = 1, \dots, r$. Then the partition function of a Gepner model describing a type IIB superstring compactification to $d + 2$ dimensions is given by

$$Z_G^{(r)}(\tau, \bar{\tau}) = \frac{1}{2} \frac{(\Im \tau)^{-\frac{d}{2}}}{|\eta(q)|^{2d}} \sum_{b_0, b_j} \sum_{\lambda, \mu}^{\beta} (-1)^{s_0} \, \chi_{\mu}^{\lambda}(q) \, \chi_{\mu + b_0 \beta_0 + b_1 \beta_1 + \dots + b_r \beta_r}^{\lambda}(\bar{q}) \quad (7.17)$$

where $\sum^{\beta}$ means that we sum only over those λ, μ in the range (7.7) which satisfy $2\,\beta_0 \bullet \mu \in 2\mathbb{Z} + 1$ and $\beta_j \bullet \mu \in \mathbb{Z}$. The summations over b_0, b_j introduce the twisted sectors corresponding to the β-restrictions. As a consequence, the Gepner partition function is non-diagonal. The sign is the usual one occurring in (spacetime) fermion one-loop diagrams. The τ-dependent factor in front of the sum accounts for the free bosons associated with the d transversal dimensions of flat external spacetime, while the $\frac{1}{2}$ is simply due to the field identification mentioned after equation (7.7). Gepner was able to prove that the partition function (7.17) is indeed modular invariant, using the following modular S-matrices for the SO(d)$_1$ and minimal model characters:

$$S_{s_0, s_0'}^{\text{f}} = \frac{1}{2} e^{-i\pi \frac{d}{2} \frac{s_0 s_0'}{2}} \, , \qquad (7.18)$$

$$S_{(l,m,s),(l',m',s')}^{k} = \frac{1}{\sqrt{2(k + 2)}} \sin \pi \frac{(l + 1)(l' + 1)}{k + 2} \, e^{i\pi \frac{mm'}{k+2}} e^{-i\pi \frac{ss'}{2}} \, . \qquad (7.19)$$

We should also remark that Gepner's partition function (7.17) can be understood as resulting from a simple current construction: the chiral algebra of the full model is extended by the fields corresponding to labels β_0 and β_j. The latter can be realised as $G^{(0)\pm} + G^{(j)\pm}$, the sum of the worldsheet supercurrents in the external and the jth internal tensor factor. The label β_0, on the other hand, is related to the spacetime supercharge; after adding superghosts, this can be realised, in the $\pm\frac{1}{2}$ picture, say, by the vertex operator $e^{\pm\frac{i}{2}\phi_{\text{gh}}}\, U_{\frac{1}{2}}^{\text{ext}}\, U_{\frac{1}{2}}^{\text{int}}\, S_\alpha$, where ϕ_{gh} bosonises the superghosts, S_α is the spacetime part of the spin field, and where the $U_{\frac{1}{2}}$ are the spectral flow operators of the external and the internal $N = 2$ SCFT – see, e.g., [83]. A more elegant formulation, avoiding the introduction of superghosts, can be given in terms of the so-called bosonic string map; see, e.g., [208].

These fields are simple currents with half-integer dimensions, and we can extend the symmetry algebra by them. The condition that $\beta_j \bullet \mu$ from equation (7.16) be integer is nothing but the familiar constraint from simple current orbifolds, namely projection onto orbits of integer monodromy charge with respect to the simple currents β_j. Likewise, projection onto states with $\beta_0 \bullet \mu \in \frac{1}{2} + \mathbb{Z}$ is the monodromy charge condition for the simple current labelled β_0 decorated with the superghost part, which accounts for an extra $2 \cdot \frac{1}{2}$ to be added to equation (7.15) – see [83].

The offset between the labels of left- and right-moving characters in equation (7.17) simply indicates that twisted sectors occur in the simple current construction; cf. equation (4.109) in Chapter 4. The multiplicities $|S_i|$ in that formula, orders of the stabilisers of certain simple current group orbits which are shorter than the generic ones, are hidden in the summations in equation (7.17). In Gepner models, these multiplicities are either 1 or 2, the latter associated to so-called "short orbits" built over representations (λ, μ) which are fixed under a certain element of the simple current group (due to divisibility properties and the field identification); these have fixed $l_i = \frac{k_i}{2}$ for all i such that $2k_i + 4$ does not divide $\text{lcm}(k_j + 2)$; all other l_i and all the m_i are arbitrary (see, e.g., [83]). In particular, there are no short orbits if all the k_i are odd. We will briefly come back to this later, when we note how the simple current point of view can be exploited to construct additional boundary states for Gepner models.

Variants of the Gepner models given above can be formulated starting from non-diagonal $\widehat{SU(2)}_k$ modular invariant partition functions. Although we will stick to the simpler case (7.17), the boundary state constructions in Section 7.2 can be extended to non-diagonal invariants, using the methods of [44, 384] – see [360] for some explicit results.

It should also be mentioned that other superstring compactifications built on $N = 2$ SCFTs have been studied in the literature, most notably Kazama-Suzuki models [310]. Boundary conditions for these theories were first analysed in [427].

7.1.2 Remarks on the target-space interpretation

There are two main reasons why Gepner models have generated quite some interest since the late 1980s. On the one hand, they provide string compactifications

to $D = 4$ (or 6 or 2) which are exactly solvable, and which display phenomenologically interesting spectra of light particles. The other reason is that these CFT
compactifications, originally formulated in purely algebraic terms, turned out to
be closely connected to σ-models on Calabi–Yau manifolds, a consequence being
that CFT and string theory triggered some fruitful new ideas in mathematics,
such as mirror symmetry.

In this subsection, we will mainly concentrate on the second aspect, concerning
the internal theory, as it is unique to this specific class of string compactifications,
but we will make some brief remarks on spacetime aspects of Gepner models first.

Closed string backgrounds based on Gepner's construction come in three ver-
sions, type IIA, type IIB, and heterotic strings. In a type IIB Gepner model, we
use the total currents

$$J_{\text{tot}} = J_{\text{ext}} + J_{\text{int}} \quad \text{and} \quad \overline{J}_{\text{tot}} = \overline{J}_{\text{ext}} + \overline{J}_{\text{int}} \tag{7.20}$$

to define the GSO projection, where the external current was defined in equation
(7.5). In a type IIA theory, we should instead start from

$$J_{\text{tot}} = J_{\text{ext}} + J_{\text{int}} \quad \text{and} \quad \overline{J}_{\text{tot}} = \overline{J}_{\text{ext}} - \overline{J}_{\text{int}} \; ; \tag{7.21}$$

see, e.g., [263] and references therein. We see that applying the mirror automorphism applied to the internal right-moving super Virasoro algebra mediates
between IIA and IIB theories. Both the external and internal U(1) currents can
be bosonised,

$$J_{\text{ext}} = i\,\partial\phi \; , \qquad J_{\text{int}} = i\sqrt{3}\,\partial H \; ,$$

where factors $\sqrt{\frac{c}{3}}$ were introduced for convenience (we are working in the light-
cone gauge and with $D = 4$ here, hence $c_{\text{ext}} = 3$). In terms of the auxiliary bosons
ϕ, H, the spectral flow operators are given as

$$U^{\text{ext}}_{-\frac{1}{2}} = e^{-\frac{i}{2}\phi} \; , \qquad U^{\text{int}}_{-\frac{1}{2}} = e^{-\frac{i\sqrt{3}}{2}H} \; , \tag{7.22}$$

and analogously for the adjoints $U_{\frac{1}{2}} = U^{*}_{-\frac{1}{2}}$. The external spectral flow operator
is just the SO(2) spin field mapping the NS to the R sector. From the total
spectral flow operator $U^{\text{tot}}_{\frac{1}{2}} := U^{\text{ext}}_{\frac{1}{2}} U^{\text{int}}_{\frac{1}{2}}$ we obtain to spacetime supercharges by

$$Q = \sqrt{p^+} \oint U^{\text{ext}}_{\frac{1}{2}} U^{\text{int}}_{\frac{1}{2}} \; , \qquad Q^* = \sqrt{p^+} \oint U^{\text{ext}}_{-\frac{1}{2}} U^{\text{int}}_{-\frac{1}{2}} \; , \tag{7.23}$$

$$S = \frac{1}{\sqrt{p^+}} \oint U^{\text{ext}}_{\frac{1}{2}} U^{\text{int}}_{\frac{1}{2}} \, \partial X \; , \qquad S^* = \frac{1}{\sqrt{p^+}} \oint U^{\text{ext}}_{-\frac{1}{2}} U^{\text{int}}_{-\frac{1}{2}} \, \partial X^* . \tag{7.24}$$

The first pair Q, Q^* are the linear, the second pair S, S^* the non-linear supercharges; see [263] and references therein. Note that the action of the total spectral
flow is very simple: it just shifts the index μ of the representations by the vector
$-\beta_0$.

As it stands, the partition function (7.17) belongs to a type IIB string compactification, because we have used the currents from equation (7.20) to define

the total U(1) charges. The different GSO projection for a type IIA Gepner model compactification requires that $\bar{q}_{\text{ext}} - \bar{q}_{\text{int}} \in 2\mathbb{Z} + 1$ and leads, at first sight, to a different partition function. However, it can be shown that the change amounts to replacing the right-moving characters $\chi_{\boldsymbol{\mu}'}^{\lambda}(\bar{q})$ in equation (7.17) by $\chi_{\natural\boldsymbol{\mu}'}^{\lambda}(\bar{q})$, where $\natural\boldsymbol{\mu} := (-s_0, m_1, \ldots, m_r, s_1, \ldots, s_r)$ for $\boldsymbol{\mu}$ as in equation (7.13) – i.e. the sign of the external charge has been reversed for the right-movers. The characters in question, however, happen to coincide with the ones in equation (7.17), because this sign reversal means passage to the conjugate sector. Thus, the same Gepner partition functions appear in IIA and IIB string theory – compare to a model of free bosons compactified at the self-dual radius, which is T-duality invariant. The transition from IIB to IIA does, however, exchange the chiral rings (c, c) and (c, a) and therefore alters the spacetime supersymmetry properties of the massless bosons in the theory (essentially, vector- and hyper-multiplets are interchanged); we refer to the literature, e.g. [263, 377], for further details.

Both type II theories (as well as the heterotic string) are tachyon-free, and their massless spectrum contains a graviton, a dilaton, and a Kalb–Ramond field along with superpartners (forming a model-independent hyper-multiplet). Furthermore, there are massless scalars in D dimensions built from CFT states of the form $\mathbf{1}_{\text{ext}} \otimes (h_{\text{int}} = \frac{1}{2})$; the number of these states depends on the concrete minimal models used in the internal theory. The distinction between the internal CFT's chiral rings (c, c) and (a, a) on the one hand and (c, a) and (a, c) on the other translates, at the level of the $N = 2$ low-energy effective theory, into vector-versus hyper-multiplets. The precise association depends on conventions, but vector- and hyper-multiplets get interchanged when switching from type IIA to type IIB. It should be noted that the "moduli" provided by the chiral rings of the internal CFT lead to parameters of the compactification, i.e. they actually reduce the predictive power of string theory. As for any full string theory, Gepner models have infinite towers of massive modes; the lowest-lying of these comes at a mass gap $\alpha' \Delta m^2 \sim (k_{\max} + 2)^{-1}$, as follows from inserting the formula for the internal conformal dimensions (7.8) into the physical state condition.

Type II superstring theories are the interesting ones when considering D-branes, but heterotic string theories provide the phenomenologically most relevant closed string backgrounds due to their inbuilt gauge symmetries. In a four-dimensional heterotic string compactification based on Gepner models, the $(c_L^{\text{int}}, c_R^{\text{int}}) = (9, 9)$ internal CFT is supplemented by four free bosons $X_\mu(z, \bar{z})$, $\mu = 0, 1, 2, 3$, along with four right-moving fermions $\bar{\psi}_\mu(\bar{z})$ and 26 free left-moving fermions $\chi^A(z)$, such that the total central charges are $(c_L, c_R) = (26, 15)$. The χ^A can be used to form an affine Lie algebra SO(26) or $E_6 \times E_8$ at level 1; see, e.g., [229]. The affine characters are such that modular invariance of the heterotic partition function is ensured, and moreover the fermions χ^A allow a GSO projection to be defined for the left-moving sector; in spite of the striking asymmetry between left- and right-movers, we obtain a consistent string theory with $N = 1$ spacetime supersymmetry, and the affine Lie algebras induce gauge symmetries for the massless fields of the effective theory. The E_6 variant in particular is interesting for particle physics phenomenology, since E_6 is a natural grand unification gauge group. It turns out the massless fields arising from the chiral rings (c, c) and (c, a) of the internal CFT transform in the 27 and $\overline{27}$ irrep of E_6, and that the number of lepton–quark generations in the low-energy theory is related to the difference of the dimensions of the (c, c) and (c, a) rings. In case the Gepner model can be related to a σ-model on a Calabi–Yau manifold M, this difference is proportional to the Euler number of M. We refer to [377] and the literature mentioned therein for more details on the spacetime properties of heterotic and type II Gepner models.

Let us now focus on the "internal part" of Gepner models, namely the tensor product of $N = 2$ super-minimal models. The most apparent feature of this type of superstring compactifications is their non-geometric nature: they are formulated in terms of rational CFTs, thus trading classical geometric intuition for computability in the string length regime. However, it has turned out that most Gepner models are intimately related to non-linear sigma models with Calabi–Yau targets.

Some Gepner models can be interpreted as a torus orbifold, and a direct translation between target geometry and CFT language is available simply because the sigma-model formulation is exact to all orders in α'. For Gepner models with $c = 6$, very explicit, and complete, relations to geometric data of K3 manifolds can be established; see [359, 445]. In the remainder of this subsection, we will outline the main steps of a correspondence that is less explicit, but applies to arbitrary Gepner models.

The correspondence entails two independent steps, the (historically) first one being to realise that $N = 2$ SCFTs arise as infrared (IR) fixed points of $N = 2$ supersymmetric Landau–Ginzburg models. For the time being, we write the classical Landau–Ginzburg action in a concise $N = 2$ covariant notation, using anticommuting supercoordinates θ, $\bar\theta$ associated with the worldsheet coordinates as well as chiral superfields $\Phi = (\Phi_1, \ldots, \Phi_n)$; the explicit expression in component fields is given in Section 7.4. The action reads

$$S_{\mathrm{LG}} = \int d^2x d^2\theta d^2\bar\theta \, K(\Phi, \overline{\Phi}) + \left(\int d^2x d^2\theta \, W(\Phi) + \text{compl. conj.} \right) \quad (7.25)$$

where K denotes the kinetic term and the potential W is a holomorphic function of the Φ_i only. Usually, it is assumed that W has no quadratic terms so that the Φ_i are massless. Moreover, W is required to be a quasi-homogeneous polynomial

$$W(\lambda^{q_1}\Phi_1, \ldots, \lambda^{q_n}\Phi_n) = \lambda^2 \, W(\Phi_1, \ldots, \Phi_n) \quad (7.26)$$

for all $\lambda \in \mathbb{C}^\times := \mathbb{C}\backslash\{0\}$ with some real q_i. This condition is necessary for the worldsheet R-symmetries to be preserved at the renormalisation group (RG) fixed point, where they are realised by the worldsheet currents J and $\bar J$ of the $N = 2$ super Virasoro algebra; the q_i become the familiar U(1) charges. A careful study of the RG flow of such Landau–Ginzburg models [288] – invoking non-renormalisation theorems for $N = 2$ theories on the way – shows that the kinetic term K in the action (7.25) is irrelevant and that the universality class of the fixed point, i.e. the resulting CFT, is determined solely by the Landau–Ginzburg potential W.

In particular, the Landau–Ginzburg model with a single superfield Φ and potential $W = \Phi^{k+2}$ flows to the kth minimal model of the super Virasoro algebra with central charge $c_k = \frac{3k}{k+2}$, and with a diagonal modular invariant partition function. The chiral primaries $(l, l, 0)$, $l = 0, \ldots, k$, of the CFT arise from the chiral fields Φ^l in the Landau–Ginzburg model. More generally, the (c, c) chiral ring can be described as the Jacobi ring $\mathbb{C}[x_1, \ldots, x_n]/(\partial_i W)$ of the Landau–Ginzburg potential. To describe tensor products of minimal models as they appear in Gepner's construction, we use sums of Landau–Ginzburg Lagrangians with the appropriate potentials.

In order to arrive at a geometric interpretation of Gepner models, it remains for us to link a certain class of Landau–Ginzburg models to non-linear sigma models with a certain class of target manifolds. The low-energy effective field theory of a sigma-model string compactification is (super)gravity on the target manifold, it can be shown that spacetime supersymmetry emerges only if the sigma-model target is a Calabi–Yau manifold, i.e. a complex Kähler manifold with vanishing first Chern class. The latter condition is equivalent to having $SU(d)$ holonomy (where d is the complex dimension of the manifold), which translates into the existence of a unique nowhere-vanishing holomorphic d-form, or into the existence of a unique covariantly constant spinor as required by unbroken spacetime supersymmetry. (See, e.g., [252, 377] for good reviews.) The Calabi–Yau condition is also equivalent to having a Kähler manifold with a Ricci-flat metric – and the latter condition arises from the vanishing of the beta function β^g; see, e.g., equation (2.14) in Chapter 2, to first order in the α'-expansion of the sigma-model perturbation theory. The papers [256] contain results on non-trivial higher-order corrections to Ricci flatness.

From the basic bosonic and fermionic fields of a Calabi–Yau sigma model and the manifold's basic geometric data, (left-moving) $N = 2$ super Virasoro generators and the spectral flow operator can be formed via

$$J \sim g_{i\bar{j}} \psi^i \psi^{\bar{j}}, \quad G^+ \sim g_{i\bar{j}} \psi^i \partial X^{\bar{j}}, \quad G^- \sim g_{i\bar{j}} \psi^{\bar{j}} \partial X^i, \quad U_{\frac{1}{2}} \sim \Omega_{i_1, \ldots, i_d} \psi^{i_1} \cdots \psi^{i_d}$$
(7.27)

where we have used the Kähler metric $g_{i\bar{j}}$ and the holomorphic d-form Ω; see, e.g., [253, 364]. The right-moving generators involve $\bar{\partial} X^j$ and the right-moving counterparts λ^j of the left-moving fermions ψ^j.

The chiral and anti-chiral primaries of this $N = 2$ super Virasoro algebra are related to the Dolbeault cohomology of the target manifold. We only sketch the main arguments here; for more details see [253, 282] and the literature quoted there. We start with R–R ground states $|\xi\rangle^{RR}_{\text{g.st.}}$, which are in one-to-one relation to (c,c) and (c,a) fields via spectral flow by $\pm\frac{1}{2}$. By definition, they satisfy $G^{\pm}_0 |\xi\rangle^{RR}_{\text{g.st.}} = \overline{G}^{\pm}_0 |\xi\rangle^{RR}_{\text{g.st.}} = 0$, which in particular implies $(L_0 - \overline{L}_0)|\xi\rangle^{RR}_{\text{g.st.}} = 0$: they have zero momentum, thus can be viewed as independent of the (worldsheet) space coordinates and treated in quantum mechanics of the zero modes. This in turn makes it easy to connect $G^{\pm}_0$ to familiar differential operators on the Calabi–Yau manifold: in quantum mechanics, the conjugate momentum $\dot{X}^\mu$ appearing in $G^{\pm}_0$ is nothing but the (covariant) derivative D^μ on the (curved) manifold, while the fermionic superpartner ψ^μ can be viewed as basic one-form dX^μ – so the $G^{\pm}_0$ become exterior derivatives. More precisely, after re-introducing the holomorphic and anti-holomorphic indices with the help of the Kähler metric $g_{i\bar{j}}$, we find that in the zero-mode sector $G^{\pm}_0$ correspond to the manifold's Dolbeault operator $\overline{\partial}$ and to the adjoint $\overline{\partial}^\dagger$, respectively. R–R ground states are annihilated by both and are therefore harmonic forms or, by Hodge's theorem, representatives of the Dolbeault cohomology $H^{\bullet,\bullet}_{\overline{\partial}}$ of the target. From the CFT perspective, the bi-grading arises from the left- and right-moving U(1) charges of the R–R ground states, and of the associated (c,c) and (c,a) primaries. The most important (anti-)chiral primaries are those with $h_L = h_R = \frac{1}{2}$, which correspond to moduli of the string compactification that appear as massless scalars in the external uncompactified directions of the target. Using the route sketched above, they turn out to be related to geometric moduli of the (complex three-dimensional) Calabi–Yau target: the (c,a) fields to elements of $H^{1,1}_{\overline{\partial}}$, also called Kähler moduli (size parameters) which can be used to deform the metric, and the (c,c) fields to elements of $H^{2,1}_{\overline{\partial}}$ which provide deformations of the complex structure.

Geometrically, the role of Kähler and complex structure moduli is very different, but at the level of the $N = 2$ SCFT they differ merely by a sign choice of the right-moving U(1) current, or by applying the mirror automorphism to the right-moving super Virasoro algebra. This fact is the origin for the *mirror symmetry* conjecture, namely that Calabi–Yau three-folds come in pairs M, $\hat{M}$ with $H^{2,1}_{\overline{\partial}}(X) \cong H^{1,1}_{\overline{\partial}}(\hat{X})$ and $H^{1,1}_{\overline{\partial}}(X) = H^{2,1}_{\overline{\partial}}(\hat{X})$.

Mirror symmetry can be seen as a generalisation of T-duality relating circle targets of inverse radii; for Calabi–Yau manifolds, however, even the topology of the dual spaces can be different. Greene and Plesser [254] suggested constructing the mirror target via an orbifold construction: the Gepner model built on minimal models $(k_1, \ldots, k_r)$ admits a group of discrete symmetries

$$ H := \left\{ \vec{n} \in \prod_{j=1}^{r} \mathbb{Z}_{k_j+2} \,\Big|\, \sum_j \frac{n_j}{k_j + 2} \in \mathbb{Z} \right\}, \tag{7.28} $$

acting on primary fields $\phi^\lambda_{\mu,\bar{\mu}}$ by multiplication with charge-dependent phases

$$ \vec{n} \cdot \phi^\lambda_{\mu,\bar{\mu}} = \exp\left\{ i\pi \sum_j \frac{n_j(m_j + \bar{m}_j)}{k_j + 2} \right\} \phi^\lambda_{\mu,\bar{\mu}} . $$

As shown in [253, 254], taking the orbifold by the group H leads to a CFT isomorphic to the starting Gepner model, with opposite right-moving U(1) charges (i.e. with $\bar{\mu} \mapsto -\bar{\mu}$). More interestingly, the construction relates whole moduli spaces of CFTs (with the role of (c, c) and (c, a) moduli interchanged), a fact which was exploited by Candelas *et al.* by writing down a "mirror map" relating complex structure moduli of X to Kähler moduli of $\hat{X}$ explicitly. The result alluded to in Subsection 5.A.1 that certain three-point functions do not receive any α' or worldsheet instanton corrections [134] plays a vital role here. The mirror map allowed numbers of rational curves in Calabi–Yau manifolds to be computed in an extremely efficient manner using string theory [93–95] – an achievement which triggered a lot of interest among mathematicians.

The Calabi–Yau manifolds that are best understood are given as varieties in weighted projective space

$$\mathbb{CP}^{n-1}[\vec{w}] := \mathbb{C}^n \setminus \{0\} / \sim \quad \text{with } (z_1, \ldots, z_n) \sim (\lambda^{w_1} z_1, \ldots, \lambda^{w_n} z_n)$$
$$\text{for all } \lambda \in \mathbb{C}^\times ; \tag{7.29}$$

where $\vec{w} = (w_1, \ldots, w_n) \in \mathbb{Q}^n$. A variety in $\mathbb{CP}^{n-1}[\vec{w}]$ is defined as the vanishing locus $\{p = 0\}$ of a polynomial which is quasi-homogeneous of some degree d, i.e. satisfies $p(\lambda^{w_1} z_1, \ldots, \lambda^{w_n} z_n) = \lambda^d p(z_1, \ldots, z_n)$ for all $\lambda \in \mathbb{C}^\times$. Certain topological data (in particular the Chern classes) of such a variety can be computed elegantly using tools from algebraic geometry, which allows one to decide whether it admits a Calabi–Yau metric (in principle; explicit Ricci-flat metrics are only known for a very few cases). The review [286] provides a good starting point for exploring these topics, but see also [282, 290, 459].

This outlines some of the general structure of Calabi–Yau sigma models and how they relate to those found in $N = 2$ SCFTs, but we still need to link specific Calabi–Yau manifolds to specific Landau–Ginzburg models. The two most prominent approaches to this problem proposed in the literature involve manipulations of path integrals for, respectively, Landau–Ginzburg models [255] and supersymmetric gauge theories interpolating between a Landau–Ginzburg and a Calabi–Yau phase [452]. We sketch both ideas in a specific example (the quintic), starting with the path integral formulation.

Consider the Landau–Ginzburg model with potential $W = \Phi_1^5 + \cdots + \Phi_5^5$. Introducing new variables $\xi_1 := \Phi_1^5$ and $\xi_k := \Phi_k/\Phi_1$ for $k = 2, 3, 4, 5$, the Jacobian is a constant, and thus performing this change of variables in the path integral appears trouble-free. Since the kinetic term in supersymmetric Landau–Ginzburg models is an irrelevant perturbation, it is ignored in the discussion. The potential becomes $W = \xi_1 (1 + \xi_2^5 + \cdots + \xi_5^5)$, and upon integrating out ξ_1, a delta constraint setting $1 + \xi_2^5 + \cdots + \xi_5^5 = 0$ arises. This is nothing but the defining equation for the quintic Calabi–Yau manifold,

$$M_{\text{quintic}} := \{ (x_1, \ldots, x_5) \in \mathbb{CP}^4 \mid x_1^5 + \cdots + x_5^5 = 0 \},$$

but written in inhomogeneous coordinates from $\mathbb{C}^5$, in the patch where $x_1 \neq 0$. The other patches needed to cover $\mathbb{CP}^4$ can be reached by integrating out other Φ_k^5 instead of Φ_1^5.

As explained in [255], this simple argument can be generalised to many other cases: if a Landau–Ginzburg model has a potential

$$W = \Phi_1^{l_1} + \cdots + \Phi_r^{l_r}$$

such that the exponents $l_i \in \mathbb{Z}_+$ satisfy

$$\sum_{i=1}^{r} \frac{1}{l_i} = 1 \qquad (7.30)$$

then the change of variables $\xi_1 := \Phi^{l_1}$, $\xi_k^{l_k} := \Phi^{l_k}/\xi_1$ for $k = 2, \ldots, r$ has a constant Jacobian, and integrating out ξ_1 in the path integral leads to the constraint $1 + \xi_2^{l_2} + \cdots + \xi_r^{l_r} = 0$, which can be read as (inhomogeneous coordinate version of) the defining equation of the hypersurface

$$M := \{(x_1, \ldots, x_r) \in W\mathbb{CP}^{r-1}[(\tfrac{l}{l_1}, \ldots, \tfrac{l}{l_r})] \mid x_1^{l_1} + \cdots + x_r^{l_r} = 0\},$$

in weighted projective space where $l = \mathrm{lcm}(l_i)$.

Using tools from algebraic geometry, it can be shown that condition (7.30) translates into $c_1(M) = 0$, i.e. the Calabi–Yau condition for the $(r-2)$-dimensional complex manifold M. At the same time, equation (7.30) means that this Landau–Ginzburg model flows to a tensor product of super Virasoro minimal models with total central charge

$$c = \sum_i \frac{3(l_i - 2)}{l_i} = 3(r - 2).$$

In general, the coordinate transformation $\Phi_i \mapsto \xi_i$ is single-valued only if the orbifold identification

$$(\Phi_1, \ldots, \Phi_r) \sim (e^{2\pi i/l_1} \Phi_1, \ldots, e^{2\pi i/l_r} \Phi_r)$$

is introduced by hand into the Landau–Ginzburg models (and likewise into the SCFT at the IR fixed point), so the path integral argument in general links sigma models on Calabi–Yau hypersurfaces to orbifolds of Landau–Ginzburg models (with the same orbifolding that is employed in Gepner's construction to ensure integral total charges). For ways to relax the condition (7.30), see [255].

The "mirror construction" via orbifolding by the group (7.28) can of course be carried over to the geometric domain of hypersurfaces in weighted projective space. However, in general the quotient M/H may have singularities even if M is a smooth Calabi–Yau manifold.

The above way of connecting Landau–Ginzburg models and Calabi–Yau targets is rather intuitive, but fraught with the usual problems of many path integral arguments. An alternative, and very influential, chain of reasoning was presented in [452] by Witten, who starts with a so-called linear sigma model in two dimensions, that is a $N = (2,2)$ supersymmetric gauge theory with abelian gauge group $U(1)_1 \times \cdots \times U(1)_{N_g}$ and chiral superfields $\Phi_1, \ldots, \Phi_{N_f}$; with respect to $U(1)_a$, the field Φ_i has charge $Q_{i,a}$. We assume that the charges satisfy

$$\sum_{i=1}^{N_f} Q_{i,a} = 0 \quad \text{for all } a = 1, \ldots, N_g, \tag{7.31}$$

to avoid anomalies in the R-symmetries.

In addition to the usual (supersymmetrised) kinetic and minimal coupling terms, the linear sigma-model Lagrangian contains a polynomial potential $W(\Phi)$ as familiar from Landau–Ginzburg models (as before, W should be quasi-homogeneous), as well as Fayet–Iliopoulos (FI) terms $\sum_{a=1}^{N_g} r^a D_a$, where D_a is the top component of the ath vector superfield and the r^a are parameters. (Actually, the full FI terms also involve a theta angle, but we will not need this here; see [452] for a very detailed discussion of the full classical action.) The D_a are auxiliary fields, i.e. their equations of motion are algebraic, and have expressions in terms of lowest components ϕ_i of the chiral superfields Φ_i.

Witten then argues that in different regimes of the parameters r^a, a non-linear sigma model with Calabi–Yau target or a Landau-Ginzburg model (both determined by W as before) can arise as "phases" of the linear sigma model. To see this, we first focus on the classical vacuum configurations of the theory, here meaning the zero set of the potential energy, which is given by

$$U = \frac{e^2}{2} \sum_a \left(r^a - \sum_i Q_{i,a} |\phi_i|^2 \right)^2 + \sum_i \left| \frac{\partial W}{\partial \phi_i} \right|^2 + \cdots.$$

Here, e is the coupling constant of the gauge group, and we have neglected terms which have to be set to zero separately in order to obtain $U = 0$ but which do not yield additional equations on the ϕ_i. The first contribution to U is often called the D-term.

Which field configurations ϕ_i minimise U depends crucially on the values of the r^a. For simplicity we restrict ourselves to the model leading to the quintic, where $N_g = 1$ and $N_f = 6$, and where the potential is $W = \Phi_6 (\Phi_1^5 + \cdots + \Phi_5^5)$. The fields $\Phi_1, \ldots, \Phi_5$ have charge $Q_1 = \ldots = Q_5 = 1$ while $Q_6 = -5$.

If $r > 0$, vanishing of the D-term implies that $\phi_6 = 0$ and $\phi_1^5 + \cdots + \phi_5^5 = 0$ under the constraint $\sum_{i=1}^5 |\phi_i|^2 = r$; due to the $U(1)$ gauge action we can identify configurations where all ϕ_i are multiplied by the same phase, and altogether the vacuum configuration for $r > 0$ can be identified with the quintic in complex projective space: in this phase, a Calabi–Yau non-linear sigma model arises from the linear sigma model.

If $r < 0$, we find that $U = 0$ implies that $\phi_1 = \cdots = \phi_5 = 0$ while the value of ϕ_6 is fixed; the vacuum manifold is the same as that of the Landau–Ginzburg model with potential $\Phi_1^5 + \cdots + \Phi_5^5$ – or rather of an orbifolded Landau–Ginzburg model, because part of the U(1) gauge invariance of the full linear sigma model survives in the form of a residual $\mathbb{Z}_5$-symmetry in the Landau–Ginzburg phase.

Witten provides strong arguments (which of course exploit the supersymmetry of the theory) that the classical vacuum configurations $\{U = 0\}$ do not receive quantum corrections, and he concludes that a Calabi–Yau sigma model and a Landau–Ginzburg orbifold appear as phases of one and the same quantum field theory (QFT). Certain quantities, such as the (Witten) index $\mathrm{tr}_{\mathcal{H}_R}(-1)^F$ are independent of Kähler moduli, allowing, e.g., the chiral rings of the two phases to be matched. In [452], Witten also discusses cases with several parameters r^a – leading to a much more complicated phase structure – as well as other generalisations. For these aspects and much more detailed explanations we refer to [253, 282, 452] and also to [273] for the open string case.

Having established a connection between Gepner models and (mirror pairs of) Calabi–Yau manifolds, let us briefly look back at the effective spacetime physics of, say, heterotic strings based on Gepner's construction. Interesting interaction terms in the low-energy effective action (LEEA) are given by the Yukawa couplings between three 27 or three $\overline{27}$ massless fields, which correspond to the (c, c) and (c, a) rings, or the holomorphic forms $H^{1,1}$ and $H^{2,1}$. It was the study of these Yukawa terms that led to surprising mathematical results [93]: supersymmetry arguments show that one class of couplings may receive instanton corrections, while the other does not; mirror symmetry interchanges the two, and in effect string theory has allowed for a very simple counting of rational curves embedded in Calabi–Yau manifolds; again, [281, 286, 290, 377, 459] give good introductions. We will return briefly to some ingredients of this (both conceptually and technically) very involved procedure in Section 7.3; for the moment it may suffice to highlight that it was a computation natural from the phenomenological point of view which led to one of the most important inputs into pure mathematics that string theory has produced so far.

7.2 Boundary states for Gepner models

Having discussed aspects of Gepner models in the bulk, we now turn towards branes for these superstring compactifications. The construction of boundary states will not rely on any geometric interpretation, but proceed in purely algebraic terms. The relation to geometric objects in Calabi–Yau manifolds will be discussed in Section 7.3.

We will restrict ourselves to constructing branes for type II string compactifications based on Gepner models and not look at unoriented worldsheets here. Gepner models in type I string models, or "open descendants" of Gepner models, have been treated in the literature; see, e.g., [20, 65, 79, 80, 244]. In these string theories, tadpole cancellation (on oriented and non-orientable worldsheets) provides the essential additional constraint and allows the determination of the Chan–Paton gauge groups that occur in such models.

Before treating Gepner models in full generality in Subsection 7.2.2, we discuss a simplified version of a $c = 3$ model, closely following [389]. On the one hand, this case displays the salient features of constructing boundary states for non-diagonal partition functions, with much lighter notation than for general Gepner models. On the other hand, the $(k = 1)^3$ Gepner model can be understood as a σ-model on a $\mathbb{Z}_3$-orbifold of a two-torus, thus it has a free field description. This fact was exploited in the treatment in [364], where the authors used the Ishibashi states of free bosons or fermions and obtained boundary states (for this specific model) which have a direct geometric interpretation. We will instead require our boundary states to preserve maximal symmetry – a condition that can be imposed for any Gepner model, whether it has a free field realisation or not. We will then show that, in this $c = 3$ example, the free field picture can be recovered from the abstract rational boundary states – which suggests that they are interesting objects to study for Gepner models in general.

7.2.1 A simplified example

We want to study the tensor product of three $k = 1$ minimal models, each restricted to the NS sector, then orbifolded to integer U(1) charge (instead of a full GSO projection to odd q_{tot}).

The irreducible representations of the $N = 2$ superconformal algebra in the NS sector of the $c = 1$ minimal model are built on the highest weight vectors

$$|(h, q)\rangle = |(0, 0)\rangle =: |0\rangle, \quad |(\tfrac{1}{6}, \tfrac{1}{3})\rangle =: |1\rangle, \quad |(\tfrac{1}{6}, -\tfrac{1}{3})\rangle =: |2\rangle .$$

Note that $|0\rangle$ and $|1\rangle$ are chiral, $|0\rangle$ and $|2\rangle$ anti-chiral states. The $N = 2$ characters in the NS sector close under the modular transformation $S : \tau \longmapsto -1/\tau$, and the elements S_{ab} in $\chi_a(\tilde{q}) = \sum_b S_{ab} \chi_b(q)$ are given by

$$S_{ab} = \frac{1}{\sqrt{3}} \, \omega^{-ab} , \tag{7.32}$$

with $\omega = e^{\frac{2\pi i}{3}}$ and $a, b = 0, 1, 2$ – as follows from equation (7.19), or more directly by diagonalising the (simple current) fusion rules of the $k = 1$ model.

We take the tensor product of three $k = 1$ minimal models (each with a diagonal partition function) and orbifold this CFT with respect to the group generated by

$$g = \exp\{2\pi i (J_0^{\text{tot}} + \overline{J}_0^{\text{tot}})\} ,$$

where $J_0 = J_0^{(1)} + J_0^{(2)} + J_0^{(3)}$ is the total left-moving charge of the $(k = 1)^3$ theory. In accordance with the general formulas given in Subsection 4.A.2, the resulting partition function is [130, 228, 254]

$$Z(q, \bar{q}) = \sum_{x=0,1,2} \sum_{\substack{\mathbf{a} \\ \Sigma a_i \equiv 0 (\text{mod } 3)}} \chi_{\mathbf{a}}(q) \, \chi_{\mathbf{a}+(x,x,x)}(\bar{q}) . \tag{7.33}$$

We have set $\chi_{\mathbf{a}} := \chi_{a_1} \chi_{a_2} \chi_{a_3}$. The (non-diagonal) terms with $x \neq 0$ are the twisted sectors, and the restriction on the $\mathbf{a} = (a_1, a_2, a_3)$ summation ensures that only states with integer charge occur in the orbifolded theory.

The maximal chiral symmetry algebra of the orbifold theory contains the diagonal ("total") $N = 2$ super Virasoro algebra with standard generators $W^{\mathrm{tot}} = \sum_i W^{(i)}$ and $\overline{W}^{\mathrm{tot}} = \sum_i \overline{W}^{(i)}$, but the bigger algebra $\mathcal{W}^{\otimes}$ generated by the three super Virasoro algebras of the individual tensor factors is also present (the g-action commutes with all these generators). While this $c = 3$ orbifold theory is rational with respect to $\mathcal{W}^{\otimes}$, it is already non-rational with respect to the diagonal super Virasoro algebra.

Therefore, if we want to construct boundary states $\|\alpha\rangle\rangle$ of the orbifolded $(k = 1)^3$ model, we can require that not only the diagonal $N = 2$ super Virasoro algebra is preserved – in the form of A-type or B-type boundary conditions, see equations (4.14, 4.15) – but that the system (bulk CFT + boundary state) in fact enjoys the full $\mathcal{W}^{\otimes}$-symmetry.

We start with boundary states satisfying A-type conditions (4.14) in each tensor factor. Out of the 27 sectors of the orbifold theory (7.33) only the nine untwisted ones ($x = 0$) can contribute because of the charge constraints $q^{(i)} = \overline{q}^{(i)}$, which follow from the zero-mode conditions $(J_0^{(i)} - \overline{J}_0^{(i)})\|\alpha\rangle\rangle_A = 0$ for $i = 1, 2, 3$.

Thus, we make the following ansatz for an A-type $\mathcal{W}^{\otimes}$-preserving boundary state of the orbifolded $(k = 1)^3$ model in the NS sector:

$$\|\alpha\rangle\rangle_A = \frac{1}{3^{\frac{1}{4}}} \sum_{\substack{\mathbf{a} \\ \Sigma a_i \equiv 0 \,(\mathrm{mod}\,3)}} \omega^{-a_1\alpha_1 - a_2\alpha_2 - a_3\alpha_3} |\mathbf{a}\rangle\rangle_A , \qquad (7.34)$$

where $\|\alpha\rangle\rangle_A = \|(\alpha_1, \alpha_2, \alpha_3)\rangle\rangle_A$ is labelled by three integers defined modulo 3. We can compare this to equation (4.100) and recognise the boundary states (7.34) as regular orbifold branes, but let us nevertheless check Cardy's conditions explicitly. Leaving out the CPT operator Θ from the cylinder amplitude for simplicity, we can compute

$$
\begin{aligned}
Z_{\alpha\tilde{\alpha}}^A(q) &\equiv {}_A\langle\langle\tilde{\alpha}\| \, \tilde{q}^{L_0^{\mathrm{tot}} - \frac{c}{24}} \, \|\alpha\rangle\rangle_A \\
&= \frac{1}{\sqrt{3}} \sum_{\substack{\mathbf{a} \\ \Sigma a_i \equiv 0 (\mathrm{mod}\,3)}} \omega^{-\mathbf{a}(\alpha - \tilde{\alpha})} \chi_{\mathbf{a}}(\tilde{q}) = \frac{1}{9} \sum_{\mathbf{c}} \sum_{\substack{\mathbf{a} \\ \Sigma a_i \equiv 0 (\mathrm{mod}\,3)}} \omega^{-\mathbf{a}(\mathbf{c} + \alpha - \tilde{\alpha})} \chi_{\mathbf{c}}(q) \\
&= \sum_{\mathbf{c}} \delta_{c_1 - c_3, \alpha_3 - \alpha_1 - \tilde{\alpha}_3 + \tilde{\alpha}_1}^{(3)} \, \delta_{c_2 - c_3, \alpha_3 - \alpha_2 - \tilde{\alpha}_3 + \tilde{\alpha}_2}^{(3)} \, \chi_{\mathbf{c}}(q) ; \qquad (7.35)
\end{aligned}
$$

in the second line, we have used

$$
{}_A\langle\langle \mathbf{a}' | \, \tilde{q}^{L_0^{\mathrm{tot}} - \frac{c}{24}} \, |\mathbf{a}\rangle\rangle_A = \delta_{\mathbf{a}', \mathbf{a}} \, \chi_{\mathbf{a}}(\tilde{q})
$$

and the modular S-matrix (7.32); finally, we have carried out the independent summations over a_1 and a_2, the third roots of unity resulting in the periodic Kronecker symbols $\delta^{(3)}$.

The ansatz (7.34) leads to a partition function which is the sum of $\mathcal{W}^\otimes$-characters with positive integer coefficients: $\mathcal{W}^\otimes$-symmetry in the open string system $Z^A_{\alpha\tilde\alpha}(q)$ is preserved, and Cardy's constraints are satisfied. But the resulting boundary CFT contains unwanted states with non-integer U(1) charge unless we require

$$\alpha_1 + \alpha_2 + \alpha_3 = \tilde\alpha_1 + \tilde\alpha_2 + \tilde\alpha_3 \,. \tag{7.36}$$

Similar compatibility conditions between two boundary states will arise later in full-fledged Gepner models; they are related to spacetime supersymmetry, while on the worldsheet they essentially arise because the terms of the partition function coupling to A-type Ishibashi states do not close under the modular S-transformation.

The partition function for $\alpha = \tilde\alpha$,

$$Z^A_{\alpha\alpha}(q) = \chi_0(q)^3 + \chi_1(q)^3 + \chi_2(q)^3 \,, \tag{7.37}$$

has contributions from highest weight states with total dimensions and charges $(0,0)$, $(\frac{1}{2},1)$, $(\frac{1}{2},-1)$. In fact, $Z^A_{\alpha\alpha}(q)$ is the vacuum character of a complex free fermion. Furthermore, the spectrum in any configuration $(\alpha\,\tilde\alpha)$ of two compatible "branes" carries a representation of this extended symmetry algebra.

This is easy to understand since the full chiral symmetry algebra of the $(k=1)^3$ model in fact contains a complex free fermion $\psi(z)$: we can, for example, identify highest weight states as $|1,1,1\rangle = \sqrt{i}\,\psi_{-\frac{1}{2}}|\text{vac}\rangle$, $|2,2,2\rangle = \sqrt{i}\,\psi^*_{-\frac{1}{2}}|\text{vac}\rangle$. With the help of the $(k=1)^3$ fusion rules it can be shown that the boundary states (7.34) satisfy

$$\psi_r\|\alpha\rangle\!\rangle_A = -i\,\omega^n\,\overline{\psi}^{\,*}_{-r}\|\alpha\rangle\!\rangle_A \tag{7.38}$$

with $n = \sum \alpha_i$. This means that the fermionic symmetry is indeed preserved by these boundary states.

This remark allows us to make contact with the results of Ooguri *et al.* [364], where the (c,c) parts of the boundary states for the $(k=1)^3$ orbifold were given in terms of free fermion and free boson coherent states. Denote by $\psi, \partial X$ the left-moving complex fermion and boson, respectively, with adjoints $\psi^*, \partial X^*$ and modes ψ_r, a_m, etc.; analogously for the right-movers. Then the coherent A-type boundary states of [364] read

$$\| B_n \rangle\!\rangle = \| B_n \rangle\!\rangle_X \, \| B_n \rangle\!\rangle_\psi \tag{7.39}$$

with

$$\| B_n \rangle\!\rangle_\psi = \exp\left\{ i\,\omega^n \sum_{r>0} \psi_{-r}\overline\psi_{-r} + i\,\omega^{-n} \sum_{r>0} \psi^*_{-r}\overline\psi^{\,*}_{-r} \right\} |\text{vac}\rangle \,, \tag{7.40}$$

$$\| B_n \rangle\!\rangle_X = \exp\left\{ -\omega^n \sum_{m>0} \frac{1}{m} a_{-m}\overline a_{-m} - \omega^{-n} \sum_{m>0} \frac{1}{m} a^*_{-m}\overline a^{\,*}_{-m} \right\} |\text{vac}\rangle \,. \tag{7.41}$$

They are obtained, in a slight generalisation of the procedure reviewed in Section 1.2, as solutions of the Ishibashi conditions

$$\left(\psi_r + i\, \omega^n\, \overline{\psi}^*_{-r} \right) \| B_n \rangle\!\rangle_\psi = 0 \,, \tag{7.42}$$

$$\left(a_m + \omega^n \bar{a}^*_{-m} \right) \| B_n \rangle\!\rangle_X = 0 \,, \tag{7.43}$$

for $n = 0, 1, 2$. The conditions for $\| B_n \rangle\!\rangle_X$ are just the superpartners of the ones for $\| B_n \rangle\!\rangle_\psi$. In fact, the bosonic part of the boundary state is hidden in the $N = 2$ families of the fermions due to $\partial X \sim G^-_{-\frac{1}{2}} \psi$.

The useful feature of this representation is that equation (7.43) has a classical geometric interpretation as Dirichlet and Neumann conditions on the torus target: the case $n = 0$ describes a one-brane extending in the direction of real X, whereas in the $n = 1, 2$ cases this one-brane is rotated by $\frac{2n\pi}{3}$ (corresponding to the $\mathbb{Z}_3$-symmetry of the torus); see [364] and also Section 7.3 below. The three worldvolumes are the simplest examples of supersymmetric cycles in the torus Calabi–Yau manifold.

It is not difficult to see that our boundary states (7.34) coincide with the coherent states (7.39) of Ooguri *et al.*: equations (7.42) and (7.38) show that they obey the same Ishibashi conditions. To compare the coefficients in front of the Ishibashi states, we use the fact that the Ishibashi state associated with an irreducible representation is unique up to normalisation, so we only need to equate the coefficients of the $N = 2$ highest weight states $|1, 1, 1\rangle \otimes |1, 1, 1\rangle = i\,\psi_{-\frac{1}{2}} \overline{\psi}_{-\frac{1}{2}} |\text{vac}\rangle$ and $|2, 2, 2\rangle \otimes |2, 2, 2\rangle = i\,\psi^*_{-\frac{1}{2}} \overline{\psi}^*_{-\frac{1}{2}} |\text{vac}\rangle$ in equations (7.34) and (7.39). In both formulas, they are given by ω^n and ω^{-n}, respectively, with $n = \sum_i \alpha_i$, establishing agreement.

In summary, our algebraic boundary state construction, which relies on the requirement that the extended algebra $\mathcal{W}^\otimes$ is preserved, automatically leads to the coherent states which are used in the (geometric) σ-model approach of Ooguri *et al.* We can take this as an encouraging sign that the "rational" boundary states constructed for arbitrary Gepner models below will be of geometric relevance. In the (non-toroidal) cases beyond $(k = 1)^3$, comparisons with geometric results are of course much less direct because the σ-model interpretation is less explicit there.

To complete our analysis of the simple $(k = 1)^3$ model, we compute, in the same way as before, boundary states satisfying B-type conditions (4.15) for each component super Virasoro algebra. Since this requires $q^{(i)} = -\bar{q}^{(i)}$ for $i = 1, 2, 3$, only three of the 27 sectors in equation (7.33) provide B-type Ishibashi states, namely those which correspond to "charge-conjugate terms" $\chi_i(q)\chi_{i^+}(\bar{q})$ in the partition function, as opposed to the diagonal terms coupling to A-type Ishibashi states. In other words, the B-type boundary states involve twisted sectors of the orbifold theory – but adjusting the general theory of orbifold boundary states to the non-standard gluing conditions would require at least as much work as a direct pedestrian computation.

Thus we regard the following as an ansatz,

$$\|\alpha\rangle\rangle_B = 3^{\frac{1}{4}} \sum_{\mathbf{b}=(b,b,b)} \omega^{-b(\alpha_1+\alpha_2+\alpha_3)} |\mathbf{b}\rangle\rangle_B, \tag{7.44}$$

and the same calculations as before lead to boundary CFT partition functions

$$Z^B_{\alpha\tilde\alpha}(q) \equiv {}_B\langle\langle\tilde\alpha\| \, \tilde{q}^{L_0^{\text{tot}} - \frac{c}{24}} \, \|\alpha\rangle\rangle_B = \sum_{\mathbf{c}=(c_1,c_2,c_3)} \delta^{(3)}_{\Sigma c_i, \Sigma\tilde\alpha_i - \Sigma\alpha_i} \, \chi_{\mathbf{c}}(q). \tag{7.45}$$

For the boundary Hilbert space to contain only states with integer U(1) charge, the same condition (7.36) as in the A-type case has to be met. The spectrum described by $Z^B_{\alpha\tilde\alpha}(q)$ is, however, different from that in $Z^A_{\alpha\tilde\alpha}(q)$.

7.2.2 Boundary states for arbitrary Gepner models

After this warm-up example, let us discuss boundary states for fully-fledged Gepner models, taking into account R sectors and the GSO projection. For definiteness, we assume that the Gepner model belongs to a type IIB compactification. We will follow the presentation given in [389], but at the end of the section make a few remarks on other approaches and on generalisations.

We first have to fix the Ishibashi conditions to be imposed on the boundary states; as above, we will choose the boundary conditions in such a way as to preserve an extended symmetry algebra $\mathcal{W}$, the obvious choice being the algebra generated by all the $N = 2$ Virasoro algebras of the r internal component theories, together with the $SO(d)_1$ of the external fermions. The external bosons will be ignored in the following; their boundary states, e.g. describing ordinary flat D-branes, just multiply the boundary states computed below.

In the generic case, i.e. if the levels k_j of the internal minimal models are pairwise different, the only way to maintain the tensor product symmetry in the presence of a boundary state is to require A-type or B-type boundary conditions as in equations (4.14, 4.15) for each subtheory separately. In special cases, when $k_{j_1} = k_{j_2}$, we can also glue the left-moving super Virasoro generators of subtheory j_1 to the right-moving generators of subtheory j_2, using permutation automorphisms; we will discuss this further at the end of this section.

We first discuss the technically simpler case where *A-type gluing conditions* (4.61) are imposed on each of the internal sub-Virasoro algebras and on the external fermions. A-type conditions imply that only left–right representations $\mathcal{H}_i \otimes \mathcal{H}_{\bar\imath}$ with $q_i = \bar{q}_{\bar\imath}$, and of course $h_i = \bar{h}_{\bar\imath}$, contribute Ishibashi states. In other words, we have to restrict to the diagonal part of the Hilbert space. Whenever a left-moving character $\chi_i(q)$ shows up at all in the Gepner model partition function (7.17) (where i labels irreducible representations of the bosonic subalgebra), then there is a term $\chi_i(q)\,\chi_i(\bar q)$ in equation (7.17). Therefore, any tensor product A-type Ishibashi state $|\boldsymbol\lambda, \boldsymbol\mu\rangle\rangle_A$ contributes to the boundary state provided $(\boldsymbol\lambda, \boldsymbol\mu)$ occurs on the (left-moving) closed string spectrum.

We use the following notation for "rational" A-type boundary states in Gepner models:

$$\|\alpha\rangle\rangle_A \equiv \|S_0; (L_j, M_j, S_j)_{j=1}^r\rangle\rangle_A = \frac{1}{\kappa_\alpha^A} \sum_{\lambda,\mu}^\beta B_\alpha^{\lambda,\mu} |\lambda,\mu\rangle\rangle_A . \qquad (7.46)$$

Here, S_0, L_j, M_j, S_j are integer labels; the summation is over states satisfying the "β-constraints" as in Gepner's partition function (7.17), κ_α^A is some normalisation constant to be determined later, and for the coefficients in front of the Ishibashi states we make the ansatz

$$B_\alpha^{\lambda,\mu} = (-1)^{\frac{s_0^2}{2}} e^{-i\pi \frac{d}{2} \frac{s_0 S_0}{2}} \prod_{j=1}^r \frac{\sin \pi \frac{(l_j+1)(L_j+1)}{k_j+2}}{\sin^{\frac{1}{2}} \pi \frac{l_j+1}{k_j+2}} e^{i\pi \frac{m_j M_j}{k_j+2}} e^{-i\pi \frac{s_j S_j}{2}} . \qquad (7.47)$$

A closer look at this formula shows that the restrictions on (l_j, m_j, s_j) – namely the field identification $(l_j, m_j, s_j) \sim (k_j - l_j, m_j + k_j + 2, s_j + 2)$ and that $l_j + m_j + s_j$ is even – lead to the same constraints on the (L_j, M_j, S_j).

Except for the phase $(-1)^{\frac{s_0^2}{2}}$, these coefficients are chosen as in Cardy's solution (4.80) for the tensor product of minimal models and external $SO(d)_1$ factor but *before* orbifolding and charge projections. Because of this, and also because the GSO projection leads out of the domain of genuine CFTs, we cannot simply rely on Cardy's general arguments and need to verify explicitly that the boundary states (7.46) lead to acceptable open string spectra $Z_{\alpha\tilde\alpha}(q) = \langle\langle\Theta\tilde\alpha\| \tilde{q}^{L_0 - \frac{c}{24}} \|\alpha\rangle\rangle$. The computations are straightforward except for the β-constraints in the summation. We have

$$Z_{\alpha\tilde\alpha}^A(q) = \frac{1}{\kappa_\alpha^A \kappa_{\tilde\alpha}^A} \sum_{\lambda,\mu}^\beta \sum_{\tilde\lambda,\tilde\mu}^\beta B_{\tilde\alpha}^{\tilde\lambda,\tilde\mu} B_\alpha^{\lambda,\mu} {}_A\langle\langle\tilde\lambda, -\tilde\mu| \tilde{q}^{L_0 - \frac{c}{24}} |\lambda,\mu\rangle\rangle_A$$

$$= \frac{1}{\kappa_\alpha^A \kappa_{\tilde\alpha}^A} \sum_{\lambda,\mu}^\beta \sum_{\lambda',\mu'}^{\text{ev}} B_{\tilde\alpha}^{\lambda,-\mu} B_\alpha^{\lambda,\mu} S_{s_0,s_0'}^{\text{f}} \prod_{j=1}^r S_{(l_j,m_j,s_j),(l_j',m_j',s_j')}^{k_j} \chi_{\mu'}^{\lambda'}(q) ,$$

where $\sum^{\text{ev}}$ denotes summation over the full range (7.7) with $l_j' + m_j' + s_j' \in 2\mathbb{Z}$ as the only constraint, and S^{f} and S^{k_j} are the modular S-matrices of the external fermions and the jth minimal model; see equations (7.18, 7.19).

In order to compute the prefactor of $\chi_{\mu'}^{\lambda'}(q)$ in $Z_{\alpha\tilde\alpha}(q)$, we introduce Lagrange multipliers ν_0 and ν_j, $j = 1, \ldots, r$, to disentangle the charge constraint and the β_j-conditions. This enables us to rewrite

$$\sum_{\lambda,\mu}^\beta = \sum_{\lambda,\mu}^{\text{ev}} \frac{1}{K} \sum_{\nu_0=0}^{K-1} e^{i\pi\nu_0(q_{\text{tot}}-1)} \prod_{j=1}^r \frac{1}{2} \sum_{\nu_j=0,1} e^{i\pi\nu_j(s_0+s_j)}$$

with $K = \text{lcm}(4, 2k_j + 4)$ and the total U(1) charge $q_{\text{tot}} = 2\beta_0 \bullet \mu$ as in equation (7.15). Now, the summations over s_0, l_j, m_j, s_j are independent of each other (except for the $l_j + m_j + s_j$ even constraint, which is easy to handle). They can be carried out with help of the identities listed in Appendix B and yield

$$Z_{\alpha\tilde{\alpha}}^{A}(q) = n_{\alpha\tilde{\alpha}}^{A} \sum_{\lambda',\mu'}{}^{\text{ev}} \sum_{\nu_0=0}^{K-1} \sum_{\nu_1,\ldots,\nu_r=0,1} (-1)^{s_0'+S_0-\tilde{S}_0} \delta_{s_0',2+\tilde{S}_0-S_0-\nu_0-2\Sigma\nu_j}^{(4)}$$

$$\times \prod_{j=1}^{r} N_{L_j\tilde{L}_j}^{l_j'} \delta_{m_j',\tilde{M}_j-M_j-\nu_0}^{(2k_j+4)} \delta_{s_j',\tilde{S}_j-S_j-\nu_0-2\nu_j}^{(4)} \chi_{\mu'}^{\lambda'}(q). \tag{7.48}$$

Here, $N_{ll'}^{l''}$ denote the fusion rules (6.11) of the SU(2) WZW model, which arise from the l_j summations via the Verlinde formula. The symbol $\delta_{r,s}^{(p)}$ means that $r \equiv s \pmod{p}$. Finally, the prefactor is

$$n_{\alpha\tilde{\alpha}}^{A} = \frac{1}{\kappa_\alpha^A \kappa_{\tilde{\alpha}}^A} 2^{\frac{r}{2}+1} \frac{2(k_1+2)\cdots(k_r+2)}{K}.$$

From equation (7.48) we see that with the (minimal) normalisation

$$\kappa_\alpha^A = 2\left(2^{\frac{r}{2}} \frac{(k_1+2)\cdots(k_r+2)}{K}\right)^{\frac{1}{2}} \tag{7.49}$$

our boundary states (7.46, 7.47) indeed satisfy Cardy's conditions – suitably modified for the supersymmetric setting: there are signs in equation (7.48) which distinguish spacetime bosons from spacetime fermions.

Similar to the constraint (7.36) in our warm-up example, there are additional string theory requirements which impose restrictions on the integers $(S_0; (L_j, M_j, S_j))$ in the ansatz (7.47): the spin structures of the component theories in $Z_{\alpha\tilde{\alpha}}(q)$ should be coupled as in the closed string case (states in $\text{NS}^{\otimes(r+1)}$ or $\text{R}^{\otimes(r+1)}$ only); therefore we must require

$$S_0 - \tilde{S}_0 \overset{!}{\equiv} S_j - \tilde{S}_j \pmod{2} \tag{7.50}$$

for all $j = 1,\ldots,r$ so as to have $\beta_j \bullet \mu' \in \mathbb{Z}$ for all states in $Z_{\alpha\tilde{\alpha}}(q)$.

In the excitation spectrum of the brane configuration $(\alpha\tilde{\alpha})$ described by $Z_{\alpha\tilde{\alpha}}(q)$ to be supersymmetric, only states with odd total charge should be present. From the δ-symbols in equation (7.48) we find that

$$2\beta_0 \bullet \mu' \equiv Q(\alpha) - Q(\tilde{\alpha}) + 1 \overset{!}{\equiv} 1 \pmod{2} \tag{7.51}$$

with

$$Q(\alpha) := -\frac{d}{2}\frac{S_0}{2} - \sum_{j=1}^{r} \frac{S_j}{2} + \sum_{j=1}^{r} \frac{M_j}{k_j+2}. \tag{7.52}$$

For $Q(\alpha) - Q(\tilde{\alpha})$ even, only states with $q_{\text{tot}} = \pm 1, \pm 3, \ldots$ contribute; since $h \geq \frac{|q|}{2}$ in unitary representations of the $N = 2$ super Virasoro algebra, all states in $Z_{\alpha\tilde{\alpha}}(q)$ have conformal weight $\frac{1}{2}$ or higher: the spectrum is tachyon-free and stable.

In particular, requirements (7.50, 7.51) are satisfied for two identical branes $\alpha = \tilde{\alpha}$ given by A-type boundary states (7.46, 7.47), which means that the excitation spectrum of a single such brane is supersymmetric and stable.

On the other hand, equations (7.50, 7.51) imply that two arbitrary boundary states may not be "compatible" with each other. Rather, we obtain groups of mutually compatible boundary states; configurations made up of two branes from different groups yield spectra which violate supersymmetry and stability. In particular, a brane–antibrane system $(\alpha\,\alpha^+)$ with $\|\alpha^+\rangle\rangle := \Theta\,\|\alpha\rangle\rangle$ has a tachyon in its spectrum – cf. equation (2.59) in Subsection 2.A.1 for the case of a flat target.

A more direct check of the fact that our boundary states describe BPS states is to apply spacetime supersymmetry charges to them. This test was carried out in [263], using the relation between supersymmetry generators and spectral flow operators, which in turn have a particularly simple realisation in $N = 2$ minimal models, as mentioned before.

The combination of left- and right-moving supercharges Q, $\overline{Q}$ that is conserved by the boundary state depends on the gluing conditions and on the quantity $Q(\alpha)$ from equation (7.52). Writing the Ishibashi states as $|\lambda, \mu\rangle\rangle_A = |\lambda; \mu, \mu\rangle\rangle_A$ to make left- and right-charges explicit, and using the action of $U_{\frac{1}{2}}^{\text{tot}}$ from Section 7.1, we have

$$Q\,\|\alpha\rangle\rangle_A = \frac{1}{\kappa_A} \sum_{\lambda,\mu}^{\beta} B_\alpha^{\lambda,\mu}\, |\lambda; \mu + \beta_0, \mu\rangle\rangle_A$$

while acting with the adjoint $\overline{Q}^*$ of the right-moving partner leads to

$$\overline{Q}^*\,\|\alpha\rangle\rangle_A = \frac{1}{\kappa_A} \sum_{\lambda,\mu}^{\beta} (-i(-1)^{s_0})\, B_\alpha^{\lambda,\mu}\, |\lambda; \mu, \mu - \beta_0\rangle\rangle_A$$

$$= \frac{1}{\kappa_A} \sum_{\lambda,\mu'}^{\beta} e^{i\pi Q(\alpha)}\, B_\alpha^{\lambda,\mu'}\, |\lambda; \mu' + \beta_0, \mu'\rangle\rangle_A \,.$$

The phases in the first line are due to the chiral CPT-operator U and to the fact that states with $s_0 = \pm 1$ are spacetime fermions so that the left-movers anticommute with $\overline{Q}^*$. The prefactors in the second line arise from the substitution $\mu \mapsto \mu + \beta_0$ in equation (7.47). Altogether,

$$\left(Q - e^{i\pi Q(\alpha)}\, \overline{Q}^*\,\right) \|\alpha\rangle\rangle_A = 0\,, \tag{7.53}$$

and it can be shown in a similar way that $\|\alpha\rangle\rangle_A$ is annihilated by three further combinations of the supersymmetry charges (7.23) and (7.24); see [263]. From the worldsheet point of view, these equations mean that our boundary states satisfy gluing conditions for the spectral flow extension of the $N = 2$ algebra.

An obvious consequence of equation (7.53) is that a configuration of two branes α and $\tilde{\alpha}$ preserves half of the supersymmetry if we have $Q(\alpha) - Q(\tilde{\alpha}) \in 2\mathbb{Z}$ – the same condition (7.51) that followed from the spectrum $Z_{\alpha\tilde{\alpha}}(q)$.

As was shown in Chapter 2, the equations (7.46, 7.47) allow us to determine the tension and R–R charges of the branes described by these boundary states, at least after taking the tensor product with the external bosonic part. We merely have to project $\|\alpha\rangle\rangle_A$ onto the massless closed string state in question, and to take into account universal prefactors from string amplitudes. While we could arrive at a closed formula for (rational) boundary states for Gepner models, the massless closed string spectrum does depend sensitively on the concrete model; we will not undertake such case-by-case studies here.

Similar remarks apply to the excitation spectrum of brane configurations, which can be extracted from $Z_{\alpha\beta}(q)$. Its massless part determines the field content of the low-energy effective field theory associated with the configuration, but varies with the boundary condition and with the underlying Gepner model.

Let us now discuss *B-type boundary states* with B-type Ishibashi conditions (4.60) imposed on each subtheory. The U(1) charges of the left- and right-moving highest weight states must satisfy $q_i = -\bar{q}_{\bar\imath}$ along with $h_i = \bar{h}_{\bar\imath}$, and the B-type Ishibashi states couple to charge-conjugate parts $\mathcal{H}_i \otimes \mathcal{H}_{i+}$ of the bulk Hilbert space. A little bit of calculation shows that a term $\chi^\lambda_\mu(q)\chi^\lambda_{-\mu}(\bar{q})$ occurs in the Gepner partition function (7.17) precisely for those μ which satisfy

$$m_j \equiv b \ (\mathrm{mod}\, k_j + 2) \tag{7.54}$$

for some $b = 0, 1, \ldots, \frac{K}{2} - 1$ and for all j simultaneously. It is those Ishibashi states that contribute to the sum in the ansatz for the B-type boundary states

$$\|\alpha\rangle\rangle_B \equiv \|S_0; (L_j, M_j, S_j)^r_{j=1}\rangle\rangle_B = \frac{1}{\kappa^B_\alpha} \sum_{\lambda,\mu}^{\beta,b} B_\alpha^{\lambda,\mu} |\lambda, \mu\rangle\rangle_B , \tag{7.55}$$

with coefficients $B_\alpha^{\lambda,\mu}$ again given by equation (7.47). Note that, generically, Gepner models possess less B-type than A-type Ishibashi states; as a consequence, the excitation spectra of B-type brane configurations will typically be richer than those of A-type branes.

We can now perform the same calculation as above in order to test this ansatz. Because of the restricted m_j range (7.54), there are slight differences compared to A-type boundary conditions, but again Lagrange multipliers can be used to disentangle all summations. For the partition function describing the excitation spectrum of a configuration of two B-type branes as in equation (7.55), we obtain

$$Z^B_{\alpha\tilde\alpha}(q) = n^B_{\alpha\tilde\alpha} \sum_{\lambda',\mu'}^{\mathrm{ev}} \sum_{\nu_0=0}^{K-1} \sum_{\nu_1,\ldots,\nu_r=0,1} (-1)^{s'_0 + S_0 - \tilde{S}_0}\, \delta^{(4)}_{s'_0, 2 + \tilde{S}_0 - S_0 - \nu_0 - 2\Sigma\nu_j}$$

$$\times\ \delta^{(K')}_{\Sigma_{m'},0} \prod_{j=1}^r N^{l'_j}_{L_j \tilde{L}_j}\, \delta^{(2)}_{m'_j + M_j - \tilde{M}_j + \nu_0, 0}\, \delta^{(4)}_{s'_j, \tilde{S}_j - S_j - \nu_0 - 2\nu_j}\, \chi^{\lambda'}_{\mu'}(q) , \tag{7.56}$$

with $K' := \mathrm{lcm}(2k_j + 4)$, $\xi_j := K'/(2k_j + 4)$ and

$$\Sigma_{m'} := \sum_{j=1}^{r} \xi_j \left(m'_j + M_j - \tilde{M}_j + \nu_0 \right) ; \tag{7.57}$$

the overall factor is $n^B_{\alpha\tilde{\alpha}} = 2^{\frac{r}{2}}/(\kappa^B_\alpha \kappa^B_{\tilde{\alpha}})$ so that we choose the normalisation in equation (7.55) to be

$$\kappa^B_\alpha = 2^{\frac{r}{4}} . \tag{7.58}$$

As in the A-type case, Cardy's conditions are always satisfied for the boundary states (7.55), and the remaining string requirements for the spectrum of a brane configuration $(\alpha\,\tilde{\alpha})$ lead to the same conditions

$$S_0 - \tilde{S}_0 \overset{!}{\equiv} S_j - \tilde{S}_j \ (\mathrm{mod}\,2) , \quad Q(\alpha) - Q(\tilde{\alpha}) \overset{!}{\equiv} 0 \ (\mathrm{mod}\,2)$$

on the labels occurring in the boundary state ansatz.

These conditions also ensure that a configuration of two B-type boundary states preserves half of the spacetime supersymmetry. The combinations of left- and right-moving charges which annihilate $\|\alpha\rangle\rangle_B$ differ from the A-type case; we have

$$\left(Q - e^{i\pi Q(\alpha)} \overline{Q} \right) \|\alpha\rangle\rangle_B = 0 . \tag{7.59}$$

Accordingly, configurations which contain both A- and B-type boundary states break spacetime supersymmetry.

Let us add some remarks on *more general rational boundary states*, which can be constructed whenever two or more of the minimal models making up the Gepner model share the same level. In these cases, we can study Ishibashi states obeying permutation gluing conditions

$$W^{(i)}(z) = \Omega_{A,B} \, \overline{W}^{(\pi(i))}(\bar{z})$$

for some permutation π; see Subsection 4.A.4. These lead to additional rational and supersymmetric boundary states with A- or B-type gluing conditions for the diagonal $N = 2$ algebra. The general theory of permutation branes was developed in [386], where applications to Gepner models were also discussed. While general features of the construction are parallel to what was outlined above, the question which terms $\chi_i(q)\,\chi_{\tilde{i}}(\bar{q})$ from the Gepner partition function (7.17) contribute to Ishibashi states is much more involved for permutation branes. The conditions that generalise equation (7.54) from the $\pi = \mathrm{id}$ B-type case turn out to depend crucially on relative divisibility properties of the number and the level of the identical minimal model factors.

In [386], permutation branes for the quintic, i.e. the Gepner model $(k = 3)^5$, were given. The formulas for the coefficients B in the boundary states are rather similar to equation (7.47), e.g. for π-B-type branes in the quintic one has

$$B^{\lambda,\mu}_{\alpha A,\pi} = (-1)^{\frac{s_0^2}{2}} e^{-\frac{i\pi}{2} s_0 S_0} e^{-\frac{i\pi}{2} \sum_{j=1}^{5} s_j S_j} \left[\prod_{\nu=1}^{P^\pi} \frac{\sin \pi \frac{(l_\nu+1)(L_\nu+1)}{5}}{\left(\sin \pi \frac{L_\nu+1}{5}\right)^{\Lambda_\nu/2}} \right] e^{\frac{i\pi}{5} \sum_{j=1}^{5} m_j M_j} ,$$

where P^π is the number of cycles of the permutation $\pi \in S_5$, and Λ_ν is the length of the νth cycle. Note that permutation boundary states have one independent label L_ν per cycle only. Excitation spectra of permutation branes and also of systems of branes corresponding to different permutations can be computed roughly along the same lines as above; see [386] and also [74] for explicit results. We will briefly come back to permutation branes later when we discuss the geometric interpretation of our algebraic constructions.

In our construction, we have focused solely on Cardy's conditions and ignored other non-linear constraints. There is, however, no doubt that the boundary states above also satisfy all the other sewing relations laid out in Section 4.4 – as long as they are elementary. The cluster condition will of course be violated as soon as a boundary state from equations (7.46, 7.55) is in fact a superposition of elementary boundary conditions (see the remarks after equation (4.93)), and this may happen if the spectral flow operator of the Gepner model has fixed points.

In principle, sewing relations can be investigated after determining the boundary OPE, but this is a very intricate task for Gepner models (see, however, [84] for some explicit results). A more conceptual approach is to view the symmetry-preserving boundary states from above as boundary conditions for a rational simple current orbifold theory; see in particular [208]. This framework makes it possible to settle the problem of sewing relations by invoking Runkel's general results. Moreover, the simple current method allows us to determine in which situations the boundary states from above are in fact superpositions, and how to resolve them into elementary branes.

The fact that Gepner model partition functions can be understood as simple current extensions was briefly mentioned at the end of Section 7.1. In fact, the extension in question is sufficiently straightforward to be able to work out the modular S-matrix of the extended characters (here given as the old S-matrix times a phase related to the monodromy charge). Therefore, rational (A-type) boundary states for the extension can be written down using Cardy's ansatz – see our general remarks in Subsection 4.A.2, and see [83, 208] and also [80] for explicit formulas. It turns out that, generically, this leads precisely to the boundary states (7.46), which contain Ishibashi states from the untwisted sector only. However, in special cases, the twisted sector of the simple current orbifold can contribute Ishibashi states, dressed with representations (cf. equation (4.102)) of the stabiliser group $\mathbb{Z}_2$ of the short orbits, cf. equation (4.109) and the remarks at the end of Section 7.1. The stabiliser representations are mere signs, and upon adding two of those elementary boundary states the twisted sector Ishibashi states may cancel out, leading to equation (7.46). However, this is rather rare

[83]: while a Gepner partition function (with at least one even k_i) may contain many short orbits under the simple current group, due to the A-type gluing condition $q_i = \bar{q}_{\bar{i}}$ only the twisted sector with $b_0 = \text{lcm}(k_i + 2)$ can contribute A-type Ishibashi states, in the notation of equation (7.17).

We conclude that sometimes the simple current construction allows us to resolve a $\mathbb{Z}_2$ fixed point under the simple current β_0. Since this simple current (or at least its internal part) is an element of the Greene–Plesser group (7.28), the additional branes can be associated with geometric singularities of the underlying Calabi–Yau hypersurfaces; we refer to [83] for some more details.

Unfortunately, it appears that the simple current point of view is not useful to discuss B-type boundary states for Gepner models or permutation branes, due to the non-standard gluing conditions. Relying on mirror symmetry, we could in principle address B-branes by passing to the mirror model first via the Greene–Plesser construction and then study A-branes there, but the relation is not very explicit. This is the main reason why we highlighted the more pedestrian, but rather adaptable construction of [389] in this section.

7.3 On the geometric and physical content of the boundary states

We have seen that the general formalism of boundary CFT, applied with some care so as to satisfy additional string requirements, can produce explicit results where geometry-inspired methods cannot. However, the boundary states (7.46, 7.55) constructed in the previous section are rather abstract, and it would be gratifying to have some more insight into the interpretation of these purely algebraic objects.

In the following, we will first outline very briefly how physical quantities in the external four-dimensional spacetime of a Gepner string compactification can be related to boundary state data; here, we apply some of the translation rules set out in Section 2.2, and also quote some results from articles such as [51, 52, 263, 264].

Afterwards, we will concentrate on the internal sector and relate Gepner model boundary states to "geometric branes" within the associated Calabi–Yau manifold. As was mentioned at the beginning of this chapter, the reduced amount of supersymmetry makes it possible that the properties of branes in the classical (large-volume) and the stringy regime differ qualitatively. Tools to attack this question were developed by Douglas and collaborators, based on some earlier results in [41, 364] and on the Gepner model boundary states constructed in [389].

We will be rather sketchy and present only some key ideas and results, as full accounts would reach far into string phenomenology and complex geometry and thus go beyond the framework of this book. Moreover, the mathematical theory behind branes in Calabi–Yau manifolds is still being developed – some further remarks on homological mirror symmetry will be collected in Section 7.4.

For a "low-energy" observer in non-compact spacetime, only the external part of the boundary state is directly visible; it gives an ordinary flat Dirichlet p-brane in the space $\mathbb{R}^d$ of $d = 8 - \frac{2c_{\text{int}}}{3}$ transversal coordinates. We are working in the light-cone gauge here, so we might speak of $(p+1)$-instantons rather than p-branes. The apparent dimension $p = -1, \ldots, d - 1$ depends on the gluing conditions for the external bosons, which by worldsheet supersymmetry are linked to those for the free fermions. Using the free field realisation (7.5), it is easy to see that $p = -1$ and $p = 1$ imply B-type conditions, whereas $p = 0$ means A-type conditions for the external $N = 2$ algebra (we focus on the most interesting case $d = 2$). Furthermore, it turns out that p is linked to the internal gluing conditions: depending on whether we start from a type IIA or type IIB string theory, different total currents (7.20, 7.21) are employed to define the GSO projection; these are related by the mirror twist $\Omega_{\text{tw}} = \text{id}_{\text{ext}} \otimes \Omega_{M,\text{int}}$ acting on the right-movers, which of course switches A-type and B-type gluing conditions for the boundary states. Altogether, the correspondences are

$$
\begin{array}{llll}
\text{IIB} & A_{\text{ext}} \otimes A_{\text{int}} & \text{D0-branes} \\[4pt]
& B_{\text{ext}} \otimes B_{\text{int}} & \text{D(}-1\text{)- or D1-branes}
\end{array}
\tag{7.60}
$$

and

$$
\begin{array}{llll}
\text{IIA} & A_{\text{ext}} \otimes B_{\text{int}} & \text{D0-branes} \\[4pt]
& B_{\text{ext}} \otimes A_{\text{int}} & \text{D(}-1\text{)- or D1-branes}
\end{array}
\tag{7.61}
$$

Below, we will see that internal A-type conditions describe odd-dimensional cycles in the Calabi–Yau three-fold, while B-type corresponds to even-dimensional cycles. Adding up dimensions, the flat-target rule that IIA (IIB) string theory admits BPS p-branes only for p even (odd) is, therefore, still obeyed in the compactifications considered here.

Having determined the branes' dimensions in the low-energy effective theory, the minimal model parts of the boundary states account for internal degrees of freedom. For example, it is natural to view the D-particles arising from A-type conditions in the external sector as extremal *black holes* with charges determined by the $B_\alpha^{\lambda,\mu}$. Here, the massless closed string states, which can couple to the "black hole boundary state", lie in vector-multiplets (for both IIA and IIB strings). On the other hand, the D-instantons and the D-strings may couple to hyper-multiplet states; see, e.g., [263].

The black hole interpretation has been pursued further for the orbifold Gepner model $T^6/\mathbb{Z}_3$ in [51, 52], where D3-branes wrapped around three-cycles in the internal orbifold were studied. In the flat-target case, a configuration of several branes is needed to "mimic" an extremal or near-extremal black hole – otherwise there are singularities in the low-energy solution (the horizon area diverges). Interestingly, indications are that in a curved space a black hole can be modelled by a single brane already, since the BPS brane may wrap supersymmetric cycles with non-vanishing self-intersections. Further discussions of properties of $N = 2$ black holes exploiting the algebraic Gepner model boundary states from above can be found in [264].

The calculation of *D-instanton effects* from these boundary states was described in some detail in [263]. As mentioned for an $\mathbb{R}^d$-target in Subsection 2.2.2, D-instantons trigger non-perturbative corrections to effective couplings of closed string fields. For example, the type IIA calculations in [263] produce a four-fermion coupling

$$\mathcal{L}_{4f} \sim \overline{\psi}^I \psi^J \overline{\psi}^K \psi^L \, R_{IJKL} \, ,$$

with $R_{IJKL} = B_\alpha^{\lambda^K, \mu^K} B_\alpha^{\lambda^L, \mu^L}$, where (λ^M, μ^M) label scalar states obtained from massless (c, c) or (a, a) fields in the Gepner model, and where ψ^I is a fermion in the corresponding hyper-multiplet. The authors of [263] also studied non-perturbative corrections to the metric on the hyper-multiplet moduli space, i.e. a kinetic term in the LEEA of the form

$$\mathcal{L}_{\text{hyp}} \sim \partial \phi^I \cdot \partial \phi^J \, g_{IJ} \, e^{-S_{\text{inst}}} \, ,$$

where ϕ^I again are massless scalars in the hyper-multiplet – in fact, the Lagrangians $\mathcal{L}_{\text{hyp}}$ and $\mathcal{L}_{4f}$ are related by spacetime supersymmetry. To leading order in the string coupling g_S, both g_{IJ} and S_{inst} can be evaluated as disk diagrams with one bulk field inserted (along with a certain number of broken supersymmetry charges – which can be tackled with functional methods; see [263] and references therein). More precisely, S_{inst} can be read off from the one-point function of the dilaton, involving $B_\alpha^{0,0}$, compare equation (2.34), while the metric is given by $g_{IJ} = B_\alpha^{\lambda^I, \mu^I} B_\alpha^{\lambda^J, \mu^J}$.

A single spacetime-filling brane (and supersymmetry-preserving configurations thereof) support a four-dimensional low-energy gauge theory with $N = 1$ supersymmetry, in general coming with a non-trivial superpotential for the scalar fields. Their type and number are determined by the internal part of the Gepner boundary states, and in principle the superpotential term can be computed from boundary field correlators (see, e.g., [72]); this is, however, hard to do in practice within the full CFT, and passage to the topologically twisted theory provides more efficient tools; see Section 7.4.

To conclude this brief discussion of the effective spacetime physics of Gepner boundary states, let us remark that they also feature in a systematic search for brane configurations which reproduce, in the low-energy limit, the Standard Model of elementary particle physics; see, e.g., [64].

For the remainder of this section, let us ignore the external flat space and turn to the geometric significance of the internal part of our D-brane boundary states. In the light of the connection between Gepner models and superstring compactifications on Calabi–Yau spaces, reviewed in Subsection 7.1.2, we expect that the boundary states from before are related to BPS branes in Calabi–Yau manifolds, but in general the connection is much less direct than in the torus example reviewed in Subsection 7.2.1.

Branes in Calabi–Yau targets were first studied in the context of 11-dimensional supergravity [41]; later, an analysis of worldsheet boundary conditions for Calabi–Yau sigma models provided a slightly refined picture [306, 364].

Supergravity in 11 dimensions possesses natural solitonic degrees of freedom (membranes), and the requirement that the effective action on the membrane worldvolume is supersymmetric translates into geometric constraints on the sub-manifold around which the membrane is "wrapped". If the (compact part of the) target is a Calabi–Yau manifold, the worldvolume has to be a *supersymmetric cycle* in the Calabi–Yau target. For definiteness, let X be a Calabi–Yau three-fold with Kähler form k and holomorphic three-form Ω. A submanifold $\gamma \subset X$ is a supersymmetric cycle if [40, 41, 270]

$$\dim_{\mathbb{R}} \gamma = 3, \qquad \iota^* k = 0, \qquad e^{i\theta}\, \iota^* \Omega = d\mathrm{vol}_\gamma \tag{7.62}$$

or

$$\dim_{\mathbb{R}} \gamma \text{ even}, \qquad \gamma \text{ holomorphic submanifold}, \tag{7.63}$$

where $\iota^* \omega$ denotes the pullback of a form ω along the embedding $\iota : \gamma \longrightarrow X$, and where $e^{i\theta}$ is an appropriately chosen constant phase. Note that, in the second case, k defines a complex structure on the cycle.

For simplicity, we have assumed that the cycles carry a trivial gauge field; not surprisingly, any gauge bundle over a holomorphic cycle has to satisfy certain holomorphicity constraints in order to preserve supersymmetry; see, e.g., [281, 306] for explicit conditions. Gauge bundles with non-zero curvature lead to generalisations of the first set of conditions, involving co-isotropic submanifolds instead of middle-dimensional cycles [306].

It can be shown that both middle- and even-dimensional supersymmetric cycles minimise the volume within their homology class: they satisfy the geometrical analogue of the Bogomolny bound.

Supergravity is supposed to describe the low-energy physics of M-theory, which in turn "includes" string theory; following those links, middle-dimensional cycles can be related to branes in type IIA and even-dimensional cycles to branes in type IIB string theory, see [41] and references therein; this correspondence holds for Calabi–Yau spaces of other complex dimensions, too, and extends that of Dirichlet $2p/(2p+1)$-branes and type IIA/B string theories from the flat case.

In the work [364], Ooguri and collaborators approached supersymmetric cycles from the perspective of $N = 2$ supersymmetric sigma models on Calabi–Yau manifolds. In particular, they used the expressions (7.27) for the $N = 2$ super Virasoro generators in terms of coordinates and geometric data of the target, and analysed the consequences of imposing A-type or B-type gluing conditions (4.14, 4.15) on those generators, including the spectral flow $U_{\frac{1}{2}}$.

Ooguri *et al.* start from an ansatz $\partial X^i = R_{ij} \bar{\partial} X^j$ with some orthogonal matrix R – which has a direct local interpretation as (rotated) Dirichlet or Neumann boundary conditions – then insert this into A- or B-type gluing conditions for the $N = 2$ generators, and find restrictions on the eigenvalues of R. These translate into conditions on the restrictions of the Kähler and the holomorphic three-form to the worldvolume, leading to equations (7.62) and (7.63).

In the paper [306], Kapustin and Orlov revisited the gluing conditions of [364] and showed that the A-type case admits a more general class of geometric realisations than middle-dimensional special Lagrangian cycles (with flat bundles), namely co-isotropic submanifolds Y. This means that the complement of TY with respect to the symplectic form is contained in TY. Furthermore, the A-type gluing conditions imply that $\dim Y = \frac{1}{2} \dim_{\mathbb{R}} X + 2n$ for some $n \in \mathbb{Z}_+$. If $n = 0$, Y is a Lagrangian submanifold.

These considerations allow us to relate classes of $N = 2$ superconformal boundary states (A-type and B-type) to classes of submanifolds (co-isotropic and holomorphic, respectively) in a Calabi–Yau manifold. They also provide a strong motivation for the homological mirror symmetry programme [321, 323], which aims at establishing an equivalence between suitable categories of sheaves on a Calabi–Yau manifold (the B-side) and the so-called Fukaya category of co-isotropic submanifolds of the mirror manifold: at the level of the CFT, applying the mirror automorphism interchanges A- and B-type boundary conditions and at the same time leads from an associated Calabi–Yau target to its mirror.

On the other hand, concentrating on gluing conditions does not allow us to identify specific boundary states with specific supersymmetric cycles. Even though it is typically difficult to give concrete supersymmetric cycles for a given Calabi–Yau manifold, enough examples are known to make it clear that manifolds may have many such cycles. For the two-torus with complex coordinate X studied in Subsection 7.2.1, the middle-dimensional cycle $\{ X \text{ real} \}$ is supersymmetric. Similarly, the fixed point set of the involution $X^i \mapsto X^{\bar{i}}$ is a supersymmetric three-cycle on the quintic $\sum_{i=1}^5 (X^i)^5 = 0$ in $\mathbb{CP}^4$ [41], and in fact this construction of A-type branes generalises to arbitrary complete intersection Calabi–Yau spaces; see [72, 144] for further details and examples.

Again, if the Calabi–Yau manifold is a torus orbifold, a direct translation between target geometry and CFT language is available simply because the sigma-model formulation is exact to all orders in α'. Identifications of individual boundary states with individual cycles in the target were in particular discussed in [263].

In more general Gepner models, it is rather intricate to extract geometric information from CFT results: the sigma-model formulation is only a large-volume approximation, and only a few of the defining quantities of a boundary state constructed at the Gepner point have direct meaning in the large-volume limit.

Among those are the one-point functions of chiral primary fields (those with $h = 2q$) in A-type boundary conditions, as shown in [364] using methods similar to those explained in Chapter 5 and exploiting major simplifications due to supersymmetry: perturbations by bulk fields that correspond to Kähler moduli of the Calabi–Yau target do not alter these one-point functions. Therefore, the value computed at the Gepner point, i.e. B_α^{λ,μ_c} from equation (7.47) with

$$(\boldsymbol{\lambda}, \boldsymbol{\mu}_c) = (\boldsymbol{\lambda}, (0; \boldsymbol{\lambda}; 0, \dots, 0))$$

(in the notation (7.13)) can be identified with its value in the large-volume (large Kähler structure) limit – where they have a geometric meaning as an integral of a differential form over the supersymmetric cycle wrapped by the brane:

$$B_\alpha^{\lambda,\mu_c} = \int_{\gamma_\alpha} \omega_{\lambda,\mu_c} . \tag{7.64}$$

Here, ω_{λ,μ_c} is a differential form uniquely associated with the chiral primary ϕ_{λ,μ_c}, e.g. $\omega_{0,0} = \Omega$; see [253, 364] and also [404] for more details. In the low-energy effective field theory, these integrals determine masses and charges of the wrapped brane.

This establishes a relation between some of our abstract CFT results and geometric objects, but it is not sufficient to extract explicit formulas for supersymmetric cycles or even topological data like Chern classes from the boundary states.

In the paper [72], Brunner *et al.* introduced much more subtle and rather far-reaching methods to relate Gepner model boundary states to D-branes in the large-volume Calabi–Yau manifold. The main tool of [72] is the *intersection form*. On a manifold, intersections between two cycles can be computed (thanks to Poincaré duality) as integrals over forms – which in turn can be related to overlaps of R ground states in the σ-model or in the associated topological field theory; see, e.g., [281] for a relatively detailed review. These R ground states correspond to open strings stretching between the two cycles (i.e. branes) α and $\tilde\alpha$ in question, and the overlap of interest is in fact the Witten index

$$I_{\alpha,\tilde\alpha} = \mathrm{tr}_{\mathcal{H}_{\alpha\tilde\alpha,\mathrm{Rg.st.}}} (-1)^F = \mathrm{tr}_{\mathcal{H}_{\alpha\tilde\alpha,\mathrm{R}}} (-1)^F q^{L_0 - \frac{c}{24}} .$$

The last inequality holds because, in the R sector of unitary $N = 2$ representations, it is precisely the ground states satisfying $L_0|\psi\rangle = \frac{c}{24}|\psi\rangle$ or equivalently $G_0^+|\psi\rangle = G_0^-|\psi\rangle = 0$ which have no partner with the same conformal weight but opposite fermion number. In its second form, the Witten index can be computed from the CFT boundary states, as the overlap of the R–R parts:

$$I_{\alpha,\tilde\alpha} = {}_{\mathrm{RR}}\langle\!\langle\alpha\|\, \tilde{q}^{L_0 - \frac{c}{24}} \,\|\tilde\alpha\rangle\!\rangle_{\mathrm{RR}}. \tag{7.65}$$

This is an integer quantity and there are no issues with moduli-dependent normalisation (as exist for the boundary state coefficients themselves), so the intersection form can be used to compare to topological data of the target space. Nevertheless, the procedure found in [72] consists of many individual steps and is rather complicated. We sketch the main steps in the following, but also recommend the rather detailed discussion given in [404].

For A-type boundary conditions, comparison to the geometric intersection form of supersymmetric cycles is relatively unproblematic since the relevant quantities are independent of (bulk) Kähler moduli; thus the large-volume limit can be performed without difficulties. On the quintic, for example, the A-type boundary states (7.46, 7.47) simply generate the R–R charge lattice for all the 204 independent three-cycles of this Calabi–Yau manifold [72].

For B-type gluing conditions, Kähler moduli do not decouple, and the comparison of target geometry and CFT data (is much more difficult but) leads to more interesting results. We concentrate on the B-type case in the following. Geometrically, a B-type brane is given by a holomorphic submanifold γ in the target X possibly carrying a bundle E – ignoring issues of continuity for the time being, which suggest the consideration of coherent sheaves instead of vector bundles. We will indeed use E to denote such a brane. We can define the Mukai vector $v(E) \in H^{\text{even}}(X)$, whose components can be expressed in terms of topological data of the bundle E, namely

$$v(E) = \left(\, r, \; c_1(E), \; \text{ch}_2(E) + \tfrac{r}{24}c_2(X), \; \text{ch}_3(E) + \tfrac{1}{24}c_1(E)c_2(X) \, \right),$$

where c_i are Chern classes, r is the rank of the bundle E and $\text{ch}_i(E)$ are its Chern characters. (Note that these depend only on the K-theory class of E.)

The Mukai vector can be used to compute the intersection between two branes E and F,

$$I(E, F) = \langle v(E), v(F) \rangle,$$

where $\langle \alpha, \beta \rangle = (-1)^{n+1} \int_X \alpha \wedge \beta \wedge \text{td}(X)$ for $\alpha, \beta \in H^{\text{even}}(X)$, and with the Todd class $\text{td}(X)$ of the Calabi–Yau manifold.

On the other hand, the Mukai vector also determines the coupling of the brane to R–R fields, because of the form of the WZ term in the low-energy brane worldvolume action: the ($N = 2$ spacetime) central charge of the brane is given by

$$Z(v(E)) = \int e^{-K(t)} \, v(E)$$

where $K = K(t)$ is the complexified Kähler form of the target X, which can be deformed by turning on the Kähler moduli t.

In transporting intersection form and central charges from the large-volume limit to the "deep interior of the moduli space", i.e. to the Gepner point where an exact CFT description is available, heavy use is made of explicit results by Candelas *et al.* [93–95]: mirror symmetry in particular relates $H^{\text{even}}(X)$ to the middle-dimensional cohomology $H^3(\hat{X})$ of the mirror. Using the holomorphic three-form $\hat{\Omega}$ on $\hat{X}$, periods Π^i can be associated with cycles $\hat{\gamma}^i \in H_3(\hat{X})$,

$$\Pi^i = \int_{\hat{\gamma}^i} \hat{\Omega} \, ,$$

which can in turn be used to write the central charge of a three-brane wrapped around the homology class $\hat{\gamma} = \sum_i n_i \hat{\gamma}^i$ as $Z = n_i \Pi^i$, where the n_i are integers; $\hat{\gamma}$ corresponds to $\gamma \subset X$ via mirror symmetry.

The periods become functions of complex structure moduli λ of $\hat{X}$ by deforming $\hat{\Omega}$, which corresponds to turning on the Kähler moduli of X via mirror symmetry. The periods from the "large complex structure" limit can be analytically continued to (the conifold and) the Gepner point. The latter is singular in that an extra "quantum symmetry" is present, namely the $\mathbb{Z}_d$ charge symmetry of the Gepner model (where we set $d := \mathrm{lcm}(2k_j = 4)$). In particular, there is a basis of periods ϖ_i at the Gepner point which form a representation of this $\mathbb{Z}_d$. (More precisely, the ϖ_i only form a spanning system, but their linear relations are under control.) The transformation m connecting the ϖ to the $\Pi = m\,\varpi$ is known explicitly for many Calabi–Yau manifolds X (with a small number of Kähler moduli); therefore the intersection form and brane charges of the large-volume limit can be related to those at the Gepner point of the Calabi–Yau moduli space,

$$I^{(L)} = m\, I^{(G)}\, m^{\mathrm{T}}, \quad n_{(L)} = n_{(G)}\, m^{-1},$$

writing $n_{(L)}$ for the row vector of integers n_i from $Z = n_i \Pi^i$.

Because of the $\mathbb{Z}_d$ symmetry at the Gepner point, $I^{(G)}$ can be written as a polynomial in a $\mathbb{Z}_d$ generator g. Not surprisingly, the same is true for the CFT Gepner model intersection form I^B of B-type boundary states as defined in equation (7.65) above. The investigations of [72] and of [127, 308, 402] result in rather concise formulas: for two rational boundary states $\|\alpha\rangle\rangle = \| L, M, 0\rangle\rangle$ and $\|\tilde{\alpha}\rangle\rangle = \| \tilde{L}, \tilde{M}, \tilde{0}\rangle\rangle$ as in [389] we have

$$I^B_{\alpha,\tilde{\alpha}} = \prod_{j=1}^{r} t_{L_j}\, I^B_{00}\, t^{\mathrm{T}}_{\tilde{L}_j}$$

$$\text{with} \quad I^B_{00} = 1 - g^{-1} \quad \text{and} \quad t_L = \sum_{l=0}^{L} g^{l-\frac{k}{2}} \, ; \tag{7.66}$$

here, rows and columns of the matrix I^B are understood as labelled by $\Delta M = M - \tilde{M}$, i.e. $g^{\frac{1}{2}}$ acts as the shift $\Delta M \mapsto \Delta M + 1$.

To proceed, we observe that in all models studied so far there is a relation of the form

$$I^B_{00} = m_1\, I^{(G)}\, m^{\mathrm{T}}_1$$

between the "geometric" and the CFT intersection forms at the Gepner point, involving some matrix m_1 which is again polynomial in g, typically given by

$m_1 = 1 - g$. This means that the $L_j = 0$ boundary states can be related to the cycles $\hat{\gamma}^i$ arising from the large-volume basis of $H^{\text{even}}(M)$. We still have to match a single CFT boundary state with a single geometric brane – e.g. the six-brane (trivial bundle over the manifold X itself), which can be followed through the whole moduli space using the results of the pioneering paper [93]. One particular finding of that work is that the mirror of the six-cycle can be identified with the three-cycle on $\hat{X}$ that vanishes at the conifold singularity. On the Gepner model side, one of the $L_i = 0$ boundary states is chosen, e.g. the one with $M = 0$, as the CFT realisation of this six-brane. This sets off the translation machine: large volume charges can now be computed from CFT charges via

$$n_{(L)} = n_B \, m_1 \, m^{-1} . \tag{7.67}$$

In order to determine n_B for an arbitrary $\pi = \text{id}$ B-type boundary state, the g-action can be used to change M and relation (7.66) to change L in the charge vector of the boundary state $\|0, 0, 0\rangle\!\rangle$.

From the large-volume charges, the topological data of the brane can be computed using the two expressions for the central charge,

$$Z = Z(v(E)) = n_i \, \Pi^i(\lambda) \; :$$

The left-hand side is a function of the Kähler moduli t of X, while the periods on the right-hand side are functions of the complex structure moduli λ of the mirror $\hat{X}$. Fortunately, the explicit map $\lambda = \lambda(t)$ has been worked out for a number of examples [93–95] using the $N = 2$ prepotential, so both sides can be written as functions of t and thus the coefficients on the left-hand side (which are the components of the Mukai vector) can be expressed in terms of the integers n_i, which in turn can be traced back to the charges at the Gepner point as above; see in particular [308] for a nice discussion.

Let us note for completeness that there is another piece of information, namely (an upper bound on) the number of brane moduli, which can be extracted in a much more straightforward way by counting the chiral primary boundary operators occurring in the spectrum $Z^B_{\alpha, \tilde{\alpha}}(q)$. The number can be encoded in formulas very similar to equation (7.66); see [72]. This information helps confirm some of the geometric interpretations reached via brane charges, and sometimes hints at the fact that some of the marginal boundary fields are not truly marginal, i.e. the corresponding moduli are lifted by a non-trivial boundary potential; see the works mentioned above for further details.

All in all, we arrive at rather explicit geometric pictures for at least some of the CFT boundary states, as collected in [72, 127, 308, 402]. For example, the papers [127, 308, 402] focus on Calabi–Yau manifolds that are elliptic or K3 fibrations, and from the topological data we can read off whether a two- or a four-brane wraps the fibre or lies in the base, which already provides a rather tight characterisation of the cycle γ.

Surprisingly, only for some Gepner models is it possible to find boundary states among those constructed in [389] which have the charges of the single D0-brane; the paper [403] contains an analysis of when this can happen. For the quintic, however, the boundary state for a single D0-brane turns out to be a permutation brane associated with $\pi = (12)(3)(4)(5) \in S_5$, as anticipated in [387] and elegantly proven in [74]. Moreover, the $\pi =$id boundary states only generate the lattice 5Λ inside the charge lattice Λ, while the permutation branes are sufficient to generate all of Λ – in the case of the quintic. It is not known whether the charge lattice can always be generated by rational boundary states of the Gepner model, or whether for some models symmetry-breaking boundary conditions need to be taken into account.

There are also boundary states whose associated topological data in the large volume limit are not those of an "ordinary" vector bundle, like rank 1 bundles with non-vanishing second Chern class. The authors of [72] speculate that such "bundles" could still exist in non-commutative geometry, which is in fact a natural explanation in view of the fact that these branes carry a B-field, which triggers a non-commutative deformation of the brane worldvolume; see Chapter 1 for the case of a flat target and Chapter 6 for group manifolds. Still, this hints at a "non-classical behaviour" of branes, and there are indeed other such features: The large-volume charges of some of the Gepner boundary states require an interpretation as superpositions of Dp-branes for different p, and frequently those p do not differ by 4. This indicates a non-supersymmetric, thus unstable, brane configuration in the large-volume limit, judging from the results of the flat-target analysis outlined in Subsection 2.A.1. On the other hand, the boundary states are clearly BPS at the Gepner point, so when varying the Kähler moduli some *line of marginal stability* must have been crossed, where stable CFT boundary states decay into non-supersymmetric configurations of branes that even repel each other; see [127, 144, 146, 301, 308, 402] for more details.

Based on these observations and deep mathematical results from algebraic geometry, and very much in line with Kontsevich's homological mirror symmetry programme, Douglas proposed a new programme for the classification of branes, using finer invariants than K-theory; namely, derived categories. Again, we refrain from attempting an outline of this very active area of research, but merely restate some of the initial observations; [25, 26, 126, 145, 146, 307] is an incomplete list of useful references.

One of the first things to realise here is that continuous vector bundles over submanifolds provide too narrow a framework to describe Calabi–Yau branes: for example, if the brane is a superposition of a D0- with a D4-brane, the rank of the bundle would have to be discontinuous. Such effects can be accommodated for by using sheaves instead of bundles; in particular there are sky-scraper sheaves supported by a single point. Within the sheaf language, it is completely natural to view (massless) open strings stretched between two branes as morphisms between two sheaves. A closer analysis shows that sequences of morphisms (involving three or more sheaves) can be viewed as describing bound state formation (an open string between two branes can "condense", leaving a new third brane behind). We are thus led to consider complexes of sheaves, which form the essential ingredient of the derived category of coherent sheaves. Douglas showed in [145] how lines of marginal stability can be encoded by a grade, which in contrast to the one usually used in the mathematical literature is real-valued and given by the argument of the $N = 2$ central charge Z – recall that its absolute value determines the mass of the BPS brane, via the $N = 2$ supersymmetry algebra. These ideas have also made it possible to perform some explicit computations concerning the behaviour of brane charges when bulk moduli are varied; see, e.g., [26].

While a complete correspondence between conformal boundary conditions and suitable derived categories is as yet difficult to formulate, the tools introduced in the following section allow us to make a precise connection for the "topological skeletons" of branes.

7.4 Topological branes and matrix factorisations

There is a procedure, called topological twisting, which leads from a full-fledged $N = 2$ superconformal field theory to a topological field theory with much smaller (often finite-dimensional) spaces of fields and states, but which nevertheless retains important information, e.g. on the (anti-)chiral ring of the original CFT [154]. We pass from the original energy-momentum tensor $T(z)$ to

$$T^{\mathrm{top}}(z) := T(z) + \frac{1}{2}\partial J(z) \,. \tag{7.68}$$

After the twist, the modes $L_n^{\mathrm{top}} = L_n - \frac{n+1}{2}J_n$ satisfy the Virasoro relations with zero central charge; $G^+(z)$ becomes a field of spin 1, called $Q(z)$, with integer modes; $G^-(z)$ acquires spin 2 and is called $G(z)$ after the twist; see, e.g., [133, 275] for the full list of commutation relations.

Alternatively we can define

$$T^{\mathrm{top}}(z) := T(z) - \frac{1}{2}\partial J(z) \,; \tag{7.69}$$

then, L_n^{top} again satisfies the Witt algebra, but now $G^+(z)$ becomes a spin 2 field (denoted $G(z)$), while $G^-(z)$ becomes the spin 1 field $Q(z)$.

For both alternatives, the zero-mode Q_0 of $Q(z)$ is nilpotent,

$$Q_0^2 = 0 \,,$$

and can be used as a BRST operator: by definition, the space of "physical" states $\mathcal{H}_{\mathrm{phys}}$ in the topological field theory arises from the full CFT state space by passing to the Q_0-cohomology.

Furthermore, $T^{\mathrm{top}}(z) = \{Q_0, G(z)\}$, i.e. the energy momentum tensor of the topological theory is BRST-exact; therefore, correlation functions of fields corresponding to physical states are not affected under local changes of the world-sheet coordinates. They can only depend on global topological properties, such as the ordering of boundary fields. The fields surviving the reduction to $\mathcal{H}_{\text{phys}}$ are as follows: for the twist (7.68), we can choose chiral primary fields as cohomology representatives. This is true since $L_0^{\text{top}} = L_0 - \frac{1}{2}J_0$ is BRST-exact and therefore vanishes on $\mathcal{H}_{\text{phys}}$; see also [133] for an alternative argument exploiting the Hodge decomposition (7.4) in the underlying unitary $N = 2$ SCFT. With the twist (7.69), the Q_0-cohomology is given by the anti-chiral primaries.

With these physical fields, the correlation functions of a topological field theory are formed. In topological string theory, additional objects are needed: string amplitudes are given by CFT correlators integrated over worldsheet insertion points; therefore, in addition to chiral (or anti-chiral) primaries ψ_c (or ψ_a) the integrated operators

$$\psi_c^{(1)} := \int_{\partial\Sigma} [G_{-1}, \psi_c] \quad \text{and} \quad \psi_a^{(1)} := \int_{\partial\Sigma} [G_{-1}, \psi_a] \tag{7.70}$$

are also needed to build topological string amplitudes.

We have so far restricted ourselves to the left- or right-moving super Virasoro algebra, which suffices to discuss open string vertex operators. In the bulk, both left- and right-moving generators are available, and they can be twisted independently. If the same sign is used for left- and right-movers, i.e. if we use equation (7.68) for both T^{top} and $\overline{T}^{\text{top}}$ or equation (7.69) for both T^{top} and $\overline{T}^{\text{top}}$, then the resulting topological model is called *B-twisted*. In an *A-twisted* theory, on the other hand, opposite signs occur in the twisting of T^{top} and $\overline{T}^{\text{top}}$. Note that if we want to add branes to such backgrounds, the gluing conditions have to be compatible with the chosen twist.

Taking the double cohomology of Q_0 and $\overline{Q}_0$ – note that they anti-commute – the physical fields of a B-twisted model are given by the (c, c)-ring, while the (c, a)-ring determines those of an A-twisted theory. In analogy to equation (7.70), we need integrated closed string vertex operators such as

$$\psi_{(c,c)}^{(1,1)} := \int_{\Sigma} [G_{-1}, [\overline{G}_{-1}, \psi_{(c,c)}]]$$

to build up topological closed string amplitudes; see, e.g., [275] for a list of formulas.

Already in view of the fact that the geometric moduli in $N = 2$ supersymmetric sigma models are (c, c) or (c, a) fields, it is expected that the topologically twisted models, in spite of the vast simplification, still capture important aspects of such string backgrounds and of their branes. In some cases, a study of bulk perturbations of amplitudes to all orders is feasible (see [133]), providing insight into how interactions depend on the moduli of the compactification. Topologically twisted bulk theories have been studied for quite some time; see, e.g., [50, 128, 133] – and of course Witten's papers [449] for a more geometrical and very fruitful approach. We will, however, not pursue these ideas here.

In the following, we are mainly interested in branes for topologically twisted theories. While we could start from CFT boundary states and then project them onto the Q_0-cohomology, a much simpler and computationally strong approach is available exploiting the link to supersymmetric Landau–Ginzburg models on worldsheets with boundary. There is no super Virasoro algebra in these theories, but topological twisting of appropriate nilpotent supercharges (corresponding to Q_0 from the SCFT) can be studied. Moreover, it is natural to ask what boundary conditions preserve the $N = 2$ supersymmetry – leading to topological branes formulated as *matrix factorisations*. The study of supersymmetric boundary conditions for Landau–Ginzburg models was initiated by Warner [442], but matrix factorisations enter into string theory following unpublished conjectures of M. Kontsevich, in the papers by Kapustin and collaborators [303–305] and in the work [78]. We will follow the main steps of the very detailed exposition in the last paper.

We start from the action (7.25) for an $N = 2$ supersymmetric Landau–Ginzburg model on a worldsheet Σ, writing out the components of the superfields. The bulk term is given by

$$S_\Sigma = \sum_j \int_\Sigma d^2x \; \partial^\mu \bar{X}^{(j)} \partial_\mu X^{(j)} - i\,\bar{\psi}_{j-} \overleftrightarrow{\partial}_+ \psi_{j-} - i\,\bar{\psi}_{j+} \overleftrightarrow{\partial}_- \psi_{j+}$$
$$+ \tfrac{1}{4} |W'^{(j)}|^2 + \tfrac{1}{2} W''^{(j)} \psi_{j+}\psi_{j-} + \tfrac{1}{2} \overline{W}''^{(j)} \bar{\psi}_{j-}\bar{\psi}_{j+}\,, \tag{7.71}$$

where $X^{(j)}$ are bosonic fields, $\psi_{j\pm}$ left- and right-moving fermions, and $W(X)$ is the Landau–Ginzburg potential, with abbreviations $W'^{(j)} := \partial W/\partial X^{(j)}$, $W''^{(j)} := \partial^2 W/\partial X^{(j)2}$ for its derivatives. (The worldsheet carries the two-dimensional version of the "mostly minus" metric.)

This action is invariant under the diagonal $N = 2$ supersymmetry transformation as long as the worldsheet has no boundary; for $\partial\Sigma \neq \emptyset$, we add boundary terms [442]

$$S_{\partial\Sigma,\psi} = \frac{i}{4} \sum_j \int_{\partial\Sigma} dx^0 \; [\bar{\theta}^{(j)}\eta^{(j)} - \bar{\eta}^{(j)}\theta^{(j)}] \tag{7.72}$$

(with $\eta := \psi_- + \psi_+,\ \theta := \psi_- - \psi_+$) as well as a term involving additional fermionic boundary degrees of freedom Π_a, namely [78]

$$S_{\partial\Sigma,\pi} = -\tfrac{1}{2} \int_{\partial\Sigma} dx^0 d^2\theta \; \overline{\Pi}_a \Pi_a - \left(\tfrac{i}{2} \int_{\partial\Sigma} dx^0 d\theta \; \Pi_a P_{1,a} + \text{compl. conj.} \right)$$
$$= \sum_{a,j} \int_{\partial\Sigma} dx^0 \left[i\,\bar{\pi}_a \partial_0 \pi_a - \tfrac{1}{2}\bar{p}_{0,a}p_{0,a} - \tfrac{1}{2}\bar{p}_{1,a}p_{1,a} \right. \tag{7.73}$$
$$\left. + \tfrac{1}{2}\pi_a(\bar{\eta}^{(j)}\bar{p}_{0,a}'^{(j)} + i\eta^{(j)}p_{1,a}'^{(j)}) - \tfrac{1}{2}\bar{\pi}_a(\eta^{(j)}p_{0,a}'^{(j)} - i\bar{\eta}^{(j)}\bar{p}_{1,a}'^{(j)}) \right].$$

Here, Π_a is a superfield with the boundary fermion π_a as its lowest component; the $p_{1,a}$ are polynomials in the bosons $X^{(j)}$ restricted to the boundary, arising as the lowest (bosonic) components of the boundary potentials $P_{1,a}$; the $p_{0,a}$

are polynomials in the bosons $X^{(j)}$ restricted to the boundary, arising as higher (bosonic) components of the Π_a.

Performing a supersymmetry variation of the action $S_\Sigma + S_{\partial\Sigma,\psi} + S_{\partial\Sigma,\pi}$ shows that in order to preserve B-type $N = 2$ supersymmetry, the polynomials $p_{i,a}(X)$ have to satisfy the factorisation condition [78]

$$\sum_a p_{0,a}\, p_{1,a} = W$$

up to a possible additive constant on the right-hand side, which will be set to zero in the following. Thus, we have a relation between factorisations of the bulk potential W and (B-type supersymmetric) boundary conditions (i.e. B-type branes).

To proceed, we restrict ourselves to the topological sector of the QFT corresponding to this boundary Landau–Ginzburg model, which in particular means that only fields that represent cohomology classes of the BRST operator are viewed as physical. It turns out that the boundary part $Q = Q_{\text{BRST},\partial\Sigma}$ of the BRST operator is determined by the polynomials $p_{i,a}(X)$; it acts on the boundary degrees of freedom as [78]

$$Q\, X = 0\,, \quad Q\, \pi = p_0\,, \quad Q\, \bar{\pi} = -ip_1\,.$$

The state space of the boundary theory is graded by the fermion number, $\mathcal{H} = \mathcal{H}^0 \oplus \mathcal{H}^1$, and using Clifford algebra anti-commutation relations

$$\{\pi_a,\, \pi_b\} = \{\bar{\pi}_a,\, \bar{\pi}_b\} = 0\,, \quad \{\pi_a,\, \bar{\pi}_b\} = \delta_{a,b}\,.$$

Q can be viewed as acting on boundary fields

$$\Phi = \begin{pmatrix} f_{00} & f_{01} \\ f_{10} & f_{11} \end{pmatrix},$$

where $f_{ij} : \mathcal{H}^j \to \mathcal{H}^i$, by graded commutator with the matrix

$$Q = \begin{pmatrix} 0 & p_0 \\ p_1 & 0 \end{pmatrix}; \tag{7.74}$$

the $\mathbb{Z}_2$-grading is given by the matrix $\text{diag}(1, 1)$.

It is straightforward to extend this to strings stretching between two different branes, where $\Phi : \mathcal{H}_\alpha \longrightarrow \mathcal{H}_{\tilde{\alpha}}$ and $D_{\alpha\tilde{\alpha}}\, \Phi = Q_{\tilde{\alpha}}\Phi \pm \Phi Q_\alpha$.

Furthermore, while the Q from above is a $2^r \times 2^r$ matrix if $a = 1,\ldots,r$, we can generalise this presentation of the BRST cohomology by allowing for matrices p_0, p_1 of arbitrary size; a very conceptual derivation using super-connections is given in [326]. In this way, we make contact to general *matrix factorisations*, which are pairs of square matrices $p_i \in \text{Mat}(k, \mathbb{C}[X^{(j)}])$ with polynomial entries, such that

$$p_0\, p_1 = W\, \mathbf{1}_k\,. \tag{7.75}$$

Altogether, the plausible conjecture is that such matrix factorisations correspond to branes in the topological Landau–Ginzburg model with superpotential W – and furthermore to topologically twisted $N = 2$ superconformal boundary states of the SCFT at the IR fixed point of this Landau–Ginzburg model.

In general, it is much easier to find matrix factorisations for a given Landau–Ginzburg potential than to construct boundary states for the corresponding CFT. In fact, there are many examples of matrix factorisations for which no associated conformal boundary state is known, see, e.g., [156]. In particular, what is lacking so far is a method to determine which CFT boundary condition a matrix factorisation of the Landau–Ginzburg potential flows to – which is perhaps hardly surprising as even the link between bulk Landau–Ginzburg models and CFTs is not very explicit, in the sense that the RG flow washes out many details of the QFT.

Up to now, correspondences between Landau–Ginzburg and CFT branes have been tested by computing intersection forms and spectra of boundary fields on both sides: passage to the topological model will not affect topological data like the Witten index considered above, which was used in [72] and subsequent works to extract target-space properties of string theory branes. Counting R ground states (with relative fermion number) in the CFT amounts to counting physical boundary fields (with relative fermion number) in the topological theory, so we merely have to compute the Euler characteristic of the BRST cohomology,

$$I_{\alpha,\tilde{\alpha}} = \dim H^{\mathrm{even}}(D_{\alpha\tilde{\alpha}}) - \dim H^{\mathrm{odd}}(D_{\alpha\tilde{\alpha}}),$$

to obtain the intersection form for two branes a and b given by two matrix factorisations with associated boundary BRST operators $Q_a^{\mathrm{BRST},\partial\Sigma}$ and $Q_b^{\mathrm{BRST},\partial\Sigma}$.

The calculations involved are relatively simple (or can be performed by a computer), and the method has the conceptual merit that only data from the topological sector are used to compute a topological quantity like the intersection number – in contrast to the CFT derivation, which relies on complete conformal boundary states.

Other data that can in principle be matched are the structure constants of the boundary chiral ring, which are straightforward to extract on the topological side, as a multiplication table of cohomology classes; but of course it is much harder to work out boundary OPEs in $N = 2$ SCFTs.

Let us give a few examples of matrix factorisations, starting with Landau–Ginzburg models with potentials $W(x) = x^{k+2}$, whose RG fixed points are given by level k super-minimal models with diagonal modular invariant partition functions. The basic topological branes are

$$Q_l = \begin{pmatrix} 0 & x^l \\ x^{k+2-l} & 0 \end{pmatrix} \quad \text{for} \quad l = 1, \ldots, k+1, \tag{7.76}$$

and they have been matched with the known elementary CFT boundary states in the papers [75, 78, 280, 305]. It is a straightforward exercise to compute the cohomologies of D_{QQ}, e.g. the bosons are found to be given by $p(x) \cdot \mathbf{1}_2$ with $p(x)$ a polynomial of degree at most $\min(l, d - l)$.

Taking tensor products, matrix factorisation is simple: if $Q^{(i)}$ are matrix factorisations of $W^{(i)}$ as in equation (7.74) for $i = 1, 2$, then the matrices

$$ p_0^{(1 \otimes 2)} = \begin{pmatrix} p_1^{(1)} \otimes 1 & 1 \otimes p_1^{(2)} \\ -1 \otimes p_0^{(2)} & p_0^{(1)} \otimes 1 \end{pmatrix}, \quad p_1^{(1 \otimes 2)} = \begin{pmatrix} p_0^{(1)} \otimes 1 & -1 \otimes p_1^{(2)} \\ 1 \otimes p_0^{(2)} & p_1^{(1)} \otimes 1 \end{pmatrix} $$

provide a factorisation of the sum $W^{(1)} + W^{(2)}$, see, e.g., [24, 156]. In these papers, it is also explained how to perform orbifolds (with respect to the charge symmetry group) of matrix factorisations, and combining the two constructions we arrive at matrix factorisations corresponding to the Gepner model branes of [389]. More recently, matrix factorisations for the more general permutation branes of [386] have been found in [74, 156]. In particular, for a permutation $\pi = (1\,2) \in S_2$ of two minimal models of level $k = d - 2$, the corresponding "matrix" factorisation is given by

$$ W(x, y) = x^d + y^d = p_0(x, y)\, p_1(x, y) $$
$$ \text{with } p_0 = \prod_{i \in I}(x - \eta_i y), \quad p_1 = \prod_{i' \in I'}(x - \eta_{i'} y) \tag{7.77} $$

where $\eta_1, \ldots, \eta_d$ is a complete set of dth roots of -1, with disjoint index sets I, I'. These factorisations were first studied in [24] and then analysed in great detail and compared to CFT permutation branes in [74]; these authors were able to, in particular, show that, in the case of the quintic, tensor products of a transposition brane (7.77) with minimal model branes (7.76) account for the long-sought-for D0-branes. The matrix factorisations considered in [156] correspond to branes induced by permutations $\pi = (1\,2 \cdots n)$ of arbitrary cycle length; for $n \geq 3$, their factors $p_{0,1}$ have rank greater than 1. Other important examples of matrix factorisations, related to elliptic surfaces and K3 surfaces, have been studied, e.g. in [76, 77, 283].

As an aside, let us remark on matrix factorisations and *topological defects* here. The link between defects separating two CFTs and boundary conditions for the tensor product CFT was already pointed out in Subsection 4.A.3. If the CFTs arise from Landau–Ginzburg models with potentials $W_1(x)$ and $W_2(y)$, a topological defect between the two can be given as a matrix factorisation of the potential $W(x; y) := W_1(x) - W_2(y)$; see in particular [81]. (The relative sign between W_1 and W_2 reflects the opposite orientation of the defect line relative to the half-planes carrying W_1 and W_2, respectively.)

The main new feature of defects is that they can be "fused" with other defects to obtain new ones, or with boundary conditions to obtain new boundary conditions. With topological defects and branes, this process does not produce singularities and, in terms of matrix factorisations, it has an explicit description: fusion is essentially nothing but taking tensor products [81]. For example, the tensor product of a defect between $W_1(x)$ and $W_2(y)$ with a boundary condition for $W_2(y)$ produces a matrix factorisation of the Landau–Ginzburg potential $W_1(x) - W_2(y) + W_2(y) = W_1(x)$, i.e. a brane for the Landau–Ginzburg model with $W_1(x)$. However, we need to remove all traces of the variables y by reducing out a (possibly infinite) number of trivial matrix factorisations from the tensor product; this can be done by equivalence (row and column) transformations that do not affect the open string spectrum.

Fusion with defects is a rather new concept still very much under investigation, but allows for various promising applications, including, e.g., the study of bulk deformations of topological branes [82].

Boundary perturbations of topological branes can be studied without invoking defects, and much more easily than in the full CFT framework. As a warm-up example, consider a topological open superstring ψ stretched between two branes corresponding to matrix factorisations Q_1 and Q_2 of a Landau–Ginzburg potential W. This can trigger a boundary RG flow ("tachyon condensation") to a new brane – a process which may be very difficult to analyse in CFT, while for the topological counterpart the picture is very simple: the system flows to the so-called cone $C\psi$ of ψ,

$$C\psi = \begin{pmatrix} Q_1 & u\,\psi \\ 0 & Q_2 \end{pmatrix}.$$

The fermionic open string $\psi \in H^{\mathrm{odd}}(D_{Q_1 Q_2})$ satisfies $Q_1\psi + \psi Q_2 = 0$, which implies that there are no u^2 terms appearing in $C\psi^2$; thus $C\psi$ is again a matrix factorisation of the same W for arbitrary values of the parameter $u \in \mathbb{C}$. The question of whether, for a chosen value of u, the cone $C\psi$ is equivalent to a previously known matrix factorisation or direct sums of such has, as yet, no general answer; instead, we have to perform a case-by-case search for equivalence transformations of the type $U\,C\psi\,U^{-1} = Q_3 \oplus Q_4 \oplus \ldots$ with an invertible polynomial matrix U which is even with respect to the $\mathbb{Z}_2$-grading.

General perturbations of a single matrix factorisation Q of W are polynomials (or even power series) in parameters u_i,

$$Q_{\mathrm{def}}(u) = Q + Q^{(1)} + Q^{(2)} + \cdots \quad \text{with} \quad Q^{(1)} = \sum_i u_i\,\psi_i, \tag{7.78}$$

where the ψ_i are a basis of $H^{\mathrm{odd}}(D_{QQ})$. One needs to determine suitable higher-order terms $Q^{(n)}$ in the u_i such that Q_{def} is again a matrix factorisation of W. To second order, we obtain the condition $Q^{(1)}Q^{(1)} = -\{Q, Q^{(2)}\}$, but usually the left-hand side is not in the image of D_{QQ}, so that we encounter obstructions, $obs^{(2)} = Q^{(1)}Q^{(1)} + \{Q, Q^{(2)}\}$.

Explicit expressions for the deformed matrix factorisation $Q_{\mathrm{def}}(u)$ as well as for the obstruction $obs(u)$ in

$$Q_{\mathrm{def}}(u)^2 = W \cdot \mathbf{1} + obs(u)$$

can be computed algorithmically [106, 316, 327], in a straightforward way or by incorporating the so-called Massey product algorithm – the latter guaranteeing that the obstruction can be expanded into a basis $\phi_j \in H^{\mathrm{even}}(D_{QQ})$ of bosonic open strings,

$$obs(u) = \sum_j f_j(u)\,\phi_j\,,$$

with polynomials (or power series) $f_j(u)$ in the parameters u_i. Their common vanishing locus $\{\,f_j(u) = 0\,\}$ parametrises families of allowed deformations, and up to identifying equivalent matrix factorisations it is the moduli space of the brane Q. In physical terms, this should correspond to the flat directions of an effective superpotential $\mathcal{W}_{\mathrm{eff}}(u)$ of the brane Q (see, e.g., [275]) so that the algorithm computing the obstructions also provides information on the interaction terms in the effective field theory describing open strings ending on Q.

Apart from matrix factorisations (or topological branes) themselves, in topological string theory we are also interested in correlation functions [274, 304], or indeed topological string amplitudes involving integrated string vertex operators,

$$B_{i_1\ldots i_n} := \langle\,\psi_{i_1}\psi_{i_2}\psi_{i_3}\psi_{i_4}^{(1)}\ldots\psi_{i_n}^{(1)}\,\rangle\,, \tag{7.79}$$

with (path-ordered) integrated fields $\psi_i^{(1)}$ as in definition (7.70). Here, we have exploited the fact that due to $\mathrm{SL}(2,\mathbb{R})$-invariance the first three of the worldsheet integrations can be traded for three fixed insertion points.

These amplitudes have to satisfy consistency relations which can be extracted from OPEs, CFT sewing relations and the BRST symmetry [275, 277] – but see also [133] for analogous studies for topological bulk models. In the boundary case, consistency conditions for open string amplitudes are especially useful, as they endow the set of physical fields with the structure of an A_∞-*algebra*: These are $\mathbb{Z}_2$-graded vector spaces A with a sequence of multilinear products $r_n : A^{\otimes n} \longrightarrow A$ for $n = 1, 2, \ldots$, such that

$$\sum_{i,j} (\pm)\, r_{n-j+1} \circ \left(\mathbf{1}^{\otimes i} \otimes r_j \otimes \mathbf{1}^{\otimes(n-i-j)}\right) = 0$$

up to some signs depending on n, i, j. Explicitly, the first two conditions are $r_1 \circ r_1 = 0$ and $r_1 \circ r_2 + r_2 \circ (r_1 \otimes \mathbf{1}) + r_2 \circ (\mathbf{1} \otimes r_1) = 0$, thus r_1 is a differential satisfying a Leibniz rule with respect to r_2; moreover, the next condition implies that the multiplication r_2 is associative only up to corrections involving r_1 and r_3 – which is why A_∞-algebras are sometimes referred to as "associative up to homotopy". We refer to [313] for precise definitions and further references.

In topological open string theory, we take $A = \mathcal{H}_{\mathrm{phys}}$ to be the space of physical boundary fields, graded by the fermion number. The A_∞-products are induced by the amplitudes (7.79) as $r_n(\psi_{i_1}, \ldots, \psi_{i_n}) := (\pm)\, B_{i_1\ldots i_n}^a\,\psi_a$, up to signs depending on the degrees of the arguments; indices are raised and lowered using the two-point function (the non-degenerate "topological metric"), which of course depends on the boundary condition, as does the "vacuum expectation value" in equation (7.79).

Existence of a metric $\langle \cdot, \cdot \rangle$, and the fact that open string insertion points can be viewed as lying on the boundary of a disk, leads to additional conditions that open string A_∞-algebras have to satisfy, namely the cyclicity property $\langle \psi_{i_0}, r_n(\psi_{i_1}, \ldots, \psi_{i_n}) \rangle = (\pm) \langle \psi_{i_1}, r_n(\psi_{i_2}, \ldots, \psi_{i_n}, \psi_{i_0}) \rangle$. As was discovered by Carqueville, the interplay of cyclicity and further A_∞-structures turns out to be surprisingly restrictive and can be exploited to solve open topological string theories (at genus 0) in the sense that all amplitudes can be computed algorithmically [105]. From the amplitudes (7.79), we can in particular obtain the effective superpotential

$$\mathcal{W}_{\text{eff}}(u) = \sum_{n \geq 3} B_{i_1 \ldots i_n} \, u_{i_1} \cdots u_{i_n}$$

of the topological brane; the u_i are deformation parameters as before, and again the formula stated is correct up to signs; see, e.g., [275] for more details.

In order to obtain a complete solution of tree-level open string theory, the work [105] draws on results from various branches of mathematics including, apart from A_∞-algebras, formal non-commutative geometry (see [237] for an introduction) and abstract category theory. For quite a while, matrix factorisations have been an important tool in commutative algebra, in particular in the study of Cohen–Macaulay modules: a factorisation (7.75) defines a periodic free resolution of the module *coker* p_1, and BRST cohomology is given the by Ext-groups of this module. More recently, matrix factorisations have proven useful in the context of the homological mirror symmetry programme initiated by Kontsevich, which was already briefly alluded to above: in [366], Orlov showed that the category $MF(W)$ of (graded) matrix factorisations of a quasi-homogeneous polynomial W is equivalent to the derived category $D^b(coh(X))$ of coherent sheaves on the variety $X = \{W = 0\}$ in projective space. The objects of $MF(W)$ are matrix factorisations as in equation (7.75); the morphisms are given by open strings between such topological branes. This equivalence of triangulated categories provides a computationally simple description of the "B-side" of homological mirror symmetry. The "A-side", given by the Fukaya category of the mirror manifold, remains much more difficult to describe.

Appendix

In this appendix, we collect formulas that are useful in connection with modular transformations of conformal characters and for the computation of spectra for Gepner model branes.

A Some omnipresent q-series

Certain q-series appear over and over again in models of conformal field theory (CFT) as building blocks of characters and partition functions. Here, we will list some of them along with formulas describing their behaviour under modular transformations.

The basic object is, of course, the *Dedekind eta function* defined by

$$\eta(q) = q^{\frac{1}{24}} \prod_{n=1}^{\infty} (1 - q^n). \tag{A.1}$$

Its inverse is the generating function of the partitions and appears as the character of Virasoro Verma modules or of irreducible representations of the U(1) current algebra.

The *classical theta functions* are of equal importance for free conformal field theories. They can be defined as follows:

$$\sqrt{\frac{\vartheta_3(q)}{\eta(q)}} = q^{-\frac{1}{48}} \prod_{n=1}^{\infty} (1 + q^{n-\frac{1}{2}}), \tag{A.2}$$

$$\sqrt{\frac{\vartheta_4(q)}{\eta(q)}} = q^{-\frac{1}{48}} \prod_{n=1}^{\infty} (1 - q^{n-\frac{1}{2}}), \tag{A.3}$$

$$\sqrt{\frac{\vartheta_2(q)}{\eta(q)}} = \sqrt{2} q^{\frac{1}{24}} \prod_{n=1}^{\infty} (1 + q^n), \tag{A.4}$$

$$\sqrt{\frac{\vartheta_1(q)}{\eta(q)}} = \frac{1}{\sqrt{2}} q^{\frac{1}{24}} \prod_{n=0}^{\infty} (1 - q^n) = 0. \tag{A.5}$$

We give the definitions in this slightly implicit form because it is those combinations which appear as characters of a free fermion ($c = \frac{1}{2}$) with various spin structures: The first two lines contain the characters $\operatorname{tr} q^{L_0 - \frac{c}{24}}$ and $\operatorname{tr} (-1)^F q^{L_0 - \frac{c}{24}}$, respectively, in the Neveu–Schwarz sector, the last two functions appear as ($2^{-\frac{1}{2}}$ times) the same traces over states in the Ramond sector. With this interpretation, $\vartheta_1(q)$ vanishes because there are two degenerate Ramond ground states with opposite fermion number $(-1)^F$.

In string theory, we often use the notations

$$f_i(q^{\frac{1}{2}}) := \sqrt{\frac{\vartheta_i(q)}{\eta(q)}} \quad \text{for} \quad i = 2, 3, 4 \quad \text{and} \quad f_1(q^{\frac{1}{2}}) := \eta(q). \tag{A.6}$$

The theta functions, and therefore the functions $f_i(q)$ from equation (A.6), satisfy the algebraic relations (for arbitrary argument q)

$$\vartheta_2(q)\,\vartheta_3(q)\,\vartheta_4(q) = 2\,\eta(q)^3, \qquad \vartheta_3(q)^4 - \vartheta_4(q)^4 - \vartheta_2(q)^4 = 0. \tag{A.7}$$

The second equation, also known as *abstruse identity*, guarantees the vanishing of one-loop partition functions in superstring theory, a fact that is also important for the investigation of BPS-brane systems.

Under the modular transformation $q \mapsto \tilde{q}$ with $q = e^{2\pi i \tau}$ and $\tilde{q} = e^{-2\pi i/\tau}$, the eta and theta functions behave in a simple way:

$$\eta(\tilde{q}) = \sqrt{-i\tau}\,\eta(q), \tag{A.8}$$

$$\vartheta_2(\tilde{q}) = \sqrt{-i\tau}\,\vartheta_4(q), \qquad \vartheta_3(\tilde{q}) = \sqrt{-i\tau}\,\vartheta_3(q), \qquad \vartheta_4(\tilde{q}) = \sqrt{-i\tau}\,\vartheta_2(q). \tag{A.9}$$

In particular, this provides a proof of abstruse identity: the combination $\vartheta_3^4 - \vartheta_4^4 - \vartheta_2^4$ is a modular form of weight two, i.e. a function $g(\tau)$ with $g(\frac{a\tau+b}{c\tau+d}) = (c\tau + d)^2 g(\tau)$ for any $\left(\begin{smallmatrix} a & b \\ c & d \end{smallmatrix}\right) \in \mathrm{SL}(2, \mathbb{Z})$. On the other hand, it is known that there are no non-trivial forms of this kind.

The *Jacobi triple product identity* is a very useful device which allows infinite products to be written as infinite sums:

$$\prod_{n=1}^{\infty} (1 - q^n)(1 + q^{n-\frac{1}{2}}\,w)(1 + q^{n-\frac{1}{2}}\,w^{-1}) = \sum_{m=-\infty}^{\infty} q^{\frac{1}{2}m^2}\,w^m\,; \tag{A.10}$$

w is an arbitrary complex argument. With the above definitions of the quotients $\vartheta_i(q)/\eta(q)$, we obtain sum-formulas for the theta functions (inserting $w = q^{-\frac{1}{2}}, 1, -1$ into the triple product identity):

$$\vartheta_2(q) = \sum_{m=-\infty}^{\infty} q^{\frac{1}{2}(m-\frac{1}{2})^2}, \quad \vartheta_3(q) = \sum_{m=-\infty}^{\infty} q^{\frac{1}{2}m^2}, \quad \vartheta_4(q) = \sum_{m=-\infty}^{\infty} (-1)^m q^{\frac{1}{2}m^2}. \tag{A.11}$$

These series show up naturally in connection with free bosons compactified to a circle and with free fermions.

The following more general theta functions are useful, e.g. in the context of branes carrying an electromagnetic field:

$$\vartheta_3(z,\tau) \equiv \sum_{m=-\infty}^{\infty} q^{\frac{1}{2}m^2} e^{2\pi imz} = (-i\tau)^{-\frac{1}{2}} e^{-i\pi\frac{z^2}{\tau}} \vartheta_3\left(\frac{z}{\tau}, -\frac{1}{\tau}\right), \qquad (A.12)$$

$$\vartheta_4(z,\tau) \equiv \sum_{m=-\infty}^{\infty} q^{\frac{1}{2}m^2} e^{2\pi im(z+\frac{1}{2})} = (-i\tau)^{-\frac{1}{2}} e^{-i\pi\frac{z^2}{\tau}} \vartheta_4\left(\frac{z}{\tau}, -\frac{1}{\tau}\right), \qquad (A.13)$$

$$\vartheta_2(z,\tau) \equiv \sum_{m=-\infty}^{\infty} q^{\frac{1}{2}(m+\frac{1}{2})^2} e^{2\pi i(m+\frac{1}{2})z} = (-i\tau)^{-\frac{1}{2}} e^{-i\pi\frac{z^2}{\tau}} \vartheta_2\left(\frac{z}{\tau}, -\frac{1}{\tau}\right), \quad (A.14)$$

$$\vartheta_1(z,\tau) \equiv \sum_{m=-\infty}^{\infty} q^{\frac{1}{2}(m+\frac{1}{2})^2} e^{2\pi i(m+\frac{1}{2})(z+\frac{1}{2})} = i\,(-i\tau)^{-\frac{1}{2}} e^{-i\pi\frac{z^2}{\tau}} \vartheta_1\left(\frac{z}{\tau}, -\frac{1}{\tau}\right).$$

$$(A.15)$$

In each line, the second equation contains the transformation rules under S : $\tau \mapsto -\frac{1}{\tau}$ and the associated change of the complex z-variable. The "specialised theta functions" defined before are obtained from the above "unspecialised" ones by putting z to zero: $\vartheta_i(q) = \vartheta_i(0,\tau)$. With the help of equation (A.10), we arrive at a product formula

$$\vartheta_1(z,\tau) = -2\,q^{\frac{1}{8}} \sin \pi z \prod_{n=1}^{\infty} (1-q^n)(1-e^{2\pi iz}q^n)(1-e^{-2\pi iz}q^n) \qquad (A.16)$$

for the last of the theta functions above. Using this representation, it is immediately shown that

$$\eta(q)^3 = -\frac{1}{2\pi}\, \partial_z \vartheta_1(0,\tau) \ .$$

Variants of the theta function (A.12) can be used to express the unspecialised characters of $\widehat{SU(2)}_k$ representations. For $k \in \mathbb{Z}_+$ and $l = 0, 1, \ldots, k$, we define

$$\Theta_l^{(k)}(z,\tau,u) = e^{-2\pi iku} \sum_{m\in\mathbb{Z}+\frac{l}{2k}} q^{k\,m^2}\, e^{-2\pi ikmz} \ .$$

With the help of $\Theta_{-l}^{(k)}(z,\tau,u) = \Theta_l^{(k)}(-z,\tau,u)$, this can be extended to negative weights l.

B Useful identities for Gepner model computations

Here, we list some simple auxiliary formulas which were used in order to find the solution of Cardy's constraints and for supersymmetry requirements in the construction of Gepner model boundary states. Each identity follows in a more-or-less straightforward way from root of unity relations; N_{ij}^l denotes the fusion rules of $\widehat{SU(2)}_k$.

$$\sum_{l=0}^{k} \sin \pi \frac{(l'+1)(l+1)}{k+2} \sin \pi \frac{(l+1)(l''+1)}{k+2} = \frac{k+2}{2} \delta_{l',l''}$$

$$\text{for } 0 \le l', l'' \le k ; \quad k = 1, 2, \ldots \quad \text{(B.1)}$$

$$\sum_{l=0}^{k} (-1)^l \sin \pi \frac{(l'+1)(l+1)}{k+2} \sin \pi \frac{(l+1)(l''+1)}{k+2} = 0$$

$$\text{for } 0 \le l', l'' \le k ; \quad k = 1, 2, \ldots \quad \text{(B.2)}$$

$$\sum_{\substack{m=0 \\ m \text{ even}}}^{2k+3} e^{i\pi \frac{mm'}{k+2}} = \frac{k+2}{2} \left(\delta_{m',0}^{(2k+4)} + \delta_{m',k+2}^{(2k+4)} \right)$$

$$\text{for } k = 0, 1, 2, \ldots \quad \text{(B.3)}$$

$$\sum_{\substack{m=0 \\ m \text{ odd}}}^{2k+3} e^{i\pi \frac{mm'}{k+2}} = \frac{k+2}{2} \left(\delta_{m',0}^{(2k+4)} - \delta_{m',k+2}^{(2k+4)} \right)$$

$$\text{for } k = 0, 1, 2, \ldots \quad \text{(B.4)}$$

$$\sum_{l=0}^{k} \frac{\sin \pi \frac{(l'+1)(l+1)}{k+2} \sin \pi \frac{(l''+1)(l+1)}{k+2} \sin \pi \frac{(l+1)(l'''+1)}{k+2}}{\sin \pi \frac{(l+1)}{k+2}} = \frac{k+2}{2} N_{l',l''}^{l'''} \quad \text{(B.5)}$$

$$\sum_{\substack{l=0 \\ l \text{ even}}}^{k} \frac{\sin \pi \frac{(l'+1)(l+1)}{k+2} \sin \pi \frac{(l''+1)(l+1)}{k+2} \sin \pi \frac{(l+1)(l'''+1)}{k+2}}{\sin \pi \frac{(l+1)}{k+2}}$$

$$= \frac{k+2}{4} \left(N_{l',l''}^{l'''} + N_{l',l''}^{k-l'''} \right) \quad \text{(B.6)}$$

$$\sum_{\substack{l=0 \\ l \text{ odd}}}^{k} \frac{\sin \pi \frac{(l'+1)(l+1)}{k+2} \sin \pi \frac{(l''+1)(l+1)}{k+2} \sin \pi \frac{(l+1)(l'''+1)}{k+2}}{\sin \pi \frac{(l+1)}{k+2}}$$

$$= \frac{k+2}{4} \left(N_{l',l''}^{l'''} - N_{l',l''}^{k-l'''} \right) \quad \text{(B.7)}$$

Note that the Verlinde formula (B.5) for the $N_{l',l''}^{l'''}$ can be applied even if l''' is outside the standard range $0 \le l''' \le k$. This was used in [72] to obtain a compact notation for the CFT intersection form of boundary states.

References

[1] A. Abouelsaood, C. G. Callan, C. R. Nappi, S. A. Yost, Open strings in background gauge fields, *Nucl. Phys.* B **280** (1987) 599

[2] I. Affleck, Universal term in the free energy at a critical point and the conformal anomaly, *Phys. Rev. Lett.* **56** (1986) 746

[3] I. Affleck, Conformal field theory approach to the Kondo effect, *Acta Phys. Polon.* B **26** (1995) 1869, cond-mat/9512099

[4] I. Affleck, A. W. W. Ludwig, Critical theory of overscreened Kondo fixed points, *Nucl. Phys.* B **360** (1991) 641; The Kondo effect, conformal field theory and fusion rules, *Nucl. Phys.* B **352** (1991) 849

[5] I. Affleck, A. W. W. Ludwig, Universal noninteger 'groundstate degeneracy' in critical quantum systems, *Phys. Rev. Lett.* **67** (1991) 161

[6] I. Affleck, A. W. W. Ludwig, Exact conformal field theory results on the multichannel Kondo effect: single-fermion Green's function, self-energy, and resistivity, *Phys. Rev.* B **48** (1993) 7297

[7] I. Affleck, M. Oshikawa, H. Saleur, Boundary critical phenomena in the three-state Potts model, cond-mat/9804117

[8] I. Affleck, Edge critical behaviour of the 2-dimensional tri-critical Ising model, *J. Phys.* A **33** (2000) 6473, cond-mat/0005286

[9] M. Aganagic, R. Gopakumar, S. Minwalla, A. Strominger, Unstable solitons in noncommutative gauge theory, *J. High Energy Phys.* **0104** (2001) 001, hep-th/0009142

[10] A. Yu. Alekseev, S. Fredenhagen, T. Quella, V. Schomerus, Non-commutative gauge theory of twisted D-branes, hep-th/0205123

[11] A. Yu. Alekseev, A. Recknagel, V. Schomerus, Generalization of the Knizhnik–Zamolodchikov equations, *Lett. Math. Phys.* **41** (1997) 169, hep-th/9610066

[12] A. Yu. Alekseev, A. Recknagel, V. Schomerus, Non-commutative world-volume geometries: branes on SU(2) and fuzzy spheres, *J. High Energy Phys.* **9909** (1999) 023, hep-th/9908040

[13] A. Yu. Alekseev, A. Recknagel, V. Schomerus, Brane dynamics in background fluxes and non-commutative geometry, hep-th/0003187

[14] A. Yu. Alekseev, V. Schomerus, D-branes in the WZW model, *Phys. Rev.* D **60** (1999) 061901, hep-th/9812193

[15] A. Yu. Alekseev, V. Schomerus, RR charges of D2-branes in the WZW model, hep-th/0007096

[16] A. Yu. Alekseev, S. Shatashvili, From geometric quantization to conformal field theory, *Commun. Math. Phys.* **128** (1990) 197; Quantum groups and WZW models, *Commun. Math. Phys.* **133** (1990) 353

[17] L. Alvarez-Gaumé, D. Z. Freedman, Geometrical structure and ultraviolet finiteness in the supersymmetric sigma model, *Commun. Math. Phys.* **80** (1981) 443

[18] L. Alvarez-Gaumé, C. Gomez, G. Sierra, Quantum group interpretation of some conformal field theories, *Phys. Lett.* B **220** (1989) 142

[19] L. Alvarez-Gaumé, C. Gomez, G. Sierra, Topics in conformal field theory, in *Physics and Mathematics of Strings*, L. Brink, D. Friedan, A. M. Polyakov (eds.), World Scientific 1990

[20] C. Angelantonj, M. Bianchi, G. Pradisi, A. Sagnotti, Y. S. Stanev, Comments on Gepner models and type I vacua in string theory, *Phys. Lett.* B **387** (1996) 743, hep-th/9607229

[21] C. Angelantonj, A. Sagnotti, Open strings, *Phys. Rept.* **371** (2002) 1, Erratum: *ibid* **376** (2003) 339, hep-th/0204089

[22] I. Antoniadis, C. Bachas, Branes and the gauge hierarchy, *Phys. Lett. B* **450** (1999) 83, hep-th/9812093

[23] F. Ardalan, H. Arfaei, M. M. Sheikh-Jabbari, Noncommutative geometry from strings and branes, *J. High Energy Phys.* **9902** (1999) 016, hep-th/9810072

[24] S. K. Ashok, E. Dell'Aquila, D. E. Diaconescu, Fractional branes in Landau–Ginzburg orbifolds, hep-th/0401135

[25] P. S. Aspinwall, The Landau–Ginzburg to Calabi–Yau dictionary for D-branes, *J. Math. Phys.* **48** (2007) 082304, hep-th/0610209

[26] P. S. Aspinwall, M. R. Douglas, D-brane stability and monodromy, *J. High Energy Phys.* **0205** (2002) 031, hep-th/0110071

[27] C. Bachas, D-brane dynamics, *Phys. Lett. B* **374** (1996) 37, hep-th/9511043

[28] C. Bachas, Lectures on D-branes, hep-th/9806199

[29] C. Bachas, On the symmetries of classical string theory, arXiv:0808.2777 [hep-th]

[30] C. Bachas, J. de Boer, R. Dijkgraaf, H. Ooguri, Permeable conformal walls and holography, *J. High Energy Phys.* **0206** (2002) 027, hep-th/0111210

[31] C. Bachas, M. R. Douglas, C. Schweigert, Flux stabilization of D-branes, *J. High Energy Phys.* **0005** (2000) 048, hep-th/0003037

[32] C. Bachas, M. R. Gaberdiel, Loop operators and the Kondo problem, *J. High Energy Phys.* **0411** (2004) 065, hep-th/0411067

[33] F. A. Bais, P. Bouwknegt, M. Surridge, K. Schoutens, Extensions of the Virasoro algebra constructed from Kac–Moody algebras using higher order Casimir invariants, *Nucl. Phys. B* **304** (1988) 348; Coset construction for extended Virasoro algebras, *Nucl. Phys. B* **304** (1988) 371

[34] V. Balasubramanian, R. G. Leigh, D-branes, moduli and supersymmetry, *Phys. Rev. D* **55** (1997) 6415, hep-th/9611165

[35] T. Banks, L. J. Dixon, D. Friedan, E. Martinec, Phenomenology and conformal field theory or Can string theory predict the weak mixing angle?, *Nucl. Phys. B* **299** (1988) 613

[36] P. Bantay, Characters and modular properties of permutation orbifolds, *Phys. Lett. B* **419** (1998) 175, hep-th/9708120; Permutation orbifolds, *Nucl. Phys. B* **633** (2002) 365, hep-th/9910079

[37] M. Bauer, Aspects de l'invariance conformé, Université Paris VII, 1990

[38] M. Bauer, P. Di Francesco, C. Itzykson, J.-B. Zuber, Covariant differential equations and singular vectors in Virasoro representations, *Nucl. Phys. B* **362** (1991) 515

[39] M. Bauer, H. Saleur, On some relations between local height properties and conformal invariance, *Nucl. Phys. B* **320** (1989) 591

[40] K. Becker, M. Becker, D. R. Morrison, H. Ooguri, Y. Oz, Z. Yin, Supersymmetric cycles in exceptional holonomy manifolds and Calabi–Yau 4-folds, *Nucl. Phys. B* **480** (1996) 225, hep-th/9608116

[41] K. Becker, M. Becker, A. Strominger, Fivebranes, membranes and non-perturbative string theory, *Nucl. Phys. B* **456** (1995) 130, hep-th/9507158

[42] R. E. Behrend, P. A. Pearce, J.-B. Zuber, Integrable boundaries, conformal boundary conditions and A–D–E fusion rules, *J. Phys. A* **31** (1998) L763, hep-th/9807142

[43] R. E. Behrend, P. A. Pearce, V. B. Petkova, J.-B. Zuber, On the classification of bulk and boundary conformal field theories, *Phys. Lett. B* **444** (1998) 163, hep-th/9809097

[44] R. E. Behrend, P. A. Pearce, V. B. Petkova, J.-B. Zuber, Boundary conditions in rational conformal field theories, *Nucl. Phys. B* **570** (2000) 525, **579** (2000) 707, hep-th/9908036

[45] A. A. Belavin, A. M. Polyakov, A. B. Zamolodchikov, Infinite conformal symmetry in two-dimensional quantum field theory, *Nucl. Phys. B* **241** (1984) 333

[46] L. Benoit, Y. Saint-Aubin, Degenerate conformal field theories and explicit expressions of some null vectors, *Phys. Lett. B* **215** (1988) 517

[47] O. Bergman, M. R. Gaberdiel, A non-supersymmetric open string theory and S-duality, *Nucl. Phys. B* **499** (1997) 183, hep-th/9701137

[48] O. Bergman, M. R. Gaberdiel, Stable non-BPS D-particles, *Phys. Lett. B* **441** (1998) 133, hep-th/9806155

[49] M. Berkooz, M. R. Douglas, R. G. Leigh, Branes intersecting at angles, *Nucl. Phys. B* **480** (1996) 265, hep-th/9606139

[50] M. Bershadsky, S. Cecotti, H. Ooguri, C. Vafa, Kodaira–Spencer theory of gravity and exact results for quantum string amplitudes, *Commun. Math. Phys.* **165** (1994) 311; hep-th/9309140

[51] M. Bertolini, P. Fré, F. Hussain, R. Iengo, C. Nuñez, C. Scrucca, Black hole – D-brane correspondence: an example, hep-th/9807209

[52] M. Bertolini, P. Fré, R. Iengo, C. Nuñez, C. Scrucca, Black holes as D3-branes on Calabi–Yau threefolds, *Phys. Lett. B* **431** (1998) 22, hep-th/9803096

[53] M. Bianchi, G. Pradisi, A. Sagnotti, Toroidal compactification and symmetry breaking in open string theories, *Nucl. Phys. B* **376** (1992) 365

[54] M. Bianchi, A. Sagnotti, On the systematics of open string theories, *Phys. Lett. B* **247** (1990) 517

[55] M. Bianchi, A. Sagnotti, Twist symmetry and open string Wilson lines, *Nucl. Phys. B* **361** (1991) 519

[56] M. Bianchi, Y. S. Stanev, Open strings on the Neveu–Schwarz pentabrane, *Nucl. Phys. B* **523** (1998) 193, hep-th/9711069

[57] M. Billó, D. Cangemi, P. Di Vecchia, Boundary states for moving D-branes, *Phys. Lett. B* **400** (1997) 63, hep-th/9701190

[58] M. Billó, B. Craps, F. Roose, Orbifold boundary states from Cardy's condition, *J. High Energy Phys.* **0101** (2001) 038, hep-th/0011060

[59] L. Birke, J. Fuchs, C. Schweigert, Symmetry breaking boundary conditions and WZW orbifolds, *Adv. Theor. Math. Phys.* **3** (1999) 671, hep-th/9905038

[60] H. W. J. Bloete, J. L. Cardy and M. P. Nightingale, Conformal invariance, the central charge, and universal finite size amplitudes at criticality, *Phys. Rev. Lett.* **56** (1986) 742

[61] R. Blumenhagen, W. Eholzer, A. Honecker, K. Hornfeck, R. Hübel, Unifying W-algebras, *Phys. Lett. B* **332** (1994) 51, hep-th/9404113

[62] R. Blumenhagen, M. Flohr, A. Kliem, W. Nahm, A. Recknagel, R. Varnhagen, W-algebras with two and three generators, *Nucl. Phys. B* **361** (1991) 255

[63] R. Blumenhagen, E. Plauschinn, Introduction to conformal field theory, *Lecture Notes in Physics*, vol. 779, Springer 2000

[64] R. Blumenhagen, T. Weigand, Chiral supersymmetric Gepner model orientifolds, *J. High Energy Phys.* **0402** (2004) 041, hep-th/0401148

[65] R. Blumenhagen, A. Wisskirchen, Spectra of 4D, N=1 type I string vacua on non-toroidal CY threefolds, *Phys. Lett. B* **438** (1998) 52, hep-th/9806131

[66] J. Böckenhauer, D. E. Evans, Modular invariants, graphs and α-induction for nets of subfactors I,II,II, *Commun. Math. Phys.* **197** (1998) 361, hep-th/9801171; *Commun. Math. Phys.* **200** (1999) 57, hep-th/9805023; *Commun. Math. Phys.* **205** (1999) 183, hep-th/9812110

[67] J. Böckenhauer, D. E. Evans, Y. Kawahigashi, On α-induction, chiral generators and modular invariants for subfactors, math.OA/9904109; Chiral structure of modular invariants for subfactors, math.OA/9907149

[68] R. Borcherds, Vertex algebras, Kac–Moody algebras, and the monster, *Proc. Natl. Acad. Sci. USA* **83** (1986) 3068; Monstrous moonshine and monstrous Lie superalgebras, *Invent. Math.* **109** (1992) 405

[69] P. Bouwknegt, V. Mathai, D-branes, B-fields and twisted K-theory, *J. High Energy Phys.* **0003** (2000) 007, hep-th/0002023

[70] P. Bouwknegt, K. Schoutens (eds.), *W-Symmetry*, World Scientific 1995

[71] I. Brunner, On orientifolds of WZW models and their relation to geometry, *J. High Energy Phys.* **0201** (2002) 007, hep-th/0110219

[72] I. Brunner, M. R. Douglas, A. Lawrence, C. Römelsberger, D-branes on the quintic, *J. High Energy Phys.* **0008** (2000) 015, hep-th/9906200

[73] I. Brunner, R. Entin, C. Römelsberger, D-branes on $T^4/\mathbb{Z}_2$ and T-Duality, *J. High Energy Phys.* **9906** (1999) 016, hep-th/9905078

[74] I. Brunner, M. Gaberdiel, Matrix factorisations and permutation branes, *J. High Energy Phys.* **0507** (2005) 012, hep-th/0503207

[75] I. Brunner, M. R. Gaberdiel, The matrix factorisations of the D-model, *J. Phys. A* **A38** (2005) 7901, hep-th/0506208

[76] I. Brunner, M. R. Gaberdiel, C. A. Keller, Matrix factorisations and D-branes on K3, *J. High Energy Phys.* **0606** (2006) 015, hep-th/0603196

[77] I. Brunner, M. Herbst, W. Lerche, B. Scheuner, Landau–Ginzburg realization of open string TFT, *J. High Energy Phys.* **0611** (2006) 043, hep-th/0305133

[78] I. Brunner, M. Herbst, W. Lerche, J. Walcher, Matrix factorizations and mirror symmetry: the cubic curve, *J. High Energy Phys.* **0611** (2006) 006, hep-th/0408243

[79] I. Brunner, K. Hori, Notes on orientifolds of rational conformal field theories, *J. High Energy Phys.* **0407** (2004) 023, hep-th/0208141

[80] I. Brunner, K. Hori, K. Hosomichi, J. Walcher, Orientifolds of Gepner models, hep-th/0401137

[81] I. Brunner, D. Roggenkamp, B-type defects in Landau–Ginzburg models, *J. High Energy Phys.* **0708** (2007) 093, arXiv:0707.0922 [hep-th]

[82] I. Brunner, D. Roggenkamp, Defects and bulk perturbations of boundary Landau–Ginzburg orbifolds, *J. High Energy Phys.* **0804** (2008) 001, arXiv:0712.0188 [hep-th]

[83] I. Brunner, V. Schomerus, D-branes at singular curves of Calabi–Yau compactifications, *J. High Energy Phys.* **0004** (2000) 020, hep-th/0001132

[84] I. Brunner, V. Schomerus, On superpotentials for D-branes in Gepner models, *J. High Energy Phys.* **0010** (2000) 016, hep-th/0008194

[85] A. O. Caldeira, A. J. Leggett, Influence of dissipation on quantum tunneling in macros, *Phys. Rev. Lett.* **46** (1981) 211; Path integral approach to quantum Brownian motion, *Physica A* **121** (1983) 587; Quantum tunnelling in a dissipative system, *Annals Phys.* **149** (1983) 374

[86] C. G. Callan, J. A. Harvey, A. Strominger, World sheet approach to heterotic instantons and solitons, *Nucl. Phys. B* **359** (1991) 611; Worldbrane actions for string solitons, *Nucl. Phys. B* **367** (1991) 60

[87] C. G. Callan, I. R. Klebanov, D-Brane boundary state dynamics, *Nucl. Phys. B* **465** (1996) 473, hep-th/9511173

[88] C. G. Callan, I. R. Klebanov, A. W. W. Ludwig, J. M. Maldacena, Exact solution of a boundary conformal field theory, *Nucl. Phys. B* **422** (1994) 417, hep-th/9402113

[89] C. G. Callan, C. Lovelace, C. R. Nappi, S. A. Yost, Adding holes and crosscaps to the superstring, *Nucl. Phys. B* **293** (1987) 83; Loop corrections to superstring equations of motion, *Nucl. Phys. B* **308** (1988) 221

[90] C. G. Callan, J. M. Maldacena, D-brane approach to black hole quantum mechanics, *Nucl. Phys. B* **472** (1996) 591, hep-th/9602043

[91] C. G. Callan, L. Thorlacius, Open string theory as dissipative quantum mechanics, *Nucl. Phys. B* **329** (1990) 117

[92] C. G. Callan, L. Thorlacius, World sheet dynamics of string junctions, *Nucl. Phys. B* **534** (1998) 121, hep-th/9803097

[93] P. Candelas, X. C. de la Ossa, P. S. Green, L. Parkes, An exactly soluble superconformal theory from a mirror pair of Calabi–Yau manifolds, *Phys. Lett. B* **258** (1991) 118; A pair of Calabi–Yau manifolds as an exactly soluble superconformal theory, *Nucl. Phys. B* **359** (1991) 21

[94] P. Candelas, X. C. de la Ossa, A. Font, S. Katz, D. R. Morrison, Mirror symmtry for two parameter models I, *Nucl. Phys. B* **416** (1994) 481, hep-th/9308083

[95] P. Candelas, A. Font, S. Katz, D. R. Morrison, Mirror symmtry for two parameter models II, *Nucl. Phys. B* **429** (1994) 626, hep-th/94030187

[96] A. Cappelli, D. Friedan, J. I. Latorre, C-theorem and spectral representation, *Nucl. Phys. B* **352** (1991) 616

[97] A. Cappelli, C. Itzykson, J.-B. Zuber, The ADE classification of minimal and $A_1^{(1)}$ conformal invariant theories, *Commun. Math. Phys.* **113** (1987) 1

[98] J. L. Cardy, Conformal invariance and surface critical behavior, *Nucl. Phys. B* **240** (1984) 514

[99] J. L. Cardy, Operator content of two-dimensional conformally invariant theories, *Nucl. Phys. B* **270** (1986) 186

[100] J. L. Cardy, Boundary conditions, fusion rules and the Verlinde formula, *Nucl. Phys. B* **324** (1989) 581

[101] J. L. Cardy, Conformal invariance and statistical mechanics, Lectures given at the Les Houches Summer School in Theoretical Physics, 1988

[102] J. L. Cardy, D. C. Lewellen, Bulk and boundary operators in conformal field theory, *Phys. Lett. B* **259** (1991) 274

[103] S. Carlip, What we don't know about BTZ black hole entropy, *Class. Quant. Grav.* **15** (1998) 3609, hep-th/9806026; Logarithmic corrections to black hole entropy from the Cardy formula, *Class. Quant. Grav.* **17** (2000) 4175, gr-qc/0005017

[104] U. Carow-Watamura, S. Watamura, Noncommutative geometry and gauge theory on fuzzy sphere, *Commun. Math. Phys.* **212** (2000) 395, hep-th/9801195

[105] N. Carqueville, Matrix factorisations and open topological string theory, *J. High Energy Phys.* **0907** (2009) 005, arXiv:0904.0862 [hep-th]

[106] N. Carqueville, L. Dowdy, A. Recknagel, Algorithmic deformation of matrix factorisations, *J. High Energy Phys.* **1204** (2012) 014, arXiv:1112.3352 [hep-th]

[107] N. Carqueville, I. Runkel, Rigidity and defect actions in Landau–Ginzburg models, *Commun. Math. Phys.* **310** (2012) 135, arXiv:1006.5609 [hep-th]

[108] M. Caselle, G. Ponzano, F. Ravanini, Towards a classification of fusion rule algebras in rational conformal field theories, *Int. J. Mod. Phys. B* **6** (1992) 2075

[109] A. S. Cattaneo, G. Felder, A path integral approach to the Kontsevich quantization formula, math.QA/9902090

[110] A. H. Chamseddine, J. Fröhlich, Some elements of Connes' non-commutative geometry, and space-time geometry, in *Chen Ning Yang, a Great Physicist of the Twentieth Century*, C. S. Liu and S.-T. Yau (eds.), International Press 1995, hep-th/9307012

[111] Y.-K. E. Cheung, M. Krogh, Noncommutative geometry from D0-branes in a background B-field, *Nucl. Phys. B* **528** (1998) 185, hep-th/9803031

[112] L. Chim, Boundary S-matrix for the tricritical Ising model, *Int. J. Mod. Phys. A* **11** (1996) 4491, hep-th/9510008

[113] P. Christe, R. Flume, The four point correlations of all primary operators of the D = 2 conformally invariant SU(2) sigma model with Wess–Zumino term, *Nucl. Phys. B* **282** (1987) 466

[114] C. Chu, P. Ho, Noncommutative open string and D-brane, *Nucl. Phys. B* **550** (1999) 151, hep-th/9812219

[115] A. Connes, *Noncommutative Geometry*, Academic Press 1994

[116] A. Coste, T. Gannon, Remarks on Galois symmetry in rational conformal field theories, *Phys. Lett. B* **323** (1994) 316

[117] B. Craps, M. R. Gaberdiel, Discrete torsion orbifolds and D branes 2, *J. High Energy Phys.* **0104** (2001) 013, hep-th/0101143

[118] A. Dabholkar, *Lectures on Orientifolds and Duality*, hep-th/9804208

[119] J. Dai, R. G. Leigh, J. Polchinski, New connections between string theories, *Mod. Phys. Lett. A* **4** (1989) 2073

[120] U. Danielsson, G. Ferretti, B. Sundborg, D-particle dynamics and bound states, *Int. J. Mod. Phys. A* **11** (1996) 5463, hep-th/9603081

[121] P. Di Francesco, P. Mathieu, D. Sénéchal, *Conformal Field Theory*, Springer 1997

[122] P. Di Francesco, J.-B. Zuber, SU(N) lattice integrable models associated with graphs, *Nucl. Phys. B* **338** (1990) 602

[123] P. Di Vecchia, M. Frau, I. Pesando, S. Sciuto, A. Lerda, R. Russo, Classical p-branes from boundary states, *Nucl. Phys. B* **507** (1997) 259, hep-th/9707068

[124] D. E. Diaconescu, M. R. Douglas, J. Gomis, Fractional branes and wrapped branes, *J. High Energy Phys.* **9802** (1998) 013, hep-th/9712230

[125] D.-E. Diaconescu, J. Gomis, Fractional branes and boundary states in orbifold theories, *J. High Energy Phys.* **0010** (2001) 001, hep-th/9906242

[126] D.-E. Diaconescu, Enhanced D-brane categories from string field theory, *J. High Energy Phys.* **0106** (2001) 016, hep-th/0104200

[127] D.-E. Diaconescu, C. Römelsberger, D-branes and bundles on elliptic fibrations, *Nucl. Phys. B* **574** (2000) 245, hep-th/9910172

[128] R. Dijkgraaf, Les Houches lectures on fields, strings and duality, in *Les Houches 1995, Quantum Symmetries*, A. Connes, K. Gawędzki (eds.) Elsevier 1995, pp. 3–147, hep-th/9703136

[129] R. Dijkgraaf, J. M. Maldacena, G. W. Moore and E. P. Verlinde, A black hole farey tail, hep-th/0005003

[130] R. Dijkgraaf, C. Vafa, E. Verlinde, H. Verlinde, Operator algebra of orbifold models, *Commun. Math. Phys.* **123** (1989) 485

[131] R. Dijkgraaf, E. Verlinde, Modular invariance and the fusion algebra, *Nucl. Phys. B Proc. Suppl.* **5B** (1988) 87

[132] R. Dijkgraaf, E. Verlinde, H. Verlinde, $C = 1$ conformal field theories on Riemann surfaces, *Commun. Math. Phys.* **115** (1988) 649

[133] R. Dijkgraaf, E. Verlinde, H. Verlinde, Toplogical strings in $D < 1$, *Nucl. Phys. B* **352** (1991) 59

[134] J. Distler, B. R. Greene, Some exact results on the superpotential from Calabi–Yau compactifications, *Nucl. Phys. B* **309** (1988) 295

[135] L. J. Dixon, Some world sheet properties of superstring compactifications, on orbifolds and otherwise, Lectures given at Trieste HEP Workshop 1987

[136] L. J. Dixon, J. A. Harvey, C. Vafa, E. Witten, Strings on orbifolds, *Nucl. Phys. B* **261** (1985) 678; Strings on orbifolds 2, *Nucl. Phys. B* **274** (1986) 285

[137] S. Doplicher, K. Fredenhagen, J. E. Roberts, The quantum structure of space-time at the Planck scale and quantum fields, *Commun. Math. Phys.* **172** (1995) 187, hep-th/0303037

[138] S. Doplicher, R. Haag, J. E. Roberts, Local observables and particle statistics I, II, *Commun. Math. Phys.* **23** (1971) 199, **35** (1974) 49

[139] S. Doplicher, J. E. Roberts, A new duality theory for compact groups, *Invent. Math.* **98** (1989) 157; Why there is a field algebra with a compact gauge group describing the superselection structure in particle physics?, *Commun. Math. Phys.* **131** (1990) 51

[140] P. Dorey, A. Pocklington, R. Tateo, G. Watts, TBA and TCSA with boundaries and excited states, *Nucl. Phys. B* **525** (1998) 641, hep-th/9712197

[141] P. Dorey, I. Runkel, R. Tateo, G. Watts, g-function flow in perturbed boundary conformal field theories, *Nucl. Phys. B* **578** (2000) 85, hep-th/9909216

[142] V. S. Dotsenko, V. A. Fateev, Conformal algebra and multipoint correlation functions in two-dimensional statistical models, *Nucl. Phys. B* **240** (1984) 312; Four point correlation functions and the operator algebra in the two-dimensional conformal invariant theories with the central charge $c \leq 1$, *Nucl. Phys. B* **251** (1985) 691

[143] M. R. Douglas, Branes within branes, hep-th/9512077

[144] M. R. Douglas, Two lectures on D-geometry and noncommutative geometry, hep-th/9901146; Topics in D-geometry, hep-th/9910170; D-branes on Calabi–Yau manifolds, math.ag/0009209

[145] M. R. Douglas, D-branes, categories and $N = 1$ supersymmetry, *J. Math. Phys.* **42**, 2818 (2001), hep-th/0011017; D-branes and $N = 1$ supersymmetry, hep-th/0105014; Dirichlet branes, homological mirror symmetry, and stability, math.ag/0207021

[146] M. R. Douglas, B. Fiol, C. Römelsberger, Stability and BPS branes, *J. High Energy Phys.* **0509** (2005) 006, hep-th/0002037; The spectrum of BPS branes on a noncompact Calabi–Yau manifold, *J. High Energy Phys.* **0509** (2005) 057 (2005), hep-th/0003263

[147] M. R. Douglas, B. R. Greene, D. R. Morrison, Orbifold resolution by D-branes, *Nucl. Phys. B* **506** (1997) 84, hep-th/9704151

[148] M. R. Douglas, C. Hull, D-branes and the noncommutative torus, *J. High Energy Phys.* **9802** (1998) 008, hep-th/9711165

[149] M. R. Douglas, D. Kabat, P. Pouliot, S. H. Shenker, D-branes and short distances in string theory, *Nucl. Phys. B* **485** (1997) 85, hep-th/9608024

[150] M. R. Douglas, G. W. Moore, D-branes, quivers, and ALE instantons, hep-th/9603167

[151] M. R. Douglas, N. A. Nekrasov, Noncommutative field theory, *Rev. Mod. Phys.* **73** (2002) 977, hep-th/0106048

[152] M. J. Duff, R. R. Khuri, J. X. Lu, String solitons, *Phys. Rept.* **259** (1995) 213, hep-th/9412184

[153] E. D'Hoker, String theory, Lecture Notes Princeton 1997; see http://www.math.ias.edu/ QFT/spring /index.html

[154] T. Eguchi, S.-K. Yang, $N = 2$ superconformal models as topological field theories, *Mod. Phys. Lett. A* **5** (1990) 1693

[155] S. Elitzur, G. Sarkissian, D-branes on a gauged WZW model, *Nucl. Phys. B* **625** (2002) 166, hep-th/0108142

[156] H. Enger, A. Recknagel, D. Roggenkamp, Permutation branes and linear matrix factorisations, *J. High Energy Phys.* **0601** (2006) 087, hep-th/0508053

[157] D. E. Evans, Y. Kawahigashi, Orbifold subfactors from Hecke algebras, *Commun. Math. Phys.* **165** (1994) 445

[158] F. Falceto, K. Gawędzki, Lattice Wess–Zumino–Witten model and quantum groups, *J. Geom. Phys.* **11** (1993) 251, hep-th/9209076

[159] B. L. Feigin, D. B. Fuchs, Invariant skew-symmetric differential operators on the line and Verma modules over the Virasoro algebra, *Funct. Anal. Appl.* **16** (1982) 114; Verma modules over the Virasoro algebra, in *Lecture Notes in Mathematics*, vol. **1060**, Springer 1984, p. 230

[160] B. L. Feigin, T. Nakanishi, H. Ooguri, The annihilating ideals of minimal models, *Int. J. Mod. Phys. A* **7** Suppl. **1A** (1992) 217

[161] G. Felder, BRST Approach to Minimal Models, *Nucl. Phys. B* **317** (1989) 215, Erratum: *ibid* B **324** (1989) 548

[162] G. Felder, J. Fröhlich, J. Fuchs, C. Schweigert, The geometry of WZW branes, *J. Geom. Phys.* **34** (2000) 162, hep-th/9909030

[163] G. Felder, J. Fröhlich, J. Fuchs, C. Schweigert, Conformal boundary conditions and three-dimensional topological field theory, *Phys. Rev. Lett.* **84** (2000) 1659, hep-th/9909140; Correlation functions and boundary conditions in RCFT and three-dimensional topology, *Compos. Math.* **131** (2002) 189, hep-th/9912239

[164] G. Felder, J. Fröhlich, G. Keller, On the structure of unitary conformal field theory I,II, *Commun. Math. Phys.* **124** (1989) 417, **130** (1990) 1

[165] G. Felder, J. Fröhlich, G. Keller, Braid matrices and structure constants for minimal conformal models, *Commun. Math. Phys.* **124** (1989) 647

[166] G. Felder, K. Gawędzki, A. Kupiainen, Spectra of Wess–Zumino–Witten models with arbitrary simple groups, *Commun. Math. Phys.* **117** (1988) 127; The spectrum of Wess–Zumino–Witten models, *Nucl. Phys. B* **299** (1988) 355

[167] P. Fendley, F. Lesage, H. Saleur, A unified framework for the Kondo problem and for an impurity in a Luttinger liquid, *J. Stat. Phys.* **85** (1996) 211, cond-mat/9510055

[168] P. Fendley, H. Saleur, N. P. Warner, Exact solution of a massless scalar field with a relevant boundary interaction, *Nucl. Phys. B* **430** (1994) 577, hep-th/9406125

[169] J. M. Figueroa-O'Farrill, S. Schrans, The spin 6 extended conformal algebra, *Phys. Lett. B* **245** (1990) 471

[170] J. M. Figueroa-O'Farrill, S. Stanciu, D-brane charge, flux quantization and relative (co)homology, *J. High Energy Phys.* **0101** (2001) 006, hep-th/0008038

[171] J. Fjelstad, J. Fuchs, I. Runkel, C. Schweigert, TFT construction of RCFT correlators. 5: Proof of modular invariance and factorisation, *Theor. Appl. Categor.* **16** (2006) 342, hep-th/0503194

[172] S. Förste, D. Ghoshal, S. Panda, An orientifold of the solitonic fivebrane, *Phys. Lett. B* **411** (1997) 46, hep-th/9706057

[173] A. Font, L. E. Ibañez, D. Lüst, F. Quevedo, Strong–weak coupling duality and nonperturbative effects in string theory, *Phys. Lett. B* **249** (1990) 35

[174] E. S. Fradkin, A. A. Tseytlin, Nonlinear electrodynamics from quantized strings, *Phys. Lett. B* **160** (1985) 69

[175] M. Frau, I. Pesando, S. Sciuto, A. Lerda, R. Russo, Scattering of closed strings from many D-branes, *Phys. Lett. B* **400** (1997) 52, hep-th/9702037

[176] K. Fredenhagen, K.-H. Rehren, B. Schroer, Superselection sectors with braid group statistics and exchange algebras I, II, *Commun. Math. Phys.* **125** (1989) 201, *Rev. Math. Phys.* Special issue (1992) 111

[177] S. Fredenhagen, Dynamics of D-branes in curved backgrounds, Ph.D. thesis (2002), available via http://www.slac.stanford.edu/spires/find/hep/www?irn=5331455

[178] S. Fredenhagen, Organizing boundary RG flows, *Nucl. Phys.* B **660** (2003) 436, hep-th/0301229

[179] S. Fredenhagen, M. R. Gaberdiel, C. A. Keller, Bulk induced boundary perturbations, *J. Phys.* A **40** (2007) F17, hep-th/0609034

[180] S. Fredenhagen, M. R. Gaberdiel, C. Schmidt-Colinet, Bulk flows in Virasoro minimal models with boundaries, *J. Phys.* A **42** (2009) 495403, arXiv:0907.2560 [hep-th]

[181] S. Fredenhagen, V. Schomerus, Branes on group manifolds, gluon condensates, and twisted K-theory, *J. High Energy Phys.* **0104** (2001) 007, hep-th/0012164

[182] S. Fredenhagen, V. Schomerus, Brane dynamics in CFT backgrounds, hep-th/0104043

[183] S. Fredenhagen, V. Schomerus, D-branes in coset models, *J. High Energy Phys.* **0202** (2002) 005, hep-th/0111189

[184] D. Friedan, The space of conformal boundary conditions for the c = 1 Gaussian model, unpublished note (1999), http://www.physics.rutgers.edu/pages/friedan/

[185] D. Friedan, A. Konechny, On the boundary entropy of one-dimensional quantum systems at low temperature, *Phys. Rev. Lett.* **93** (2004) 030402, hep-th/0312197

[186] D. Friedan, E. Martinec, S. H. Shenker, Conformal invariance, supersymmetry and string theory, *Nucl. Phys.* B **271** (1986) 93

[187] D. Friedan, Z. Qiu, S. H. Shenker, Conformal invariance, unitarity and two-dimensional critical exponents, *Phys. Rev. Lett.* **52** (1984) 1575

[188] J. Fröhlich, New superselection sectors ('soliton states') in two-dimensional Bose quantum field theories, *Commun. Math. Phys.* **47** (1976) 269

[189] J. Fröhlich, Statistics of fields, the Yang–Baxter equation and the theory of knots and links, in *Non-perturbative Quantum Field Theory*, G. t'Hooft *et al.* (eds.), Plenum 1988

[190] J. Fröhlich, The non-commutative geometry of two-dimensional supersymmetric conformal field theory, in *PASCOS, Proceedings of the Fourth International Symposium on Particles, Strings and Cosmology*, K. C. Wali (ed.), World Scientific 1995

[191] J. Fröhlich, J. Fuchs, I. Runkel, C. Schweigert, Kramers–Wannier duality from conformal defects, *Phys. Rev. Lett.* **93** (2004) 070601, cond-mat/0404051; Defect lines, dualities, and generalised orbifolds, arXiv:0909.5013 [math-ph]

[192] J. Fröhlich, F. Gabbiani, Braid statistics in local quantum theory, *Rev. Math. Phys.* **2** (1990) 251

[193] J. Fröhlich, K. Gawędzki, Conformal field theory and the geometry of strings, *CRM Proceedings and Lecture Notes*, Vol. 7, CRM 1994, 57, hep-th/9310187

[194] J. Fröhlich, O. Grandjean, A. Recknagel, Supersymmetric quantum theory, non-commutative geometry, and gravitation, in *Les Houches 1995*, Elsevier 1995, *Quantum Symmetries*, A. Connes, K. Gawędzki (eds.), hep-th/9706132

[195] J. Fröhlich, O. Grandjean, A. Recknagel, V. Schomerus, Fundamental strings in Dp–Dq brane systems, *Nucl. Phys.* B **583** (2000) 381, hep-th/9912079

[196] J. Fröhlich, T. Kerler, Quantum groups, quantum categories and quantum field theory, *Lecture Notes in Mathematics*, vol. 1542, Springer 1993

[197] J. Fröhlich, C. King, The Chern–Simons theory and knot polynomials, *Commun. Math. Phys.* **126** (1989) 167; Two-dimensional conformal field theory and three-dimensional topology, *Int. J. Mod. Phys.* A **4** (1989) 5321

[198] J. Fuchs, *Affine Lie Algebras and Quantum Groups: An Introduction, with Applications in Conformal Field Theory*, Cambridge University Press 1992

[199] J. Fuchs, Fusion rules in conformal field theory, *Fortsch. Phys.* **42** (1994) 1, hep-th/9306162

[200] J. Fuchs, A. Klemm, C. Scheich, M. G. Schmidt, Gepner models with arbitrary affine invariants and the associated Calabi–Yau spaces, *Phys. Lett.* B **232** (1989) 317; Spectra and symmetries of Gepner models compared to Calabi–Yau compactifications, *Ann. Phys.* **204** (1990) 1

[201] J. Fuchs, I. Runkel, C. Schweigert, TFT construction of RCFT correlators. 1: Partition functions, *Nucl. Phys. B* **646** (2002) 353, hep-th/0204148; TFT construction of RCFT correlators. 2: Unoriented world sheets, *Nucl. Phys. B* **678** (2004) 511, hep-th/0306164;TFT construction of RCFT correlators. 3: Simple currents, *Nucl. Phys. B* **694** (2004) 277, hep-th/0403157; TFT construction of RCFT correlators 4: Structure constants and correlation functions, *Nucl. Phys. B* **715** (2005) 539, hep-th/0412290

[202] J. Fuchs, A. N. Schellekens, C. Schweigert, A matrix S for all simple current extensions, *Nucl. Phys. B* **473** (1996) 323, hep-th/9601078

[203] J. Fuchs, C. Schweigert, *Symmetries, Lie Algebras and Representations: A Graduate Course for Physicists*, Cambridge University Press 1997

[204] J. Fuchs, C. Schweigert, A classifying algebra for boundary conditions, *Phys. Lett. B* **414** (1997) 251, hep-th/9708141

[205] J. Fuchs, C. Schweigert, Branes: from free fields to general conformal field theories, *Nucl. Phys. B* **530** (1998) 99, hep-th/9712257

[206] J. Fuchs, C. Schweigert, Completeness of boundary conditions for the critical three-state Potts model, *Phys. Lett. B* **441** (1998) 141, hep-th/9806121

[207] J. Fuchs, C. Schweigert, Orbifold analysis of broken bulk symmetries *Phys. Lett. B* **447** (1999) 266, hep-th/9811211; Symmetry breaking boundaries I. General theory, *Nucl. Phys. B* **558** (1999) 419, hep-th/9902132; Symmetry breaking boundaries II. More structures; examples, *Nucl. Phys. B* **568** (2000) 543, hep-th/9908025

[208] J. Fuchs, C. Schweigert, J. Walcher, Projections in string theory and boundary states for Gepner models, *Nucl. Phys. B* **588** (2000) 110, hep-th/0003298

[209] P. Furlan, G. M. Sotkov, I. T. Todorov, Two-dimensional conformal quantum field theory, *Riv. Nuovo Cim.* **12** (1989) 1

[210] F. Gabbiani, J. Fröhlich, Operator algebras and conformal field theory, *Commun. Math. Phys.* **155** (1993) 569

[211] M. R. Gaberdiel, Fusion in conformal field theory as the tensor product of the symmetry algebra, *Int. J. Mod. Phys. A* **9** (1994) 4619, hep-th/9307183

[212] M. R. Gaberdiel, An introduction to conformal field theory, *Rept. Prog. Phys.* **63** (2000) 607, hep-th/9910156

[213] M. R. Gaberdiel, Discrete torsion orbifolds and D branes, *J. High Energy Phys.* **0011** (2000) 026, hep-th/0008230

[214] M. R. Gaberdiel, P. Goddard, Axiomatic conformal field theory, *Commun. Math. Phys.* **209** (2000) 549, hep-th/9810019

[215] M. R. Gaberdiel, A. Recknagel, Conformal boundary states for free bosons and free fermions, *J. High Energy Phys.* **0111** (2001) 016, hep-th/0108238

[216] M. R. Gaberdiel, A. Recknagel, G. M. T. Watts, The conformal boundary states for SU(2) at level 1, *Nucl. Phys. B* **626** (2002) 344, hep-th/0108102

[217] M. R. Gaberdiel, A. Konechny, C. Schmidt-Colinet, Conformal perturbation theory beyond the leading order, *J. Phys. A* **42** (2009) 105402, arXiv:0811.3149 [hep-th]

[218] T. Gannon, The classification of affine SU(3) modular invariant partition functions, *Commun. Math. Phys.* **161** (1994) 233, hep-th/9212060; The classification of SU(3) modular invariants revisited, *Annales Henri Poincaré: Phys. Theor.* **65** (1996) 15, hep-th/9404185; The level 2 and 3 modular invariants of SU(n), *Lett. Math. Phys.* **39** (1997) 289, hep-th/9511040

[219] T. Gannon, Integers in the open string, *Phys. Lett. B* **473** (2000) 80, hep-th/9910148

[220] T. Gannon, Boundary conformal field theory and fusion ring representations, *Nucl. Phys. B* **627** (2002) 506, hep-th/0106105

[221] M. R. Garousi, R. C. Myers, Superstring scattering from D-branes, *Nucl. Phys. B* **475** (1996) 193, hep-th/9603194

[222] E. Gava, J. F. Morales, K. S. Narain, G. Thompson, Bound states of type I D-strings, *Nucl. Phys. B* **528** (1998) 95, hep-th/9801128

[223] K. Gawędzki, Quadrature of conformal field theories, *Nucl. Phys. B* **328** (1989) 733; Coulomb gas representation of the SU(2) WZW correlators at higher genera, *Lett. Math.*

Phys. **33** (1995) 335, hep-th/9404012; SU(2) WZW theory at higher genera, *Commun. Math. Phys.* **169** (1995) 329, hep-th/9402091

[224] K. Gawędzki, Lectures on conformal field theory, Lecture Notes Princeton 1996; see http://www. math.ias.edu/QFT/fall/index.html

[225] K. Gawędzki, Conformal field theory: a case study, hep-th/9904145

[226] K. Gawędzki, Boundary WZW, G/H, G/G and CS theories, *Annales Henri Poincaré* **3** (2002) 847, hep-th/0108044

[227] K. Gawędzki, A. Kupiainen, G/H conformal field theory from gauged WZW model, *Phys. Lett. B* **215** (1988) 119; Coset construction from functional integrals, *Nucl. Phys. B* **320** (1989) 625

[228] D. Gepner, Space-time supersymmetry in compactified string theory and superconformal models, *Nucl. Phys. B* **296** (1988) 757

[229] D. Gepner, Lectures on N=2 string theory, Lectures at the Trieste Spring School on Superstrings 1989

[230] D. Gepner, Z. Qiu, Modular invariant partition functions for parafermionic field theories, *Nucl. Phys. B* **285** (1987) 423

[231] D. Gepner, E. Witten, String theory on group manifolds, *Nucl. Phys. B* **278** (1986) 493

[232] A. A. Gerasimov, S. L. Shatashvili, On exact tachyon potential in open string field theory, *J. High Energy Phys.* **0010** (2000) 034, hep-th/0009103

[233] G. W. Gibbons, N. S. Manton, Classical and quantum dynamics of BPS monopoles, *Nucl. Phys. B* **274** (1986) 183

[234] E. G. Gimon, J. Polchinski, Consistency conditions for orientifolds and D-manifolds, *Phys. Rev. D* **54** (1996) 1667, hep-th/9601038

[235] P. Ginsparg, Applied conformal field theory, Lectures given at the Les Houches Summer School in Theoretical Physics 1988

[236] P. Ginsparg, Curiosities at $c = 1$, *Nucl. Phys. B* **295** (1988) 153

[237] V. Ginzburg, *Lectures on Noncommutative Geometry*, math.AG/0506603

[238] A. Giveon, D. Kutasov, Brane dynamics and gauge theory, *Rev. Mod. Phys.* **71** (1999) 983, hep-th/9802067

[239] P. Goddard, Meromorphic conformal field theory, in *Infinite-dimensional Lie Algebras and Lie Groups*, V. G. Kac (ed.), World Scientific 1989

[240] P. Goddard, A. Kent, D. I. Olive, Virasoro algebras and coset space models, *Phys. Lett. B* **152** (1985) 88; Unitary representations of the Virasoro and Supervirasoro algebras, *Commun. Math. Phys.* **103** (1986) 105

[241] P. Goddard, D. I. Olive, Kac–Moody and Virasoro algebras in relation to quantum physics, *Int. J. Mod. Phys. A* **1** (1986) 303

[242] J. Gomis, D-branes on orbifolds with discrete torsion and topological obstruction, *J. High Energy Phys.* **0005** (2000) 006, hep-th/0001200

[243] R. Gopakumar, S. Minwalla, A. Strominger, Noncommutative solitons, *J. High Energy Phys.* **0005** (2000) 020, hep-th/0003160

[244] S. Govindarajan, J. Majumder, Crosscaps in Gepner models and type IIA orientifolds, *J. High Energy Phys.* **0402** (2004) 026, hep-th/0306257

[245] K. Graham, I. Runkel, G. M. T. Watts, Minimal model boundary flows and $c = 1$ CFT, *Nucl. Phys. B* **608** (2001) 527, hep-th/0101187

[246] K. Graham, G. M. T. Watts, Defect lines and boundary flows, *J. High Energy Phys.* **0404** (2004) 019, hep-th/0306167

[247] M. B. Green, A gas of D-instantons, *Phys. Lett. B* **354** (1995) 271, hep-th/9504108

[248] M. B. Green, M. Gutperle, Symmetry breaking at enhanced symmetry points, *Nucl. Phys. B* **460** (1996) 77, hep-th/9509171

[249] M. B. Green, M. Gutperle, Light-cone supersymmetry and D-branes, *Nucl. Phys. B* **476** (1996) 484, hep-th/9604091

[250] M. B. Green, M. Gutperle, D-instanton partition functions, *Phys. Rev. D* **58** (1998) 046007, hep-th/9804123

[251] M. B. Green, J. A. Harvey, G. Moore, I-brane inflow and anomalous couplings on D-branes, *Class. Quant. Grav.* **14** (1997) 47, hep-th/9605033

[252] M. B. Green, J. H. Schwarz, E. Witten, *Superstring Theory I, II*, Cambridge University Press 1987

[253] B. R. Greene, String theory on Calabi–Yau manifolds, TASI lectures, hep-th/9702155

[254] B. R. Greene, M. R. Plesser, Duality in Calabi–Yau moduli spaces, *Nucl. Phys. B* **338** (1990) 14

[255] B. R. Greene, C. Vafa, N. P. Warner, Calabi–Yau manifolds and renormalization group flows, *Nucl. Phys. B* **324** (1989) 371

[256] M. T. Grisaru, A. E. M. van de Ven, D. Zanon, Four loop beta function for the $N = 1$ and $N = 2$ supersymmetric nonlinear sigma model in two dimensions, *Phys. Lett. B* **173** (1986) 423; Two-dimensional supersymmetric sigma models on Ricci flat Kahler manifolds are not finite, *Nucl. Phys. B* **277** (1986) 388

[257] D. J. Gross, N. A. Nekrasov, Monopoles and strings in noncommutative gauge theory, *J. High Energy Phys.* **0007** (2000) 034, hep-th/0005204

[258] H. Grosse, C. Klimčík, P. Prešnajder, Towards finite quantum field theory in non-commutative geometry, *Int. J. Theor. Phys.* **35** (1996) 231, hep-th/9505175; Field theory on a supersymmetric lattice, *Commun. Math. Phys.* **185** (1997) 155, hep-th/9507074; Simple field theoretical models on noncommutative manifolds, Lecture Notes Clausthal 1995, hep-th/9510177

[259] S. S. Gubser, A. Hashimoto, I. R. Klebanov, J. M. Maldacena, Gravitational lensing by p-branes, *Nucl. Phys. B* **472** (1996) 231, hep-th/9601057

[260] S. S. Gubser, I. R. Klebanov, A. M. Polyakov, Gauge theory correlators from non-critical string theory, *Phys. Lett. B* **428** (1998) 105, hep-th/9802109

[261] S. Gukov, I. R. Klebanov, A. M. Polyakov, Dynamics of $(n, 1)$ strings, *Phys. Lett. B* **423** (1998) 64, hep-th/9711112

[262] M. Gutperle, Aspects of D-instantons, hep-th/9712156

[263] M. Gutperle, Y. Satoh, D-branes in Gepner models and supersymmetry, *Nucl. Phys. B* **543** (1999) 73, hep-th/9808080

[264] M. Gutperle, Y. Satoh, D0-branes in Gepner models and $N = 2$ black holes, *Nucl. Phys. B* **555** (1999) 477, hep-th/9902120

[265] R. Haag, *Local Quantum Physics*, Springer 1992

[266] M. Hamermesh, *Group Theory and its Applications to Physical Problems*, Addison-Wesley 1962

[267] A. Hanany, E. Witten, Type IIB superstrings, BPS monopoles, and three-dimensional gauge dynamics, *Nucl. Phys. B* **492** (1997) 152, hep-th/9611230

[268] J. A. Harvey, Komaba lectures on noncommutative solitons and D-branes, hep-th/0102076

[269] J. A. Harvey, P. Kraus, F. Larsen, Exact noncommutative solitons, *J. High Energy Phys.* **0012** (2000) 024, hep-th/0010060

[270] R. Harvey, H. B. Lawson, Calibrated geometries, *Acta Math.* **148** (1982) 47

[271] A. Hashimoto, I. R. Klebanov, Decay of excited D-branes, *Phys. Lett. B* **381** (1996) 437, hep-th/9604065; Scattering of strings from D-branes, *Nucl. Phys. B Proc. Suppl.* **55B** (1997) 118, hep-th/9611214

[272] K. Hashimoto, K. Krasnov, D-brane solutions in non-commutative gauge theory on fuzzy sphere, *Phys. Rev. D* **64** (2001) 046007, hep-th/0101145

[273] M. Herbst, K. Hori, D. Page, Phases of $N = 2$ theories in 1+1 dimensions with boundary, arXiv:0803.2045 [hep-th]

[274] M. Herbst, C. I. Lazaroiu, Localization and traces in open–closed topological Landau-Ginzburg models, *J. High Energy Phys.* **0505** (2005) 044, hep-th/0404184

[275] M. Herbst, C. I. Lazaroiu, W. Lerche, Superpotentials, A-infinity relations and WDVV equations for open topological strings, *J. High Energy Phys.* **0502** (2005) 071, hep-th/0402110

[276] Y. Hikida, M. Nozaki, Y. Sugawara, Formation of spherical D2-brane from multiple D0-branes, *Nucl. Phys. B* **617** (2001) 117, hep-th/0101211

[277] C. Hofman, On the open–closed B-model, *J. High Energy Phys.* **0311** (2003) 069, hep-th/0204157

[278] J. Hoppe, Diffeomorphism groups, quantization and SU(∞), *Int. J. Mod. Phys. A* **4** (1989) 5235

[279] P. Hořava, E. Witten, Heterotic and type I string dynamics from eleven dimensions, *Nucl. Phys. B* **460** (1996) 506, hep-th/9510209; Eleven-dimensional supergravity on a manifold with boundary, *Nucl. Phys. B* **475** (1996) 94, hep-th/9603142

[280] K. Hori, Boundary RG flows of $N = 2$ minimal models, hep-th/0401139

[281] K. Hori, A. Iqbal, C. Vafa, D-branes and mirror symmetry, hep-th/0005247

[282] K. Hori, S. Katz, A. Klemm, R. Pandharipande, R. Thomas, C. Vafa, R. Vakil, E. Zaslow (eds.), *Mirror Symmetry*, Clay Mathematics Monographs 2003

[283] K. Hori, J. Walcher, F-term equations near Gepner points, *J. High Energy Phys.* **0501** (2005) 008, hep-th/0404196

[284] G. T. Horowitz, The origin of black hole entropy in string theory, gr-qc/9604051

[285] G. T. Horowitz, A. Strominger, Black strings and p-branes, *Nucl. Phys. B* **360** (1991) 197

[286] S. Hosono, A. Klemm, S. Theisen, Mirror symmetry, mirror map and applications to Calabi–Yau hypersurfaces, *Commun. Math. Phys.* **167** (1995) 301, hep-th/9308122; *Lectures on Mirror Symmetry*, hep-th/9403096

[287] B.-Y. Hou, K.-J. Shi, P. Wang, R.-H. Yue, The crossing matrices of WZW SU(2) model and minimal models with the quantum 6j symbols, *Nucl. Phys. B* **345** (1990) 659

[288] P. S. Howe, P. C. West, $N = 2$ Superconformal models, Landau–Ginzburg Hamiltonians and the epsilon expansion, *Phys. Lett. B* **223** (1989) 377

[289] Y. Z. Huang, Vertex operator algebras and the Verlinde conjecture, math.qa/0406291; Vertex operator algebras, the Verlinde conjecture and modular tensor categories, *Proc. Nat. Acad. Sci. USA* **102** (2005) 5352, math.qa/0412261; Rigidity and modularity of vertex tensor categories, math.qa/0502533; Vertex operator algebras, fusion rules and modular transformations, math.qa/0502558

[290] T. Hubsch, *Calabi–Yau Manifolds: A Bestiary for Physicists*, World Scientific 1992

[291] L. R. Huiszoon, A. N. Schellekens, N. Sousa, Klein bottles and simple currents, *Phys. Lett. B* **470** (1999) 95, hep-th/9909114

[292] F. Hussain, R. Iengo, C. Nuñez, C. A. Scrucca, Interaction of moving D-branes on orbifolds, *Phys. Lett. B* **409** (1997) 101, hep-th/9706186; Interaction of D-branes on orbifolds and massless particle emission, hep-th/9711021; Aspects of D-brane dynamics on orbifolds, hep-th/9711020; Closed string radiation from moving D-branes, *Nucl. Phys. B* **517** (1998) 92, hep-th/9710049

[293] K. A. Intriligator, Bonus symmetry in conformal field theory, *Nucl. Phys. B* **332** (1990) 541

[294] N. Ishibashi, The boundary and crosscap states in conformal field theories, *Mod. Phys. Lett. A* **4** (1989) 251

[295] N. Ishibashi, T. Onogi, Conformal field theories on surfaces with boundaries and crosscaps, *Mod. Phys. Lett. A* **4** (1989) 161

[296] C. Itzykson, H. Saleur, J.-B. Zuber (eds.), *Conformal Invariance and Applications to Statistical Mechanics*, World Scientific 1988

[297] C. Itzykson, J. B. Zuber, Two-dimensional conformal invariant theories on a torus, *Nucl. Phys. B* **275** (1986) 580

[298] R. A. Janik, Exceptional boundary states at $c = 1$, *Nucl. Phys. B* **618** (2001) 675, hep-th/0109021

[299] D. P. Jatkar, G. Mandal, S. R. Wadia, K. P. Yogendran, Matrix dynamics of fuzzy spheres, *J. High Energy Phys.* **0201** (2002) 039, hep-th/0110172

[300] D. Kabat, P. Pouliot, A comment on zero-brane quantum mechanics, *Phys. Rev. Lett.* **77** (1996) 1004, hep-th/9603127

[301] S. Kachru, J. McGreevy, Supersymmetric three-cycles and (super)symmetry breaking, *Phys. Rev. D* **61** (2000) 026001, hep-th/9908135

[302] A. Kapustin, D-branes in a topologically nontrivial B-field, *Adv. Theor. Math. Phys.* **4** (2000) 127, hep-th/9909089

[303] A. Kapustin, Y. Li, D-branes in Landau–Ginzburg models and algebraic geometry, *J. High Energy Phys.* **0312** (2003) 005, hep-th/0210296

[304] A. Kapustin, Y. Li, Topological correlators in Landau–Ginzburg models with boundaries, *Adv. Theor. Math. Phys.* **7** (2004) 727, hep-th/0305136

[305] A. Kapustin, Y. Li, D-branes in topological minimal models: the Landau–Ginzburg approach, *J. High Energy Phys.* **0407** (2004) 045, hep-th/0306001

[306] A. Kapustin, D. Orlov, Remarks on A branes, mirror symmetry, and the Fukaya category, *J. Geom. Phys.* **48** (2003) 84, hep-th/0109098

[307] A. Kapustin, D. Orlov, Lectures on mirror symmetry, derived categories, and D-branes, math.AG/0308173

[308] P. Kaste, W. Lerche, C. A. Lütken, J. Walcher, D-branes on K3-fibrations, *Nucl. Phys.* B **582** (2000) 203, hep-th/9912147

[309] H. Kausch, G. M. T. Watts, A study of W-algebras using Jacobi identities, *Nucl. Phys.* B **354** (1991) 740

[310] Y. Kazama, H. Suzuki, New $N = 2$ superconformal field theories and superstring compactification, *Nucl. Phys.* B **321** (1989) 232

[311] R. Kedem, T. R. Klassen, B. M. McCoy, E. Melzer, Fermionic quasiparticle representations for characters of $G_1^{(1)} \times G_1^{(1)} / G_2^{(1)}$, *Phys. Lett.* B **304** (1993) 263, hep-th/9211102; Fermionic sum representations for conformal field theory characters, *Phys. Lett.* B **307** (1993) 68, hep-th/9301046

[312] R. Kedem, B. M. McCoy, Construction of modular branching functions from Bethe's equations in the 3-state Potts chain, hep-th/9210129

[313] B. Keller, Introduction to A-infinity algebras and modules, *Homology, Homotopy Appl.* **3** (2001) 1, math.RA/9910179

[314] A. N. Kirillov, N. Y. Reshetikhin, Representations of the algebra U(q)(sl(2), q-orthogonal polynomials and invariants of links, in *New Developments in the Theory of Knots*, T. Kohno (ed.), World Scientific 1990

[315] C. Klimčík, A nonperturbative regularization of the supersymmetric Schwinger model, *Commun. Math. Phys.* **206** (1999) 567, hep-th/9903112

[316] J. Knapp, H. Omer, Matrix factorizations, minimal models and Massey products, *J. High Energy Phys.* **0605** (2006) 064, hep-th/0604189

[317] V. G. Knizhnik, A. B. Zamolodchikov, Current algebra and Wess–Zumino model in two dimensions, *Nucl. Phys.* B **247** (1984) 83

[318] A. Konechny, g function in perturbation theory, *Int. J. Mod. Phys.* A **19** (2004) 2545, hep-th/0310258

[319] A. Konechny, A. Schwarz, Introduction to M(atrix) theory and noncommutative geometry, *Phys. Rept.* **360** (2002) 353, hep-th/0012145

[320] A. Konechny, A. Schwarz, Introduction to M(atrix) theory and noncommutative geometry, Part II, *Phys. Rept.* **360** (2002) 353, hep-th/0107251

[321] M. Kontsevich, Homological algebra of mirror symmetry, alg-geom/9411018

[322] M. Kontsevich, Deformation quantization of Poisson manifolds I, *Lett. Math. Phys.* **66** (2003) 157, q-alg/9709040

[323] M. Kontsevich, Y. Soibelman, Homological mirror symmetry and torus fibrations, math.SG/0011041

[324] D. Kutasov, M. Marino, G. W. Moore, Some exact results on tachyon condensation in string field theory, *J. High Energy Phys.* **0010** (2000) 045, hep-th/0009148; Remarks on tachyon condensation in superstring field theory, hep-th/0010108

[325] O. A. Laudal, Matric Massey products and formal moduli I, in *Lecture Notes in Mathematics*, vol. 1183, Springer 1986, p. 218

[326] C. I. Lazaroiu, On the structure of open–closed topological field theory in two dimensions, *Nucl. Phys.* B **603** (2001) 497, hep-th/0010269

[327] C. I. Lazaroiu, On the boundary coupling of topological Landau–Ginzburg models, *J. High Energy Phys.* **0505** (2005) 037, hep-th/0312286

[328] R. G. Leigh, Dirac–Born–Infeld action from Dirichlet sigma model, *Mod. Phys. Lett.* A **4** (1989) 2767

[329] W. Lerche, Recent developments in string theory, hep-th/9710246

[330] W. Lerche, B. Schellekens, N. P. Warner, Lattices and strings, *Phys. Rept.* **177** (1989) 1

[331] W. Lerche, C. Vafa, N. P. Warner, Chiral rings in $N = 2$ superconformal theories, *Nucl. Phys. B* **324** (1989) 427

[332] F. Lesage, H. Saleur, Boundary conditions changing operators in non conformal theories, *Nucl. Phys. B* **520** (1998) 563, hep-th/9801089

[333] F. Lesage, H. Saleur, P. Simonetti, Boundary flows in minimal models, *Phys. Lett. B* **427** (1998) 85, hep-th/9802061

[334] D. C. Lewellen, Sewing constraints for conformal field theories on surfaces with boundaries, *Nucl. Phys. B* **372** (1992) 654

[335] M. Li, Boundary states of D-branes and Dy-strings, *Nucl. Phys. B* **460** (1996) 351, hep-th/9510161

[336] A. W. W. Ludwig, Field theory approach to critical quantum impurity problems and applications to the multi-channel Kondo effect, *Int. J. Mod. Phys. B* **8** (1994) 347; Methods of conformal field theory in condensed matter physics: an introduction to nonabelian bosonization, in: *Low-dimensional Quantum Field Theories for Condensed Matter Physicists*, S. Lundqvist, G. Morandi, Y. Lu (eds.), World Scientific 1995

[337] D. Lüst, S. Theisen, Lectures on string theory, *Lecture Notes in Physics*, vol. 346, Springer 1989

[338] G. Mack, V. Schomerus, Quasi-Hopf quantum symmetry in quantum theory, *Nucl. Phys. B* **370** (1991) 185; Action of truncated quantum groups on quasi-quantum planes and a quasi-associative differential geometry and calculus, *Commun. Math. Phys.* **149** (1992) 513

[339] J. Madore, The fuzzy sphere, *Class. Quant. Grav.* **9** (1992) 69

[340] J. Madore, *An Introduction to Noncommutative Differential Geometry and its Physical Applications*, Cambridge University Press 1999

[341] J. Majumder, A. Sen, 'Blowing up' D-branes on non-supersymmetric cycles, *J. High Energy Phys.* **9909** (1999) 004, hep-th/9906109

[342] J. M. Maldacena, Black holes in string theory, hep-th/9607235

[343] J. M. Maldacena, The large N limit of superconformal field theories and supergravity, *Adv. Theor. Math. Phys.* **2** (1998) 231, hep-th/9711200

[344] J. M. Maldacena, G. W. Moore, N. Seiberg, Geometrical interpretation of D-branes in gauged WZW models, *J. High Energy Phys.* **0107** (2001) 046, hep-th/0105038

[345] J. M. Maldacena, G. W. Moore, N. Seiberg, D-brane instantons and K-theory charges, *J. High Energy Phys.* **0111** (2001) 062, hep-th/0108100

[346] N. S. Manton, A remark on the scattering of BPS monopoles, *Phys. Lett. B* **110** (1982) 54

[347] D. Matalliotakis, H. P. Nilles, S. Theisen, Matching the BPS spectra of heterotic – type I–type I' strings, *Phys. Lett. B* **421** (1998) 169, hep-th/9710247

[348] K. Matsubara, V. Schomerus, M. Smedbäck, Open strings in simple current orbifolds, *Nucl. Phys. B* **626** (2002) 53, hep-th/0108126

[349] G. Moore, N. Reshetikhin, A comment on quantum group symmetry in conformal field theory, *Nucl. Phys. B* **328** (1989) 557

[350] G. Moore, N. Seiberg, Polynomial equations for rational conformal field theories, *Phys. Lett. B* **212** (1988) 451; Classical and conformal quantum field theory, *Commun. Math. Phys.* **123** (1989) 177; Lectures on rational conformal field theory, http://www.physics.rutgers.edu/~gmoore/LecturesRCFT.pdf

[351] J. E. Moyal, Quantum mechanics as a statistical theory, *Proc. Cambridge Phil. Soc.* **45** (1949) 99

[352] R. C. Myers, Dielectric-branes, *J. High Energy Phys.* **9912** (1999) 022, hep-th/9910053

[353] W. Nahm, Lie group exponents and SU(2) current algebras, *Commun. Math. Phys.* **118** (1988) 171

[354] W. Nahm, Quantum field theories in one and two dimensions, *Duke Math. J.* **54** (1987) 579; Chiral algebras of two-dimensional chiral field theories and their normal ordered

products, Proceedings of the Trieste Conference on Recent Developments in Conformational Field Theories, Trieste, October 1989

[355] W. Nahm, A proof of modular invariance, *Int. J. Mod. Phys. A* **6** (1991) 2837

[356] W. Nahm, Quasi-rational fusion products, *Int. J. Mod. Phys. B* **8** (1994) 3693, hep-th/9402039

[357] W. Nahm, Conformal quantum field theories in two dimensions, in preparation

[358] W. Nahm, A. Recknagel, M. Terhoeven, Dilogarithm identities in conformal field theory, *Mod. Phys. Lett. A* **8** (1993) 1835, hep-th/9211034

[359] W. Nahm, K. Wendland, A Hiker's guide to K3: aspects of $N = (4, 4)$ superconformal field theory with central charge $c = 6$, *Commun. Math. Phys.* **216** (2001) 85, hep-th/9912067

[360] M. Naka, M. Nozaki, Boundary states in Gepner models, *J. High Energy Phys.* **0005** (2000) 027, hep-th/0001037

[361] N. Nekrasov, A. Schwarz, Instantons on noncommutative $\mathbb{R}^4$ and (2,0) superconformal six dimensional theory, *Commun. Math. Phys.* **198** (1998) 689, hep-th/9802068

[362] N. A. Obers, B. Pioline, U-duality and M-theory, *Phys. Rept.* **318** (1999) 113, hep-th/9809039; Eisenstein series and string thresholds, *Commun. Math. Phys.* **209** (2000) 275, hep-th/9903113

[363] A. Ocneanu, Quantized groups, string algebras and Galois theory for algebras, in *Operator Algebras and Applications II, London Mathematical Society*, Cambridge University Press 1989; Quantum symmetry, differential geometry of finite graphs and classification of subfactors, University of Tokyo Seminary Notes 45, recorded by Y. Kawahigashi, July 1990

[364] H. Ooguri, Y. Oz, Z. Yin, D-branes on Calabi–Yau spaces and their mirrors, *Nucl. Phys. B* **477** (1996) 407, hep-th/9606112

[365] H. Ooguri, Z. Yin, TASI lectures on perturbative string theories, hep-th/9612254

[366] D. Orlov, Derived categories of coherent sheaves and triangulated categories of singularities, in *Algebra, Arithmetic, and Geometry: in Honor of Yu. I. Manin*, vol. II, Birkhäuser 2009, math.AG/0503632

[367] M. Oshikawa, I. Affleck, Boundary conformal field theory approach to the critical two-dimensional Ising model with a defect line, *Nucl. Phys B* **495** (1997) 533, cond-mat/9612187

[368] B. Ovrut, $N = 1$ supersymmetric vacua in heterotic M-theory, hep-th/9905115

[369] V. Pasquier, Operator content of the ADE lattice models, *J. Phys. A* **20** (1987) 5707; Two-dimensional critical systems labeled by Dynkin diagrams, *Nucl. Phys. B* **285** (1987) 162; Etiology of IRF models, *Commun. Math. Phys.* **118** (1988) 355

[370] V. B. Petkova, J.-B. Zuber, On structure constants of sl(2) theories, *Nucl. Phys. B* **438** (1995) 347, hep-th/9410209

[371] V. B. Petkova, J.-B. Zuber, From CFT to graphs, *Nucl. Phys. B* **463** (1996) 161, hep-th/9510175; Conformal field theory and graphs, hep-th/9701103

[372] V. B. Petkova, J.-B. Zuber, Generalized twisted partition functions, *Phys. Lett. B* **504** (2001) 157, hep-th/0011021

[373] V. B. Petkova, J.-B. Zuber, The many faces of Ocneanu cells, *Nucl. Phys. B* **603** (2001) 449, hep-th/0101151

[374] J. Polchinski, Combinatorics of boundaries in string theory, *Phys. Rev. D* **50** (1994) 6041, hep-th/9407031

[375] J. Polchinski, Dirichlet branes and Ramond-Ramond charges, *Phys. Rev. Lett.* **75** (1995) 4724, hep-th/9510017

[376] J. Polchinski, TASI lectures on D-branes, hep-th/9611050

[377] J. Polchinski, *String Theory I, II*, Cambridge University Press 1998

[378] J. Polchinski, Y. Cai, Consistency of open superstring theories, *Nucl. Phys. B* **296** (1988) 91

[379] J. Polchinski, S. Chaudhuri, C. V. Johnson, Notes on D-Branes, hep-th/9602052

[380] J. Polchinski, L. Thorlacius, Free fermion representation of a boundary conformal field theory, *Phys. Rev. D* **50** (1994) 622, hep-th/9404008

[381] A. P. Polychronakos, Flux tube solutions in noncommutative gauge theories, *Phys. Lett. B* **495** (2000) 407, hep-th/0007043

[382] G. Pradisi, A. Sagnotti, Open string orbifolds, *Phys. Lett. B* **216** (1989) 59

[383] G. Pradisi, A. Sagnotti, Y. S. Stanev, Planar duality in SU(2) WZW models, *Phys. Lett. B* **354** (1995) 279, hep-th/9503207; The open descendants of non-diagonal SU(2) WZW models, *Phys. Lett. B* **356** (1995) 230, hep-th/9506014

[384] G. Pradisi, A. Sagnotti, Y. S. Stanev, Completeness conditions for boundary operators in 2d conformal field theory, *Phys. Lett. B* **381** (1996) 97, hep-th/9603097

[385] A. Pressley, G. Segal, *Loop Groups*, Clarendon 1988

[386] A. Recknagel, Permutation branes, *J. High Energy Phys.* **0304** (2003) 041, hep-th/0208119

[387] A. Recknagel, On Permutation branes, *Fortsch. Phys.* **51** (2003) 824

[388] A. Recknagel, D. Roggenkamp, V. Schomerus, On relevant boundary perturbations in unitary minimal models, *Nucl. Phys. B* **588** (2000) 552, hep-th/0003110

[389] A. Recknagel, V. Schomerus, D-branes in Gepner models, *Nucl. Phys. B* **531** (1998) 185, hep-th/9712186

[390] A. Recknagel, V. Schomerus, Boundary deformation theory and moduli spaces of D-branes, *Nucl. Phys. B* **545** (1999) 233, hep-th/9811237

[391] A. Recknagel, V. Schomerus, Moduli spaces of D-branes in CFT-backgrounds, *Fortsch. Phys.* **48** (2000) 195, hep-th/9903139

[392] K.-H. Rehren, Markov traces as characters for local algebras, *Nucl. Phys. B Proc. Suppl.* **18B** (1990) 259; Braid group statistics and their superselection rules, in *The Algebraic Theory of Superselection Sectors. Introduction and Recent Results*, D. Kastler (ed.), World Scientific 1990; Quantum symmetry associated with braid group statistics, in *Lecture Notes in Physics*, vol. 370, Springer 1990; Quantum symmetry associated with braid group statistics II, in: *Quantum Symmetries* Doebner et al. (eds.), World Scientific 1993

[393] K.-H. Rehren, B. Schroer, Einstein causality and Artin braids, *Nucl. Phys. B* **312** (1989) 715

[394] S.-J. Rey, The confining phase of superstrings and axionic strings, *Phys. Rev. D* **43** (1991) 526

[395] A. Rocha-Caridi, Vacuum vector representations of the Virasoro algebra, in *Vertex Operators in Mathematics and Physics*, J. Lepowsky et al. (eds.), Springer 1985

[396] D. Roggenkamp, K. Wendland, Limits and degenerations of unitary conformal field theories, *Commun. Math. Phys.* **251** (2004) 589, hep-th/0308143

[397] I. Runkel, Boundary structure constants for the A-series Virasoro minimal models, *Nucl. Phys. B* **549** (1999) 563, hep-th/9811178

[398] I. Runkel, Structure constants for the D-series Virasoro minimal models, *Nucl. Phys. B* **579** (1999) 561, hep-th/9908046

[399] A. Sagnotti, Open strings and their symmetry groups, in *Non-perturbative Methods in Field Theory*, G. Mack et al. (eds.), Lecture Notes Cargèse 1987

[400] A. Sagnotti, Some properties of open string theories, hep-th/9509080

[401] A. Sagnotti, Surprises in open-string perturbation theory, *Nucl. Phys. B Proc. Suppl.* **56B** (1997) 332, hep-th/9702093

[402] E. Scheidegger, D-branes on some one- and two-parameter Calabi–Yau hypersurfaces, *J. High Energy Phys.* **0004** (2000) 003, hep-th/9912188

[403] E. Scheidegger, D0-branes in Gepner models, *J. High Energy Phys.* **0208** (2002) 001, hep-th/0109013

[404] E. Scheidegger, D-branes on Calabi–Yau spaces, Ph.D. thesis, Ludwig-Maximilians-Universität, Munich (2001), available at http://edoc.ub.uni-muenchen.de/archive/00000445

[405] A. N. Schellekens, S. Yankielowicz, Extended chiral algebras and modular invariant partition functions, *Nucl. Phys. B* **327** (1989) 673; Modular invariants from simple currents: an explicit proof, *Phys. Lett. B* **227** (1989) 387

[406] A. N. Schellekens, S. Yankielowicz, Simple currents, modular invariants and fixed points, *Int. J. Mod. Phys. A* **5** (1990) 2903

[407] A. N. Schellekens, S. Yankielowicz, Field identification fixed points in the coset construction, *Nucl. Phys. B* **334** (1990) 67

[408] V. Schomerus, Construction of field algebras with quantum symmetry from local observables, *Commun. Math. Phys.* **169** (1995) 193, hep-th/9401042

[409] V. Schomerus, Non-compact string backgrounds and non-rational CFT, *Phys. Rept.* **431** (2006) 39, hep-th/0509155.

[410] V. Schomerus, D-branes and deformation quantization, *J. High Energy Phys.* **9906** (1999) 030, hep-th/9903205

[411] M. Schottenloher (ed.), A mathematical introduction to conformal field theory, *Lecture Notes in Physics*, vol. 759, Springer 2008, p. 1

[412] J. H. Schwarz, Superstring theory, *Phys. Rept.* **89** (1982) 223

[413] A. Schwimmer, N. Seiberg, Comments on the $N = 2, N = 3, N = 4$ superconformal algebras in two dimensions, *Phys. Lett. B* **184** (1987) 191

[414] G. Segal, The definition of conformal field theory, in *Differential Geometrical Methods in Theoretical Physics*, K. Bleuler, M. Werner (eds.), Kluwer 1988

[415] N. Seiberg, E. Witten, String theory and noncommutative geometry, *J. High Energy Phys.* **9909** (1999) 032, hep-th/9908142

[416] A. Sen, Tachyon condensation on the brane antibrane system, *J. High Energy Phys.* **9808** (1998) 012, hep-th/9805170; Stable non-BPS bound states of BPS D-branes, *J. High Energy Phys.* **9808** (1998) 010, hep-th/9805019; Stable non-BPS states in string theory, *J. High Energy Phys.* **9806** (1998) 007, hep-th/9803194

[417] A. Sen, SO(32) spinors of type I and other solitons on brane–antibrane pair, *J. High Energy Phys.* **9809** (1998) 023, hep-th/9808141

[418] A. Sen, Developments in superstring theory, hep-ph/9810356

[419] A. Sen, Type I D-particle and its interactions, *J. High Energy Phys.* **9810** (1998) 021, hep-th/9809111

[420] A. Sen, Descent relations among bosonic D-branes, *Int. J. Mod. Phys. A* **14** (1999) 4061, hep-th/9902105

[421] A. Sen, Non-BPS states and branes in string theory, hep-th/9904207

[422] A. Sen, Moduli space of unstable D-branes on a circle of critical radius, *J. High Energy Phys.* **0403** (2004) 070, hep-th/0312003

[423] A. Sen, Tachyon dynamics in open string theory, *Int. J. Mod. Phys. A* **20** (2005) 5513, hep-th/0410103

[424] S. L. Shatashvili, Comment on the background independent open string theory, *Phys. Lett. B* **311** (1993) 83, hep-th/9303143; On the problems with background independence in string theory, *Alg. Anal.* **6** (1994) 215, hep-th/9311177.

[425] M. M. Sheikh-Jabbari, Classification of different branes at angles, *Phys. Lett. B* **420** (1998) 279, hep-th/9710121; More on mixed boundary conditions and D-branes bound states, *Phys. Lett. B* **425** (1998) 48, hep-th/9712199

[426] S. H. Shenker, Another length scale in string theory?, hep-th/9509132

[427] S. Stanciu, D-branes in Kazama–Suzuki models, *Nucl. Phys. B* **526** (1998) 295, hep-th/9708166

[428] S. Stanciu, D-branes in group manifolds, *J. High Energy Phys.* **0001** (2000) 025, hep-th/9909163

[429] S. Stanciu, A note on D-branes in group manifolds: flux quantization and D0-charge, *J. High Energy Phys.* **0010** (2000) 015, hep-th/0006145

[430] S. Stanciu, A. Tseytlin, D-branes in curved spacetime: Nappi–Witten background, *J. High Energy Phys.* **9806** (1998) 010, hep-th/9805006

[431] Y. S. Stanev, talk given at the Workshop on Conformal Field Theory of D-Branes, DESY, Hamburg, September 1998. http://www.desy.de/~jfuchs/CftD-s.html

[432] K. S. Stelle, Lectures on supergravity p-branes, hep-th/9701088; BPS branes in supergravity, hep-th/9803116

[433] A. Strominger, Open p-branes, *Phys. Lett. B* **383** (1996) 44, hep-th/9512059

[434] A. Strominger, C. Vafa, Microscopic origin of the Bekenstein–Hawking entropy, *Phys. Lett. B* **379** (1996) 99, hep-th/9601029

[435] W. Taylor, D2-branes in B-fields, *J. High Energy Phys.* **0007** (2000) 039, hep-th/0004141

[436] J. Teschner, Remarks on Liouville theory with boundary, hep-th/0009138; Liouville theory revisited, *Class. Quant. Grav.* **18** (2001) R153, hep-th/0104158

[437] P. K. Townsend, The eleven-dimensional supermembrane revisited, *Phys. Lett. B* **350** (1995) 184, hep-th/9501068

[438] A. A. Tseytlin, Ambiguity in the effective action in string theories, *Phys. Lett. B* **176** (1986) 92

[439] A. A. Tseytlin, Born–Infeld action, supersymmetry and string theory, hep-th/9908105

[440] C. Vafa, Modular invariance and discrete torsion on orbifolds, *Nucl. Phys. B* **273** (1986) 592

[441] E. Verlinde, Fusion rules and modular transformations in 2-d conformal field theory, *Nucl. Phys. B* **300** (1988) 360

[442] N. P. Warner, $N = 2$ Supersymmetric integrable models and topological field theories, Lectures at the Trieste Summer School on High Energy Physics and Cosmology, 1992, hep-th/9301088

[443] N. P. Warner, Supersymmetry in boundary integrable models, *Nucl. Phys. B* **450** (1995) 663, hep-th/9506064

[444] G. M. T. Watts, unpublished TCSA computations (February 2000)

[445] K. Wendland, Orbifold constructions of K3: a link between conformal field theory and geometry, hep-th/0112006

[446] H. Weyl, Quantum mechanics and group theory, *Z. Phys.* **46** (1927) 1

[447] E. T. Whittaker, G. N. Watson, *A Course of Modern Analysis*, Cambridge University Press 2002

[448] E. Witten, Non-abelian bosonization in two dimensions, *Commun. Math. Phys.* **92** (1984) 455

[449] E. Witten, Topological quantum field theory, *Commun. Math. Phys.* **117** (1988) 353; Topological sigma models, *Commun. Math. Phys.* **118** (1988) 411

[450] E. Witten, Quantum field theory and the Jones polynomial, *Commun. Math. Phys.* **121** (1989) 351

[451] E. Witten, On background independent open string field theory, *Phys. Rev. D* **46** (1992) 5467, hep-th/9208027; Some computations in background independent off-shell string theory, *Phys. Rev. D* **47** (1993) 3405, hep-th/9210065

[452] E. Witten, Phases of $N = 2$ theories in two dimensions, *Nucl. Phys. B* **403** (1993) 159, hep-th/9301042

[453] E. Witten, String theory dynamics in various dimensions, *Nucl. Phys. B* **443** (1995) 85, hep-th/9503124

[454] E. Witten, Bound states of strings and D-branes, *Nucl. Phys. B* **460** (1996) 335, hep-th/9510135

[455] E. Witten, Solutions of four-dimensional field theories via M theory, *Nucl. Phys. B* **500** (1997) 3, hep-th/9703166

[456] E. Witten, Anti de Sitter space and holography, *Adv. Theor. Math. Phys.* **2** (1998) 253, hep-th/9802150

[457] E. Witten, D-branes and K-theory, *J. High Energy Phys.* **9812** (1998) 019, hep-th/9810188

[458] E. Wong, I. Affleck, Tunneling in quantum wires: a boundary conformal field theory approach, *Nucl. Phys. B* **417** (1994) 403

[459] S. T. Yau (ed.), *Essays on Mirror Manifolds*, International Press 1992

[460] S. A. Yost, Bosonized superstring boundary states and partition functions, *Nucl. Phys. B* **321** (1989) 629

[461] A. B. Zamolodchikov, Infinite additional symmetries in two-dimensional conformal quantum field theory, *Theor. Math. Phys.* **65** (1985) 1205

[462] A. B. Zamolodchikov, "Irreversibility" of the flux of the renormalization group in a 2-d field theory, *JETP Lett.* **43** (1986) 730

[463] A. B. Zamolodchikov, V. A. Fateev, Operator algebra and correlation functions in the two-dimensional Wess–Zumino $SU(2) \times SU(2)$ chiral model, *Sov. J. Nucl. Phys.* **43** (1986) 657

[464] Y. Zhu, Modular invariance of characters of vertex operator algebras, *J. Amer. Math. Soc.* **9** (1996) 237

[465] J.-B. Zuber, Graphs, algebras, conformal field theories and integrable lattice models, *Nucl. Phys. B Proc. Suppl.* **18B** (1990) 313; C-algebras and their applications to reflection groups and conformal field theories, hep-th/9707034

[466] J.-B. Zuber, talk given at the Workshop on Conformal Field Theory of D-Branes, DESY, Hamburg, September 1998

Index

Printed in the United States
by Baker & Taylor Publisher Services